CONCEPTUAL DATA MODELING AND DATABASE DESIGN: A FULLY ALGORITHMIC APPROACH

Volume 1

The Shortest Advisable Path

CONCEPTUAL DATA MODELING AND DATABASE DESIGN:
A FULLY ALGORITHMIC APPROACH

Volume 1

The Shortest Advisable Path

Christian Mancas

*Mathematics and Computer Science Department,
Ovidius State University, Constanta, Romania*

*Computer Science Taught in English Department,
Politehnica University, Bucharest, Romania*

*Software R&D Department, Asentinel International and
DATASIS ProSoft, Bucharest, Romania*

Apple Academic Press Inc. | Apple Academic Press Inc.
3333 Mistwell Crescent | 9 Spinnaker Way
Oakville, ON L6L 0A2 | Waretown, NJ 08758
Canada | USA

Library and Archives Canada Cataloguing in Publication

Mancas, Christian, author
Conceptual data modeling and database design : a fully algorithmic approach / Christian Mancas (Mathematics and Computer Science Department, Ovidius State University, Constanta, Romania, Computer Science taught in English Department, Politehnica University, Bucharest, Romania, Software R&D Department, Asentinel International and DATASIS ProSoft, Bucharest, Romania).

Includes bibliographical references and index.
Contents: Volume 1. The shortest advisable path.
Issued in print and electronic formats.
ISBN 978-1-77188-124-1 (v. 1: bound).--ISBN 978-1-4987-2844-7 (v. 1: ebook)
1. Database design. 2. Computer algorithms. 3. Conceptual structures (Information theory). I. Title.

QA76.9.D26M35 2015   005.74'3   C2015-904816-8   C2015-904882-6

Library of Congress Cataloging-in-Publication Data

Mancas, Christian, author.
Conceptual data modeling and database design : a fully algorithmic approach. Volume 1, The shortest advisable path / Christian Mancas.

pages cm
Includes bibliographical references and index.
ISBN 978-1-77188-124-1 (alk. paper)
1. Database design. 2. Computer algorithms. 3. Conceptual structures (Information theory)
I. Title. II. Title: Shortest advisable path.

QA76.9.D26M36 2015   005.74'3--dc23   2015025172

To the loving memory of my parents,
to my revered Professors,
to my students, colleagues, love, children, and friends.

Our standards in databases (and not only) should always be at least equal to
the maximum of the corresponding highest academic and customer ones.

Laudamus veteres, sed nostris ultimur annis:
mos tamen est aeque dignus uterque coli.
—Publius Ovidius Naso

We cannot have a future without respecting our past.
—King Michael I of Romania

What we think, we become.
All that we are arises with our thoughts.
With our thoughts, we make the world.
—The Buddha

My prayer:
Lord, do not ever let me be satisfied with myself.
—Lucian Blaga

THINK.
—Thomas J. Watson

As we look ahead into the next century,
leaders will be those who empower others.
—Bill Gates

I believe people have to follow their dreams—I did.
—Larry Ellison

The cave you fear to enter holds the treasure you seek.
—Joseph Campbell

The Key
The whole Key
and nothing but the Key,
so help me Codd.
—David J. Emmick

CONTENTS

Dedication .... *v*

About the Author .... *ix*

Foreword by Professor Bernhard Thalheim .... *xi*

Foreword by Professor Dan Suciu .... *xv*

Preface .... *xvii*

Acknowledgments .... *xxiii*

List of Symbols .... *xxv*

1. **Data, Information and Knowledge in the Computer Era** 1

2. **The Quest for Data Adequacy and Simplicity:**
   The Entity-Relationship Data Model (E-RDM) 35

3. **The Quest for Data Independence, Minimal Plausibility,**
   and Formalization: The Relational Data Model (RDM) 115

4. **Relational Schemas Implementation and Reverse Engineering** 341

5. **Conclusion** 595

   Appendix: Mathematic Prerequisites for the Math Behind 615

   Index 635

ABOUT THE AUTHOR

Christian Mancas, PhD, is currently an Associate Professor with both the Mathematics and Computer Science Department of Ovidius University, Constanta, Romania, and the Engineering Taught in Foreign Languages Department (Computer Science and Telecommunications in English stream) of Politehnica University, Bucharest, Romania (as an invited Professor). Since 2012, he is also a database architect with Asentinel International srl, Bucharest, a subsidiary of Asentinel LLC, Memphis, Tennessee. His specialties include university teaching, R&D, business analysis, conceptual data and knowledge modeling and querying, client-server and hierarchical software architecture, object-oriented and event-driven design, structured development, complex projects and small IT company management, Datalog, SQL, C#, XML programming, etc.

Professor Christian Mancas has published dozens of scientific papers (in Romania, USA, Austria, and Greece), which have been indexed by ACM Digital Library, Zentralblatt MATH, Scopus, DBLP, Arnetminer, Researchr, TDGS, SCEAS, etc. He has also published three books in Romanian and dozens of reviews (mostly in USA, including *ACM Reviews*). He was a program committee member and session chairman for several software conferences in the USA, Austria, and Romania, and he is a member of several associations (including ACM, the Romanian Mathematics Sciences Society, and the International Who's Who of Professionals). Since 2006, his biography is included in *Marquis Who's Who in the World* and *Who's Who in Science and Technology* and *Hubners' Who's Who in Romania.*

Since 1990, he also worked for several IT startups, including his own DATASIS Consult srl (co-owned with his good friend and faculty colleague Ion Draghicescu) and DATASIS ProSoft srl (which had 25 programmers working for the design and development of several ERP-type database applications for customers from France, UK, Switzerland, USA, Israel, Greece, and Romania).

His main research areas are the conceptual data and knowledge modeling and querying, database design, implementation, and optimization, as well as the architecture, design, development, fine-tuning, and maintenance of data and knowledge base management systems.

Dr. Mancas graduated in 1977 from the Computers Department at Politehnica University of Bucharest, Romania, with a thesis on Generating Parsers for LR(k) Grammars, under the supervision of Professor Dan Luca Serbanati. Up until the fall of communism in 1990, he worked as a software engineer and, since 1980, had been R&D manager of a state-owned computer center in Bucharest (contributing to the design, development, and maintenance of a dedicated ERP), and conducted (from time to time) computer programming labs at Politehnica University of Bucharest, but, for political reasons, he was not accepted for PhD studies. He started this program under the supervision of Professor Cristian Giumale in 1992 and obtained his PhD degree in 1997 from the same as above department, with a thesis on Conceptual Data and Knowledge Modeling.

FOREWORD BY
PROFESSOR BERNHARD THALHEIM

What Christian Mancas wanted to do is to write the best possible book on real, pragmatic database design available, bar none. He succeeded. Christian had most of these tricks and approaches in his head when he planned this book more than a dozen years ago. When you work on a new approach, you tend to pick up things that have been developed in the past by researchers. Mimicking accepted approaches without reconsideration, reorganization and renovation is a common unculture in computer science. The area of database and information system design is not an exception. There are at least three-score books at present that follow the main agenda and repeat what has been done in the past with some minor changes in scope and order. However, if you are smart and clever, you collect main results and ideas, work on them, and figure out whether the classical approach is really the way how topics must be given in a course book. But what if the classical approach is not the most appropriate one? What if the classical approach leaves pragmatism completely out of scope? What if the classical approach is mainly based on syntax? It is very hard for a person new to a language to walk into a discipline knowing only syntax and some few integrity constraints and design a complex database schema. Ever try to get along in a foreign country with only a small dictionary that gives word-to-word transformations and additionally a small rulebook?

Okay, now we have the purpose of this book. In order to develop database applications one must have a methodological background, a number of well-developed techniques, and the best possible theory in the background. The next step is how to write so that everybody can use the book as a main source for database design and development. It seems that in the age of the Internet someone has only to look into the ocean of sources in the Internet and will succeed in this task. The agility of criticism on the Internet seems to provide the path for improvement of any book. Anybody can propose new topics that must be covered, can add what has been left out, can correct typos and real errors, and propose some kind of insight how things should be developed for real applications.

So, we have now a glimpse into the real need of practical database development. One must start with a good understanding of the topic and the application, must use the best language, must be supported by most appropriate tools for development, and must be completely sure that the approach is well founded and will not add errors due to the partial background of the language. The classic approach uses some extension of the entity-relationship modeling language for introduction to conceptual database design, allows to transfer schemata developed in such language to another language—say some abstraction of relational database management system language, and finally derives the description within a system language. This book follows such order as well. Whether the given extension is a most appropriate is a matter of culture, education and background. Whether the transfer or mapping works well is a matter of insight. This book is, however, based on methods—called algorithms—for database design and on stewardship by good guidelines—called good practices. I know only four books in our area that follow such approach. Practical people wrote all four books during the last almost three decades. And that is it. All four books provide an insight what should be done for database development. They do, however, not provide an insight how database design should be backed by an appropriate language and theoretical underpinning.

Since I have a reputation for expertise in object-relational database design and development, in database theory, in database programming and in methodology development for this area, I was one of the people Christian Mancas asked for a review of the manuscript. And I was surprised and satisfied while discovering stuff and tricks I have not seen before. While improving the co-design methodology to information systems development with the goal to reach maturity at SPICE level 3, we had to realize that database design is an engineering activity far beyond an art and must be based on a well-founded technology of design. Many areas of knowledge begin as an art and only later become scientific. The maturity of such an area can be measured by the existence of well-defined methods, by a chance for being repeatable, by becoming manageable, and by logistics on the basis of methods for continuous optimization. Currently, database design is still an art. An artisan or design mechanic is supported by constructs mainly at the syntactic level. So, a database developer knows templates for structure description and templates for coherence specification. The second kind of templates is called integrity constraints and considered to provide semantics. It is, however, only the starting point for semantics description. Artisans apply myriad rules of thumb, each rule with only a narrow scope. So, we require maturation of the area. We need a small number of small powerful methods that can be applied systematically. Systematic application must be guided.

This book will find its way into the literature on database design and development. It has a good number of ideas that must be considered in any design task. It uses a sample-based approach and is thus easy to understand. It supports digestion due to nice exercises. And, finally it discusses in details also the result of a design in different DBMS languages. So, a reader can be sure that the book guides to the right track.

Enjoy it! Use it!

Bernhard Thalheim

Professor with the Department of Computer Science,
Christian-Albrechts-University Kiel, Germany
E-mail: thalheim@is.informatik.uni-kiel.de

October 2014

FOREWORD BY PROFESSOR DAN SUCIU

For a computer professional today, data management skills are no longer just nice to have, but a necessity. Companies collect ever-increasing amounts of data, and make critical business decisions based on sophisticated analysis of this data. They need experts that understand how to model and store the data, how to query and analyze it, and how to be advanced users of modern database management systems. Professionals with strong data management skills are in high demand today, including not only the traditional database administrator (DBA), but also newer skills like the data analyst, the data scientist, the data steward, or the data enthusiast.

The first volume of this book, by Prof. Christian Mancas, covers the classical data management topics that any computer professional should master. Based on courses taught by Prof. Mancas over several years at the University Ovidius in Constanta and the Polytechnic University in Bucharest, Romania, this volume is a gentle yet rigorous and extensive introduction to the main topics in data management, with concrete examples on several popular database systems. There are lots of detailed examples, and each concept is covered in detail, and from several perspectives, using alternative definitions or notations where needed. The book ensures that no reader is left behind, and all potential questions are answered. Basic yet fundamental data management tasks like designing an E/R diagram, converting it into relations, checking for normal forms, are described in full detail and with lots of examples. The book covers all major normal forms, 1st, 2nd, 3rd, BCNF, 4th, and 5th, which one rarely finds described in an introductory textbook to databases. Special treatment is given to the relational algebra, which the book defines rigorously, using mathematical concepts, then describes with the same rigor the connection between the relational algebra and SQL. The SQL chapter has case studies using several major commercial and open source database management systems.

The first volume of this book is best suited for the practitioner who wants to achieve a thorough understanding of the fundamental concepts in data management. The increasingly central role that data, sometimes called *big data*, plays in our society today makes it imperative for every computer professional to be skilled in the topics covered by the book. One should think of this first volume as an important first step in understanding the complexities of data today.

Dan Suciu
Professor
University of Washington, Seattle, Washington, USA
September 2014

PREFACE

You can have the vastest horizons from the mountains
that obliterated them until lately.
—Lucian Blaga

There are those who believe that databases (dbs) do not need architecture and design. In the best such case, they consider that it is enough to build and design your software application, then develop it accordingly by using a CASE[1] ORM[2] development environment, like, for example, Java's Hibernate, which automatically generates the needed underlying db scheme too.

If the application's architecture is correct, then architecture of the corresponding db is also correct; however, generally, even the best software architects who are not also db ones are extremely tempted, for example in the case of commercial applications, to treat cities, states, and sometimes even countries as character strings, instead of objects, and/or to semantically overload classes, which will result in db instance anomalies and/or semantically overloaded corresponding db tables, both of them prone to implausible data storing.

Moreover, even if the db architecture is perfect, it is not enough: in order to guarantee data plausibility, its design should also take into consideration all existing business rules (constraints) in the given universe of discourse.

Consequently, even when perfectly using such CASE ORM tools, the generated db scheme should then be refined.

In the worst such case, they simply launch a database management system (DBMS), like SQLite, MySQL, MS Access, or even IBM DB2, Oracle Database (Oracle, for short), Sybase Advantage Database Server (former SQL Server), or MS SQL Server, etc. and start creating tables and filling them with data.

---

[1] The acronym for Computer Aided Software Engineering.

[2] The acronym for Object-Relational Mapping.

Generally, such persons believe that dbs are only some technological artifacts and almost completely ignore the old software adage "garbage in, garbage out", which is equivalent to, for example, building not only houses, but even skyscrapers, without architecture and design.

Unfortunately, in dbs there is no law prohibiting such "constructions", and the day when mankind will realize that data is, after health, education, and love, perhaps our most valuable asset (and so it deserves legal regulations too) is not yet foreseeable.

Then, there are those who consider db architecture and design as being rather an art, or even a "mystery", but definitely not a science.

This book advocates a dual approach: conceptual data modeling, strengthened with CASE algorithms, is db architecting, while db design and querying is applied mathematics, namely the (semi-naïve) algebraic theory of sets, relations, and functions, plus the first-order predicate logic.

The main four chapters of this book are devoted to four data models that I'm considering as being cornerstones in this field: the relational, the entity-relationship (E-R), the logical, and a(n elementary) mathematical ones.

Two chapters are devoted to implementing optimized relational dbs (rdbs), by using five of the currently most widely used relational DBMSs (RDBMSs): IBM DB2, Oracle Database and MySQL, Microsoft SQL Server and Access.

Essentially, this book emphasizes that using RDBMSs, so, generally, technology mastering is a must, but much more important are the previous two steps of db architecture and design: it almost does not matter *what pen* (and RDBMSs too are mainly pens…) you write with, it is *what you write* that matters the most.

The proposed approach is fully algorithmic: Entity-Relationship (E-R) diagrams and business rule (restriction) sets are obtained in the conceptual data modeling architectural phase by using an assistance algorithm; they are next algorithmically translated into mathematical schemes, which are then refined by applying other five algorithms; next, mathematical schemes are algorithmically translated into relational ones and associated sets of nonrelational constraints, which, in their turn, are then algorithmically translated into actual RDBMSs rdbs. Finally, rdbs are optimized with a couple of other algorithms.

Moreover, this book also includes "shortcut" algorithms, like, for example, the one for directly translating E-R diagrams and business rule sets into relational schemas (or even dbs on a particular RDBMS version) and associated sets of nonrelational constraints, as well as reverse engineering (*RE*) algorithms, like, for example, the duals of the above mentioned forward (*FE*) ones.

Except for those in the case studies from the Appendices of the second volume, all algorithms are written in an easy to understand and commented pseudocode.

Almost no R, S, a, b, X, Y, etc. type (i.e., purely syntactic) examples are included: except for some exercises from the third chapter of this first volume that are borrowed from other books, all other exercises, examples, and case studies provided in this book are everyday life ones, full of semantics.

Informal and formal aspects are always clearly isolated: informal ones always precede corresponding formal ones, as (only relevant) math, if any, is presented in dedicated sections placed at the end of the chapters; needed math prerequisites are grouped in a dedicated Appendix of the first volume.

No other theoretical or technological prerequisites are needed; however, operating systems basic notions and (especially object-oriented and event-driven) computer programming experience, as well as math prerequisite knowledge are very useful for a faster and deeper understanding of this book.

For founding solid architecture, design, implementation, manipulation, and optimization, as well as criteria for objective comparisons between possible alternative solutions, a set of db axioms are provided too.

Exemplified best practice rules are also provided for all levels: those that are only searching for such practical receipts might skip everything else from this book, perhaps except for some examples and case studies.

Equivalences between informal, E-R, relational, math, and RDBMS jargons are also provided. Bibliographic notes are adding optional concepts and results, as well as retracing domain mainstream history. References include both up-to-date and seminal relevant papers and books.

On one hand, given the immense quantity of published work in this field, it is impossible to exhaustively cite them; on the other, online searchable bibliographies cover essentially all recent ones. Thus, in this book we rather mention only the papers of historical importance, major secondary sources, and useful surveys.

There are some well-known proposed exercises, especially in the third chapter, but most of them are original. Those marked with a star are more difficult; those that have two stars are very difficult. For all chapters, at the beginning of the exercises section, problems of all types are solved.

The subject part of the index is, generally, built dually: for example, both "function, domain" and "domain, function" are included; moreover, the index also includes abbreviations and relational and math symbols (thus serving as a legend too), as well as paper, book, and motto authors. Mottos without authors come from previous books by this book's author.

Essentially, I am using the first volume (together with my book on *SQL ANSI-92 Programming*) for my undergraduate database lectures, and the second one (together with my books on *Conceptual Data and Knowledge Modeling and Querying* and on *Database Management Systems*) for the postgraduate *Conceptual Data and Knowledge Modeling and Querying* and *Architecture, Design and Development of Database Applications* lectures, at both Ovidius State University Math and Computer Science Dept., Constanta (Romanian and English) and Bucharest Politehnica University Computer Science taught in Foreign Languages Department, English stream.

The first volume describes the shortest advisable path for conceptual data modeling, db design and implementation, which completely skips both the intermediate mathematical level (except for algorithm $A7/8$–3, an adaptation of $A3$—see Figs. 1.1 and 1.2 in Section 1.7) and the additional logic one.

The second volume of this book, scheduled to appear next year, contains advanced topics (on mathematic schemas refinement, recursive *SQL*, knowledge bases, nonrelational constraint enforcement and db optimizations) and is also made up of five chapters:

1.   The quest for knowledge: Recursive *SQL*, Datalog, and knowledge bases (KBs, also including Datalog and deductive dbs).

2.   The quest for rigor: The (Elementary) Mathematical Data Model (E)MDM—including math db schemas and instances, algorithms $A1$ to $A7$, $REA1$ and $REA2$ –see Figs. 1.1 and 1.2 in Section 1.7, (E) MDM models of the E-RDM, RDM, and itself, E-R and relational data models of the (E)MDM, the equivalence between E-R and

functional data modeling, as well as a presentation of the corresponding prototype *KDBMS MatBase*.

3. The quest for ultimate data plausibility: Non-relational constraints
   enforcement (which includes all (E)MDM constraints related to
   sets, mappings, homogeneous binary relations, and objects, as well
   as the family of algorithms *AF*9/9′—see Figs. 1.1 and 1.2 in Section 1.7).

4. The quest for speed: Optimizing schemas and data manipulation
   (including the families of algorithms *AF*10 and *AF*11—see Figs.
   1.1 and 1.2 in Section 1.7).

5. Conclusion.

Naturally, second volume chapters also include corresponding best
practice rules, math behind, conclusion, past, present, references, and
solved, as well as proposed exercises sections.

Moreover, besides its index, the second volume also includes two
appendices, each one containing a full case study: one for a reigns and the
other one for a geographical db software applications.

There are, however, sections that are only optional supplementary
reading for my students: from the first volume, 2.14, 3.10, 4.4, and 4.12;
from the second one, all ones on the past, present and references.

Overall, this book embeds more than 40 years of db industry and
scientific research, as well as 20 years of university teaching experience. Besides the author's original results it contains (e.g., most of the
25 proposed (family of) algorithms, formalizations of the Key Propagation Principle and referential integrity constraints (typed inclusion
dependencies/foreign keys), E-RDM, and RDM, the (E)MDM, their
metamodels (i.e., E-R, relational, and mathematical data models of the
E-RDM, RDM, and (E)MDM), the equivalence between the E-R and
the functional data modeling (which proves that the only fundamental
mathematical relations for data modeling are the functions), relationship hierarchies, restriction sets, and descriptions associated to E-R diagrams, the relational Domain-Key-Referential-Integrity Normal Form
(*DKRINF*), the (E)MDM Semantic NF, the *MatBase* KDBMS proto-
types, the proof that any table having n columns may have at most $C(n,$
$[n/2])$ keys, the orthogonality of keys and foreign keys, the formalization
of the *SQL* GROUP BY clause with quotient sets computed by the kernel

of mapping products equivalence relations, case studies, examples, and exercises, many database axioms and best practice rules, etc.), the main difference between it and the most of the other similar books is the completely algorithmic approach it advocates.

The intended audience is diverse: computer science students and professors; db architects, designers, developers and, partially, administrators; and, as it also contains (mainly in the second volume) open problems and new points of view in this field, db scientific researchers as well.

Obviously, some might wish to study and use only the 25 algorithms and the almost 100 best practice rules (many of them having several sub-rules), others only the examples, or only the math behind, or only the plethora of relevant technological details on the latest versions of the 5 flagship RDBMSs considered, etc.

Although I proofread this first volume several times, I fear that it still may contain both typo errors and inaccurate details. Thanks in advance to all those who will contribute to make subsequent editions better and better!

ACKNOWLEDGMENTS

The wise man does not lay up his own treasures.
The more he gives to others, the more he has for his own.
—Lao Tzu

Acknowledgements are firstly due to those who seeded in me the love for mathematics and computer science—my wonderful parents, Elena and Ioan, and professors, chronologically, who were the most influential: Lucia Țene, Octavian Stănășilă, Gheorghe Șabac, Alexandru Dincă, Ana Burghelea, Luca Dan Șerbănați, Pierre (Petre) Dimo, Irina Athanasiu, and Christian Giumale.

Among the colleagues with whom I had the privilege to learn from and work with, in the same order, I'm especially indebted to Professors Val Tannen (Valeriu Breazu), Victor Markowitz (Mănescu), and Dan Suciu; to my teaching assistants Lecturer Dr. Elena Băutu and PhD student Alina Iuliana Dicu; as well as to Lavinia Crasovschi, Edward Sava-Segal, Adrian Lungu, Vladimir Ivanov, Vadim Leandru, and Adrian Raicu.

I will always recall with great pleasure the effervescent atmosphere and fruitful discussions that I had in the 80s with the outstanding members of the Miller seminar of the mathematics and computer science faculty of the Jassy (Iași) "A. I. Cuza" University (Professors Sorin Istrail, Gabriel Ciobanu, Christian Masalagiu, Victor Felea, etc.); moreover, they offered me too access to the seminar's extraordinary library, an incredible one in those dark communist years, when we were almost completely isolated from the rest of the world.

Thanks to my students from the mathematics and computer science department of the Ovidius State University at Constanta and from the computer-science-taught-in-foreign-languages department, English stream, at the Bucharest Politehnica University who inspired my teaching: for convincing them of the beauty of conceptual data modeling and querying, I always had to find better and better algorithms, examples, and exercises.

Thanks to God for the hundreds of customers' projects on which I worked and which provided me with a wealth of such interesting examples.

Thanks for their patience and kindness to Sandra Sickels and Ashish Kumar at Apple Academic Press.

Last only chronologically, but first in my mind and heart, special thanks and all my gratitude to Professors Dan Suciu and Bernhard Thalheim, who had the kindness and patience to be the first readers of this first volume and to write the forewords that honor it, as well as to all of its reviewers, and especially to Professors Cristian Calude and Andra Hugo.

Bucharest, Romania, October 1, 2014

LIST OF SYMBOLS

∀	*universal quantifier* (*for any*)
∃	*existential quantifier* (*there exists*)
∈	(element) *belongs to* (set)
∉	(element) *does not belong to* (set)
Σ	*sum* operator symbol
Π	*function product* operator
∃!	*there exists a unique* (element of a set)
&	string *concatenation* operator
[x]	*integer part* (of x)
\	*difference* set operator
—	*difference* (set, numbers) operator
|P|	*arity* (of P)
π	*projection* (RA) operator
ρ	*renaming* (RA) operator
σ	*selection* (RA) operator
|><|	(*inner*) *join* (RA) operator
^{L}|><|	*left* (*outer*) *join* (RA) operator
|><|^P	*right* (*outer*) *join* (RA) operator
^{L}|><|^R	*full* (*outer*) *join* (RA) operator
|>	*left anti-semijoin* (RA) operator
<|	*right anti-semijoin* (RA) operator
|><	*left semi-join* (RA) operator
><|	*right semi-join* (RA) operator
÷	*division* (RA) operator
,	*SQL set Cartesian product* operator
×	(math) *set Cartesian product* operator
ε, λ	*empty string*
|−	logic *implication* (between sets of formulas)
○	*function composition* operator
•	*function* (*Cartesian*) *product* operator

$\oplus$	*direct sum* set operator
$\emptyset$	*empty set*
$\cap$	*intersection* set operator
$\cup$	*union* set operator
$\subseteq$	(set) *inclusion* relation
$\subset$	(set) *strict inclusion* relation
$\not\subset$	(set) *not inclusion* relation
$\neg$	(logical) *not* operator
$\wedge$	(logical) *and* operator
$\vee$	(logical) *or* operator
$\Rightarrow$	(logical) *implication* operator
$\neq$	*not equal*
$<>$	*not equal*
$x!$	*factorial of x*
ASCII(n)	the set of all strings of ASCII characters of length at most n
C(n,j)	*n choose j* (*combinations of n taken j*)
card	the (math) *cardinal* function
COUNT	the (*SQL* aggregation) *cardinal* function
f^{-1}	(mapping) f's *inverse*
iff	*if and only if*
Im	*function image*
INT(n)	the set of all integers having at most n digits
ker	(mapping) *kernel* operator
S$_{/\sim}$	the quotient set of S with respect to the equivalence relation $\sim$
max	the *maximum* function
NAT(n)	the set of all naturals having at most n digits
RAT(n,m)	the set of all rationals having at most n digits before the decimal dot (comma, in Europe) and m after it
UNICODE(n)	the set of all strings of UNICODE characters of length at most n

CHAPTER 1

DATA, INFORMATION AND KNOWLEDGE IN THE COMPUTER ERA

Data is not information, Information is not knowledge,
Knowledge is not understanding, Understanding is not wisdom.
—Clifford Stoll

CONTENTS

1.1   Data, Information and Knowledge.. 2
1.2   Data Analysis and Conceptual Modeling.. 6
1.3   Data and Knowledge Bases 9
1.4   Constraints (Business Rules) 11
1.5   Data and Knowledge Base Management Systems
      (DBMS, KBMS) 15
1.6   Static and Dynamic Aspects of Databases 17
1.7   Conclusion and Book Overview 18
1.8   Past and Present 25
Keywords 29
References... 31

After studying this chapter you should be able to:

- Concisely define informally the following key terms: *data* (analysis, model, aggregation, generalization, factorization, plausibility, positive, negative, incomplete, semistructured, big, homogeneous, heterogeneous), *metadata*, *information*, *knowledge*, *assumption*

(closed world, closed domain, unique objects), *object set* (property), *database* (fundamental, derived, scheme, instance—both consistent and inconsistent), *inference rule, knowledge base, business rule* (constraint) (fundamental, implied, trivial, redundant, satisfaction, violation, plausible, implausible), *constraint set* (closure, equivalence, minimal, redundant, coherent, incoherent), *implication problem, conceptual data modeling, data models* (Relational (*RDM*), Entity-Relationship (*E-R*) (*E-RDM*), elementary mathematical (*(E)MDM*), logical (*LDM*)), *database management system* (*DBMS*), *knowledge-based management system* (*KBMS*), *Structured Query Language* (*SQL*) (*stored methods, Data Definition Language* (*DDL*), *Data Manipulation Language* (*DML*)), *Datalog, NoSQL* systems.
- Give examples of both commercial and open source DBMSs and KBMSs of all existing types.
- Outline the steps involved in data analysis, conceptual data modeling, database design, implementation, and optimization, as well as the proposed corresponding forward and reverse engineering (families of) algorithms.

1.1 DATA, INFORMATION AND KNOWLEDGE

The goal is to turn data into information, and information into insight.
—Carly Fiorina

Data comes from the Latin *datum*, which means both a fact (known from direct observation) and a premise (from which conclusions are drawn). Information also comes from Latin: *informatio* (meaning formation, conception, education) is the participle stem of *informare* (from where "to inform" descends).

Generally, in Information Technology (*IT*), we mean by *data* facts' (objects, concepts, etc.) property descriptions (through some values, generally of type character strings, numbers, calendar dates, etc.) which are worth being noticed and (electronically) stored for later retrieving and processing, while by *information*—increments of knowledge that can be derived from data.

For example, all cells in Table 1.1 contain data: those in the first row (of type character string) store the names of the corresponding columns (the so-called *table header* or *scheme*), which in fact, together with the table name, is *metadata*—that is, data on data—, while all of the others (the *table instance* or *content*) contain data on some Carpathians peaks.

Obviously, for example, line two of this table embeds the information that in the (Carpathians) "Făgăraşi" mountains there is a peak called "Moldoveanu" whose altitude (in meters) is 2544. Generally, each (data, not metadata) line of a table embeds the information that there is (in the corresponding set of objects) an element (object, item, etc.) having corresponding values for the corresponding properties.

But more information, of higher degrees, may also be derived from this table, for example:

1. There are 3 mountains and 5 peaks for which data is stored, out of which 2 from the "Făgăraşi" and "Retezat", and 1 from the "Piatra Craiului" mountains.
2. "La Om (Piscul Baciului)" is less high than "Retezat", which is less high than "Peleaga", which is less high than "Negoiu", which is less high than "Moldoveanu"; "Moldoveanu" is the highest, while "La Om (Piscul Baciului)" is the lowest.
3. Only 3 peaks are at least 2500 m, with two from "Făgăraşi" and one from "Retezat".
4. Overall average altitude is 2461.6 m; for "Făgăraşi" it is 2539.5 m, for "Retezat" it is 2495.5 m, and for "Piatra Craiului" it is 2238 m.
5. There is only one peak ("Retezat") that gives its name to its mountain.

Traditionally, data storage uses some communication means (e.g., images, languages, sculptures, shrines, computers, etc.) and media (lime stone, granite, papyrus, paper, HDD/CD/DVD/SSD, etc.). Getting

TABLE 1.0 Carpathians Peaks

MOUNTAIN_PEAKS

Mountain	Peak	Altitude (m)
Făgăraşi	Negoiu	2535
Făgăraşi	Moldoveanu	2544
Retezat	Peleaga	2509
Retezat	Retezat	2482
Piatra Craiului	La Om (Piscul Baciului)	2238

information generally needs retrieving and processing data (e.g., arithmetically, statistically, logically, etc.).

Note that such stored data is always positive (i.e., we always store the fact that somebody has a child, and never that somebody does not have children) and that the following three implicit assumptions are governing both storage and processing:

a. *closed world*: facts that are not known to be *true* (i.e., for which there is no stored data) are implicitly considered as *false*;

b. *closed domains*: there are no other objects than those for which there is stored data;

c. *unique objects*: objects are uniquely identifiable by their corresponding stored data.

Moreover, conventional IT information is never obtained through recursion.

Our minds, our books, our libraries have finite capabilities of both storing data and deriving information. Our computers, even if some of them have today more storage capacity and faster processing speeds than humans, are and will always be in the foreseeable future finite too.

Most of the object sets we are currently dealing with (e.g., people, organizations, services, products, etc.) are, fortunately, finite; even when they might be infinite (as, for example, the number of universe's particles) most of them are countable and we know that we can only store and process finite subsets of them.

We know how to define and characterize infinite not countable sets too (like, for example the rational, real, and complex numbers), but when it comes to manipulating we have to digitize them. Probably for all practical purposes, from measuring it to real-time management systems (be them nuclear power plants or missiles batteries), even Time is digitized.

Most of the real and complex mappings (as, for example, sine, cosine, tangent, arc cotangent, exponential, logarithm, fractals, etc.) are, first of all, only ideally encountered (e.g., when no friction, no gravity, no obstacles, etc. exist) and, secondly, are, practically, only approximately computable, with, at most, a maximum number of digits both before and after the digital dot, using discrete computations.

Fortunately, just like for sets, the vast majority of currently encountered mappings are finite too. Moreover, most of the fundamental ones (e.g., names, birth dates, e-mail addresses, population, surface, latitude and longitude, altitude, minimum and maximum temperatures, etc.) do not have any other computation "law" than their graphs (i.e., the corresponding set of pairs $<x, f(x)>$).

Consequently, we may use any finite data structure for storing data: (independent) variables, records, lists (vectors), stacks, matrices, trees, acyclic or cyclic graphs, etc. Not only for their extremely powerful intuitive appeal or algebraic naturalness, but especially as manipulating data (both for adding, modifying, removing, and querying it for deriving information) is, by far, as we will see, the simplest possible for tables (which are matrices extended with headers for storing metadata too), data is since decades and will continue to be in the foreseeable future stored in and processed from tables, even if, for example, graphs have lately reemerged as a very powerful alternative.

Knowledge (coming from old English) generally means facts, information and skills acquired through experience and/or education, as well as theoretical and/or practical understanding of a subject. In IT, *knowledge* includes positive data and information, but the latter is also obtainable from data through recursion, by applying *inference rules* (e.g., anybody who is a child of one of somebody's descendants is that somebody's descendant too, for any somebody); moreover, data may be also stored as first-order logic predicates, which allow for *negative* (e.g., "Iosif Visarionovici loved nobody") and *incomplete data* too.

For example, incomplete data is used for asserting that:
- something has a given property even if no values are available for it (e.g., "John Adam has a child");
- all elements of a set, which have not to be enumerated, have a certain property (e.g., "all children have parents");
- something has either one or another value for a property, without having to be sure which one (e.g., "Mary Eve is 29 or 30");
- two distinct expressions (e.g., "Mary Adam" and "Adam's wife") are referring or not to a same object.

Generally, positive data is *homogeneous* (i.e., for all objects of interest, same types of data are desired, like, for example, in Table 1.1), while negative and incomplete one is not. Both theory and experience proved that, for obtaining best possible querying speed and storage size, positive, homogeneous data should be stored in tables, leaving to logic predicates only negative and incomplete data, as well as inference rules. When not otherwise stated, this book deals with homogeneous data.

1.2   DATA ANALYSIS AND CONCEPTUAL MODELING

> *Given a large mass of data, we can by judicious selection*
> *construct perfectly plausible unassailable theories—*
> *all of which, some of which, or none of which may be right.*
> —Paul Arnold Srere

As usual, we are not happy only seeing trees (data): we want to see especially forests (information). As such, some data interpretation is always needed that should be enough abstract and powerful in order to understand not only data values, but especially data interconnections.

Generally, *models* are such intellectual tools, intensively used in science for abstracting details and outlining structure; from mathematical and physical ones (e.g., Big Bang, Big Crunch, relativity, quantum, strings, unification theories, etc.), economics (e.g., Nash equilibrium), geography and meteorology (e.g., maps), and up to everyday life (e.g., phone books, yellow pages, time schedules and tables, etc.) models allow us to at least partially derive and understand information.

Data models existed long before computers: from maps to phone books and all color pages, from Alumni to cook books, and from ships/trains/planes/buses time tables to accounting worksheets mankind stored data according to several data models since at least the ancient Egyptians, Babylonians, Mayans, and Chinese. We will, however, focus here only on computer manageable ones.

Such models are structuring elementary data items on which a predefined set of operations is allowed. Elementary data items are of type *<object set, object unique id, object properties, properties' values, timestamp>*, as objects descriptions are finite and characterized by sets of aggregated properties' values that may vary in time. For example, *MOUNTAIN_PEAKS* objects from Table 1.1 are aggregated from the *Mountain*, *Peak*, and *Altitude* properties. *PEOPLE* might be aggregated from *SSN*, *FirstName*, *LastName*, *Sex*, *BirthDate*, *BirthPlace*, and *e-mail*.

As in most of the cases we are only interested in current values or, at most, in the history of few some of them (e.g., hire/marriage/divorce/start/ end dates), generally timestamps are dropped or replaced by calendar date/ time values.

The sets of allowed operations generally include *create*, *alter*, and *drop*—for object set collections, objects, users, and user group sets, etc., *select*, *insert*, *update*, and *delete*—for elementary data items aggregated in instances of above collections and sets, *grant* and *revoke* rights on object collections and sets to/from users and groups, etc.

Besides properties *aggregation*,[3] data structuring is obtained by hierarchies of *categorizations/classifications/generalizations*[4] and *factorization* (of common properties, on appropriate levels). For example, we might characterize persons as the aggregation of SSN[5], first and last name, birth date, birth place, sex, and e-mail address; some catalog items may be classified as water, wine, beer, cheese, vegetables, women cosmetics, men cosmetics, etc.; then, the first two above may be categorized as beverages, the following two as food, and the last two as cosmetics; beverages and food may then be generalized as aliments, aliments and cosmetics may be generalized as products, etc.; corresponding classes, categories, types, families, etc. names and properties should be factorized on their appropriate levels of this hierarchy.

This is why it is no wonder that the first data model was the *hierarchical* one, especially as—from states, armies, and organizations to biology, botany, and zoology—we are living in a world of hierarchies.

Then, hierarchies (trees) were extended to the more general graphs in a *network* data model, where nodes are elementary data items and edges are connections (links, relationships, etc.) between them.

The first mathematical data model was the *relational* (RDM) one, presented in the third chapter of this first volume: initially, although several equivalent interpretations are possible (and we will provide one more in Section 3.12), it was introduced as sets of representatives of mathematical relations (defined over data domains/types) equivalence classes (where

---

[3] Mathematically, aggregations are function (Cartesian) products.
[4] Mathematically, such hierarchies are function compositions; in Artificial Intelligence (AI), categorizations/classifications/generalizations are IS-As.
[5] The acronym for Social Security Number.

each such class contains all possible permutations of underlying sets of a relation), interconnected by common data domains. Ultimately, although, again, several equivalent representations are possible, the one consisting of sets of tables (each made out of a collection of columns) interconnected by foreign keys prevailed, as it is the most natural and simple one for representing finite data.

Essentially, RDM is purely syntactic: for example, theoretically, no table should correspond to any actual set of objects. This is why designing RDM schemas proved almost intractable: which column should belong to which table and why? Which tables should we add and why? This is why, hundreds of semantic data models aroused in these last four decades and this will continue to happen.

One of the most still used is graphical, so very powerful if correctly used: the *entity-relationship* data model (*E-RDM*), presented in the next chapter, uses rectangles for atomic object sets, diamonds for mappings and relations between them, and ellipses for their base properties; due to their simplicity, its *E-R diagrams* (*E-RDs*) can be best understood by our customers too.

However, E-RDM is fine only for depicting the structure of data (which object sets are of interest, what are their properties and interconnections between them), but not also for embedding all of the business rules that are governing any given subuniverse of discourse. This is why hundreds of higher level (semantic) more sophisticated data models were proposed too.

For example, this book proposes the (*elementary*) *mathematical* one (*(E)MDM*), presented in the second chapter of its second volume, for refining the conceptual data modeling such as to guarantee not only the best suited structure, but also the highest possible level of data plausibility.

In (E)MDM, for example, one can also express constraints like "any capital of a country must be a city of that country" and "any hotel room may be reserved for any day by only one person", which are not expressible either in E-RDM or in RDM.

Before using any data model, the first and one of the most important steps is *data analysis*: understand exactly what data is needed in order to correctly model a given subuniverse of discourse (e.g., an e-commerce, a bank, a supermarket, etc.) and what are the business rules that govern it.

The next chapter is devoted to both data analysis and its deliverables: an exhaustive, concise and clear informal description of the needed data and existing business rules, as well as a corresponding set of E-RDs.

The shortest advisable path for db design and implementation needs then an algorithm to translate E-RDs and associated business rules into rdb schemas and (possibly empty) sets of associated nonrelational constraints (i.e., business rules that are not expressible in the RDM framework and, consequently, need to be enforced by software applications built on top the corresponding database).

However, the best approach to conceptual data analysis and modeling, as well as database design and implementation is to first translate E-RDs and associated business rules into a higher level semantic data model scheme, refine it as much as possible, and only afterwards translating this refined scheme into RDM schemas and (possibly empty) sets of associated nonrelational constraints.

Finally, RDM schemas can be implemented as databases (through corresponding translation algorithms presented in the fourth chapter), by using any desired (or imposed) *relational database management system* (*RDBMS*) version.

Knowledge is stored both using RDM and logic (for negative and incomplete data) and inferred by *logical data models* (presented in the first chapter of the second volume), based on first order logic.

1.3 DATA AND KNOWLEDGE BASES

> *It is a capital mistake to theorize before one has data.*
> *Insensibly one begins to twist facts to suit theories,*
> *instead of theories to suit facts.*
> —Sir Arthur Conan Doyle

Any data collection structured according to a data model is called a *database* (*db*). This is why, on one hand, there are hierarchical, network, relational (*rdb*), object-oriented (*O-O*), etc. dbs and, on the other, not any data collection (e.g., not even a "relational database" storing aberrant data) is a db. As such, dbs are models representing subuniverses of interest to at least some of us.

For example, a geographic rdb could be made of the interrelated tables *CONTINENTS, ARCHIPELS, ISLANDS, MOUNTAIN_RANGES, MOUNTAINS, PEAKS, RIVERS, LAKES, SEAS, OCEANS, COUNTRIES, STATES, CITIES*, etc.

Databases are three-dimensional objects: their *schemes* store the structure and the associated constraints (i.e., their metadata), their *instances* store actual data, whereas their *SQL stored programs* (views, triggers, functions, procedures, etc.) are modifying and/or querying instances.

For example, such a stored view could compute, in alphabetic order, the set of triples <country name, state name, city name>; a stored trigger could automatically add to *STATES* an "UnknownState" whenever a new country is added to *COUNTRIES*; a stored procedure might compute the number of states, the corresponding number of cities, as well as their total number for any country given to it as a parameter.

Beware of notational abuses, as, for example, instances are often referred to also as "databases".

Modern *knowledge bases* (*kb*) extend databases with *inference rules* (*IR*), adding to them the power of recursiveness and negation.

For example, a genealogic rdb, containing a *PEOPLE* table also including columns *Mother* and *Father*, could be extended with *Datalog*[6] programs for computing the set of all genealogic ancestor and/or descendent relationships[7], or all (male and/or female) ancestors and/or descendants of a given person (*see* Exercise II.1.1 in the second volume).

Note that *conventional dbs* do not embed inference rules, so can only infer information, not knowledge.

Also note that there are *fundamental dbs* (which store data which cannot be retrieved from other dbs) and *derived dbs* (like, for example, the Online Analytical Processing (*OLAP*) and/or Business Intelligence (*BI*) ones, which store only data that can be obtained from other dbs). Except when otherwise explicitly stated, this book deals only with fundamental ones.

---

[6] *Datalog* (see the first chapter of the second volume) is a happy marriage between rdbs, the programming language based on first order logic *Prolog*, and a much more efficient computational semantics than *Prolog*'s one.

[7] The so-called *transitive closure* of *Mother* and/or *Father*.

1.4   CONSTRAINTS (BUSINESS RULES)

> *Mathematics may be compared to a mill of exquisite workmanship,*
> *which grinds you stuff of any degree of fineness; but, nevertheless, what*
> *you get out depends upon what you put in; and as the grandest mill in the*
> *world will not extract wheat-flour from peasecods, so pages of formulae*
> *will not get a definite result out of loose data.*
>   —Thomas Henry Huxley

Besides interrelated tables simplicity and naturalness, besides the extremely powerful associated querying formalism also introduced, RDM had the major contribution of introducing and formalizing *constraints* (*business rules*) too: logic formulas that are added to dbs schemas and enforced by RDBMSs in order to reject attempts to store undesired data in their instances.

For example, if we leave *Altitude* from Table 1.1 above to get integer values (as most computer programmers, but also too many database designers are still doing), end-users might enter peaks having either negative, less than 1000, or greater than 8850 ones too.

Some are still speaking about data correctness in this respect, although not only me is inclined to not ever use this syntagm: even today, for example, there is still debate even on mountain peaks' altitudes, as it depends on measurement instruments, meteorological conditions, Earth tectonic plates' movements, etc. Moreover, even if we, humans, were in the possession of only relatively correct altitude values for all peaks, how could an automaton (e.g., a computer program) reject human data entering mistakes (e.g., typo ones or swapping between altitudes of two peaks)?

But even if we do not know (or agree on the fact) whether, for example, peak Moldoveanu has 2543 m or 2545 m (as some newer measurements reportedly suggest), so we cannot instruct our software to accept or reject any such value, we can at least instruct it that only values between 1000 (by geographical convention, the minimum altitude in meters for a peak in order to consider it a mountain one) and 8850 (Everest's altitude) are acceptable, as, otherwise, they are not plausible. Of course that, for enhancing plausibility, we could even store maximum elevations per mountains and instruct software to dynamically compute maximum thresholds mountain per mountain.

In order to always guarantee *data plausibility*, data model schemas should also include all needed constraint types. For example, the above "peaks should have altitudes between 1000 and 8850" is formalized in RDM as the "domain" constraint "*Altitude* between [1000, 8850]".

When enforced, constraints are always *satisfied* by (or *hold in*) any corresponding db instance, as any attempt to *violate* them is rejected. For example, above constraint is satisfied by Table 1.1's instance. When not enforced, some instances may *violate* (or *do not satisfy*) them. For example, if we replace in Table 1.1 any altitude with 900 or 9000 (or any other natural less than 1000 or greater than 8850), then above constraint would be violated.

An instance is said to be *consistent* (*valid*) if it satisfies all (table/db) constraints; otherwise, it is called *inconsistent* (*invalid*).

Note that constraints exist independently of whether we know and/ or ignore them and that only adding all existing ones in our data models may guarantee data plausibility: "*garbage in, garbage out*" applies in this context too if and only if there is at least one missing constraint from our corresponding model.

1.4.1 IMPLIED CONSTRAINTS AND CONSTRAINT SET CLOSURES

> *You can have data without information,*
> *but you cannot have information without data.*
> —Daniel Keys Moran

As constraints are first order logic formulas,[8] one of them or a set of constraints may *imply* other constraints as well: a set of constraints C implies a constraint c if c holds in all instances in which C holds (dually, C does not imply a constraint c if there is an instance for which c does not hold, but C holds).

The standard logical notation for formulas implication is $C \vdash c$; c is called an *implied constraint*. For example, in Table 1.1, constraint set

---

[8] In particular, *closed* ones, that is, formulas whose variable occurrences are bound to at least one logic quantifier (be it "for any" or "there is"). Dually, *open* formulas have at least one occurrence of a variable free (i.e., not bound to any logic quantifier), which, in dbs, are formalizing queries.

{*Altitude* > 1000, *Altitude* < 2550} implies constraint *Altitude* $\in$ (1000, 2550); trivially, the vice-versa is also true. Constraints that are not implied are called *fundamental*.

We should never enforce implied constraints in db schemas: it would only be superfluously time consuming.

Constraint implication is not important only for minimizing constraint sets, but also for defining constraint sets closures that are needed for defining many important db theoretical notions, including the highest RDM normal form: given any constraint class CC, its subset containing all constraints implied by any of its subsets C is called the *closure* of C (with respect to that class) and is denoted by $(C)_{CC}^{+}$ (or, if CC is understood from the context, simply C^{+}). For example, if $C = \{S \subseteq T, T \subseteq U\}$, its transitive closure is $C^{+} = \{S \subseteq T, T \subseteq U, S \subseteq U\}$.

Closures of a set with respect to different classes are generally different (e.g., the reflexive closure of the above C is $C^{+} = \{S \subseteq T, T \subseteq U, S \subseteq S, T \subseteq T, U \subseteq U\}$).

Note that any set is included in its closure and that the closure of a closure is itself (i.e., for any C and CC, $(C^{+})^{+} = C^{+}$). Moreover, the closure of a subset is a subset of the closure of the corresponding superset (i.e., for any C, D, and CC, $C \subseteq D \Rightarrow C^{+} \subseteq D^{+}$), whereas the union of two closures is distinct of the closure of the corresponding union (i.e., for any C, D, and CC, $C^{+} \cup D^{+} \neq (C \cup D)^{+}$).

A set which is equal to its closure is called *closed with respect to implication* (e.g., $C = \{S \subseteq T, T \subseteq U, S \subseteq U\}$ is transitively closed).

Two sets having same closures are called *equivalent* and are said to *cover* each other (e.g., C and C' above are equivalent or covering each other).

1.4.2 REDUNDANT CONSTRAINTS AND MINIMAL CONSTRAINT SETS

> *There is an enormous redundancy in every well-written book.*
> *With a well-written book I only read the right-hand page*
> *and allow my mind to work on the left-hand page.*
> *With a poorly written book I read every word.*
> —Marshall McLuhan

A constraint is said to be *redundant* in a set when the set's closure remains the same after removing it (e.g., $S \subseteq U$ is redundant in C' above).

Any cover not containing redundant constraints is said to be *minimal* (e.g., C above is minimal under both transitivity and reflexivity).

1.4.3 THE IMPLICATION PROBLEM

> *To think out in every implication the ethic of love for all creation...*
> *this is the difficult task which confronts our age.*
> —Albert Schweitzer

Testing for constraint sets equivalence and/or constraints redundancy can be reduced to the well-known *implication problem*: given any constraint set C and constraint c, does C imply c? (e.g., does C above imply $S \subseteq U$? does it imply $T \subseteq U$ too?)

Note that, depending on the constraint classes, this problem may be solved very quickly (i.e., linearly), very slowly (i.e., exponentially), or be impossible to solve (even undecidable[9]).

1.4.4 TRIVIAL CONSTRAINTS

> *Science is an organized pursuit of triviality.*
> *Art is a casual pursuit of significance.*
> *Let's keep it in perspective.*
> —Vera Nazarian

A very particular case of implied constraints are obtained when C is empty: a constraint implied by the empty set is said to be *trivial*, as it always holds in any db instance.

For example, $(A = B \lor A \neq B)$ is trivial for any sets A and B.

---

[9] A problem is said to be *undecidable* if it is neither provable, nor refutable or for which it is impossible to design an algorithm that always (i.e., in any context) correctly answers to its questions (see Appendix).

1.4.5 PLAUSIBLE AND IMPLAUSIBLE CONSTRAINTS: COHERENT AND INCOHERENT CONSTRAINT SETS

Civilization is a progress from an indefinite, incoherent homogeneity toward a definite, coherent heterogeneity.
—Herbert Spencer

Given any db and its associated set of constraints C, C is *coherent* (with respect to the db scheme) if for any table of the db there is at least a non-void instance that satisfies C and *incoherent* otherwise.

For example, if a db scheme contains the above *MOUNTAIN_PEAKS* table and constraints C_1: *Altitude* $\geq$ 1000 and C_2: *Altitude* < 1000 (or, equivalently, constraint C: *Altitude* $\geq$ 1000 $\wedge$ *Altitude* < 1000), then the only possible instance for *MOUNTAIN_PEAKS* is obviously the empty set, which means that the corresponding set of constraints C is incoherent.

Trivially, from any incoherent set a coherent one may be obtained by removing any pair of contradicting (sub)constraints (C_2 in the above example).

Note that, depending on the context, constraints too may be *plausible* (i.e., corresponding to an existing business rule governing that context) or *implausible* (*dictatorial*, *aberrant*: that is not corresponding to an existing business rule governing that context): for example, C_1 above is plausible in any mountains context, while C_2 is not. Obviously, we must always include in our data models only plausible constraints.

1.5 DATA AND KNOWLEDGE BASE MANAGEMENT SYSTEMS (DBMS, KBMS)

The only source of knowledge is experience.
—Albert Einstein

Any software tool providing db schemas definition facilities (i.e., a *data definition language* (*DDL*), including *create*, *add*, *alter*, *drop*, *grant*, and *revoke* statements), querying and processing instances facilities (i.e., a *data manipulation language* (*DML*), including *select*, *insert*, *update*, and *delete* statements), as well as other needed facilities (e.g., at least, *con-*

straint enforcement (*CE*), *data access control* (*DAC*), *concurrency control* (*CC*), and *data and programs backup and restore* (*DPBR*)) is said to be a *database management system* (*DBMS*).

For example, IBM *DB2*, Oracle *Database* (*Oracle*, for short) and *MySQL*, Sybase *Advantage Database Server*, Microsoft *SQL Server* and *Access*, Tandem *NonStop SQL Server*, Apache *Hadoop*[10] (actually, its DBMSs components *Cassandra*, *HBase*, and *Hive*), etc. are all DBMSes. All of them provide a variant of the ANSI[11] standard relational language *SQL*[12], which include both a *DDL* and a *DML*. The relational DBMS (RDBMS) cores (i.e., *SQL* parsers, compilers/interpreters, optimizers, evaluators, etc.) are called *SQL engines*.

Note that binary, assembly, and high level programming languages (*PLs*) are manipulating data as bits, bytes, words, records, record sets, and/or files (stored and managed by operating systems (*OS*) file systems); hierarchical, network, and relational DBMSs are storing and manipulating data values; O-ODBMSs are storing and manipulating objects, while *knowledge base management systems* (*KBMS*, for example: *LDL*, *Coral++*, etc.) are storing and manipulating knowledge. There are also hybrid products, like, for example, *MatBase*, based on both *Datalog*, (E) MDM, E-RDM and RDM that is storing and manipulating both objects and knowledge (which is why they are called *knowledge and DBMS*, *KDBMS* for short).

Please note that, just like for OSs, PLs, or any other system software, (K)DBMSs are just tools, let's say pens, and that it is important to know how to use them the best possible, but do not ever forget that *it does not matter with what pen you write: what matters is what you write*.

Moreover, for all tools, including the above mentioned ones, keep in mind that, during your professional lifetime, they are continuously evolving, sometimes dying, others appearing and that, generally, you will have to switch several times to other positions, asking from you to work with other tools than those you were already acquainted with: consequently, the

---

[10] Open source framework developed in *Java* that stores, manages and processes gigantic amounts of data in a highly parallel manner, on clusters of commodity machines, and supporting large scale data analysis by allowing question decomposition into discrete chunks that can be executed independently very close to slices of the data in question and, ultimately, reassembled into an answer to the posed question; used by *Facebook* too.

[11] The acronym for American National Standards Institute.

[12] The acronym for Structured Query Language.

most important things are to master conceptual data analysis, db design, implementation and optimization, as well as querying and manipulations principles, algorithms and best practice rules, and mastering technologies comes only last (even if this is a must too).

This is why, although necessarily having to exemplify with some of the most currently used RDBMS versions, this book is not about IBM *DB2* or *Oracle* or *MySQL* or MS *SQL Server* or *Access* or whatever other such tool.

Also note that using a DBMS does not guarantee at all that your outcome is a database, just like using wood and appropriate carpenter tools does not guarantee at all that your outcome is, at least, furniture. Without proper data analysis, conceptual and physical db design, implementation optimization, one cannot equal in this field a wood artist (be it sculptor or cabinetmaker) and not even a good furniture manufacturer, but simply remains a woodcutter.

1.6   STATIC AND DYNAMIC ASPECTS OF DATABASES

> *Hiding within those mounds of data is knowledge*
> *that could change the life of a patient, or change the world.*
> —Atul Butte

Let us consider stores also selling CDs/DVDs/Blu-rays and potential buyers trying to get answers (either directly, from a db application provided on public available computers, or indirectly, through store employees) to their following possible questions:
- "What recordings of Dinu Lipatti do you have?"
- "Who else is also costarring in these recordings?"
- "How much do they cost?"
- "What are the opuses contained by each of them?"
- "What are their corresponding recording houses?"
- "What are their recording quality ratings?"

This typical example involves two subtopics:
- a *static* one: what data should be stored and how, such that answers to these queries (and any other similar ones) may be provided correctly and promptly (keeping in mind too that stocks

are varying in real time)? (i.e., what should be the scheme of the needed db and how should it be implemented on a DBMS? What columns and constraints in what tables?) For example, Table 1.1 should be split into two tables, one for mountains and the other for peaks, and all existing constraints should be added to both of them.

- a *dynamic* one: how should the corresponding db be queried for getting correct and prompt answers? (i.e., what should be the architecture, design, and development of the needed application built on top of the above db?)

Generally, dbs have four dimensions: design, implementation, optimization, and usage (i.e., programming for querying and modifying data). The design, implementation and optimization are static, while usage is dynamic.

Note that, generally, it is true that daily db tasks involve in average some 90% usage, 3% design and implementation, and 7% optimization; consequently, at a first glance, it might seem that mastering only *SQL*'s *DML* is almost the only key to success in the db field.

However, both design, implementation and optimization are crucial and can make the huge difference between a very poor and cumbersome set of interrelated tables and a true database: who would not love to decorate and live in a beautifully and intelligently architectured, designed, and built spacious home full of light and splendid views around, but prefer instead a dark underground labyrinth shared with some minotaur?

1.7   CONCLUSION AND BOOK OVERVIEW

> *There are two modes of acquiring knowledge,*
> *namely, by reasoning and experience.*
> *Reasoning draws a conclusion and makes us grant the conclusion,*
> *but does not make the conclusion certain, nor does it remove doubt*
> *so that the mind may rest on the intuition of truth*
> *unless the mind discovers it by the path of experience.*
> —Roger Bacon

Data was stored since the dawn of humanity on various media in order to infer from it information and knowledge. Most of the data is discrete, which makes tables and graphs ideal candidates to store it, with tables having great advantages on both storing and querying, which are still the simplest and fastest.

Especially in the computer era, where the amount of stored data is huge (think, for example, of the astronomical one) and there may be several thousands of people simultaneously updating it, but especially as crucial global and local decisions, be them in politics, intelligence, military, science, health, commerce, etc., are more and more heavily based on data, guaranteeing at least its plausibility is of paramount importance.

Consequently, on one hand, more and more sophisticated data models were introduced, even if, commercially, from the implementation point of view, the vast majority of the existing database management systems are relational. This means that only part of the existing business rules may be enforced through them and for the rest we have to develop software applications built on top of the corresponding databases.

On the other hand, just as starting to directly write *Java* (or *C#* or whatever other PL) code, immediately after getting some desired software application requirements, without proper problem analysis, software architecture and design is a huge mistake, starting to directly create tables in a database on a RDBMS is an even greater one: databases are the foundations of db software applications.

Just like no building resists earthquakes, tsunamis or even significant winds or rains if its foundations are not solid enough, no db software application may be easily designed, developed, maintained, and used if its underlying db was not properly designed and implemented.

As such, whenever a new database is needed or an existing one has to be expanded, first of all, besides establishing exactly what are the corresponding data needs, data analysis has to also discover all business rules that govern the corresponding subuniverse.

Then, these findings should be structured according to a data model that, even if it is not as powerful as needed in order to guarantee data plausibility, is comprehensible by the customers too. We will see in the next chapter that *E-R data models* (i.e., E-RDs accompanied by restriction sets and informal descriptions of the corresponding subuniverse) are an excellent candidate for this first step.

Chapter 3 of this first volume provides enough evidence that RDM is indeed excellently suited for elegantly and efficiently data both storing and manipulating.

Finally, Chapter 4 hopefully proves not only that for any RDBMS version there is an algorithm that translates any relational scheme into a corresponding db, but also that it is worthwhile first thoroughly analyzing the corresponding subuniverse, obtaining its E-RDs and associated restriction set, translating it into a relational scheme, and only then translating this scheme into a db.

The second volume first tackles recursive queries, a logic data model for also getting knowledge from data, not only information, and the highest relational normal forms that we are proposing; then, it introduces an elementary mathematical data model, for refining as much as possible E-RDs and restriction sets before translating them into relational schemas. Its final two chapters deal with enforcing nonrelational constraints and optimizations of dbs, respectively.

Obviously, forward engineering algorithms should be applied whenever new dbs need to be designed, implemented, optimized, and extended. However, much more frequently than in almost any other industry, we are dually confronted with the need to exploit, maintain, and extend legacy dbs, which, generally, lack either completely or at least partially corresponding design, implementation, and optimization adequate documentation. This is why corresponding dual reverse engineering algorithms are very useful (and sometimes even crucial) too.

Figure 1.0 presents the 25 algorithms proposed in this book. Figure 1.1 shows an overview of the proposed fully algorithmic data analysis, conceptual data modeling, database design, and optimization forward engineering process, as well as its dual reverse engineering one.

$A7/8$–3 needs to be applied at the RDM level only if neither $A3$, nor $A3'$ were previously applied (i.e., when the shortcut $A1$–7 was taken instead of the full normal path $A1$, $A2$, $A3$, $A4$, $A5$, $A6$, $A7$); similarly, it needs to be applied at the RDBMS level only if neither $A3$ nor $A3'$ were previously applied and if it was not applied at the RDM level either (i.e., when the shortcut $AF1$–8 was taken instead of the full normal path $A1$, $A2$, $A3$, $A4$, $A5$, $A6$, $A7$, $A8$, $AF8'$).

*A*0—The algorithm for assisting data analysis and initial conceptual data modeling (in order to obtain E-R data models)

*A*1—The algorithm for translating E-R data models into mathematical schemas

*A*2—The algorithm for assisting validation of the initial E-R data model

*A*3—The algorithm for assisting keys discovery (mathematical level)

*A*3′—The improved algorithm for assisting keys discovery (mathematical level)

*A*4—The algorithm for assisting analysis of E-RD cycles

*A*5—The algorithm for guaranteeing the coherence of constraint sets

*A*6—The algorithm for minimization of constraint sets

*A*7—The algorithm for translating mathematical schemas into relational ones and associated nonrelational constraint sets

*A*7/8–3—The algorithm for assisting keys discovery (relational/ RDBMS levels)

*A*8—The algorithm for translating RDM schemas into *SQL* ANSI *DDL* scripts

*AF*8′—The family of algorithms for translating ANSI *DDL* scripts into rdbs

*AF*9—The family of algorithms for assisting enforcement of nonrelational constraints through *SQL* triggers

*AF*9′—The family of algorithms for assisting enforcement of nonrelational constraints through high level PL embedding *SQL* trigger-type methods

*AF*10—The family of algorithms for assisting initial optimization of rdbs

*AF*11—The family of algorithms for assisting regular optimizations of rdbs

*A*1–7—The algorithm for translating E-R data models into relational schemas and associated nonrelational constraint sets*

*AF*1–8—The family of algorithms for translating E-R data models into rdbs and associated nonrelational constraint sets*

*REAF*0′—The family of algorithms for translating rdb schemas into *SQL* ANSI *DDL* scripts

---

*REA*0—The algorithm for translating *SQL* ANSI *DDL* scripts into relational schemas

*REA*1—The algorithm for translating relational schemas into mathematical ones

*REA*2—The algorithm for translating mathematic schemas into E-R data models

*REAF*0–1—The family of algorithms for translating rdb schemas into mathematical ones

*REAF*0–2—The family of algorithms for translating rdb schemas into E-R data models

*REA*1–2—The algorithm for translating relational schemas into E-R data models

*: *without the refinements brought by the algorithms A2 to A6.*

FIGURE 1.0   Proposed algorithms for data analysis, db design, implementation, and optimization, as well as their dual reverse engineering ones.

In fact, out of these 25 algorithms 9 (*AF*8', *AF*9, *AF*9', *AF*10, *AF*11, *AF*1–8, as well as *REAF*0', *REAF*0–1, and *REAF*0–2) are family of algorithms (with the 'F' in their name from "Family"), each of them including one corresponding algorithm per RDBMS version. For example, all of them include the subfamilies *DB2*, *Oracle*, *SQL Server*, *Access*, etc.; in their turn, for example, all *Oracle* subfamilies include algorithms for its versions 12c, 11g, 10g, 9i, etc. For each such family, this book actually presents, as an example, only a couple of algorithms and invites its readers to similarly design their own ones for the rest of the RDBMS versions.

18 algorithms are of forward and 7 of reverse engineering types. Reverse engineering algorithms are represented with upwards dashed arrows and their names are prefixed with "REA"; forward ones have downwards not dashed arrows and their names are prefixed with "A".

19 algorithms are fundamental and 6 are derived ones: these latter "shortcut" ones are composed: *A*1–7 is the composition of *A*1 followed by *A*7; any algorithm from *AF*1–8 is the composition of *A*1–7 followed by *A*8, and one from *AF*8'; *REA*1–2 is the composition of *REA*1 followed by *REA*2; any algorithm from *REAF*0–1 is the composition of one algorithm

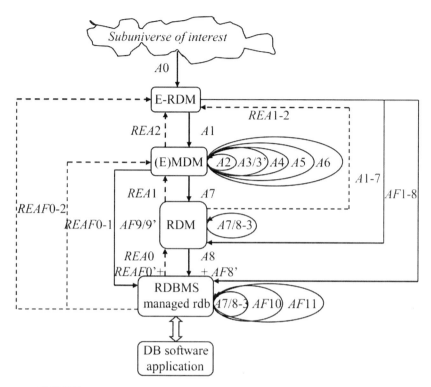

FIGURE 1.1 Overview of proposed algorithms for data analysis, db design, implementation, and optimization, as well as their dual reverse engineering ones.

from *REAF0'* followed by *REA0* and *REA1*; finally, any algorithm from *REAF0–2* is the composition of one algorithm from *REAF0–1* followed by *REA2*.

16 algorithms are of translation and 9 of refinement types: *A2*, *A3*, *A3'*, *A4*, *A5*, *A6*, *A7/8–3*, *AF10*, and *AF11* are refining-type algorithms; all of the others are translation-type (from one formalism to another) ones. 15 algorithms need not human intervention, while 10 (the assisting ones) do need it: *A0*, *A2*, *A3*, *A3'*, *A4*, *A7/8–3*, *AF9*, *AF9'*, *AF10*, and *AF11* are assisting algorithms, as only humans may take corresponding needed decisions.

All of these 25 algorithms may and should be embedded in any powerful DBMS. Unfortunately, today's RDBMSs are only providing A8, *AF8'* and, partially, *REAF0–2* and/or *AF10/AF11*; only a couple of them (e.g., *DB2* and *Oracle*) are also offering *AF1–8* (but without the refinements brought by the algorithms *A2* to *A6* or, at least, by *A7/8–3*).

Theoretically, E-RDM might be dropped (in which case an $A0$–1 algorithm should replace $A0$ and $A1$), but, in my opinion, it would not only be a pity, but neither users and nor even business analysts and db designers would be prepared or at least ready to get prepared for it in the foreseeable future.

(E)MDM might be replaced with any other higher-level conceptual data model that provides at least equivalent to $A1$, $A2$, $A3/A3$', $A4$, $A5$, $A6$, $A7$, $AF9$, and $AF9'$ algorithms. Dropping this level would generally result in semantic overloading of object sets and/or much, much poorer data plausibility, as well as possible constraint sets incoherence and/or redundancy.

RDM might be replaced as well (e.g., with a modern graph-based data model, an O-ODM, a semistructured, or a heterogeneous[13] one), provided that $A7$, $A7/8$–3, if needed, and $AF8'$ are replaced with corresponding equivalent ones. Dropping this level would compromise the portability of the conceptual data modeling solutions on any number of existing and to come RDBMS versions. Moreover, learning RDM is for RDBMSs similar, for example, to learning formal languages for PLs: you may then extremely easily navigate between any versions of any manufacturer, only having to pay attention to their minor syntactical and technological peculiarities.

RDBMS versions might be replaced, for example, with O-ODBMS or *NoSQL*[14] ones, provided that $AF8$', $AF9$, $AF10$, and $AF11$ are replaced with corresponding equivalent ones.

This first volume introduces only the E-RDM, RDM, RDBMS dbs, and (the family of) algorithms $A0$, $A1$–7, $A7/8$–3, $A8$, $AF8$', $AF1$–8, $REAF0$', $REA0$, $REA1$–2, and $REAF0$–2; the (E)MDM and the rest of the algorithms are presented in the second volume.

In my opinion, the shortest forward advisable path is $A0$, $AF1$–8, $A7/8$–3, $AF10$, while the professional one is $A0$, $A1$, $A2$, $A3/A3$', $A4$, $A5$, $A6$, $A7$, $A8$, $AF8$', $AF10$, $AF11$.

Unfortunately, the most frequent path taken today (if any!) is $AF1$–8, with some pros also considering some partial implicit variants of $A0$, $AF10$ and $AF11$.

---

[13] Data is *heterogeneous* when has several format types (e.g., a mixture of at least two of the following: binary, contiguous, graph, image, map, audio, video, ranking, URL, etc.). *Heterogeneous DBMS* (*HDBMS*) are systems able to integrate heterogeneous data managed by other DBMSs.

[14] *NoSQL DBMS* (*NoSQL* = Not Only *SQL*) are systems that replaced RDM with some other types of less restrictive data models (e.g., based on pairs of type <key, value>, or graphs, or documents, etc.), but still use *SQL*-like languages (for both declaring and processing data).

1.8 PAST AND PRESENT

I like the dreams of the future better than the history of the past.
—Thomas Jefferson

Real generosity toward the future lies in giving all to the present.
—Albert Camus

In my opinion, modern data modeling era opened with Mealy (1967) and Quillian (1968).

Codd (1970) introduced the RDM.

Chen (1976) introduced the E-RDM.

Smith and Smith (1977) was the first to explore using generalization and aggregation as data modeling tools.

Gallaire and Minker (1978) first married logic and databases: knowledge bases (in particular, deductive databases) sprung from it.

Hammer and McLeod (1978) proposed a first semantic data model.

Mancas (1985) introduced the (E)MDM, and Mancas (1990, 1997, 2002) extended it.

The two main prototype implementations that immensely contributed to the technology of the RDBMS field were the *System R* project at IBM Almaden Research Center (Astrahan, 1976) and the *INGRES* project at Berkeley University (Stonebraker et al., 1976). Both of them were crucial for establishing this type of DBMS as the dominant db technology of today, including the relational languages *SQL* (Chamberlin and Boyce, 1974), the *de facto* standard one today, and its rival *Quel* (Stonebraker et al., 1976).

IBM *DB2* (http://www-01.ibm.com/software/data/db2/), *System R*'s direct descendant, celebrated a couple of years ago its 30 years of success and superior innovation (Campbell et al., 2012).

DB2's main competitor is another *System R*'s descendant, the *Oracle Database* (http://www.oracle.com/us/products/database/overview/index.html), or, simply, *Oracle*, which managed to reach the market some two years before *DB2* and which continued ever since to challenge it in a beautiful race of innovation and excellence (Ashdown and Kyte, 2014).

INGRES too spawned several very successful commercial RDBMSs: the current open source commercial *Ingres* supervised by Actian Corporation (http://www.actian.com/products/operational-databases/), Sybase (now a SAP company) *SQL Server* (currently known as the *Advantage Database Server*, http://www.sybase.com/products/databasemanagement/advantagedatabas-eserver, after Microsoft bought its license in 1992), Tandem *NonStop SQL* (http://h17007.www1.hp.com/ro/en/enterprise/servers/integrity/nonstop/nonstop-sql.aspx#.U8BUe7EhHCI), now owned by Hewlett-Packard, the post-*Ingres* family of products *Postgres*, which evolved into the current open source *PostgreSQL* (http://www.postgresql.org/), and *Informix Database*, now IBM owned (http://www-01.ibm.com/software/data/informix/) are the best known ones.

Besides the *SQL Server* (http://www.microsoft.com/en-us/server-cloud/products/sql-server/#fbid=o6jljjvm1J0, http://www.sqlservercen-tral.com/), now the biggest competitor for *DB2* and *Oracle* (all three of them currently having cloud-empowered solutions available too), Micro-soft also still offers its genuine smaller scale RDBMS *Access* (http://office.microsoft.com/en-us/access/).

MySQL (http://www.mysql.com/), one of the most popular open source small scale RDBMS, is now under Oracle Corp.'s umbrella too.

Although they use their powerful in-house developed DBMS *BigTable* (https://developers.google.com/appengine/articles/storage_breakdown), not distributed outside the company, but to which users of the *Google App Engine* cloud platform as a service (https://developers.google.com/appen-gine/) may have access however, Google is offering their *MySQL*'s *Cloud SQL* version too (https://developers.google.com/cloud-sql/).

Nowadays, on one hand, even smartphones are equipped with RDBMSs (e.g., Android is using *SQLite*, http://www.sqlite.org/, an open source small RDBMS engine sponsored mainly by Oracle, Bentley, Mozilla, Nokia, Adobe, and Bloomberg, which, probably, is the current most widely deployed mobile db engine worldwide).

On the other, the so-called *big data*[15] companies are using, just like Google, so-called *NoSQL* DBMSs, like Apache *Cassandra* (http://cas-sandra.apache.org/, developed by Facebook when *MySQL* proved to not be satisfactory anymore and currently used by more than other 1,500

---

[15] *Big data* is denoting data sets generally unstructured and having sizes of petabytes (PB, 10^{15} bytes = 10^3 terabytes), which are consequently beyond the ability of commonly used software tools to manage and process within user tolerable periods of time.

companies too, among which, for example, are CERN, the Swiss-based European Council for Nuclear Research, which built and uses the Large Hadron Collider, eBay, and The Weather Channel), *HBase* (http://www.webuzo.com/sysapps/databases/HBase), the main *Hadoop* db engine (http://hadoop.apache.org/, whose underlying technology was invented by Google and then developed by Yahoo), and *Hive* (http://hive.apache.org/, of *SQL*-type, born within the *Hadoop* project too, but now being a project of its own), all of them open source and best suited for mission-critical data that is replicated in hundred worldwide scattered data centers.

For example, (Redmond and Wilson, 2012) presents 7 *NoSQL* open source solutions (including *HBase* and *Postgres*). Moreover, there are also bridges built between the complementary *SQL* and *NoSQL* solutions, for bringing speed and BI capabilities to querying heterogeneous big data, like, for example, the one between the MS *SQL Server 2012* and *Hadoop* (Sarkar, 2013).

O-ODBMS were introduced starting with the 1982 *Gemstone* proto-type (see http://gemtalksystems.com/index.php/products/gemstones/). Even if they only managed to conquer niche areas (CAD, multimedia pre-sentations, telecom, high energy physics, molecular biology, etc.), their influence on RDM and RDBMSs, at least for the hybrid *object-relational* (O-R) systems was significant. Out of them, probably the most famous, especially after IBM bought it (from Illustra), is *Informix* (http://www-01.ibm.com/software/data/informix/).

Only some 10 years after the first commercial RDBMSs were launched, Chang (1981) described a first system, which is marrying rdbs and infer-ence rules. Most probably, the name *Datalog* was first coined to the cor-responding language in Maier (1983).

It then took almost another decade until Naqvi and Tsur (1989), after a 5 years efforts of MCC, introduced *LDL*, a language extending *Datalog* with sets, negation, and data updating capabilities, which was the founda-tion of the first widely known homonym KDBMS (Chimenti et al., 1990).

The University of Wisconsin at Madison started in 1992 the *CORAL* project (Ramakrishnan et al., 1992), experimenting a marriage between *SQL* and *Prolog*, which was implemented in a corresponding homonym KDBMS one year later (Ramakrishnan et al., 1992). *CORAL++* (Srivas-tava et al., 1993) is its object-oriented extension.

Today, there are lot of prototype KDBMS and very few commercial ones. Among the former, for example, there are the notable open source semantic web framework *Jena* (http://jena.apache.org//), developed (first by HP, but then by Apache) in *Java* and also embedding *Datalog* and *OWL*, the *Datalog* educational system *DES* (http://www.fdi.ucm.es/ profesor/fernan/des/), licensed by the Free Software Foundation, and the Apache *Clojure* and *Java* library for querying data stored on *Hadoop* clusters *Cascalog* (http://cascalog.org/).

Among the latter, for example, there are *QL* (de Moor et al., 2008), a commercial object-oriented variant of *Datalog* created by Semmle Ltd. (http://semmle.com/), a free-of-charge database binding for *pyDatalog* (a logic extension for *Python*, see https://sites.google.com/site/pydatalog/) from FoundationDB (https://foundationdb.com/key-value-store/documentation/datalog.html), and a distributed db system for scalable, flexible and intelligent applications running on cloud architectures, which uses *Datalog* as the query language, called *Datomic* (http://www.datomic.com/ rationale.html), from Cognitect (http://cognitect.com/).

Even Microsoft Research developed and provides a security policy language based on *Datalog* called *SecPal* (http://secpal.codeplex.com/).

MatBase has several prototype versions (MS *C* and Borland *Paradox*, MS *Access*, C# and *SQL Server*) and was presented in a series of papers: Mancas et al. (2003), Mancas and Dragomir (2004), and Mancas and Mancas (2005, 2006). Its users may either work in (E)MDM (and the system automatically translates everything into both E-RDM and RDM), which also includes a *Datalog* engine, in E-RDM (and the system automatically translates everything into both (E)MDM and RDM), or in RDM (and the system automatically translates everything into both E-RDM and (E) MDM).

The excellent (Tsichritzis and Lochovsky, 1982) and the more recent (Hoberman, 2009; Hoberman et al., 2009; Simsion and Witt, 2004; Simsion, 2007), etc. are devoted to data modeling.

The recently revived graph-based database models are discussed in depth, for example, in Robinson et al. (2013).

For data model examples (even if not all of them correct and/or optimal), see, for example, http://www.databaseanswers.org/data_models/ index.htm.

Not only in my opinion, the current "bible" version for database theory still remains (Abiteboul et al., 1995), while (Garcia-Molina et al., 2014) is the one for DBMSs.

Other remarkable books in this field are, for example, Date (2003, 2011, 2012, 2013), Churcher (2012), Hernandez (2013), Lightstone et al. (2011), Ullman, (1988, 1989), and Ullman and Widom (2007).

Especially in O-ODBMSs, but not only, data is frequently stored in *XML*; as such data is somewhere at the half-road between dbs and plain text documents it is called *semistructured*; again not only in my opinion, the current "bible" in this field remains (Abiteboul et al., 2000).

DBLP (http://www.sigmod.org/dblp/db/index.html), started and still managed by Michael Ley, lists now more than 1,200,000 publications in the db field. A smaller searchable index of db research papers (only some 40,000 entries, but out of a total of more than 3,000,000 ones in computer science, with more than 600,000 also having links to the full papers), http://liinwww.ira.uka.de/bibliography/Database/, is maintained by Alf-Christian Achilles.

All URLs mentioned in this section were last accessed on July 11th, 2014.

KEYWORDS

- **big data**
- **business rule (BR)**
- **closed domain assumption**
- **closed world assumption**
- **coherent constraint set**
- **conceptual data modeling**
- **consistent database instance**
- **constraint**
- **constraint satisfaction**
- **constraint set**
- **constraint set closure**

- **constraint sets equivalence**
- **constraint violation**
- **data**
- **data aggregation**
- **data analysis**
- **Data Definition Language (DDL)**
- **data factorization**
- **data generalization**
- **Data Manipulation Language (DML)**
- **data model**
- **data plausibility**
- **database (db)**
- **database instance**
- **database management system (DBMS)**
- **database scheme**
- **database stored methods**
- **Datalog**
- **derived database**
- **E-R data model**
- **(Elementary) Mathematical Data Model ((E)MDM)**
- **Entity-Relationship (E-R) Data Model (E-RDM)**
- **fundamental constraint**
- **fundamental database**
- **heterogeneous data**
- **homogeneous data**
- **implausible constraint**
- **implication problem**
- **implied constraint**
- **incoherent constraint set**
- **incomplete data**
- **inconsistent database instance**
- **inference rule**

- **information**
- **knowledge and database management system (KDBMS)**
- **knowledge base (kb)**
- **knowledge-based management system (KBMS)**
- **Logical Data Model (LDM)**
- **metadata**
- **minimal constraint set**
- **negative data**
- **NoSQL DBMS**
- **object set**
- **object set property**
- **plausible constraint**
- **positive data**
- **redundant constraint**
- **redundant constraint set**
- **Relational Data Model (RDM)**
- **semistructured data**
- **Structured Query Language (SQL)**
- **trivial constraint**
- **unique objects assumption**

REFERENCES

No finite point has meaning without an infinite reference point.
—Jean-Paul Sartre

Abiteboul, S., Buneman, P., Suciu, D. (2000). *Data on the Web. From Relations to Semistructured Data and XML*. Morgan Kaufman: San Francisco, CA.

Abiteboul, S., Hull, R., Vianu, V. (1995). *Foundations of Databases;* Addison-Wesley: Reading, MA.

Ashdown, L., Kyte, T. (2014). *Oracle Database Concepts, 12c Release 1 (12.1)*. Oracle Corp. (http://docs.oracle.com/cd/E16655_01/server.121/e17633.pdf).

Astrahan, M. M., et al. (1976). System R: A Relational Approach to Database Management. In ACM TODS, 1(2), 97–137.

Campbell, J., Haderle, D., Parekh, S., Purcell, T. (2012). *IBM DB2: The Past, Present and Future: 30 Years of Superior Innovation.* MC Press Online, LLC, Boise, ID.

Chamberlin, D. D., Boyce, R. F. (1974). SEQUEL—A Structured English QUEry Language. In SIGMOD Workshop, 1, 249–264.

Chang, C. C. (1981). On the Evaluation of Queries Containing Derived Relations in a Relational Database. In H. Gallaire, J. Minker, and J. Nicolas, eds., *Advances in Database Theory,* Vol. I, Plenum Press.

Chen, P. P. (1976). The entity-relationship model: Toward a unified view of data. *ACM TODS* 1(1), 9–36.

Chimenti, D., Gamboa, R., Krishnamurti, R., Naqvi, S., Tsur, S., Zaniolo, C. (1990). The LDL system prototype. *IEEE* TKDE 2(1):76–90.

Churcher, C. (2012). *Beginning Database Design: From Novice to Professional,* 2nd ed., Apress Media LLC: New York, NY.

Codd, E. F. (1970). A relational model for large shared data banks. *CACM* 13(6), 377–387.

Date, C. J. (2003). *An Introduction to Database Systems, 8th ed.,* Addison-Wesley: Reading, MA.

Date, C. J. *SQL and Relational Theory: How to Write Accurate SQL Code, 2nd ed.;* Theories in Practice; O'Reilly Media, Inc.: Sebastopol, CA, 2011.

Date, C. J. *Database Design & Relational Theory: Normal Forms and All That Jazz;* Theories in Practice; O'Reilly Media, Inc.: Sebastopol, CA, 2012.

Date, C. J. *Relational Theory for Computer Professionals: What Relational Databases are Really All About;* Theories in Practice; O'Reilly Media, Inc.: Sebastopol, CA, 2013.

De Moor, O. et al., (2008). QL: Object-Oriented Queries Made Easy. Lammel, R., Visser, J., Saraiva, J. (Eds.): GTTSE 2007, LNCS 5235: 78–133, Springer-Verlag, Berlin Heidelberg.

Gallaire, H., Minker, J. (1978). *Logic and Databases.* Plenum Press, New York, U.S.A.

Garcia-Molina, H., Ullman, J. D., Widom, J. (2014). *Database Systems: The Complete Book,* 2nd ed., Pearson Education Ltd.: Harlow, U.K., (Pearson New International Edition)

Hammer, M., McLeod, D. (1978). The Semantic Data Model: a Modeling Mechanism for Database Applications. In *ACM SIGMOD Int. Conf. on the Manag. of Data.*

Hernandez, M. J. (2013). *Database Design for Mere Mortals: A Hands-on Guide to Relational Database Design,* 3rd ed., Addison-Wesley: Reading, MA.

Hoberman, S. (2009). *Data Modeling Made Simple* 2nd Edition. Technics Publications LLC, Bradley Beach, NJ.

Hoberman, S., Blaha, M., Inmon, B., Simsion, G. (2009). *Data Modeling Made Simple: A Practical Guide for Business and IT Professionals 2nd Edition.* Technics Publications LLC, Bradley Beach, NJ, (Fourth Printing).

Lightstone, S. S., Teorey, T. J., Nadeau, T., Jagadish, H. V. (2011). *Database Modeling and Design: Logical Design,* 5th ed., Data Management Systems; Morgan Kaufmann: Burlington, MA.

Maier, D. (1983). *The Theory of Relational Databases.* Computer Science Press: Rockville, MD.

Mancas, C. (1985). Introduction to a data model based on the elementary theory of sets, relations and functions (in Romanian). In Proc. of INFO IASI'85, 314–320, A.I. Cuza University, Iasi, Romania.

Mancas, C. (1990). A Deeper Insight into the Mathematical Data Model. Proc. 13th Intl. Seminar on DBMS, ISDBMS'90, 122–134, Mamaia, Romania.

Mancas, C. (1997). *Conceptual data modeling.* (in Romanian) Ph.D., Thesis: *Politehnica* University, Bucharest, Romania.

Mancas, C. (2002). On Knowledge Representation Using an Elementary Mathematical Data Model. In Proc. IASTED IKS 2002. Conf. on Inf. and Knowledge Sharing, 206–211, Acta Press, St. Thomas, U.S. Virgin Islands, U.S.A.

Mancas, C., Dragomir S., Crasovschi, L. (2003). On modeling First Order Predicate Calculus using the Elementary Mathematical Data Model in *MatBase* DBMS. In Proc. IASTED AI 2003. MIT Conf. on Applied Informatics, 1197–1202, Acta Press, Innsbruck, Austria.

Mancas, C., Dragomir S. (2004). *MatBase Datalog¬* Subsystem Metacatalog Conceptual Design. In Proc. IASTED SEA 2004. MIT Conf. on Software Eng. and App., 34–41, Acta Press, Cambridge, MA.

Mancas, C., Mancas, S. (2005). *MatBase* E-R Diagrams Subsystem Metacatalog Conceptual Design. In Proc. IASTED DBA 2005. Conf. on DB and App., 83–89, Acta Press, Innsbruck, Austria.

Mancas, C., Mancas, S. (2006). *MatBase* Relational Import Subsystem. In Proc. IASTED DBA 2006. Conf. on DB and App., 123–128, Acta Press, Innsbruck, Austria.

Mealy, G. H. (1967). Another Look at Data. In *AFIPS Fall Joint Comp. Conf., 31*, 525–534.

Naqvi, S., Tsur, S. (1989). *A Logical Language for Data and Knowledge Bases.* Computer Science Press: Rockville, MD.

Quillian, H. R. (1968). Semantic memory. In Semantic Inf. Processing, Minskz, M. ed., M.I.T. Press.

Ramakrishnan, R., Srivastava, D., Sudarshan, S. (1992). CORAL: Control, Relations and Logic. In VLDB.

Ramakrishnan, R., Srivastava, D., Sudarshan, S., Sheshadri, P. (1993). Implementation of the {CORAL} deductive database system. In SIGMOD.

Redmond, E., Wilson, J. R. (2012) *Seven Databases in Seven Weeks: A Guide to Modern Databases and the NoSQL Movement.* Pragmatic Bookshelf.

Robinson, I., Webber, J., Eifrem, E. (2013). *Graph Databases.* O'Reilly Media Inc., Sebastopol, CA.

Sarkar, D. (2013) *Microsoft SQL Server 2012 with Hadoop.* Packt Publishing, Birmingham, Mumbai.

Simsion, G. C. (2007). *Data Modeling: Theory and Practice.* Technics Publications LLC, Bradley Beach, NJ.

Simsion, G. C., Witt, G. C. (2004). *Data Modeling Essentials, 3rd Edition.* Morgan-Kaufmann Publishers, Burlington, MA.

Smith, J. M., Smith, D. P. C. (1977). Database abstractions: Aggregation and generalization. ACM TODS 2(2):105–133.

Srivastava, D., Ramakrishnan, R., Sudarshan, S., Sheshadri, P. (1993). CORAL++; Adding Object-orientation to a Logic Database Language. In Proc. *VLDB* Dublin, Ireland, 158–170.

Stonebraker, M., Wong, E. A., Kreps, P., Held, G. (1976) The design and implementation of Ingres. In ACM TODS, 1(3), 189–222.

Tsichritzis, D. C., Lochovsky, F. (1982). *Data models.* Prentice-Hall: Upper Saddle River, NJ.

Ullman, J. D. (1988). *Principles of Database and Knowledge-Based Systems—Volume 1: Classical Database Systems.* Computer Science Press: Rockville, MD.

Ullman, J. D. (1989). *Principles of Database and Knowledge-Based Systems—Volume 2: The New Technologies.* Computer Science Press: Rockville, MD.

Ullman, J. D., Widom, J. (2007). *A First Course in Database Systems*, 3rd ed. Prentice Hall: Upper Saddle River, NJ.

CHAPTER 2

THE QUEST FOR DATA ADEQUACY AND SIMPLICITY: THE ENTITY-RELATIONSHIP DATA MODEL (E-RDM)

The most incomprehensible thing about the world is that
it is comprehensible.
—Albert Einstein

CONTENTS

2.1   Entity and Relationship Type Object Sets 37

2.2   Attributes and Surrogate Keys 38

2.3   Entity-Relationship Diagrams (E-RDs) 40

2.4   Functional Relationships and the Key Propagation
Principle (*KPP*) 42

2.5   Relationship Hierarchies 44

2.6   Higher Arity Non-Functional Relationships 48

2.7   Restriction Sets 49

2.8   Case Study: A Public Library (Do We Know Exactly
What a Book Is?) 60

2.9   The Algorithm for Assisting the Data Analysis and Modeling
Process (A0): An E-R Data Model of the E-RDM 69

2.10   Best Practice Rules 75

2.11   The Math Behind E-RDs and Restriction Sets—The Danger
of "Many-To-Many Relationships" and the Correct E-RD
of E-RDM 92

2.12   Conclusion 98

2.13   Exercises 99

2.14   Past and Present 110

Keywords 111

References 112

After studying this chapter you should be able to:
- Concisely and precisely define, and exemplify the following key terms: *entity*, *relationship* (one-to-one, one-to-many, many-to-one, many-to-many, functional, recursive, non-functional), *attribute*, *surrogate key*, *role*, *Entity-Relationship* (*E-R*) *Diagram* (*E-RD*), their equivalent (Chen, Barker, and Mancas) *notations*, *E-R data model*, the *Key Propagation Principle* (*KPP*), *relationship hierarchy*, *restriction set*, *restriction* (attribute range, compulsory data, uniqueness, other types), *controlled data redundancy*.
- Apply algorithm *A1* (assistance of data analysis and conceptual modeling) and the best practice rules for correctly solving at least all exercises found at the end of the chapter.

Generally, our customers (be them companies, or our team leaders, professors, etc.) are providing us a brief informal description of the dbs they would like us to design and implement for them. These descriptions are very rarely complete, accurate, clear, and not ambiguous; sometimes they are even only oral ones.

Starting to implement corresponding dbs is a wrong decision even for the very small and simple ones of them, just like starting to type code before analyzing corresponding software applications usage, architecting, and designing them: without data analysis, conceptual, and physical design, dbs will not be correct, scalable, and optimal.

For each db, data analysis should come up with an *E-R data model*, made out of the following data modeling deliverables, all of them agreed by both the customer and the db designer:
1. A complete, accurate, clear, not ambiguous, well structured, and concise written *description* of the desired db (that may refer to the documents in 2 and 3 below).

2. A set of corresponding *E-R diagrams* (E-RDs).
3. An associated *restriction set* (that refers to 1 and 2 above).

Moreover, a "contract" (be it juridical or not) between the customer and the db designer (that should have as its appendixes all of the above documents) stipulating requirements, deadlines, costs, etc., would be desirable too.

Obviously, the "contract" is not within the scope of this book, whereas 1 is fairly simple (and we too will provide examples in the sequel), as it is merely the result of an informal reverse engineering process applied to the results of 2 and 3; this is why we are only focusing on obtaining the second and third above components, which are the first and, generally, the most important steps towards both formalizing the customer db requirements and needs, as well as understanding undocumented legacy dbs.

2.1 ENTITY AND RELATIONSHIP TYPE OBJECT SETS

> *I love bringing the inanimate object to life.*
> —John Lasseter

E-RDM considers two types of object sets:

- *Entity*: atomic, independent ones (e.g., *PEOPLE, EMPLOYEES, STUDENTS, COURSES, CUSTOMERS, CITIES, COUNTRIES, PRODUCTS, PRODUCT_TYPES, SERVICES*, etc.), graphically represented by rectangles.
- *Relationship*: complex, nonindependent, mathematically relation type (i.e., Cartesian products subsets; for example, *MARRIAGES, STOCKS, COAUTHORS, BIRTH_PLACES, HOME_TOWNS, CAPITALS*, etc.), graphically represented by diamonds. Initially, their underlying sets were only of entity type. Depending on their *cardinalities* (that are labeling corresponding edges), they are classi-fied into *many-to-many*, *one-to-many*, *many-to-one*, and *one-to-one* types (e.g., *STOCKS, MARRIAGES, CO-AUTHORS* are many-to-many, *BIRTH_PLACES* is either one-to-many, if *CITIES* is repre-sented at its left side, or many-to-one, if *PEOPLE* is represented at its left side, whereas *CAPITALS* is one-to-one). Cardinalities are

also used for representing compulsory type restrictions: by using 0 instead of 1, you are asserting that the corresponding property value might not always be known or applicable (e.g., see *BIRTH_ PLACES* in Fig. 2.1). Relationship edges should be labeled by *roles* (e.g., *Husband*, *Wife*, *Product*, *Warehouse*, etc.). We extended relationships by allowing them to be defined on other relationships as well, thus being possible to build relationship hierarchies too.

2.2   ATTRIBUTES AND SURROGATE KEYS

> *Painting is concerned with all the 10 attributes of sight, which are: Darkness, Light, Solidity and Color, Form and Position, Distance and Propinquity, Motion and Rest.*
> —Leonardo da Vinci

Both types of these object sets have associated *attributes* (e.g., *Sex, First-Name, LastName, e-mailAddr, BirthDate, MarriageDate, HireDate, CustomerName, CityName, ZipCode, Address, CountryName, CountryCode, CountryPopulation, WarehouseName, ProductTypeName, StockQuantity, EndDate*, etc.), graphically represented by ellipses attached to the corresponding rectangles or diamonds.

Please note that in any such data model, names of the object sets (be them entity or relationship type) must be unique. Similarly, all attributes of any given object set should have distinct names (but there may be attributes of different object sets having same names).

Mathematically, attributes are mappings defined on corresponding associated sets and taking values into corresponding value sets (i.e., subsets of data types; for example, {'F', 'M'}, [01/01/1900, 31/12/2012], {*True, False*}, [30, 250], ASCII(32), UNICODE(32), etc.).

In the original E-RDM, attributes may take values into Cartesian product sets as well; we do not allow it and restrict their codomains to atomic sets only.

The E-RDM also generally recommends adding for each object set (be it of type entity or relationship) a distinguished attribute called *surrogate key*, having no other semantics except for numeric unique identification of the elements of the corresponding set. Their names generally contain either "ID" (most commonly) or "#"; for example, *ID_Discipline* or *DisciplineID*, or *Discipline#*.

However, we do not recommend adding surrogate keys at this highest data modeling level, first of all in order not to confuse users (for whom they are irrelevant), but also in order to simplify business analysts work, as they may be automatically added later. When they are added, their names should be underlined.

Please note that same concepts should be correctly modeled either as (entity-type) object sets or as attributes, depending on the context. For example, for people, their eyes or hair colors may be modeled as attributes (*EyeColor* and *HairColor*) of the *PEOPLE* object set; in a physics context, where colors are defined by very many properties (e.g., Kelvin grades temperature, saturation, hue, etc.) you should abstract an object set *COLORS* instead.

Generally, whenever several properties are needed to define an object you should abstract a corresponding object set; if there is only one property, then you might not do so, but simply declare that property (e.g., *Color*) as being a property of a corresponding related object set (like *EyeColor* and *HairColor* above).

However, sometimes, for optimality, it is preferable to abstract object sets even if they have only one property. For example, in order not to store several times the names of eye and hair colors, but especially in order not to impose to users to type them every time (but provide them a drop-down list from where to choose the desired values), the best modeling solution for this subuniverse is to consider the entity-type object sets *PEOPLE*, *EYE_COLORS* and *HAIR_COLORS* and the many-to-one relationship-type sets *PEOPLE_EYE_COLORS* and *PEOPLE_HAIR_COLORS* between them.

Context dependency is also dictating the needed data modeling refinement level, from which we should similarly decide too whether a given concept should be modeled as an (entity-type) object set or as a property (attribute).

For example, in most usual subuniverses where people addresses are also needed, besides the entity-type object sets *PEOPLE* and *CITIES* and the many-to-one relationship-type one *PEOPLE_HOMETOWNS*, all that is needed is an attribute *Address* of *PEOPLE*.

However, again, when street addresses risk to repeat themselves a lot, like, for example, in a postal or electoral subuniverse, addresses should not be modeled anymore as (text) attributes, but split on several refinement levels, by abstracting object sets *STREETS, BUILDINGS,* and *APPARTMENTS* (and in big cities also *NEIGHBORHOODS* and *STAIRCASES*), as well as the corresponding many-to-one relationships between them and *PEOPLE.*

2.3   ENTITY-RELATIONSHIP DIAGRAMS (E-RDS)

> *Each relationship nurtures a strength or weakness within you.*
> —Mike Murdock

Figure 2.1 presents an example of an entity-relationship diagram (E-RD).

Many other graphical conventions exist and very many extensions were added to E-RDM, mainly in order to represent other constraint types. For example, Oracle uses the so-called *"crow's foot"* (or Barker) notation instead of diamonds and corresponding cardinalities; for example, *MARRIAGES* and *BIRTH-PLACES* are represented as in Fig. 2.2.

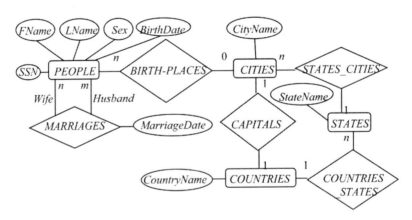

FIGURE 2.1   An example of an entity-relationship diagram (E-RD).

FIGURE 2.2   An example of an Oracle-style E-RD.

Similarly, mainly because, mathematically, both one-to-many, many-to-one, and one-to-one relations are functions, we prefer a mathematical type convention and represent these mappings by arrows, double arrows for one-to-one and onto ones and with reverse arrows not tangent to the domain set for only one-to-one and not onto ones, like for *Capital* in Fig. 2.3.

Note that this notation is not only simpler and familiar since elementary school, but also more natural; for example, in contrast with Fig. 2.1, *BIRTH-PLACES* is replaced by *BirthPlace* (anybody has a birth place and only one), *STATES_CITIES* by *State* (any city belongs to a state and to only one), *COUNTRIES_STATES* by *Country* (any state belongs to a country and to only one), and *CAPITALS* by *Capital* (any country has a capital and only one and, dually, no city may simultaneously be the capital of several countries, but of only one). Also note that one-to-one attributes are represented similar to one-to-one relationships (e.g., see *SSN* in Fig. 2.5).

Moreover, in our notation, inclusions[16] are represented consistently too (and do not need names[17]), like in Fig. 2.4.

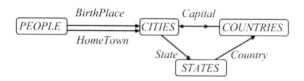

FIGURE 2.3   An example of a mathematical-style E-RD.

FIGURE 2.4   Representation of the inclusion EMPLOYEES ⊆ PEOPLE.

Computed objects, be them entity or relationship type object sets or attributes, are represented by dotted rectangles, diamonds, arrows, and ellipsis, respectively; in Fig. 2.5, **MALES* and **FEMALES*, as well as **Age* and **HomeCountry* are computed objects.

The table in Fig. 2.6 summarizes contrastively the original (Chen), Barker (Oracle), and our (Mancas) E-RD notations.

---

[16] IS-A relationships from AI.

[17] As associated mappings are the corresponding canonical inclusion ones.

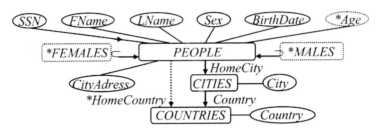

FIGURE 2.5 An example of an E-RD also including a one-to-one attribute and computed object sets, structural mappings, and attributes.

Equivalences between three E-RD notational conventions

Items	Chen	Barker	Mancas
Entity-type set			
Relationship-type set *m:n* (many-to-many)	*m* *n*		
Relationship-type set 1:*n* (one-to-many)	1 *n*		
Relationship-type set *n*:1 (many-to-one)	*n* 1		
Relationship-type set 1:1 (one-to-one)	1 1		(dom) (co-dom)
Set inclusion: A ⊆ B		A / B	A → B

FIGURE 2.6 Chen, Barker, and Mancas E-RD notational equivalences.

2.4 FUNCTIONAL RELATIONSHIPS AND THE KEY PROPAGATION PRINCIPLE (*KPP*)

> *Failure comes only when we forget our ideals and objectives and principles.*
> —Jawaharlal Nehru

The relationship examples from Figs. 2.3 and 2.4 suggest that not all relationship types need to be associated with object sets: all the one-to-one,

one-to-many and many-to-one ones may be modeled instead as properties of some object sets, as they are *functional* (i.e., they have at least an "one" part); only the many-to-many (which are *non-functional*) ones need adding to the corresponding data models such object sets.

For example, if we are to store all marriages we need the many-to-many relationship type object set from Fig. 2.1 (as almost anybody may get married several times during their lifetimes).

Dually, for example, we could also use a set of pairs <city, country> for storing the countries where cities are located, according to Fig. 2.7, although this is a many-to-one relationship (any city belonging to only one country and each country having, generally, several cities):

However, it is simpler and optimal to consider a *Country* property of *CITIES* instead, like in Fig. 2.8.

Obviously, *Country* should store corresponding values of any property of *COUNTRIES* that uniquely identifies corresponding countries; as we will see in the next chapter, surrogate keys are always the best optimal choice for it.

Generally, the *Key Propagation Principle* (*KPP*) states, at this conceptual data modeling level, that functional relationship types may be modeled as object set properties and that it is better to model them like that, instead of modeling them as object sets. We will deal in details with KPP's implementation in the next chapter.

Applying KPP for all functional relationship types is guaranteeing minimality of your data models in terms of the total number of

FIGURE 2.7   Functional relationship CITIES_LOCATIONS modeled as a relationship-type object set.

FIGURE 2.8   CITIES_LOCATIONS modeled as a functional relationship Country.

needed object sets and mappings, one of the main measures of model complexity.

From this point of view too, but not only, please note that the cardinalities of relationship types are not absolute, but highly depending on the context.

For example, if we are only interested in current marriages (and not in their history too) for a subuniverse where only one marriage is allowed at a time, then *MARRIAGES* from Fig. 2.1 is not anymore a many-to-many, but a one-to-one relationship type.

In such cases, obviously, by applying KPP, we should model *MARRIAGES* rather as the one-to-one *Spouse*, like in Fig. 2.9 (which is an example of a so-called *recursive relationship*[18], which is any functional relationship between a set and itself); please also note that, in such contexts, *MarriageDate* is not anymore a property of *MARRIAGES*, but of *PEOPLE*.

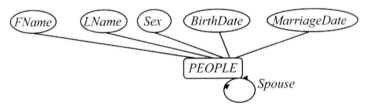

FIGURE 2.9   An example of a recursive relationship.

2.5   RELATIONSHIP HIERARCHIES

> *Hierarchies are celestial. In hell all are equal.*
> —Nicolas Gomez Davila

E-RDM was introduced with very many simplifying restrictions. In an attempt to make it THE only needed conceptual data model, hundreds of extensions were later added to it. From our point of view, however, only two of its restrictions are not justified, either theoretically or practically, namely:

---

[18] Mathematically, recursive relationships are autofunctions (functions having same domain and codomain).

- relationship-type sets should be based only on entity-type sets;
- there are only four types of business rules that apply for E-RDs data instances (namely the RDM ones: uniqueness, attribute value ranges, their existence, and tuple constraints). Moreover, the original E-RDM did not propose any formalism for embedding business rules; later extensions embedded primary keys and existence constraints.

In what follows, we consider examples of our E-RDM main extensions: relationship hierarchies and restriction sets.

For example, consider first any high-school, college, or university sub-universe. Figure 2.10 presents the best possible data modeling solution for students' attendance to classes and their grades, which makes use of a four level hierarchy of both relationship and entity type object sets.

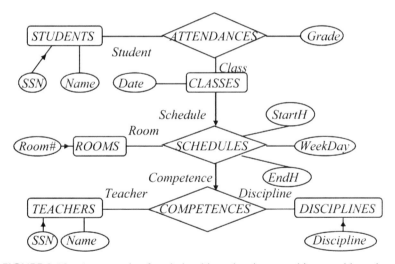

FIGURE 2.10   An example of a relationship and entity type object sets hierarchy.

Please first note that all relations between a relationship-type object and any of its underlying object sets need to be labeled by their corresponding *roles* (the "role" of that object set in that relationship-type object set)[19]; in this case, they are *Student* and *Class* for *ATTENDANCES*, *Room* and *Competence* for *SCHEDULES*, and *Teacher* and *Discipline* for *COMPETENCES*. All role names of a relationship-type set should be unique not

---

[19] Formally, roles are the names of the corresponding canonical Cartesian projections.

only between them: moreover, there may not be an attribute of a relation-ship-type set having a same name as one of its roles.

Then, note that *Schedule* stores for each class an element of the *SCHED-ULES* relationship-type set, which, in its turn, is built upon the entity-type set *ROOMS* and the relationship-type set *COMPETENCES*. Please also note that, in order to graphically distinguish between relationships and their underlying sets in such cases, the normal line for the corresponding role (here *Competence*) has to be replaced by an arrow pointing from the higher level relationship-type set (here *SCHEDULES*) to its corresponding underlying relationship-type set (here *COMPETENCES*).

In this solution, *COMPETENCES* stores teachers competency for various disciplines; as *SCHEDULES* is defined on *ROOMS* and *COM-PETENCES*, it is impossible to schedule classes for teachers that are not competent for the corresponding disciplines; next, *CLASSES* are the actual daily instances of *SCHEDULES*, factorizing out dates (note that if it were missing, dates would have to be stored for each corresponding student attendance); finally, *ATTENDANCES* takes full advantage of this four levels hierarchy, only implicitly storing student attendances (to classes that were scheduled in rooms according to teachers competences in cor-responding disciplines) and explicitly only the grades they got (if any).

A poorer solution would alternatively use only one flat relationship *STUDENTS_ACTIVITIES* of arity 4 (see Fig. 2.11), overloaded with all of the four semantics above (those of *CLASSES* and of the three relationships *COMPETENCES*, *SCHEDULES*, and *ATTENDANCES*), which would

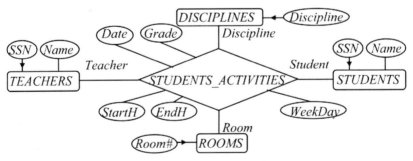

FIGURE 2.11   Alternative flat solution instead of using relationship and entity types object sets hierarchy.

obviously need many additional restrictions in order to reject implausible instances.

This solution seems to be simpler (as it only needs 5 object sets instead of 8), but, in fact, it is much more complicated, as it needs the following additional associated restrictions to be enforced, for any element of *STUDENTS_ACTIVITIES*:

- Teachers should teach corresponding disciplines.
- The disciplines should be scheduled with the corresponding teachers in the corresponding rooms in order for students to be able to attend corresponding classes.

Without enforcing them, implausible data might be entered; for example, teachers of music only could be stored as teaching and giving marks also for algebra, biology, and physics.

Moreover, disciplines, teachers, rooms, week day, start and end hours, as well as dates are stored (also meaning that they have to be entered) lots of times redundantly: for example, assuming that there are 10 teachers, teaching a total of 16 disciplines, during a 30 weeks year, 5 days a week, 6 h daily, for some 100 students, even if we simplify assuming that each discipline is taught only once every week for 2 h each, all seven above cells would need to be filled with data $16 \times 30 \times 5 \times 3 \times 100 = 720{,}000$ times each, that is a grand total of $720{,}000 \times 7 = 5{,}040{,}000$.

For the solution in Fig. 2.7 one would only need to fill 16 lines in *COMPETENCES* (i.e., for teachers and disciplines: $16 \times 2 = 32$ cells), $3 \times 5 = 15$ lines in *SCHEDULES* (i.e., for competences, rooms, weekdays, start and end hours: $15 \times 5 = 75$ cells), $30 \times 15 = 450$ lines in *CLASSES* (i.e., for dates and schedules: $2 \times 450 = 900$ cells), and $450 \times 100 = 45{,}000$ lines in *ATTENDANCES* (i.e., for students and classes: $2 \times 45{,}000 = 90{,}000$ cells), that is a grand total of only $32 + 75 + 900 + 90{,}000 = 91{,}007$, that is only some 1.8% of the 5,040,000 for *STUDENTS_ACTIVITIES*!

Even up to 20 years ago, dbs had to be more carefully designed as limited disk and memory available (both in terms of size, price, and speed) did not allow for such a waste; never think please that, as disk and memory nowadays are both affordable and much faster, the flat solution is an option: always think that you should never ask your end-users to work more than 50 times more (and, even worse, for repetitive redundant tasks)!

Otherwise, rather sooner than later, you will lose them, as competition in this field too is merciless.

Consequently, abstract relationship hierarchies whenever this is possible in the considered subuniverse of discourse. Please note that this process is, essentially, just another type of factorization of data as much as possible, so that each data is placed on the highest possible generalization level, in order to avoid storing any duplicates.

2.6   HIGHER ARITY NON-FUNCTIONAL RELATIONSHIPS

The depth of any story is proportionate to the protagonist's commitment to their goal, the complexity of the problem, and the grace of the solution.
—Steve House

The above example should not mislead us: properly defined, without semantic overloading, relationships having arity greater than two may be successfully used too. For example, in order to store various options for flying between pairs of airports, the following ternary relationship might be used, like in Fig. 2.12.

However, we will see in the second volume that (E)MDM includes both a theorem proving that the only relationships fundamentally needed for data modeling are the functions (i.e., the arrows in E-RDM), which means that we could never use relationship-type sets, as well as counterexamples proving that, generally, it is not a good idea to use relationships of arity greater than two.

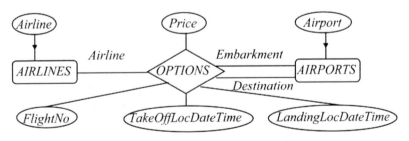

FIGURE 2.12   An example of a ternary relationship.

Consequently, we would recommend only usage of binary non-functional (i.e., many-to-many) relationship-type sets, just to draw the attention on their non-functionality.

2.7 RESTRICTION SETS

> *The golden rule for every business man is this:*
> *"Put yourself in your customer's place".*
> —Orison Swett Marden

All business rules of any data subuniverse to be modeled, be them explicit or implicit, should be discovered and added to our data models, in order to be able to guarantee that only plausible data is stored in the corresponding dbs.

Obviously, some of them are embedded in E-RDs; for example, any attribute attached to an object set implies that each element of that object set may have at most one value for that attribute (in Fig. 2.12, for example, each airline has only one name); similarly, each arrow implies that to each element of the object set from which the arrow starts corresponds at most one element from the one to where the arrow ends (in Fig. 2.5, for example, each person has only one home city); moreover, a backward arrow implies that the dual functional property holds too (in Fig. 2.12, for example, the backward arrow that links *Airline* to *AIRLINES* means that there may not be several airlines having same name), etc.

Note that we are not even representing compulsory type restrictions on E-RDs[20] as, anyhow, it is either impossible to graphically represent all restriction types or doing it even only partially complicates them too much: it is true that a good picture is worth hundreds of words, but it is also true that, as the number of objects in a picture increases, it is becoming increasingly uninteresting.

Moreover, recall that our customers should agree with our E-RDs, so let's keep them simple, only presenting the structure of the corresponding subuniverses, the results of our data abstraction and aggregation efforts.

---

[20] For example, Barker (Oracle) notation uses dotted half-lines for indicating that no element might correspond to the corresponding object set.

Consequently, we prefer attaching restriction sets to every E-RD; another argument would be that, anyhow, restrictions' not only formalizations, but especially informal descriptions should be managed apart, in order to fully understand and agree with our customers on them too.

There are always explicit business rules that we have to discover. In order to do it, it is simpler to first search for those of the types that are provided by the vast majority of the available commercial and open source DBMSs, which are the following three:

- *Attribute ranges* (or *allowable/permitted values*)
- *Compulsory attributes* (*data*) (or *mandatory values*, or *not null values*)
- *Elements* (*objects*) *uniqueness*

Next, we should add a fourth type for *other business rule types*, whenever needed.

As "business" sounds inappropriate in noncommercial contexts and as "constraint" is not that prized by most of our customers, I've tried several other words for "business rule" in almost 40 years of experience, out of which "*restriction*" was the most widely accepted.

Notationally, it is very convenient to uniquely label all restrictions associated to the E-RDs of a db (for ease of subsequent references to them) and to group them by object set and then by their type.

2.7.1 RANGE RESTRICTIONS

> *Wisdom begins with sacrifice of immediate pleasures*
> *for long-range purposes.*
> —Louis Finkelstein

The most trivial restriction type is the attributes' ranges: what should be, for example, the plausible values for the attributes in Fig. 2.10?

For some of them, the answer is obvious to most humans; for example, *Weekday* should probably take values from {1, 2, 3, 4, 5}, as there are no classes either on Saturdays or Sundays. Note that, even for these attributes, their ranges are not absolute, but relative to the corresponding context: for example, under the communism, in all east European countries almost everybody was working Saturdays too, including students; probably today too there are still countries where students have classes on Saturdays (or even Sundays).

Similar considerations apply for *StartH* and *EndH* (most probably the former always ranges from 8 to 19 and the latter from 9 to 20).

Even for text-type attributes we should indicate what is the longest possible corresponding text string; for example, how long could be a person (be it student or teacher) or discipline name? Obviously, it is not plausible that they might have, for example, 4 billion characters each.

Similarly, it is trivial that *Date* should not take values between, for example, January 1st, 4712 BC and December 31st, 9999 (as *Oracle* is letting us to store), but only between something like October 1st 2010 and today.

For other attributes, the answer is not that obvious; for example, should *Room#* take natural or text values? Probably, as at least some schools are labeling rooms by using letters too, in order to accommodate them too (for having as many customers as possible), we would decide to restrict them to text strings (e.g., of at most 8 or 16 ASCII characters), even if numbers are stored and processed more conveniently by computers.

Similar considerations apply for *Grade* (in some countries varying from F to A+, in others from 1 to 7, or to 10, or to 20 etc.).

The nine digits social security number (*SSN*) is used for uniquely identifying U.S. citizens and residents; in Romania, for example, its equivalent (called *personal numeric code*), has 13 digits (and embeds birthdate and sex); other countries have other types of equivalent codes that include letters too (e.g., Greece, Italy, Hong Kong, etc.), while others (e.g., U.K.) do not have such codes.

Last, but not least, we should always also specify the maximum cardinality of each object set. For example, corresponding values might be 1,000 for *DISCIPLINES*, *TEACHERS* and *ROOMS*, 10,000 for *COMPETENCES* and *CLASSES*, 100,000 for *STUDENTS* and *SCHEDULES*, and 1,000,000,000 for *ATTENDANCES*.

The restriction on the maximum cardinality of the corresponding set is the range of its surrogate key: for example, the range of *ID_Discipline* would be [1; 1,000].

Specifying maximum cardinality restrictions should be done too, not only for later implementation considerations, but also for having the most precise idea possible on the complexity of the considered subuniverse.

Range restrictions are the basic ones for guaranteeing data plausibility. Consequently, for any attribute (ellipsis) of any E-RD you should add a corresponding range restriction.

Unfortunately, there are still very many of those believing that it is enough to specify as ranges data types: numbers (generally divided in implementation subtypes: naturals between 0 and 256, integers between –2,147,483,648 and 2,147,483,647, etc.), calendar dates, etc. Obviously, this is not generally guaranteeing data plausibility (except for rare cases, like the Boolean data type): for example, *StockQty* should never be negative and should always be less than a maximum value that depends on the context, but which is never that large as 2,147,483,647 (also see *Altitude* from Table 1.1).

To conclude with, to any attribute of any E-RD you should add a sensible range restriction, such that its values are plausible in the corresponding context.

Notationally, each range restriction may be represented by the name of the corresponding attribute, followed by a column (or by the keywords "in" or "∈"), the corresponding range set, and the (unique) restriction label (possibly enclosed in parentheses) (see R02, R03, etc.).

Distinction between integer and noninteger ranges should be done by corresponding minimum and maximum values: for example, $[1, 5]$ is a subset of the naturals, while $[1.0, 5.0]$ is one of the reals (rationals only for computers).

$ASCII(n)$, $UNICODE(n)$, etc. represent the sets of strings made out of characters from the corresponding alphabets and having at most n (natural) such characters.

Similarly, $NAT(n)$ represents the subset of naturals having at most n digits, $INT(n)$ the corresponding one of integers, $RAT(n.m)$ the one of rationals having at most n digits before the decimal dot (or comma in Europe) and m after it, and $RAT^+(n.m)$ its subset of positive values (obviously, $NAT(n) = INT^+(n)$).

$CURRENCY(n)$ is a subset of $RAT^+(n.2)$.

For expressing the particular case of cardinality constraints, we may use the standard algebraic mappings *card*, which returns the number of elements in a set and *max*, which computes the maximum value of an expression (see R01, R04, R07, etc.).

For example, the following might be (for Romania) the range restrictions associated to the E-RD in Fig. 2.10 (where *SysDate*() is a system function returning current OS calendar date):

STUDENTS
    max(*card*(*STUDENTS*)) = 10^5 (R01)
    SSN: [1000101000000, 8991231999999]   (R02)
    Name: ASCII(255)   (R03)
TEACHERS
    max(*card*(*TEACHERS*)) = 10^3 (R04)
    SSN: [1000101000000, 8991231999999]   (R05)
    Name: ASCII(255)   (R06)
DISCIPLINES
    max(*card*(*DISCIPLINES*)) = 10^3 (R07)
    Discipline: ASCII(128)   (R08)
ROOMS
    max(*card*(*ROOMS*)) = 10^3 (R09)
    Room#: [1, 10^4]   (R10)
CLASSES
    max(*card*(*CLASSES*)) = 10^4 (R11)
    Date: [01/10/2010, *SysDate*()]   (R12)
SCHEDULES
    max(*card*(*SCHEDULES*)) = 10^5 (R13)
    Weekday: [1, 5]   (R14)
    StartH: [8, 19]   (R15)
    EndH: [9, 20]   (R16)
ATTENDANCES
    max(*card*(*ATTENDANCES*)) = 10^9 (R17)
    Grade: [1, 10]   (R18)
COMPETENCES
    max(*card*(*COMPETENCES*)) = 10^4 (R19)

Obviously, as already discussed, the above range restrictions might have different minimum and maximum values and/or even different data types for other countries. Of course that if you want to deliver a universal solution, you should choose the more general data types (e.g.,

ASCII(2) for *Grade*) and the corresponding widest intervals (e.g., [1, 7] for *Weekday*).

Please note that, generally, besides choosing the right data type set for every attribute, it is crucial that all of your range restrictions specify as narrow as possible intervals (explicitly or implicitly of the type [*minValue*, *maxValue*]), such as to actually guarantee that only plausible data will be accepted for all attributes.

As a counterexample, many RDBMSs (including *DB2*, MS *SQL Server*, *Access*, and *MySQL*) chose for the DATE data type ranges starting in our era and ending on Dec. 31st, 9999 (most probably dreaming that their products will last unchanged, from this point of view, at least another 7,986 years from now); obviously, we generally need no more than 100 years in the future, but, whenever confronted with mankind history data, we badly need dates before our era, which are very hard to store and process with these RDBMSs.

For example, the famous battle of Actium took place on September 2, 31 BC; the almost equally one of Kadesh, in May 1274 BC; the oldest recorded one (on the walls of the Hall of Annals from the Temple of Amun-Re in Karnak, Thebes—today's Luxor) is the one of Megiddo, which took place on April 16, 1457 BC (or, after other calculations, 1479 BC or even 1482 BC), etc.

Consequently, a much more intelligent choice for DATE would have been something like [1/1/–7000, 12/31/2999], as, almost surely, 985 years from now on are enough from any practical point of view, and, for example, the oldest intact bows discovered (in Holmegaard, Denmark) have some 8000 years, the Goseck Circle astronomic observatory discovered in Germany has some 7000 years, the oldest discovered (near Ljubljiana, Slovenia) wheel has some 5200 years, etc.

2.7.2 *COMPULSORY DATA RESTRICTIONS*

> *Learning is not compulsory ... neither is survival.*
> —W. Edwards Deming

Sometimes we either do not temporarily know the values of some attributes or they do not apply to some objects. Dually, it is obviously useless to store no data on any object, be it only temporarily.

Consequently, for each object set there should be at least one property whose values should be known for all of its objects. For example, taking once more into consideration Fig. 2.10, as we cannot even refer to a discipline without knowing its name, *Discipline*'s values should be compulsory; similarly, all other attributes, except for *Grade*, should have all their values known: for example, how could both students and teachers attend classes without knowing where (*Room#*), when (*WeekDay*, *StartH*), and for how long (*EndH*) to go and remain for?

Grade is an example of attribute whose values should not be mandatory, as it is not applicable for all elements of *ATTENDANCES*: there may be (lot of) students that do not get any grade during (many) classes.

Similarly, if, for example, we were to add *BirthDate* to both *STUDENTS* and *TEACHERS*, in most contexts we should not add corresponding compulsory data restrictions, as temporarily we might not know some of their values.

Arrows (i.e., many-to-one, one-to-many, and one-to-one relationships) also should or should not be compulsory, depending on the context.

For example, *Schedule* from Fig. 2.10 should be mandatory, as no class may be scheduled at no matter what date without knowing exactly in what room, starting with what hour, up to what end hour, etc.

All roles should be compulsory too, because no relationship may be defined without knowing all of its components.

For example, both *Student* and *Class* must be mandatory, as otherwise we might not know what classes attended each students or/and what students attended each class.

Similarly, *Room*, *Competence*, *Teacher*, and *Discipline* should be compulsory too.

Notationally, each compulsory data restriction may be represented by the name of the corresponding object set, followed by a column, and then the corresponding property (attribute, arrow, role) name (see R22 and R23 bellow); for simplicity, all such restrictions for a same object set may be merged into a single compound one, by factoring out the name of the object set and separating the names of the properties by columns (see R20, R21, R24, R25, R26, and R27).

For example, the following might be the compulsory data restrictions associated to the E-RD in Fig. 2.10.

STUDENTS: *SSN, Name*	(R20)
TEACHERS: *SSN, Name*	(R21)
DISCIPLINES: *Discipline*	(R22)
ROOMS: *Room#*	(R23)
CLASSES: *Date, Schedule*	(R24)
SCHEDULES: *Weekday, StartH, EndH, Room, Competence*	(R25)
ATTENDANCES: *Student, Class*	(R26)
COMPETENCES: *Teacher, Discipline*	(R27)

2.7.3 UNIQUENESS RESTRICTIONS

> *Always remember that you are absolutely unique.*
> *Just like everyone else.*
> —Margaret Mead

Just like for (algebraic) sets, db object sets do not allow for duplicates: it does not make sense to store data on objects that are not uniquely identifiable.

For example, in the subuniverse modeled by the E-RD from Fig. 2.10, we should be able to uniquely distinguish between any two students, teachers, disciplines, rooms, etc.

Any time when we do not need to uniquely distinguish between elements of a same set of objects (like, for example, two kilos of sugar that we are buying) for which we would like to store data, we should abstract a corresponding higher conceptual level set instead (e.g., *PRODUCT_TYPES*, having "sugar" as one of its elements and a *Quantity* property, besides the *ProductTypeName* that should be unique).

Uniqueness is a constraint that applies either on properties or on sets of properties. For example, by definition, *SSN* values are unique, but in order to uniquely distinguish between world's states we need both their names and the corresponding country (as there may not be states having same names in a same country, but, worldwide, there may be states of different countries having same names: for example, both Spain and Argentina have states called Cordoba and La Rioja).

Object sets may have several uniqueness constraints; for example, as in no room may simultaneously either start or end more than one class, *SCHEDULES* above has two such triple constraints: one built with *Room*, *Weekday* and *StartH*, and the other (its dual one) with *Room*, *Weekday* and *EndH*.

At this conceptual level, however, in order to be sure that object sets are correctly defined (i.e., from this point of view, they do not allow for duplicates), it is enough to add to any of them at least one uniqueness constraint; of course that anytime we discover even at this stage several ones, we should add them all to the corresponding uniqueness restriction set.

There are only a couple of exception types among object sets that might not have associated uniqueness restrictions: subsets, people genealogy-type sets, and poultry/rabbit/etc. cages type sets.

For example, the subset *DRIVERS* of *EMPLOYEES* (that we might abstract in order to store only for them the otherwise nonapplicable properties *LicenseType*, *LicenseDate*, etc.) might not have any uniqueness restriction. For subsets, unique identification may be done through the uniqueness restrictions of the corresponding supersets: all subsets of any set inherit all of its restrictions.

PEOPLE in a genealogical db might not have uniqueness restrictions either: for example, there may be several persons having same names on which we do not know anything else (but their names). For such sets, the only way to uniquely distinguish between their elements is by the values of their surrogate keys.

Poultry/rabbit/etc. *CAGES* in a (farming) db might not need any other uniqueness than the one provided by surrogate keys: you can use their system generated numbers as cage labels too (surrogate keys thus getting semantics too).

Notationally, we may simply specify uniqueness restrictions just like the compulsory data ones; however, as they are not that obvious, we should add the corresponding reason in parenthesis after each of them.

Compound uniqueness restrictions are specified by using the '•' separator[21] between the corresponding property names.

For example, the following are the uniqueness restrictions associated to the E-RD in Fig. 2.10.

STUDENTS: *SSN* (there may not be two students having same SSN)   (R28)

TEACHERS: *SSN* (there may not be two teachers having same SSN)   (R29)

DISCIPLINES: *Discipline* (there may not be two disciplines having
                                                    same name)   (R30)

ROOMS: *Room#* (there may not be two rooms having same number)   (R31)

CLASSES: *Date • Schedule* (there may not be two classes scheduled at
                                                the same date)   (R32)

SCHEDULES:

    Room • Weekday • StartH (in no room may simultaneously start more
                                        than one class) (R33)

    Room • Weekday • EndH (in no room may simultaneously end more
                                        than one class) (R34)

ATTENDANCES: *Student • Class* (there is no use to store more than once
                            the fact that a student attended a class) (R35)

COMPETENCES: *Teacher • Discipline* (there is no use to store more
than once the fact that a teacher has the competence to teach a discipline) (R36)

2.7.4. OTHER TYPES OF RESTRICTIONS

> *Progress imposes not only new possibilities for the future,*
> *but new restrictions.*
> —Norbert Wiener

Even a very simple subuniverse, like, for example, the one in Fig. 2.10 may have other type of business rules than the three above that are governing it. Some of them are of common sense types (like, for example,

---

[21]Mathematically, '•' is the Cartesian product operator on mappings.

nothing may end before it starts, nobody can simultaneously be present in different places, no resource that can accommodate only one object at a time may do it simultaneously for several ones, etc.), others may be applicable strictly to that particular subuniverse (e.g., in order to be eligible, anybody who would like to run for an office should have a minimum age: in Romania, for example, currently they are 23 for local or national representatives and Europarlimentaries, 25 for senators, and 35 for presidents).

As not adding any business rule as a restriction to our data models allow for storing implausible data, each existing business rule must be always added.

Notationally, we may specify such restrictions at this level even in plain English (or any other natural language that is understood by our customers and by all other implied persons).

Of course that, when appropriate, we may use simple (i.e., propositional) logic formulas (like, for example, $StartH < EndH$), but it is a good practice also in this case to provide an explanation (enclosed in parenthesis) for each such restriction too.

When such a restriction involves properties of several object sets, it is better to include them at the end of the restriction set or, at least, at the end of the subset of the corresponding last object set involved (in order not to reference properties not yet defined).

For example, the following are additional restrictions associated to the E-RD in Fig. 2.10.

SCHEDULES:

$StartH < EndH$ (no class may end before it starts)	(R37)
No teacher may be simultaneously present in more than one room.	(R38)
No student may be simultaneously present in more than one room.	(R39)
No room may simultaneously host more than one class.[22]	(R40)
There may not be two people (be them teachers or students) having same SSN.[23]	(R41)

---

[22] Note that (R40) implies both (R33) and (R34).
[23] Note that (R41) implies both (R28) and (R29).

2.8   CASE STUDY: A PUBLIC LIBRARY (DO WE KNOW EXACTLY WHAT A BOOK IS?)

> *A room without books is like a body without a soul.*
> —Cicero

As the public libraries subuniverse is familiar to almost everyone, the only description for it in this case is a very simple one: the final db application should assist librarians in managing books and subscribers.

Most of my undergraduate, but also very many graduate students will come up in some 5′ with the entity-type object sets *BOOKS* (having properties *Title, ISBN, Price, BooksNo*), *AUTHORS* (*FName, LName*), and *SUBSCRIBERS* (*FName, LName, e-mail*), the relationship-type one *BORROWS* (*BorrowDate, ReturnDate*), and the ER-D in Fig. 2.13.

First of all I'm generally drawing their attention on the fact that among authors there are lot of still living persons and that at least some of them might also be subscribers (writers are among the most assiduous readers!); consequently, each such author should be stored twice in their db, once as author and a second time as subscriber (without even being possible to tell whether or not these two object set elements correspond to a same person). Moreover, by consolidating *AUTHORS* and *SUBSCRIBERS* (e.g., into *PERSONS*), their solution becomes not only conceptually better, but also simpler and faster in terms of implementations.

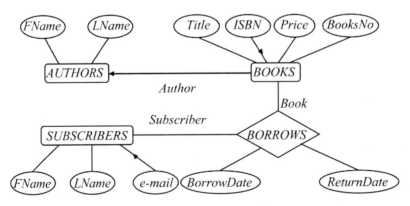

FIGURE 2.13   Public library first data analysis iteration E-RD.

Secondly, I remind them that there are plenty of books (generally scientific, technologic, but not only) having more than one author: consequently, *Author* should be replaced by a relationship-type set (e.g., *CO-AUTHORS*); moreover, for retrievals (but also for accurately modeling this subuniverse), their positions in the coauthor lists should also be stored.

Next, for finishing with "minor" issues, I remind them too that return dates are generally not the same for all books that you borrow together and that actual return dates should also be stored for each book, first of all for keeping track of still not returned, although overdue ones, but not only (e.g., tracking both best subscribers, those that are never late, and less or much less accountable ones, as they are frequently returning books late, very late, etc.).

Finally, I'm closing this first round of comments and suggestions by asking them the following questions (for which, unfortunately, I'm not getting correct, generally even satisfactory answers):

- What is, for example, the ISBN of Shakespeare's *King Lear*?
- What's the price of Homer's *Odyssey*?
- Is the books number (by which they mean number of copies of a same book from the library) computable or not?
- How exactly do they ask for a book in a public library (do they generally know its ISBN)?
- What do they exactly mean by "book"?

Here are then my answers:

- When written, no book has associated ISBN; moreover, generally there are several editions of a book (even by a same publisher!), each one having different ISBNs.
- Like any other masterpiece, Homer's *Odyssey* is priceless! Even ordinary books do not have prices when written: only their editions have prices.
- The number of copies of an edition, as well as the total one for all editions of any book is computable, so it may be eliminated from the data model.
- Except for insignificant number of cases, subscribers (and even librarians) do not know ISBNs: generally, you ask for a book by title, authors, publishing house, and edition year.

- They mean at least two things by "book": written, conceptual works, on one hand and their (printed or electronic) copies on the other.

Consequently, they are replacing the previous E-RD with the much better one in Fig. 2.14.

When trying then to establish uniqueness for *BOOKS*, another issue arises, as there may be several books having same title (e.g., "Logic", "Algebra", "Euclidean Geometry", "Poems", etc.) and I'm suggesting them to add a *FirstAuthor* functional relationship (and to consequently store coauthors, if any, only starting with the second position in coauthor lists); as there are, however, few exceptions of books having same first author and title, I'm also suggesting them to also add the year of book finishing (which is also interesting *per se*).

Next, focusing on their meaning for *COPIES*, I'm asking them why are they breaching first db axiom ("*any db should store any fundamental data only once*") with both *ISBN* and *Price*, as there are, generally, several identical copies of any edition; moreover, which is worse, I'm asking them how about several editions of a book (by different publishers, but even by a same one) that have different ISBNs? The answer is that *COPIES* is still overloaded with several semantics, so we identify together the entity-type object set *EDITIONS* that is also needed. Note that, besides publisher, title, and year, as these three properties do not uniquely identify editions, similar to *BOOKS*, *EDITIONS* should have an additional functional relationship *FirstBook* (as no publisher would publish more than one edition per year of a book under a same edition title).

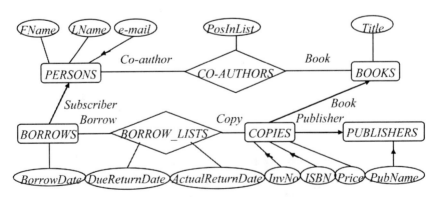

FIGURE 2.14   Public library second data analysis iteration E-RD.

In order to skip redrawing the E-RDs too many times, I'm then also drawing their attention to other two relevant facts, proving that, if left like that, *EDITIONS* is also overloaded semantically:

- Often (possibly regardless of the initial (co)author(s) book volumes structuring, which is of no great interest to libraries), editions are split into several volumes, each with their own distinct ISBN and possibly having their own title.
- Sometimes, even if not that often, volumes may contain several books (e.g., "*Complete Shakespeare's Dramatic and Poetical Works*").

This proves that their initial *BOOKS* object set was overloaded with the semantics of five object sets, namely: *BOOKS, VOLUMES_CONTENTS, VOLUMES, EDITIONS*, and *COPIES*.

Finally, I'm also suggesting them to add a Boolean *Author?* to *PEOPLE*, which will prove very useful for both users and db application developers.

2.8.1 PUBLIC LIBRARY E-RD

> *I have always imagined that Paradise will be a kind of library.*
> —Jorge Luis Borges

The final corresponding E-RD is presented in Fig. 2.15, immediately followed by its associated restriction set and by its full description.

2.8.2 PUBLIC LIBRARY RESTRICTION SET

> *Never trust anyone who has not brought a book with them.*
> —Lemony Snicket

1. **PERSONS** (The set of book (co)authors and subscribers of interest to the library.)
   a. *Cardinality: max(card(PERSONS)) = 10^9*   (*RP0*)
   b. *Data ranges:*
      FName: ASCII(128)   (*RP1*)
      LName: ASCII(64)   (*RP2*)

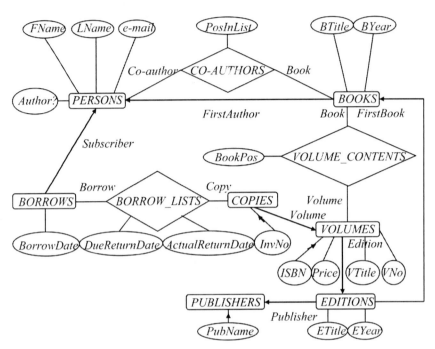

FIGURE 2.15   Public library E-RD.

Author?: BOOLEAN	*(RP3)*
e-mail: ASCII(255)	*(RP4)*
c. Compulsory data: LName	*(RP5)*

d. Uniqueness: e-mail • FName • LName (there may not be two persons having same first and last name, as well as same e-mail)   *(RP6)*

e. Other type restrictions:
  FName and *e-mail* should be compulsory for subscribers.   *(RP7)*

2. **BOOKS** (The set of written works of interest to the library.)
a. Cardinality: max(card(BOOKS)) = 10^6   *(RB0)*
b. Data ranges:
  BTitle: ASCII(255)   *(RB1)*
  BYear: [−5000, current year]   *(RB2)*

c. *Compulsory data: BTitle, FirstAuthor*   (RB3)

d. *Uniqueness: FirstAuthor • BTitle • BYear* (no author
   writes two books with a same title in a same year)   (RB4)

3. **CO-AUTHORS** = (*PERSONS, BOOKS*) (The set of
   pairs <*p, b*> storing the fact that book *b* was (also)
   written by person *p*.)

a. *Cardinality: max(card(CO-AUTHORS)) = 2 × 10⁶*   (RCA0)

b. *Data ranges: PosInList: [2, 16]* (*p*'s position in *b*'s
   co-authors list)   (RCA1)

c. *Compulsory data: Co-author, Book, PosInList*   (RCA2)

d. *Uniqueness: Co-author • Book* (no author should appear
   more than once in any book co-authors list)   (RCA3)

4. **PUBLISHERS** (The set of publishers of interest to
   the library.)

a. *Cardinality: max(card(PUBLISHERS)) = 10⁴*   (RPB0)

b. *Data ranges: PubName: ASCII(128)*   (RPB1)

c. *Compulsory data: PubName*   (RPB2)

d. *Uniqueness: PubName* (there may not be two publishers
   having same name)   (RPB3)

5. **EDITIONS** (The set of book editions of interest to
   the library.)

a. *Cardinality: max(card(EDITIONS)) = 10⁶*   (RE0)

b. *Data ranges:*

       ETitle: ASCII(255)   (RE1)

       EYear: [−2500, current year]   (RE2)

c. *Compulsory data: Publisher, FirstBook, ETitle*   (RE3)

d. *Uniqueness: Publisher • FirstBook • EYear* (no publisher
   publishes more than one edition of any given book in any
   given year)   (RE4)

6. **VOLUMES** (The set of edition volumes of interest to
   the library.)

a. *Cardinality: max(card(VOLUMES)) = 2 × 10⁶*   (RV0) *b. Data ranges:*

    ISBN: ASCII(16)   (*RV1*)

    VTitle: ASCII(255)   (*RV2*)

    VNo: [1, 255]   (*RV3*)

    Price: [0, 100000]   (*RV4*)

c. Compulsory data: Edition, VNo, Price   (*RV5*)

d. Uniqueness: ISBN (by definition, *ISBN*s are unique for any volume)   (*RV6*)

7. **VOLUME_CONTENTS** = (*BOOKS, VOLUMES*) (The set of pairs <*b, v*> storing the fact that book *b* is included in volume *v*.)

a. Cardinality: max(card(VOLUME_CONTENTS)) = 4×10^6 (*RVC0*)

b. Data ranges: BookPos: [1, 16] (*b*'s position in *v*'s table of contents)   (*RVC1*)

c. Compulsory data: Book, Volume, BookPos   (*RVC2*)

d. Uniqueness: Volume • Book (no book should be included more than once in any volume)   (*RVC3*)

e. Other type restrictions:

    For any edition, its first book should be the first one published in its first volume.   (*RVC4*)

    No edition may contain same book more than once.   (*RVC5*)

8. **COPIES** (The set of volume copies that the library ever possessed.)

a. Cardinality: max(card(COPIES)) = 32×10^6 (*RC0*)

b. Data ranges: InvNo: ASCII(32)   (*RC1*)

c. Compulsory data: InvNo, Volume   (*RC2*)

d. Uniqueness: InvNo: (by definition, inventory numbers are unique for any copy)   (*RC3*)

9. **BORROWS** (The set of copy borrows by subscribers.)

a. Cardinality: max(card(BORROWS)) = 10^{11} (*RBR0*)

b. *Data ranges: BorrowDate:* [1/6/2000, *SysDate*()]
   (assuming, for example, that first borrow date of interest
   is Jan. 6th, 2000)   (*RBR*1)

c. *Compulsory data: BorrowDate, Subscriber*   (*RBR*2)

d. *Uniqueness: BorrowDate • Subscriber*
   (no subscriber may simultaneously borrow more
   than once)   (*RBR*3)

10. **BORROW_LISTS** = (*BORROWS, COPIES*) (The set of
    pairs <*b, c*> storing the fact that volume copy *c* was
    (also) borrowed in borrow *b*.)

a. *Cardinality: max*(*card*(*BORROW_ LISTS*)) = 10^{12} (*RBL*0)

b. *Data ranges:*
   DueReturnDate: [1/6/2000, *SysDate*() + 300]
   (assuming, for example, that maximum borrow
   period is 300 days)   (*RBL*1)
   ActualReturnDate: [1/6/2000, *SysDate*()]   (*RBL*2)

c. *Compulsory data: Borrow, Copy, DueReturnDate*   (*RBL*3)

d. *Uniqueness: Borrow • Copy* (no copy may be
   simultaneously borrowed more than once)   (*RBL*4)

e. *Other type restrictions:*
   No copy may be borrowed less than 0 days or
   more than 300 days.   (*RBL*5)
   No copy may be simultaneously borrowed to
   more than one subscriber.   (*RBL*6)
   No copy may be returned before it was
   borrowed or after 100 years since corresponding
   borrow date.   (*RBL*7)

2.8.3 PUBLIC LIBRARY CLEAR, CONCISE AND COMPREHENSIVE DESCRIPTION

> *It is what you read when you don't have to*
> *that determines what you will be when you can't help it.*
> —Oscar Wilde

The db should store data on books (title, writing year, first author, coauthors, and their order), people (authors and/or library subscribers) e-mail addresses, first and last names, whether they are authors or not, as well as (book copies) borrows by subscribers (borrow, due, and actual return dates).

Books are published by publishing houses, possibly in several editions, even by same publishers (in different years). Each edition may contain several volumes, each having a number, a title, a price, and an ISBN code. Volumes may contain several books.

The library owns several copies (uniquely identified by an inventory code) of any volume and may lend several copies for any borrow; not all borrowed copies should be returned at a same date; maximum lending period is 300 days.

Last names, book titles, first authors, publisher names, edition publishers, first books and titles, as well as volume numbers and prices, copy inventory codes, borrow subscribers, dates, and due return dates are compulsory; for subscribers, first names and e-mail addresses are compulsory too.

People are uniquely identified by their first and last names, plus e-mail address; books by their first author, title, and writing year; publishers by their names; editions by their first book, publisher, title, and year; volumes by corresponding edition and volume number, as well as by their ISBN; copies by their inventory number.

There may be at most 1,000,000 persons, books, and editions, 2,000,000 coauthoring and volumes, 10,000 publishers, 4,000,000 books publishing, 32,000,000 copies, 100,000,000,000 borrows, and 1,000,000,000,000 copy borrows of interest. For example, first borrow date of interest is 01/06/2000.

No author should appear more than once in any book coauthors list.

For any edition, its first book should be the first one published in its first volume. No edition may contain same book more than once.

No copy may be borrowed less than 0 days or more than 300 days.

No copy may be simultaneously borrowed to more than one subscriber.

No copy may be returned before it was borrowed or after 100 years since corresponding borrow date.

2.9   THE ALGORITHM FOR ASSISTING THE DATA ANALYSIS AND MODELING PROCESS (A0): AN E-R DATA MODEL OF THE E-RDM

The secret of getting ahead is getting started. The secret of getting started is breaking your complex overwhelming tasks into small manageable tasks, and starting on the first one.

—Mark Twain

The intuition gained with the above case study should make easier the task of always applying first the algorithm for assisting data analysis and conceptual modeling processes (*A0*) presented in Fig. 2.16[24].

Algorithm A0 (*Data analysis and conceptual modeling process assistance*)

Input: an informal description of a subuniverse of interest (provided by the customer, team leader, professor, etc.);

Output: a corresponding initial E-R data model;

Strategy: *done = false*;

while not done repeat

  rewrite input description in order to make it crystal-clear, comprehensive, concise, structured, with highlighted keywords (that is all words susceptible of defining and characterizing object sets, their properties, as well as the rules governing the corresponding subuniverse of discourse; for example, customers, contracts, people, cities, services, e-mails, quantities, addresses, uniquenesses, etc.);

  repeat for all keywords in the above description

    select keyword

      case: object set

        if current E-RD is already too complicated *then*

          start another E-R subdiagram;

        if object set is linking two other existing object sets *then*

          add a corresponding diamond in the current E-RD (correspondingly labeled);

          else add a rectangle in the current E-RD (correspondingly labeled);

---

[24] Whose complexity is linear in the sum of object set collection, property and restriction sets cardinals (see Exercise 2.9)

if current E-R subdiagram is a new but not the first one *then*
link the new object set with at least another existing one (from previous E-R subdiagrams);
   case: object set property
      if the corresponding object set does not exist in the data model *then*
         add a corresponding object set like above;
         if property is a functional relationship *then*
         add corresponding arrow (correspondingly labeled);
         else add corresponding ellipse (correspondingly labeled);
   case: business rule
      add corresponding restriction to the restriction set (note: sometimes it may be the case that, in order to do it, you should first add new object sets and/or properties);
   end select;
   end repeat;
   meet again customers and ask them about everything which is still not clear or missing, if any, or whether they agree with your deliveries;
   if everything is clear and you got customers' OK *then done* = *true*;
   end while;
   End algorithm A0;

FIGURE 2.16   Algorithm *A0* (Data analysis and modeling process assistance).

Note that such algorithms are sometimes referred as being of type *forward engineering* (as opposed to their reverse engineering dual ones—see, for example, Sections 3.5, 4.4, and 4.6).

By applying A0 to our variant of the E-RDM, the corresponding E-R data metamodel from Fig. 2.17 is obtained.

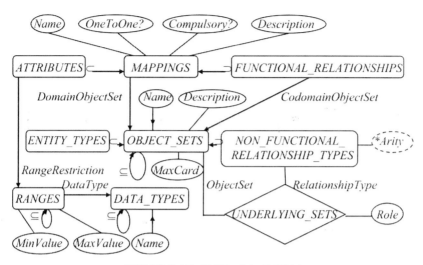

FIGURE 2.17   E-RD of the E-RDM.

2.9.1   E-RD OF E-RDM

> *He who thinks less is much more wrong.*
> —Lucian Blaga

Figure 2.17 presents the corresponding E-RD.

2.9.2   AN EXAMPLE OF POSSIBLE RESTRICTION SET ASSOCIATED TO E-RDM

> *After having discovered that life is senseless,*
> *the only thing that remains to be done is to give it a sense.*
> —Lucian Blaga

The following example uses some arbitrary chosen maximum values, which may actually differ depending on the context (e.g., possible implementation).

1. OBJECT_SETS (The finite nonempty collection of object sets of interest in any subuniverse of discourse, partially ordered by set inclusion)
   a. Cardinality: $max(card(OBJECT\_SETS)) = 10^{38}$ (RO1)

b. Data ranges:
   Name, Description: ASCII(255)   (*RO2*)
   MaxCard: NAT(38) (maximum cardinality)   (*RO3*)
c. Compulsory: *Name, MaxCard, Description*
   (*RO4*)
d. Uniqueness: *Name* (There may not be two object sets
   having same names in a same subuniverse of discourse.) (*RO5*)
e. Other type restrictions:
   There are no other types of object sets than entity and
   relationship ones and any object set is either of entity
   or of relationship-type.[25]   (*RO6*)

2. **ENTITY_TYPES ⊆ OBJECT_SETS** (The finite
nonempty object sets subcollection of atomic sets)

3. **NON_FUNCTIONAL_RELATIONSHIP_TYPES ⊆ OBJECT_
SETS** (The finite object sets subcollection of Cartesian
product type sets)
   *Arity: corresponding Cartesian product arity

4. **UNDERLYING_SETS** = (*RelationshipType* → *NON_FUNC-
TIONAL_RELATIONSHIP_TYPES, ObjectSet* → *OBJECT_SETS*)
(The finite set of pairs of type <*r, o*>, storing the fact that the non-
functional relationship-type object set *r* has object set *o* as one of
its underlying sets)
   a. Cardinality: *max(card(OBJECT_SETS))* = 10^{38} (*RU1*)
   b. Data ranges: *Role*: ASCII(255) (the role of *o* in *r*)   (*RU2*)
   c. Compulsory: *RelationshipType, ObjectSet, Role*   (*RU3*)
   d. Uniqueness: *RelationshipType • Role* (There may not
     be two underlying object sets of a same non-functional
     relationship-type object set having same roles.)   (*RU4*)
   e. Other type restrictions:
   For each non-functional relationship-type object set
   there should be at least two underlying object sets.   (*RU5*)
   The graph of *UNDERLYING_SETS* should be acyclic
   (that is no non-functional relationship-type object set

---

[25] Mathematically, *OBJECT_SETS* = *ENTITY_TYPES* ⊕ *NON_FUNCTIONAL_RELATIONSHIP_
TYPES*.

may be defined on itself, neither directly, nor indirectly); equivalently: recursive definitions of non-functional relationship-type object sets is not allowed. (*RU6*)

5. **MAPPINGS** (The finite nonempty set of mappings defined on object sets of interest)

   a. *Cardinality*: *max*(*card*(*MAPPINGS*)) = 10^{38} (*RM1*)

   b. *Data ranges*:

   Name, *Description*: ASCII(255) (*RM2*)

   OneToOne?: Boolean (is mapping uniquely identifying the object set elements?) (*RM3*)

   Compulsory?: Boolean (*RM4*) (should mapping have known all of its values?) (*RM4*)

   c. *Compulsory*: *Name*, *DomainObjectSet*, *OneToOne?*, *Compulsory?* (*RM5*)

   d. *Uniqueness*: *Name* • *DomainObjectSet* (There may not be two mappings defined on a same object set and having same names.) (*RM6*)

   e. *Other type restrictions*:

   There are no other types of mappings than attributes and functional relationship ones and any mapping is either of attribute or of functional relationship-type.[26] (*RM7*)

6. **ATTRIBUTES ⊆ MAPPINGS** (The finite nonempty mapping subset of attributes)

   a. *Compulsory*: *RangeRestriction* (*RA1*)

7. **FUNCTIONAL_RELATIONSHIPS ⊆ MAPPINGS** (The finite mapping subset of functional relationships)

   a. *Compulsory*: *CodomainObjectSet* (*RF1*)

8. **DATA_TYPES** (The finite nonempty collection of needed data type sets, partially ordered by set inclusion)

   a. *Cardinality*: *max*(*card*(*DATA_TYPES*)) = 100 (*RD1*)

   b. *Data ranges*: *Name*: ASCII(255) (*RD2*)

---

[26] Mathematically, *MAPPINGS* = *FUNCTIONAL_RELATIONSHIPS* ⊕ *ATTRIBUTES*.

      c. Compulsory: Name   *(RD3)*

      d. Uniqueness: Name (There may not be two data type sets
          having same names in a same subuniverse of discourse.) *(RD4)*

9. **RANGES** (The finite nonempty collection of needed attribute
    data range sets, partially ordered by set inclusion)
    a. Cardinality: max(card(RANGES)) = 10^4 *(RR1)*
    b. Data ranges: MinValue, MaxValue: ASCII(255)   *(RR2)*
    c. Compulsory: DataType, MinValue, MaxValue   *(RR3)*
    d. Uniqueness: DataType • *MinValue* • *MaxValue* (There
      may not be two data range sets of a same data type and
      having same minimum and maximum values.)   *(RR4)*

2.9.3 DESCRIPTION OF E-RDM

> *Reality is the ruin of a fairy tale.*
> —Lucian Blaga

There are at most 10^{38} (for example) object sets of interest in any sub-universe of discourse; some of them may be included in others; they are characterized by the compulsory properties name (an ASCII string of maximum 255 characters, for example), uniquely identifying each object set in any given subuniverse of discourse, maximum cardinal (a natural having at most 38 digits, for example) and a description (an ASCII string of maximum 255 characters, for example).

Object sets may be either atomic (of the so-called "entity" type) or of Cartesian product type (the so-called "relationship").

The Cartesian product type ones obviously have arity greater than one and are nonrecursively defined on other underlying object sets. For any such relationship-type set, each canonical Cartesian projection has a compulsory name (of at most, for example, 255 ASCII characters), unique within the set and called "role".

Each subuniverse of interest may also need at least one and at most 100 (for example) so-called (programming) "data types", which are finite subsets of the naturals, integers, rationals, strings of characters over some alphabet (generally ASCII or UNICODE), calendar dates, Boolean truth values, etc. They are uniquely identified by their compulsory name (an

ASCII string of maximum 255 characters, for example) and some of them may be included in other ones.

There are at most 10^{38} (for example) mappings of interest in any sub-universe of discourse defined on object sets; they are characterized by the compulsory properties name (an ASCII string of maximum 255 characters, for example), uniquely identifying each mapping within those defined on the same object set, and a description (an ASCII string of maximum 255 characters, for example); it is also compulsory to specify for any mapping whether or not its values are compulsory (or may be sometimes left unspecified)—the so-called "compulsory restriction"—and are uniquely identifying the elements of its domain object set (i.e., if the mapping is one-to-one or not)—the so-called "uniqueness restriction".

Mappings may take values in either object sets (in which case they are called "functional relationships", which are formalizing existing links between object sets) or data ranges (in which case they are called "attributes", which are formalizing object set properties).

Attribute codomains (the so-called "range restrictions") are subsets (called "ranges") of data types, uniquely identified by the triple made out of their compulsory data type superset, minimum and maximum values (both being, for example, ASCII strings of maximum 255 characters); some ranges may be included into other ones.

In order to keep things simpler, this E-R data metamodel does not include restrictions of other types and uniquenesses made out of several mappings (see Exercise 2.6, as well as Section 3.3).

2.10   BEST PRACTICE RULES

> *The Golden Rule finds no limits of application in business.*
> —James Cash Penney

Here is a set of best practice rules in data analysis and conceptual modeling.

2.10.0   *NO SEMANTIC OVERLOADING*

> *Information overload is a symptom of our desire*
> *to not focus on what's important.*
> —Brian Solis

R-DA-0. No object set, be it of entity or relationship type, should be semantically overloaded: any object set should have only one simple and clear semantics. On the other hand, somewhat dually, obeying to this paramount best practice rule should not diminish your power of abstraction.

For example, in a medical db, do not abstract sets of *PATIENTS* and *DOCTORS*, but only one set of *PEOPLE* (also having a Boolean attribute *Doctor?*—you will see why in the second volume): MDs may also get sick, do have the right to also be patients, and if you were choosing the former solution you should duplicate data for each such MD, without ever being able to conclude only from the corresponding db instance that any two persons from these object sets, even if having exactly same data, do represent a same actual person.

Similarly, do not abstract sets of *SUPPLIERS* and *CUSTOMERS* (why not starving to make at least some of your suppliers your customers too?), but a set of *PARTNERS* instead (and, by the way, for example, banks are also your partners, so there is no need to add a *BANKS* object set either: abstract banks too as partners).

2.10.1   NAMING CONVENTIONS

> *I named my album today. It was like naming a child for me.*
> *Such a hard decision.*
> —Carly Rae Jepsen

R-DA-1.
 a. All software objects (e.g., dbs, object sets, properties, tables, columns, constraints, applications, classes, variables, methods, libraries, etc.) should be named consistently, adequately, in a uniform manner, and their names should be the same at all involved levels (data analysis, E-RDs, restriction sets, mathematical, relational, and implemented db schemas, *SQL*, high level programming languages code, as well as both global/external and local/embedded in-line documentation).
 b. Object set names should be plurals and uppercase; property and restriction names should be singular and using generally lowercase, except for acronyms and letters starting words. Examples:

PEOPLE, CITIES, COUNTRIES, SSN, FName, LName, Birth-Place, Capital, CitiesKey.

c. Object set and constraint names should be unique in any data model, whereas property names should be unique within the set of all properties of the corresponding object set.

d. Do not ever use the ID suffix or prefix for properties relating object sets (i.e., functional relationships): not only that, for db people, it should be trivial that their implementation will reference a surrogate key, whereas for customers implementation details should remain transparent; moreover, it is absolutely unnatural (when asking somebody in what city they live, we do not ever ask them in what cityID or IDcity they do!), but it is also a very bad idea for any object set for which there are several properties relating them to a same object set—and not only conceptually, but also relationally (as for any table its columns should have distinct names). For example, *BirthPlace, HomeTown, WorkingPlace*, etc., all of them relating *PEOPLE* to *CITIES*, are much, much better than *CityID1*, *CityID2*, and *CityID3*, respectively.

2.10.2   OBJECT SET TYPES

> *I'm more of an adventurous type than a relationship type.*
> —Bob Dylan

R-DA-2. Object sets are either atomic (i.e., of "entity" type) or not (i.e., of non-functional "relationship" type) that is, mathematically, relations (i.e., subsets of Cartesian products).

a. For each abstracted object set, you should always first determine its type in the corresponding context.

b. In different contexts, object sets might have different types.

c. Obviously, if a set is not atomic, then you should also exactly determine and specify its underlying object sets.

For example, considering a logistic subuniverse where several product types might be stored in a same warehouse and warehouses may store several product types, we should abstract three related object sets:

FIGURE 2.18   Many-to-many (nonfunctional) STOCKS E-RD fragment.

- *PRODUCTS* (atomic),
- *WAREHOUSES* (atomic), and
- *STOCKS* (nonatomic, linking the former two atomic ones, which are its underlying sets, through two corresponding roles: *Product* and *Warehouse*).

The attribute *StockQty* is a property of *STOCKS*, as there may be, for example, q_1 products of type p_1 in warehouse w_1, q_2 in w_2, and so on.

The corresponding E-RD fragment should then be as in Fig. 2.18.

When there is only one warehouse, both *WAREHOUSES* and *STOCKS* are no more needed, and attribute *StockQty* is associated to *PRODUCTS* instead.

2.10.3 THE KEY PROPAGATION PRINCIPLE

> *Change your opinions, keep to your principles;*
> *change your leaves, keep intact your roots.*
> —Victor Hugo

R-DA-3. Always model functional relationships, according to *KPP*, as (relating/referencing) properties, instead of relationship-type object sets.

2.10.4 OBJECT SETS ABSTRACTION

> *I paint objects as I think them, not as I see them.*
> —Pablo Picasso

R-DA-4.
  a. Objects should be always abstracted as set elements, not as character strings, numbers, calendar dates, etc., that is object property

values (e.g., people, companies, cities, states, countries, invoices, payments, etc. are all object sets; for example, a city is an object, element of an object set *CITIES*, not the character string denoting its name).

b. Dually, you should never add to your data models sets that either are not relevant (e.g., subset of calendar dates when invoices were issued), or should conceptually be merged with other ones (e.g., passports set should be merged with the corresponding people set), or are equipotent with other ones, even if implementation would then distribute them apart (e.g., HR, salary, competence data, etc.), and so on.

c. Moreover, whenever sets become obsolete they should either be dropped from the corresponding data models or, at least, renamed accordingly.

d. To all of the abstracted object sets there should consequently be added corresponding entity or relationship-type sets in the E-R data models conceptually modeling the corresponding subuniverse of interest.

e. Make full use of abstraction: conceptual data models should have a minimum possible number of object sets, none of them semantically overloaded though. Always inflate instances and deflate schemes.

For example, you should not abstract sets *NOTEBOOKS*, *DESK-TOPS*, *MONITORS*, *MOTHERBOARDS*, *MEMORIES*, *HARD_DISKS*, etc.; instead, you should abstract sets:

- *PRODUCTS*
- *PRODUCT_TYPES*—whose elements include notebook, desktop, monitor, motherboard, memory, hard disk, etc.
- *CHARACTERISTICS*—whose elements include monitor dimension, CPU speed, hard disk size, hard disk speed, memory size, speed, etc.
- *MEASURE_UNITS*, and
- *PROD_TYPES_CHARACTERISTICS*—storing data on what characteristics has each product type, as well as the values for these characteristics as presented in the E-RD from Fig. 2.19.

f. Data models too may contain computed sets (e.g., *CAPITALS* ⊆ *CITIES*, *STATE_CAPITALS* ⊆ *CITIES*, *MEN* ⊆ *PEOPLE*,

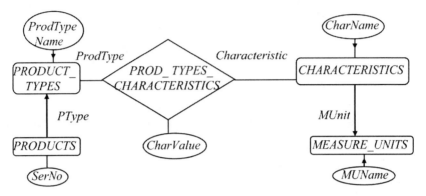

FIGURE 2.19   Example of using abstraction in data modeling for reducing models complexity.

WOMEN ⊆ PEOPLE, etc.), provided they are consistently named, documented, read-only for users, and automatically recomputed whenever needed.

2.10.5 OBJECT SET BASIC PROPERTIES

A theory is the more impressive the greater the simplicity of its premises is, the more different kinds of things it relates, and the more extended is its area of applicability. Therefore the deep impression which classical thermodynamics made upon me. It is the only physical theory of universal content concerning which I am convinced that within the framework of the applicability of its basic concepts, it will never be overthrown.
—Albert Einstein

R-DA-5. Object sets should be described by all of their relevant properties of interest, properties that should also make possible unique identification of their elements.

For example (also see Fig. 2.18), *PName* (unique) and *PPrice* could characterize *PRODUCTS*, *WName* (unique)—*WAREHOUSES*, and *Stock-Qty*—*STOCKS* (obviously, stocks are uniquely identified by the product of the corresponding *Product* and *Warehouse* values), as in Fig. 2.20.

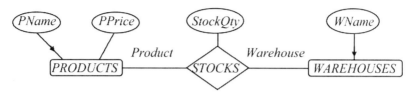

FIGURE 2.20 Example of property aggregation for uniquely identifying elements of a non-functional relationship-type object set.

2.10.6 OBJECT SET PROPERTY TYPES

I bought a two story house. ... One story before I bought, and another after.
                                                        —(anonymous author)

R-DA-6. Object set properties are mappings defined on the corresponding object sets and taking values either in some finite subset of some data types (for the basic ones, represented by ellipsis in E-RDs) or in other (but also possibly the same) object sets (for the functional type relationships—see R-DA-3 above and R-DA-7 below). As such, we should always ask ourselves for each of them whether or not, in the corresponding subuniverse, they are one-to-one or not.

For example, in R-DA-5 above, *PName* and *WName* are taking values in subsets of the ASCII strings (e.g., having maxim length of 64 for the former and 16 for the latter) and, being one-to-one, both are linked to *PRODUCTS* by arrows, while *PPrice* and *Qty* are taking numerical values (e.g., in [0, 1000]) and are not one-to-one (as several products may have same price and as several stocks may have same quantity, respectively).

In Fig. 2.21, both *CountryName* and *CityName* are taking values, for example, in ASCII(64); *CountryName* is one-to-one (as there may not be two countries having a same name), *CityName* is not (as there may be several cities, even in a same country, but in different counties, having same names), whereas *Capital* is one-to-one too (as no city may simultaneously be the capital of more than one country).

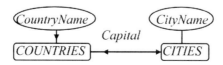

FIGURE 2.21 Example of one-to-one and not one-to-one properties.

2.10.7 CONTROLLED DATA REDUNDANCY: COMPUTED DATA

> *Redundancy is my favorite business strategy.*
> —Amit Kalantri

R-DA-7.

a. Fundamental dbs should store any fundamental data only once. Any computable data is redundant; hence, generally, it should not be stored.

b. Exceptions justified by querying speeding-up needs should not only be documented as such (starting with prefixing their name with a special character, like '*', indicating the corresponding expression for computing it, as well as the reasons why it was added to the model), but always be made read-only for end-users in any application built on that db, and their values be updated only automatically, by trigger-like methods, immediately when needed (which is referred to as *controlled data redundancy*).

c. Generally, such exceptions should not be introduced at this level (the only conceptual one available to customers too), as it may confuse customers (possible exception: when you want to justify to your customers higher costs for faster solutions), but only starting with the next conceptual level.

d. Before introducing such exceptions, you should make sure that they are either critical or that significant querying speed-up advantages justify additional costs in developing, db size, and corresponding automatic updates needed time.

e. You might add both computed sets, possibly having associated properties and computed properties of fundamental sets.

f. Computed restrictions and fundamental objects uncontrolled redundancy should be banned!

For example, let us consider the E-RD from Fig. 2.22 (see also, for example, the standard MS *Northwind* demo db), consisting of object sets:

- *CUSTOMERS*,
- *CITIES*,
- *STATES*, and

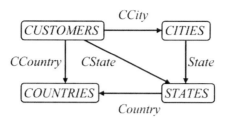

FIGURE 2.22   E-RD of part of the MS Northwind demo db.

- *COUNTRIES*,

where:
- every customer has its headquarters in only one city,
- any city belongs to only one state, and
- any state belongs to only one country, and, consequently,
- any headquarter is located in only one state and country.

Both *CState* and *CCountry* are computable, hence redundant:
CState = State ° CCity and
CCountry = Country ° CState = Country ° State ° CCity.[27]

This means that, whenever you also add them to your solution, you should also add two corresponding additional constraints (*CState = State ° CCity* and either *CCountry = Country ° CState* or *CCountry = Country ° State ° CCity*) to it: otherwise, implausible instances would be possible.

For example, even if, in *CITIES*, Paris belongs to Ile-de-France, in *STATES* Ile-de-France belongs to France, and Tennessee to U.S., in *CUSTOMERS* Sephora, whose headquarters are actually in Paris, may otherwise, according to such a db instance, be located, for example, in the Greenland's unknown state from Romania!

Obviously, such a solution is either very costly (and not that much because of the needed extra db space for storing these two computed columns and time to update their values too—be it automatically, as it should, or manually, as it shouldn't—, but especially for enforcing these two extra needed constraints) or an incorrect one (whenever these two constraints are not added too).

---

[27] Recall please that "°" is the algebraic symbol for function composition; for example, for any city *c*, the country to which *c* belongs may be computed by composing *Country* and *State*: (*Country ° State*) (*c*) = *Country*(*State*(*c*)).

Normally, all cycles in this E-RD should be broken, by dropping both *CState* and *CCountry* (which may be computed anytime when needed), implying that no extra constraints are needed anymore, like in Fig. 2.23.

Whenever querying speed is critical or updates to *CState* and *CCountry* are much less frequent than queries involving them (and additional needed db space is affordable), they may be added as computed properties *CState = State ° CCity* and/or *CCountry = Country ° State ° CCity* (with controlled redundancy, that is being read-only to end-users and automatically updated whenever changes occur in *CCity*, *State*, and/or *Country* corresponding values, by trigger-like methods); note that this solution (see Fig. 2.24) is as costly as the initial one with the two extra constraints added too, since plausibility is always guaranteed).

Moreover, in such cases, even if costlier, in fact it is often better that these two redundant properties store not pointers to the corresponding object sets, but plain names for customer cities, states, and countries (i.e., *CCity = CityName ° CCity*, *CState = StateName ° State ° CCity*, and *CCountry = CountryName ° Country ° State ° CCity*), like in Fig. 2.25: with this solution, when querying data about customers, only *CUSTOMERS* have to be accessed (whereas even in the previous one, from Fig. 2.24, for getting corresponding city, state, and country names too, all four object sets have to be accessed).

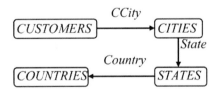

FIGURE 2.23   Resulting E-RD after breaking the cycles in Fig. 2.22.

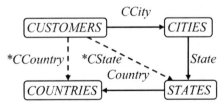

FIGURE 2.24   Resulting E-RD after adding controlled redundancies to the one in Fig. 2.23.

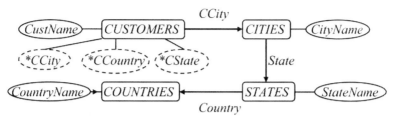

FIGURE 2.25   Resulting E-RD after optimizing controlled redundancies of the one in Fig. 2.24.

2.10.8   DATA COMPUTABILITY

> *If you don't want to be replaced by a computer, don't act like one.*
> —Arno Penzias

R-DA-8. Fundamental dbs should not ever store computable data instead of corresponding fundamental one.

For example, you should not ever store (for people/living beings/spiritual periods, etc.) *Age* instead of, say, *BirthDate/Year*: trivially, *Age* is computable from *BirthDate/Year*; moreover, if you would add *Age* instead of *BirthDate/BirthYear*, you are forcing users each day/year to check whether or not one or more of its values should be incremented and, whenever it is the case, to do it!

2.10.9   RESTRICTIONS NECESSITY

> *Computers are like Old Testament gods; lots of rules and no mercy.*
> —Joseph Campbell

R-DA-9. Any fundamental (i.e., non-implied, non redundant) restriction governing a subuniverse should be added in all corresponding data models. Not including such a constraint might lead to implausible db instances.

For example, consider (from an academic subuniverse) object sets (of entity-type) *STUDENTS*, *COURSES*, and (of relationship-type) *ENROLLMENTS* and *PREREQUISITES*, like in Fig. 2.26.

FIGURE 2.26   E-RD for a fragment of the student enrollment subuniverse.

PREREQUISITES should be acyclic, as no course should be a prerequisite of itself (neither directly, nor indirectly). If this restriction is not added to the data model of such a subuniverse, then, obviously, no student may ever be enrolled in any course belonging to a cycle in the prerequisites graph.

For example, suppose that in order to be enrolled for *Databases*, you need as prerequisites *Computer Programming*, *Algebra*, and *Logic*; for being enrolled in *Computer Programming* you need as a prerequisite *Operating Systems*, but *Operating Systems* would have *Databases* as a prerequisite; then, trivially, no student may ever enroll either for *Operating Systems*, or *Computer Programming*, or *Databases*; moreover, he/she might never be enrolled for any other courses having any one of these three as a prerequisite!

Here is another, simpler and somewhat dual example: assume that *SSN* is not declared as one-to-one; obviously, implausible instances might be possible, because any number of people might then have stored same SSNs, which should not be acceptable.

2.10.10 NO IMPLAUSIBLE RESTRICTIONS

> *It has been claimed at times that our modern age of technology*
> *facilitates dictatorship.*
> —Henry A. Wallace

R-DA-10. No implausible (dictatorial, not existing, aberrant) restrictions should ever be added to db schemas and/or applications (and not only: in fact, to anything else in the Universe!). Any such restriction is not only time consuming, but is aberrantly rejecting possible plausible instances.

For example, in a worldwide context, adding the restriction that city names, or zip codes, or state telephone prefixes, or vehicle plate codes, etc. should be (worldwide) unique is aberrant and would prevent, for example, storing more than one city having any given name (when, in fact, for example, besides France's capital, there are plenty of other cities in the world named Paris, from Tennessee, Texas, and Kentucky, to Illinois, Ontario, etc.).

2.10.11 NO RESTRICTIONS ON COMPUTED OBJECTS, EXCEPT FOR KEYS

> *Any kind of restrictions put on free speech would have worse consequences than bullying.*
> —Colleen Martin (aka Lady Starlight)

R-DA-11. No restrictions should ever be added to computed object sets or properties: as their instances/values must be computed only from plausible instances, they should also be intrinsically plausible (of course, provided that their computation expressions are at least plausible too!), except for computed columns added especially in order to be able to define keys (See Section 3.2.3.3).

For example, no constraints should be added either on computed properties of fundamental object sets or on computed (derived, non fundamental) sets, be them temporary or not, except for prime properties (see Fig. 3.5).

2.10.12 MAXIMUM CARDINALITIES

> *The more minimal art, the more maximum the explanation.*
> —Hilton Kramer

R-DA-12. For each object set its maximum possible number of elements should be added as its first associated restriction.

For example, *#Cities* values should be between 0 and 1,000 for states/regions/departments/lands/etc., or 100,000 for countries, or 100,000,000 worldwide, whereas *#Countries* values should be between 0 and 250.

Generally, reasonable margins should be provided in order to let room for foreseeable evolutions (e.g., 16 billion people worldwide by 2300, as current U.N. forecast is of some 9 billion).

2.10.13 RANGE RESTRICTIONS

I admire how she protects her energy and understands her limitations.
—Terry Tempest Williams

R-DA-13. For each attribute you should associate a corresponding range restriction, that is, a plausible finite subset of a data type.

For example, *FName* and *LName* might both take values in ASCII(64), the subset of all ASCII strings having length less or equal to 64; *BirthDate* and *HireDate* might have [1/1/1945, SysDate()] ⊂ DATE as range restrictions; *Sex* might take values into {'F', 'M'}, *Quantity* into [0.0, 100.0], *MarriageDate* into [1/1/-4000, SysDate() + 1year], *Grade* into [0, 20], *StartH* into [8, 18], etc.

Please note that one of the most typical programming runtime errors arises when programs assign implausible values to variables (e.g., out of range subscripts) and that the syntagm "garbage in, garbage out" would not perhaps be still that actual nowadays if programs were always checking both their input and output values for plausibility. Consequently, do not ever range restrict properties with implausible sets: for example, letting *BirthDate* to take any DATE value for *EMPLOYEES* would allow end-users to enter implausible data, resulting, for example, in employees that are either 2,000, 2, or −2,000 years old!

2.10.14 MANDATORY (COMPULSORY) TYPE RESTRICTIONS

Getting to the top is optional. Getting down is mandatory.
—Ed Viesturs

R-DA-14. For each object set, there should be at least one compulsory property (i.e., one that should not accept NULL values), besides the corresponding surrogate key. Do not forget to also ban dirty nulls.

For example, *SSN, FName, LName, CityName, StateName, Country-Name, State, Country, Precondition, Course, Sex,* and *Quantity* above should all be compulsory, while *BirthDate, BirthPlace, HomeTown, WorkingPlace,* and *Capital* above might allow null values too.

Generally, names, surrogate keys, and roles should always be compulsory.

Dually, it is obvious that it would be useless, incomprehensible, and implausible to store no data for all of the properties of an element of any object set (even if it would have an associated surrogate key value).

2.10.15   UNIQUENESS RESTRICTIONS

> *Feeling unique is no indication of uniqueness.*
> —Douglas Coupland

R-DA-15.

a.   For each object set, except for very rare cases, we should be able to uniquely identify any of its elements through at least one semantic (i.e., nonsurrogate) attribute, functional relationship, or aggregation of attributes and/or functional relationships.

For example, countries should be uniquely identified by their *CountryName, CountryCode,* and *Capital,* states should be uniquely identified by the product *StateName • Country* (as no country may simultaneously have two states having same name), people should be uniquely identified by their *SSN,* etc.

b.   There are only three exception types to "a." above:

•   Subsets: some of them might not have any semantic key (please note that they may, however, be uniquely identified by the semantic uniquenesses of their corresponding objects supersets, as they inherit them all). For example, *DRIVERS ⊆ EMPLOYEES* is not generally having any semantic uniqueness of its own (but it inherits *EMPLOYEES'* ones, for example *SSN*).

•   Object sets for which it makes sense to overload their surrogate (syntactic) keys with some semantics (and, consequently, it does not make sense anymore to duplicate its values in an identical semantic property). For example, *RABBIT_CAGES* might actually too be uniquely identified by their (generally, automatically generated number) surrogate key values (e.g., cage 1, cage 2, cage 3, etc.) and only by these values.

- Object sets for which we may not have enough data for some of their elements (like people that lived in ancient times and are ancestors of people for which we have enough data). However, note that even in such cases we might add some one-to-one *Notes* property (e.g., "ancestor of x from the Xth century BC who distinguished himself in the battle of y").

c. Please note that, even if most of the object sets have several semantic uniquenesses (that we should eventually discover and enforce in the end—see next chapter), in this first stage discovering only one (generally, the most used one by your customers) is enough; however, if you already discovered several ones, you should add them all to the corresponding restriction set.

2.10.16 OTHER RESTRICTION TYPES

> *The golden rule is that there are no golden rules.*
> —George Bernard Shaw

R-DA-16. Very frequently, apart from cardinality, range, mandatory, and uniqueness restrictions, data subuniverses are also governed by restrictions of other types as well: obviously, all of them should always be added to our corresponding models (see R-DA-9 above). Do not ever forget, in particular, that trees are acyclic, so you should add this constraint to any autofunction modeling a tree.

For example, in R-DA-6 above, a restriction enforcing that, for each country, its capital should be one of its cities (and not one of another country) should be added; in R-DA-9 above, we already seen that *PREREQUISITES* should be acyclic; in R-DA-13 above, a restriction stating that minor young ones cannot be employed should be added too: for each employee, $HireDate \geq BirthDate + 18years$; for the above case study, a restriction should ban first authors from also appearing in the corresponding coauthor lists.

2.10.17 OBJECT SETS AND THEIR RESTRICTIONS ORDERING AND NAMING

> *To create architecture is to put in order.*
> *Put what in order? Function and objects.*
> —Le Corbusier

R-DA-17.
    a. You should consider object sets in the descending order of their importance in the corresponding subuniverse.
    b. For each object set you should specify its restrictions grouped by their types (and always order types consistently; above R-DA-12 to R-DA-16 rule types order is preferable).
    c. For other restriction types involving several object sets (e.g., *BorrowDate ≤ DueReturnDate* and *BorrowDate ≤ ActualReturnDate* from the case study above), you should place them either in the restrictions subset of the last involved object set or at the end of the restriction set, in a dedicated multiset restrictions section.
    d. All restrictions of any data model should be uniquely named within it.

2.10.18 OBJECT SETS AND THEIR PROPERTIES AND RESTRICTIONS DOCUMENTATION

> *Documentation is like sex: when it is good, it is very, very good;*
> *and when it is bad, it is better than nothing.*
> —Dick Brandon

R-DA-18. Except for cases when (preferably generally, not only in that particular subuniverse) object sets, properties, and restriction names are unambiguous and self-explanatory (e.g., *PEOPLE*, *FirstName*, *SSN*, *PEOPLE_SSNuniqueness*, etc.), you should always associate substantial, comprehensive, concise, clear, nonambiguous descriptions for both object sets, properties, and restrictions.

2.10.19 E-RDS ARCHITECTURE FOR COMPLEX DATA MODELS

> *The art of simplicity is a puzzle of complexity.*
> —Douglas Horton

R-DA-19.

a. First establish atomic subdiagrams, only containing one object set and its basic properties (i.e., one rectangle or one diamond – with its underlying object sets too – and all associated ellipsis).

b. All subsequent E-RDs should be structural diagrams (i.e., only containing rectangles, diamonds, and arrows, without ellipsis).

c. Structural E-RDs should be established by (sub)domains of the universe of discourse (HR, Salaries, Accounting, Finance, Customers, Orders, etc.).

d. Links between these structural E-RDs are naturally ensured by commonly shared object sets (e.g., *EMPLOYEES* will appear both in HR, Salaries, and Orders).

e. Each E-RD should fit on one page.

f. Rectangles, diamonds, ellipsis, arrows, and lines should not intersect each other.

For example, the E-RD from Fig. 2.15 could be split into nine atomic subdiagrams and one structural E-RD, as presented in Figs. 2.27–2.37.

2.11 THE MATH BEHIND E-RDS AND RESTRICTION SETS: THE DANGER OF "MANY-TO-MANY RELATIONSHIPS" AND THE CORRECT E-RD OF E-RDM

> *If people do not believe that mathematics is simple,*
> *it is only because they do not realize how complicated life is.*
> —John Louis von Neumann

2.11.1 THE MATH BEHIND E-RDS AND RESTRICTION SETS

> *The person who reads too much and uses his brain too little will fall*
> *into lazy habits of thinking.*
> —Albert Einstein

FIGURE 2.27   PERSONS subdiagram.

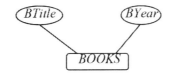

FIGURE 2.28   BOOKS subdiagram.

FIGURE 2.29   CO-AUTHORS subdiagram.

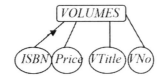

FIGURE 2.30   VOLUMES subdiagram.

FIGURE 2.31   PUBLISHERS subdiagram.

FIGURE 2.32   EDITIONS subdiagram.

FIGURE 2.33   VOLUME_CONTENTS subdiagram.

FIGURE 2.34   BORROWS subdiagram.

FIGURE 2.35   COPIES subdiagram.

FIGURE 2.36   BORROW_LISTS subdiagram.

Formally, any E-RD D is a pair $D = <S, \mathcal{M}>$, where S is a finite nonempty collection of sets and $\mathcal{M}$ a finite nonempty set of mappings defined on and taking values from sets in S; $(S, \subseteq)$ is a poset.

$S = \Omega \oplus \mathcal{V}$, where Ω is a nonempty collection of object sets and $\mathcal{V}$ is one of value sets; sets of Ω are explicitly figured (as rectangles or diamonds) in any E-RD, while those in $\mathcal{V}$ are implicitly assumed (never figured).

$\Omega = \mathcal{E} \oplus \mathcal{R}$, where $\mathcal{E}$ is a nonempty collection of atomic sets (represented as labeled rectangles and whose elements are called *entities*) and $\mathcal{R}$ is a (possibly empty) collection of mathematical relations (represented as labeled diamonds and whose elements are called *relationships*).

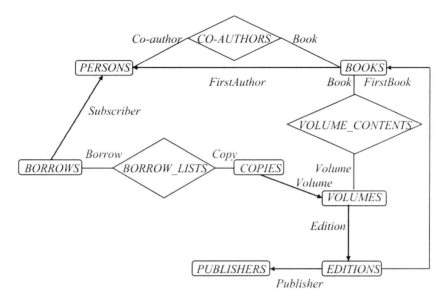

FIGURE 2.37   Public library structural E-RD.

Sets in $\mathcal{V}$ (generally called *data ranges*) are generally finite subsets of the so-called data types from computer programming, like naturals, integers, rationals, calendar dates, freely generated monoids over some alphabets (usually ASCII and UNICODE), Boolean values, etc., but may also be user defined, like, for example, the set of the rainbow colors.

$\mathcal{M} \subseteq (S \cup \text{NULLS})^{\Omega}$, where NULLS is a distinguished countable set of *nulls* (or *null values*); $\mathcal{M} = \mathcal{A} \oplus \mathcal{F}$, where $\mathcal{A} \subseteq (\mathcal{V} \cup \text{NULLS})^{\Omega}$ is a nonempty set of *attributes* (represented as labeled ellipsis) and $\mathcal{F} \subseteq (\Omega \cup \text{NULLS})^{\Omega}$ is a (possible empty) set of *functional relationships* (or *structural functions*, represented as labeled arrows). $\mathcal{R} \cap \mathcal{F} = \varnothing$.

Roles are names of canonical Cartesian projections; for example, in Fig. 2.25, *Prerequisite* and *Course* are the names of the corresponding (first and second) such projections of *PREREQUISITES* on *COURSES*.

The only unlabeled arrows (represented in E-RDs as ⊆_) are canonical injections; for example, Fig. 2.4 above represents the canonical injection i : *EMPLOYEES* → *PEOPLE*.

Obviously, both canonical Cartesian projections and canonical inclusions are structural functions (functional relationships).

Surrogate keys are natural injections defined on object sets (not more than one per set) having no semantics other than uniquely identifying their elements: $\forall O \in \Omega$, there may be only one surrogate key $x : O \rightarrow NAT(n)$, n natural; other authors are naming them ID or *OID* or ID*O* or #*O* etc. Obviously, surrogate keys are attributes.

Any *restriction set* C associated to an E-RD D is a nonempty set of closed Horn clauses whose first order predicates only refer to sets and mappings of D.

Any *(attribute) range restriction* $c \in C$ is, in fact, specifying the codomain of a mapping $a \in \mathcal{A}$ (c = "range a is C", where $dom(a) = O \in \Omega$, is equivalent to $a : O \rightarrow C$).

Any *(attribute) compulsory restriction* $c \in C$ is stating that the codomain of a mapping $m \in \mathcal{M}$ is disjoint from NULLS (c = "compulsory m" is equivalent to $codom(m) \cap$ NULLS $= \emptyset$).

Any *uniqueness restriction* $c \in C$ is stating that a mapping $m \in \mathcal{M}$ is one-to-one or that a mapping product $m_1 \bullet \ldots \bullet m_n$, $m_i \in \mathcal{M}$, $\forall i \in \{1, \ldots, n\}$, $n > 1$ natural, is minimally one-to-one. Note that, unfortunately, some authors and db "experts" wrongly consider one-to-one "relationships" as also being onto.

2.11.2   THE DANGER OF "MANY-TO-MANY RELATIONSHIPS" AND THE CORRECT E-RD OF E-RDM

Discovery consists of seeing what everybody has seen and thinking what nobody has thought.
—Albert Szent–Gyorgyi

From the E-RDM point of view, the relationship-type object set *UNDERLYING_SETS* from Subsection 2.9.2 above is correctly used: any object set may be the underlying set of several non-functional relationship-type sets and any non-functional relationship-type set has several (at least two) underlying object sets, so this is a "typical" so-called many-to-many relationship-type set.

In fact, modeled as such, it is not even a set, as it should accept duplicates: any non-functional relationship-type set may have a same object set as underlying one as many times as needed—and the most frequently studied and encountered such type of math relation is the homogeneous binary ones (i.e., those defined on two occurrences of a same set, like, for example, *PREREQUISITES* from Fig. 2.26).

Generally, as it is shown in Section 2.15 of the second volume (dedicated to the math behind (E)MDM), for any math relation the product of its canonical Cartesian projections should be one-to-one (and, in particular, for the binary ones, minimally one-to-one).

As a same object set may occur several times among the underlying sets of a non-functional relationship-type object set, the product *RelationshipType • ObjectSet* is not one-to-one, which means that, mathematically, *UNDERLYING_SETS* is ill-defined: the correct data model of it should use an entity-type set and not a relationship-type one.

Figure 2.38 shows the corresponding correct E-RD for the E-RDM.

From this example (and other similar ones are provided in Section 2.3 of the second volume), it is clear that "many-to-many" relationships should be used very cautiously if taking the shortest advisable path of conceptual data modeling and db design presented in this volume, as they are dangerously imprecise.

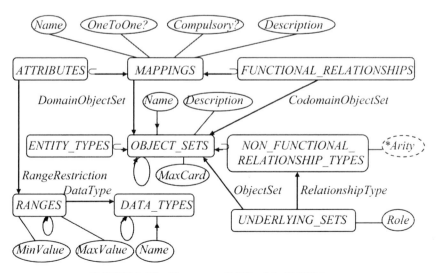

FIGURE 2.38   The correct E-RD of the E-RDM.

Fortunately, as formally proved in Section 2.10 of the second volume, you can always replace relationship-types object sets with entity-type ones (and "downgrading" their canonical Cartesian projections to ordinary functional relationships), just as in Fig. 2.38 as compared to the E-RD from Subsection 2.9.2 (Fig. 2.17).

2.12   CONCLUSION

> *The effective exploitation of his powers of abstraction*
> *must be regarded as one of the most vital activities of a competent*
> *programmer.*
> —Edsger W. Dijkstra

Db design should start with a thorough data analysis and modeling, for correctly identifying all needed object sets, their properties, as well as the rules governing the corresponding subuniverse of discourse. This process should result in an *E-R data model*—that is a set of E-RDs, an associated restriction set, and an informal but clear, concise, and comprehensive description of the corresponding subuniverse.

Not only that you should never start creating the tables of a db without a proper data analysis, but, moreover, during data analysis you should never think in terms of db tables, but in terms of object sets, their properties, the relationships between them, and the business rules that are governing the corresponding subuniverse.

Algorithm *A*0 from Section 2.9, together with the best practice rules from Section 2.10, provide a framework for assisting this process. As also seen from the case study presented in Section 2.8, this process is an iterative one. In the end, customers and db designers should agree (preferable formally) on all three deliverables of this process.

Obviously, this process is greatly favored by abstraction power, knowledge of the subuniverses of discourse semantics and governing rules, communication skills (in order to both find out from customers exactly what they need and everything relevant you don't know, but they do, as well as to adequately present your deliverables as to convince and get from your customers the final agreements), and experience (i.e., a library of best solutions to as many subuniverses as possible: contributing to this library is also one of the main goals of this book).

Two golden rules are crucial to the success or failure of data analysis:

- Never overload object sets with several semantics.
- Discover and include in your models not only all needed object sets, properties of and functional relationships between them, but also all business rules that are governing the corresponding subuniverses of discourse.

Although further refinements are generally needed, as shown in the second volume, a corresponding relational db scheme and an associated (possibly empty) set of non-relational constraints (that will have to be enforced either through db triggers or by db applications) may already be obtained from these deliverables by applying algorithm *A1-7* introduced in Section 3.3 and also exemplified in Section 3.4.

Dually, when you have to work with an unknown db, for which there is no such documentation, by using the reverse engineering algorithm *REA*1-2 introduced and exemplified in Section 3.5 you will benefit a lot especially from the structural E-RD and restriction set that you will obtain, which is always providing a very useful overview on both the set of all existing interobject sets connections, as well as on the business rules governing that subuniverse.

Hopefully, the E-R data model of the E-RDM introduced in Section 2.9 makes understanding of this data model simpler. The math behind this E-RDM kernel (presented in Section 2.11) is very straightforward, while providing a deeper understanding of this simple, yet powerful data model.

As many are wrongly assuming that one-to-one relationships are also onto, as using many-to-many ones proves dangerous (see Subsection 2.11.2) and not fundamental (see Section 2.10 of the second volume), and as many-to-one and one-to-many ones may be misleading too (as, in fact, they are both ordinary mappings with only the arrow direction being different between them), we would strongly recommend to get rid of them all and always think in data modeling only in terms of atomic object sets and mappings defined on them.

2.13   EXERCISES

> *In any situation, the best thing you can do is the right thing;*
> *the next best thing you can do is the wrong thing;*
> *the worst thing you can do is nothing.*
> —Theodore Roosevelt

2.13.1 SOLVED EXERCISES

> *He who asks a question may be a fool for five minutes,*
> *but he who never asks a question remains a fool forever.*
> —Thomas Martin Connelly Jr.

2.0. Apply algorithm *A0* and the above best practice rules to design the E-RDs, restriction set, and description for the following simplified payments subuniverse: payments to suppliers (names and locations) by customers (names and locations) are done for invoices (supplier, customer, number, date, total amount) either cash, with credit/debit cards (type, credit/debit, number, expiration month and year, holder, security code, account), or through bank wires; payment documents are, correspondingly, receipts, bank payment confirmations and orders; for all of them, supplier, customer, number, date, and total amount are needed; moreover, the paid total amount of a payment may represent the sum of several partial invoice payments; for bank payments also needed are involved bank accounts (bank subsidiaries' addresses and cities, holders, number, IBAN code, available amount); suppliers may be customers too; there may not be two cities of a same country having same zip code.

Solution:

Banks may also be suppliers and/or customers, so we abstract them too as partners; as not all suppliers and/or customers are banks, we add to *PARTNERS* a Boolean property *Bank?* too.

For simplicity, we assume that each bank account has only one holder and at most one associated card valid in any given time period, and that any card is valid at least one year.

Obviously, total amounts on both invoices and payment documents are computable (for payment documents, from corresponding amounts in payments; for invoices, from corresponding subtotal amounts in invoice lines, not figured here as it does not matter from the payments point of view); however, in order to speed-up querying, we have added two computed **TotAmount* properties, one for *INVOICES* and the other for *PAYM_DOCS*.

The corresponding description of this subuniverse is as follows:

1. Payments to suppliers by customers are done for invoices (supplier, customer, number, date, all of them compulsory) either cash, with

credit/debit cards (type, credit/debit, number, expiration month, holder, security code, account, all of them mandatory), or through bank wires; payment documents are, correspondingly, receipts, bank payment confirmations, and corresponding orders; for all of them, debtor, number, date, and type are needed (all of them mandatory); for bank wires, supplier and customer corresponding accounts (bank subsidiaries, holders, IBAN code, available amount, all of them mandatory) are needed too; for card payments, card is needed too; moreover, the paid total amount of a payment may represent the sum of several partial invoice payments.

2. Suppliers and banks may be customers too; consequently, suppliers, customers, and banks are considered as partners, for which names, cities, and addresses within cities (all of them mandatory) are stored; some partners may be subsidiaries of others; obviously, no partner may be a subsidiary of its own, either directly, or indirectly; only bank type partners may manage bank accounts. There may not be two partners having same name, city, and address. There may be at most 100,000 partners of interest.

3. There may not be two invoices issued by a same supplier and having same number and date. Self-invoicing is not allowed. There may be at most 10,000,000 invoices of interest.

4. There may not be two payments issued by a same debtor and having same type, number, and date. No payment for an invoice can be done before the invoice is established or after more than 100 years after corresponding establishing date. The sum of all paid amounts for different invoices should be equal to the total amount of the corresponding payment document. The sum of payments for an invoice may not be greater than the corresponding invoice total amount. For cash payments, there are no associated bank accounts or cards. For card ones, card and supplier account are mandatory, while the debtor's one should not be stored (as it is the corresponding card account); no payment may be done with an expired card; there may not be two cards of a same type and having same number. For bank wire ones, there may be no associated card, but the suppliers' and the debtors' accounts are mandatory; no self-payment is allowed. There may not be two payments for a same invoice on a

same payment document. There may be at most 100,000,000 payment documents and details, 1,000,000 cards, and 100 card types of interest.

5. There may not be two bank accounts having same IBAN code. There may be at most 1,000,000 bank accounts of interest.

6. There may not be two cities of a same country having same zip code. City names, corresponding countries, and zip codes are mandatory. There may be at most 100,000 cities of interest.

7. There may not be two countries having same name. Country name is mandatory. There may be at most 250 countries of interest.

Figure 2.39 presents the corresponding E-RD. Corresponding restriction set is the following:

1. **PARTNERS** (The set of all involved partners: suppliers, customers, banks)

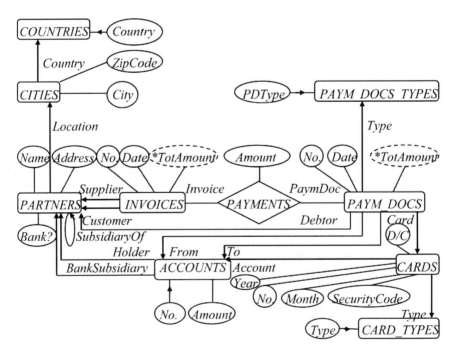

FIGURE 2.39   E-RD for the "payments" subuniverse.

a. *Cardinality: max(card(PARTNERS)) = 100,000*   (*P0*)
b. *Data ranges*:
   Bank?: BOOLE   (*P1*)
   Name: ASCII(128)   (*P2*)
   Address: ASCII(255)   (*P3*)
c. *Compulsory data: Bank?, Name, Address, Location*   (*P4*)
d. *Uniqueness: Name • Location • Address* (there may not
   be two partners having same name, city, and address)   (*P5*)
e. *Other type restrictions*:
   SubsidiaryOf should be acyclic (no partner can be a
   subsidiary of itself, either directly or indirectly).   (*P6*)
   All and only the subsidiaries of a bank should be of
   bank type.   (*P7*)

2. **INVOICES** (The set of all involved invoices issued by
   suppliers to customers)
a. *Cardinality: max(card(INVOICES)) = 10,000,000*   (*I0*)
b. *Data ranges:*
   No.: ASCII(16)   (*I1*)
   Date: [1/1/2000, SysDate()]   (*I2*)
   *TotAmount = Sum(INVOICE_DETAILS.Amount)
   (Exercise 2.8)   (*I3*)
c. *Compulsory data: No., Date, Supplier, Customer*   (*I4*)
d. *Uniqueness: No. • Date • Supplier* (there may not be
   two invoices issued by a same supplier and having
   same number and date)   (*I5*)
e. *Other type restrictions: Supplier* and *Customer* values
   should not ever be equal for a same invoice.   (*I6*)

3. **PAYM_DOCS** (The set of all involved payment documents
   issued by customers according to supplier invoices)
a. *Cardinality: max(card(PAYM_DOCS)) = 100,000,000*   (*PD0*)
b. *Data ranges:*
   No.: ASCII(16)   (*PD1*)
   Date: [1/1/2000, SysDate()]   (*PD2*)
   *TotAmount = Sum(PAYMENTS.Amount)   (*PD3*)

    c. *Compulsory data: No., Date, Type, Debtor*   (*PD4*)

    d. *Uniqueness: No. • Date • Debtor • Type* (there may not
    be two payments issued by a same debtor and having
    same type, number, and date)   (*PD5*)

    e. *Other type restrictions:*

      No payment for an invoice can be done before the
      invoice is established or after more than 100 years after
      corresponding establishing date.   (*PD6*)

      Whenever *Type* is '*c*' (cash), *To*, *From*, and *Card*
      should be null.   (*PD7*)

      Whenever *Type* is '*C*' (card), *From* should be null,
      whereas *To* and *Card* should not be null.   (*PD8*)

      Whenever *Type* is '*w*' (bank wire), *Card* should be null
      and *To* and *From* should not be null.   (*PD9*)

      To and *From* values should not ever be equal for a same
      payment document.   (*PD10*)

4. **PAYMENTS** (The set of all involved, possibly partial
and/or consolidated, payments made by customers for their
invoices)

    a. *Cardinality: max(card(PAYMENTS)) = 100,000,000*   (*PM0*)

    b. *Data ranges:*

      Amount: CURRENCY(6)   (*PM1*)

    c. *Compulsory data: Amount, Invoice, PaymDoc*   (*PM2*)

    d. *Uniqueness: Invoice • PaymDoc* (there may not be two
    payments for a same invoice on same payment document) (*PM3*)

    e. *Other type restrictions:*

      Sum of all payments for a same invoice should not be
      greater than the corresponding invoice total amount.   (*PM4*)

5. **PAYM_DOCS_TYPES** (The set of all 3 possible payment types:
cash, card, and bank wire)

    a. *Cardinality: max(card(PAYM_DOCS_TYPES)) = 3*   (*PDT0*)

    b. *Data ranges:*

      PDType: {"c", "C", "w"}   (*PDT1*)

    c. *Compulsory data: PDType*   (*PDT2*)

   d. Uniqueness: PDType (there may not be two payment
      types having same name)   *(PDT3)*

6. **ACCOUNTS** (The set of all involved bank accounts hold
   by partners)
   a. Cardinality: max(card(ACCOUNTS)) = 1,000,000   *(A0)*
   b. Data ranges:
      No.: ASCII(32)   *(A1)*
      Amount: CURRENCY(32) (available amount)   *(A2)*
   c. Compulsory data: BankSubsidiary, Holder, No., Amount   *(A3)*
   d. Uniqueness: No. (there may not be two bank accounts
      having same IBAN code)   *(A4)*
   e. Other type restrictions:
      Bank subsidiaries should always be bank type partners. *(A5)*

7. **CARDS** (The set of all involved debit/credit cards used
   by partners for their payments)
   a. Cardinality: max(card(CARDS)) = 1,000,000   *(C0)*
   b. Data ranges:
      No.: $[10^{15}, 10^{16} - 1]$ *(C1)*
      Month: [1, 12]   *(C2)*
      Year: [*Year*(*SysDate*()) — 2, *Year*(*SysDate*()) + 2]   *(C3)*
      SecurityCode: [0, 999]   *(C4)*
      D/C: {'D', 'C'} (debit or credit card)   *(C5)*
   c. Compulsory data: Account, No., Type, Month,
      Year, SecurityCode, D/C   *(C6)*
   d. Uniqueness: No. • *Type* • *Year* (there may not be two cards of
      a same type and having same number and expiration year) *(C7)*
   e. Other type restrictions:
      No card payment may be done with an expired card.   *(C8)*

8. **CARD_TYPES** (The set of all involved card types;
   e.g. American Express, Visa, Visa Electron, Maestro, etc.)
   a. Cardinality: max(card(CARD_TYPES)) = 64   *(CT0)*
   b. Data ranges: CardType: ASCII(16)   *(CT1)*

c. *Compulsory data: CardType*   (*CT2*)

d. *Uniqueness: CardType* (there may not be two card types
   having same name)   (*CT3*)

9. **CITIES** (The set of all involved cities, where partners reside)

a. *Cardinality: max(card(CITIES))* = 100,000   (*CS0*)

b. *Data ranges:*

    City: ASCII(64)   (*CS1*)

    ZipCode: ASCII(16)   (*CS2*)

c. *Compulsory data: Country, City, ZipCode*   (*CS3*)

d. *Uniqueness: Country • ZipCode* (there may not be two
   cities of a same country having same zip code)   (*CS4*)

10. **COUNTRIES** (The set of all involved countries, where
    partners reside)

a. *Cardinality: max(card(ACCOUNTS))* = 250   (*CN0*)

b. *Data ranges:*

    Country: ASCII(128)   (*CN1*)

c. *Compulsory data: Country*   (*CN2*)

d. *Uniqueness: Country* (there may not be two countries
   having same name)   (*CN3*)

2.13.2 PROPOSED EXERCISES

How you think about a problem is more important than the problem itself
– so always think positively.
—Norman Vincent Peale

Apply algorithm *A*0 to the following subuniverses:

2.1. Add to the public library db data on people birth dates, phone numbers, and addresses, other types of both authors and subscribers (e.g., companies, governments, legislators, nongovernmental organizations, etc.), books topics, publisher headquarters locations, edition types (e.g., hard cover, paperback, electronic), volumes' number of pages and format (e.g., B5, A4, A3, etc.), and copies' status (e.g., brand new, new, satisfactory, deteriorated, etc.). Establish E-R subdiagrams for each object set and one or several structural ones, and update correspondingly restriction set, and description of this augmented subuniverse.

2.2. Extend the public library db for also managing journals, magazines, newspapers, movies, music, etc., both in printed and electronic editions. Update correspondingly E-RDs, restriction set, and description of this extended subuniverse.

2.3. The subuniverses from Figs. 2.5, 2.12, 2.17, 2.18, 2.23–2.25, as well as the one mentioned in Section 1.6.

2.4. The following subuniverses:
   a. public schools
   b. TV stations and their programs
   c. Internet web sites and their contents
   d. e-commerce sites (e.g., eBay)
   e. music and film sites (e.g., YouTube)
   f. social networking sites (e.g., Facebook)
   g. pictures and albums sites (e.g., Picassa)
   h. e-mail clients (e.g., Gmail)

2.5. The following subuniverses:
   a. (*"supplies"*): manufacturers (name, headquarters, plants' countries), vendors (name, headquarters, subsidiaries), and products (name, manufacturer, type, vendor quotations, serial no., manufacturer code, vendor code, stocks, characteristics, conditioning); for countries, only their name and capital are needed; for cities, only name and country; for characteristics, only name and measurement unit; for measurement units, only names and immediately upper ones (e.g., for KB, 1024 KB = 1 MB).
   b. (*"orders"*): customers (types, e.g., people, ltd/Inc./etc. company, governmental agency, etc., names, contact persons, and locations), products (name, manufacturer, type, manufacturer code, vendor code, own code, stocks, characteristics, conditioning), orders

(number, date, deadline, customers, products, quantities, prices, employees in charge).

c. ("*marketing*"): products/services (name, type), campaigns (time interval, types, e.g., internet, TV, radio, banners, etc., sites, managers, costs, sale increases per periods).

d. ("*sales*"): products/services (name, type, conditioning), discount percentages and sold quantities per period and conditioning.

e. ("*invoices*"): customers (types, e.g., people, ltd/Inc./etc. company, governmental agency, etc., names, invoice and delivery locations, and orders), invoices (number, date, and total amount), products/services (code, name, quantity, and price).

f. ("*deliveries*"): carriers (names, locations, costs), customers (types, e.g., people, ltd/Inc./etc. company, governmental agency, etc., names, invoices, and delivery locations), invoices (number, date, total amount), products/services (code, name, quantity, price).

g. ("*human resources*"): employees and candidates (first and last name, birth place and date, home and e-mail addresses, sex, parents, spouse, children, education, previous and current jobs, hire and leaving dates).

h. ("*salaries*"): employees (first and last name, job, base salary, taxes, due and actual working hours, bonuses, retention clause, bank account).

i. ("*postal*"): continents (names), countries (names, abbreviations, phone prefixes, state names—e.g. state, department, land, region, etc.), states (name, country), cities (name, state, subordinated cities), and zip codes.

j. ("*time tables*"): departures, stops, and destinations (name, city) weekdays and hours, per seasons, companies (type, e.g., airline, railroad, bus, ship, etc., name, country), prices per comfort classes.

k. ("*football leagues*"): countries (names), cities (names, countries), stadiums (name, city, owner club), clubs (name, city), matches (home and host teams, season, date, score, referees, attendees number), players (names, origin countries, clubs, numbers, positions, starting and ending seasons, buying and selling amounts), goals (match, player, minute), coaches (names, origin countries, clubs, salaries, won nonleague competitions), referees (names, origin countries).

l. ("*school schedules*"): disciplines (titles, credits), pre and coreq-uisites, teachers (first and last names, competences, sex), rooms (numbers, buildings floors), schedules (semesters, weekdays, dis-ciplines, teachers, start and end hours, rooms), students (first and last names, disciplines enrollments, sex).

m. ("*movies*"): industry people (e.g., directors, actors, composers, cameramen, sound engineers, etc.) first and last names, birth and passed away places and dates, movies (title, producers, year, lan-guage, novel/play/movie adapted, casting, prizes, category—e.g., drama, western, musical, etc.).

n. ("*music*"): composers (first and last names, birth and passed away places and dates), their works (title, opus and catalog number, tonality, soloist instruments, parts, lyrics language), interpreters (types – e.g., person, duo, trio, quartet, choral, chamber/symphonic orchestra–, first and last names, birth and passed away places and dates), interpretations (year, works, interpreters, hall/studio), recordings (publisher, release year, interpretations, media type – e.g., blu-ray, dvd, cd, vhs–, price).

2.6. Add to the E-R data metamodel presented in Section 2.9 unique-ness restrictions made out of several mappings, as well as all existing restrictions of other types.

2.7. Design an E-R data metamodel for the original E-RDM.

2.8. Add to Exercise 2.0 products (unique name and unit price) and invoice details (invoice, line number, product, unit price, quantity, and total amount).

2.9. Characterize algorithm $A0$, proving that its complexity is linear in the sum of object set collection, property and restriction sets cardinals.

2.10. Analyze the data models proposed by http://www.databas-eanswers.org/data_models/index.htm and establish E-R data models for each of them.

2.11. Extend problem 2.0 considering that a bank account can have several holders and several associated cards simultaneously valid, one for each such holder.

2.14   PAST AND PRESENT

> *Employ your time in improving yourself by other men's writings,*
> *so that you shall gain easily what others have labored hard for.*
> —Socrates

> *If you are depressed, you are living in the past.*
> *If you are anxious, you are living in the future.*
> *If you are at peace, you are living in the present.*
> —Lao Tzu

The E-RDM introduced by Chen (1976), was the first hugely successful higher level conceptual data model, not only still in use today, but, very probably, to be at least as used and revered in the foreseeable future too.

Each year there are dozens of E-RDM international conferences, mainly proposing new extensions to it (e.g., the International Conferences on Conceptual Modeling (ER), see, for example, http://www.informatik. uni-trier.de/~ley/db/conf/er/index.html).

Besides the thousands of papers on it, there are dozens of good books too; for example, one of the "bibles" in E-RDM is Thalheim (2000).

A recent book in this field is Bagui and Earp (2012).

The math behind E-RDM section 2.10 of this chapter follows the E-RDM algebraic formalization from Mancas (1985), which also extended E-RDM with hierarchies of relationships, refined in Mancas (2005).

Even if E-RDM is so simple and with very limited constraint expressive power, it is still heavily used in conceptual data modeling, generally as a first step, due to its graphical suggestive power.

E-RDs (also abbreviated ERDs) too are ancestors of UML diagrams (see http://www.iso.org/iso/iso_catalog/catalog_tc/catalog_detail. htm?csnumber=32624 and http://www.omg.org/spec/UML/2.4.1/Super-structure/PDF/), both being types of organograms.

Most of the RDBMS are offering their variants of E-RDs, at least for visualizing links between db tables. For example, MS *Access* provides among its db tools the so-called *Relationships* window; its MS *SQL Server* equivalent is the *Database Diagrams*; IBM *Data Studio* (http://www.ibm. com/developer works/downloads/im/data/index.html) and Oracle *SQL*

Developer Data Modeler (http://www.oracle.com/technetwork/developer-tools/datamodeler/overview/index.html) empower their users to also create and modify db schemas through their E-R-type diagrams.

Dually, there are also E-R data modeling tools that can connect to several RDBMSs like, for example, CA *Erwin* (http://erwin.com/products/data-modeler).

Algorithm *A*0 and the E-R data metamodel of the E-RDM (see Section 2.9), the public library case study (Section 2.8), as well as the best practice rules (Section 2.10) and the above exercises (Section 2.13) come from Mancas (2001).

All URLs mentioned in this section were last accessed on July 17th, 2014.

KEYWORDS

- **attribute**
- **attribute range restriction**
- **Barker E-RD notations**
- **Chen E-RD notations**
- **compulsory data restriction**
- **computed data**
- **computed objects**
- **controlled data redundancy**
- **data analysis**
- **data computability**
- **data modeling**
- **E-R data model (E-RDM)**
- **E-RD architecture**
- **entity**
- **Entity-Relationship (E-R)**
- **Entity-Relationship Diagram (E-RD)**
- **functional relationship**
- **implausible restriction**
- **Key Propagation Principle (KPP)**

- **Mancas E-RD notations**
- **many-to-many relationship**
- **many-to-one relationship**
- **non-functional relationship**
- **object set basic properties**
- **object set property types**
- **object sets abstraction**
- **one-to-many relationship**
- **one-to-one relationship**
- **Oracle E-RD notations**
- **other restriction types**
- **plausible restriction**
- **recursive relationship**
- **relationship**
- **relationship hierarchy**
- **restriction**
- **restriction set**
- **role**
- **semantic overloading**
- **surrogate key**
- **uniqueness restriction**

REFERENCES

> *Principles and rules are intended to provide a thinking man*
> *with a frame of reference.*
> —Carl von Clausewitz

Bagui, S., Earp, R. (2012). *Database Design Using Entity-Relationship Diagrams*, 2nd Edition: CRC Press, Boca Raton, FL.

Chen, P. P. (1976). The entity-relationship model: Toward a unified view of data. *ACM TODS* 1(1), 9–36.

Mancas, C. (1985). A Formal Viewpoint on the Entity-Relationship Data Model: *Proc. 6th Int. Conf. on Control Syst. and Comp. Sci.*, 2, 174–178, *Politehnica* University, Bucharest, Romania.

Mancas, C. (2001). *Databases for Undergraduates Lecture Notes*: Computer Science Department, *Ovidius* State University, Constanta, Romania.

Mancas, C., Mancas, S. (2005). *MatBase* E-R Diagrams Subsystem Metacatalog Conceptual Design. Proc. IASTED DBA 2005. Conf. on DB and App., 83–89, Acta Press, Innsbruck, Austria.

Thalheim, B. (2000). *Fundamentals of Entity-Relationship Modeling*. Springer-Verlag, Berlin.

CHAPTER 3

THE QUEST FOR DATA INDEPENDENCE, MINIMAL PLAUSIBILITY, AND FORMALIZATION: THE RELATIONAL DATA MODEL (RDM)

Facts and theories are different things, not rungs in a hierarchy of increasing certainty. Facts are the world's data. Theories are structures of ideas that explain and interpret facts. Facts do not go away while scientists debate rival theories for explaining them. Einstein's theory of gravitation replaced Newton's, but apples did not suspend themselves in mid-air pending the outcome.

—Stephen Jay Gould

CONTENTS

3.1 First Normal Form Tables, Columns, Constraints, Rows, Instances..118

3.2 The Five Basic Relational Constraint Types............................. 123

3.3 The Algorithm for Translating E-R Data Models into Relational Schemas and Non-Relational Constraint Sets (*A*1-7): An RDM Model of the E-RDM 146

3.4 Case Study: The Relational Scheme of the Public Library Data Model.. 152

3.5 The Reverse Engineering Algorithm for Translating Relational Schemas into E-R Data Models (*REA*1-2)................................. 156

3.6 The Algorithm for Assisting Keys Discovery (*A*7/8-3) 164

3.7   RDBMS Metacatalogs. Relational and E-R Data Models
        of the RDM 177

3.8   Relational Schemas Definition. *SQL DDL* 193

3.9   Relational Instances Manipulation. *SQL DML*. Relational
        Calculi and Algebra... 196

3.10   Higher and the Highest RDM Normal Forms............................... 227

3.11   Best Practice Rules.. 237

3.12   The Math behind RDM... 247

3.13   Conclusion 256

3.14   Exercises 263

3.15   Past and Present 313

Keywords.. 325

References.. 332

After studying this chapter you should be able to:

- Concisely and precisely define, and exemplify the following key terms: *relational db* (rdb) (name, scheme, instance), *First Normal Form* (1NF), *table* (*relation*) (name, nested, header (scheme), column (*attribute*, field), arity, row, instance, cardinal, cursor, record set, fundamental, computed, temporary, universal), *cross tabulation*, *relation* (name, scheme, instance, tuple, arity, cardinal), *null* (value, not applicable, temporarily unknown, no information, dirty, marked), *attribute* (name, prime, nonprime, domain (range), not null (mandatory, compulsory, required, total, computed), unique, concatenation (product), minimally uniqueness), *constraint* (dependency) (name, nonrelational, relational, primitive, domain, (unique) key, referential integrity (typed inclusion), check (tuple), existence, not null (totality), (partial, transitive) functional, multivalued (complementary property), join, acyclicity, object, data, equalities/tuple generating, preservation (representation)), *key* (surrogate (syntactic), autonumber, primary, candidate (semantic), simple, concatenated (composed), superkey, foreign, Propagation Principle), *dangling pointer*, *metadata catalog* (metacatalog), *relational calculus* (domain (*DRC*), *QBE*, tuple (*TRC*), SQL, Quel, domain independence, type/value domain, CH- completeness (computable queries, *QL*), *IsNULL* predicate), *relational algebra* (union, intersection, complementation, Cartesian product, division, *selection*, (total) *projection*, *renaming*, inner/outer (left, right, full) natural/equi/theta/complete/semi/anti *join*, augmentation, kernel equivalence relation, quotient set, expressive power, lossless decomposition, completeness), *SQL* (query, extended, embedded, static, dynamic, *DDL*, CREATE/ALTER/DROP SCHEMA/ TABLE, CONSTRAINT, NOT NULL, UNIQUE, FOREIGN KEY, CHECK, GRANT, REVOKE, *DML*, INSERT, UPDATE, DELETE, SELECT, *aggregation functions* (COUNT, SUM, MIN, MAX, AVG), DISTINCT, DISTINCTROW, TOP, AS, FROM, INNER/OUTER (LEFT, RIGHT, FULL) JOIN, WHERE, GROUP BY, HAVING, ORDER BY, ANY, SOME, ALL, (NOT) IN, (NOT) EXISTS, UNION (ALL), BEGIN/START/COMMIT/ROLLBACK TRANSACTION, *subquery* (static, dynamic), standards), *view*, *stored procedure/function/trigger*, *transaction*, *ACID*ity, *transitive closure*, *data manipulation anomalies* (insert, update, delete), 2nd *Normal Form* (2NF), 3rd *Normal Form* (3NF), *Boyce-Codd Normal Form* (BCNF, 3.5NF), 4th *Normal Form* (4NF), 5th *Normal Form* (5NF), *Projection-Join Normal Form* (PJNF), *Domain-Key Normal Form* (DKNF), *Domain-Key Referential Integrity Normal Form* (DKRINF), *normalization problem* (synthesis, decomposition), *derivation rules* ((limited) complete axiomatization, soundness, completeness, finite implication, decidability/undecidability), *chase* (tableaux).

- Correctly apply algorithm A1–7 (translation of E-RDs and restriction sets into relational schemas and associated nonrelational constraint sets) to any E-R data model.
- Correctly apply algorithm REA1–2 (translation of relational schemas into E-R data models) to any rdb scheme.
- Correctly apply algorithm A7/8–3 (assistance of existing keys validation and discovery of all keys) to any rdb table scheme.
- Use provided key terms definitions, A1–7, REA1–2, A7/8–3 algorithms, and/or the best practice rules for correctly solving at least all exercises found at the end of this chapter.

3.1 FIRST NORMAL FORM TABLES, COLUMNS, CONSTRAINTS, ROWS, INSTANCES

> *He is able who thinks he is able.*
> —The Buddha

Informally, a *relational db* (*rdb*) is a finite set of interlinked tables, each of which has to have a unique name within the rdb and be in the so-called *First Normal Form* (*1NF*)—that is all of their columns should be atomic (i.e., not subtables; tables that are not in 1NF are called *nested*) and there should not be duplicate rows. Both RDM and RDBMSs are generally considering only 1NF rdbs.[28]

Note that, formally, atomic *cross tabulations* like the one in Table 3.1a are in 1NF; in fact, as cross-tabulations are computable, in RDM they are never stored: Table 3.1b shows how data from Table 3.1a is stored instead (as generally, only strictly positive data is stored, while negative one it is not).

TABLE 3.1A   Example of a Cross Tabulation

Supplier	Part 1	Part 2	Part 3
1	3	2	1
2	4	0	2
3	0	2	0

TABLE 3.1B   RDM 1NF Table Storing Data from Which Cross-Tabulation in Table 3.1a May Be Computed

Supplier	Part	Qty.
1	1	3
1	2	2
1	3	1
2	1	4
2	3	2
3	2	2

[28] But there are both theoretical (e.g., O-O extensions of RDM) and implementation (e.g., MS *Access* 2010 "multiple values" combo and list boxes) approaches of nested table schemes too.

Also note that RDM's *relations* (derived from the homonym sets algebra ones) may be also represented using graphs or lists, etc., but, fortunately, their most common representation uses tables. Historically, in RDM tables are called *relations*, columns are called *attributes*, and rows are called *tuples* (or *n*-tuples).

Table *headers* (*schemas*) contain both the corresponding finite set of *column* names (which should be unique across any table) and a finite set of relational constraints. The number of columns of a table (its *arity*) is a natural strictly positive number. Table rows (if any) are making up corresponding table (*data*) *instances*. Note that, abusively, *relation* sometimes designates in RDM only the graph of a relation (i.e., its corresponding table instance). The number of rows of a table (or its *cardinal*) is a natural number. The union of all table headers of an rdb constitutes its *scheme*; the union of all table instances of an rdb constitutes its *instance*.

Figure 3.1 summarizes basic notion equivalences between these three formalisms' notational conventions.

For example, in Fig. 3.2, *COMPOSERS* has arity 5 and cardinality 8; *MUSICAL_WORKS* has arity 6 and cardinality 9; *TONALITIES* has arity 2 and cardinality 6; in total, this rdb instance has cardinality $8 + 9 + 6 = 23$.

Parenthesis after tables names, as well as second and third table headers rows contain the associated *relational constraints*:

> ➤ in the parenthesis after the table names are listed the *key* (*uniqueness*) *constraints*, stating that corresponding columns or concatenation[29] of columns do not accept duplicate values (in this example: in table *COMPOSERS* there may not be two rows having same values in columns *FirstName* and *LastName*; in table *MUSICAL_WORKS* there may not be two rows having same values in columns *Composer*, *Opus* and *No*; in table *TONALITIES* there may not be two rows having same values in column *Tonality*); consequently, they may be used for unique identification of corresponding instance rows;

> ➤ note that each table has a so-called *primary key*, denoted *x* in this example (and almost always in this book: Subsection 3.2.3.1 explains why); primary key names are underlined; unfortunately, RDM allows to declare as primary any key, which you should never

---

[29] Formally, $f \cdot g$ denotes the *mapping* (*Cartesian*) *product* of mappings f and g (see Appendix). In RDM, it is usually abbreviated as fg.

informal	RDM	Math
table instance (content)	relation instance (content)	function (Cartesian) product image
table header (scheme)	relation scheme	function (Cartesian) product definition ⊕ associated (relational) constraint definitions
(table) column, field	attribute	(Cartesian product) member function
(table) row	(n-)tuple	function (Cartesian) product image element
db instance (content)	rdb instance (content)	union of all function (Cartesian) product images
(table) number of columns	relation arity	function (Cartesian) product arity
(table) number of rows	(relation) instance cardinal	function (Cartesian) product image cardinal
non-nested table	1NF (First Normal Form)	atomic (that is no member is a function product in its turn or has a Cartesian product as its codomain) Cartesian function product
business rule	constraint, dependency	logical first order predicate, closed formula
cursor (recordset) definition	query	logical first order predicate, open formula

FIGURE 3.1   Equivalences between informal, RDM, and math terminology.

do (see why in Subsection 3.2.3.1); also note that all tables should have all existing constraints, so all existing keys, which means that, with a couple of exceptions to be explained and exemplified in Subsection 3.2.3.1, all tables should have at least one non primary key as well: for example, if *Tonality* were not a key in *TONALITIES*, instances of this table might store, for example, "G major" twice (or any number of times, thus violating axiom A-DI17 from Section 5.1.2);

➤ after these parenthesis are listed the *check constraints*, each of which is stating that the corresponding formula should hold for every corresponding instance row (in this example, *BirthYear* + 3 < *PassedAway*

COMPOSERS ($\underline{x}$, *FirstName* • *LastName*) *BirthYear* + 3 < *PassedAwayYear*,
PassedAwayYear – *BirthYear* < 120

$\underline{x}$	FirstName	LastName	BirthYear	PassedAway Year
auto-num(8) NOT NULL	ASCII(32)	ASCII(32) NOT NULL	[1200, currentYear() – 3] NOT NULL	[1200, currentYear()]
1	Antonio	Vivaldi	1678	1741
2	Johann Sebastian	Bach	1685	1750
3	Joseph	Haydn	1732	1809
4	Wolfgang Amadeus	Mozart	1756	1791
5	Ludwig van	Beethoven	1770	1827
6	Franz	Schubert	1797	1828
7	Frédéric	Chopin	1810	1849
8	Johannes	Brahms	1833	1897

MUSICAL_WORKS ($\underline{x}$, *Composer* • *Opus* • *No*)

$\underline{x}$	Composer	Opus	No	Tonality	Title
auto-num(12) NOT NULL	COMPO-SERS.x NOT NULL	ASCII(16) NOT NULL	[1, 255]	TONALI-TIES.x	ASCII(255)
1	1	8 RV 269		4	La primavera
2	2	BWV 846–893			The Well-Tempered Clavier
3	3	H 1/94		5	The "Surprise" symphony
4	4	K 527			Don Giovanni
5	5	125		3	The 9th symphony
6	6	D 759		2	The "Unfinished" symphony
7	7	37	2	5	Nocturne
8	8	83		1	The 2nd piano concerto
9	8	115		2	The clarinet quintet

TONALITIES (_x_, *Tonality*)

_x_	*Tonality*
autonum(2)	ASCII(13)
NOT NULL	NOT NULL
1	B-flat major
2	B minor
3	D minor
4	E major
5	G major
6	G minor

FIGURE 3.2   A classical music rdb example.

Year states that every composer has to live at least 3 years –as it is not plausible that somebody could compose before being 3–, while *Passed-AwayYear—BirthYear* < 120 states that no composer may live longer than 120 years);

➢ in the first header line are listed the names of the columns (attributes, mappings);

➢ in the second header line are listed both the domain constraints and the foreign keys:

  ✓ *domain constraints* specify column range restrictions (in this example, all *x* column values are naturals auto-generated by the system: for *COMPOSERS* having at most 8 digits, for *MUSICAL_WORKS* at most 12, and for *TONALITIES* at most 2; *FirstName* and *LastName* may only be character strings using the ASCII alphabet, having length at most 32; *PassedAwayYear* may only be a natural number between 1200 and the current year; etc.);

  ✓ *foreign keys* specify that corresponding columns should take values only among the values of other columns (in this example, there are only two such foreign keys, both in the *MUSICAL_WORKS* table: *Composer* has to take values only among those of column *x* in table *COMPOSERS*, while *Tonality* has to take values only among those of column *x* in table *TONALITIES*); foreign keys are interrelating tables or same table's instances; if intelligently used, always only like in the *MUSICAL_WORKS* table (and everywhere throughout this book), namely always referencing numeric, single primary keys, especially when coupled with combo boxes (drop down lists), preferably hiding the stored numerical pointers to primary keys and replacing them with corresponding nonprimary key values (e.g., in column *Tonality* of table *MUSICAL_WORKS* with corresponding *Tonality* values from table *TONALITIES*), foreign keys are reducing rdbs size (by factorizing data according to axiom A-DI10 in Section 5.1.2), simplifying end-user task (selecting from combo boxes desired values, instead of repeatedly typing them), and guaranteeing *referential integrity*: any pointer should always point to an existing row; dually, note that, unfortunately, the RDM allows for foreign keys to reference any other columns, not only primary keys, which you should never do; note that, in order to avoid ambiguity, all referential integrity constraints specify referenced columns with their full name, consisting of the name of the table separated by a dot from the name of the column (e.g., here, *COMPOSERS.x* and *TONALITIES.x*);

➢ finally, in the third header line are listed the *not-null constraints*, each of which stating that values in the corresponding column are mandatory (in this rdb example, end-users should always enter values for *LastName* and *BirthYear* of table *COMPOSERS*, *Composer* and *Opus* in table *MUSICAL_WORKS*, as well as for *Tonality* in table *TONALITIES*, whereas values in all other not primary keys columns might always miss; for example, obviously, if we were to declare *PassedAwayYear* as not-null too, still alive composers could not ever been stored); note that, by definition, values for primary keys are always mandatory, so, for example, the not-null constraints in all three *x* columns from Fig. 3.2 might have been omitted, as they are implicitly assumed; also note that, always, each table should have at least one not-null non primary key column (even if RDM does not request it). Conventionally, not known and/or not applicable values (like, for example, those in column *No* of table *MUSICAL_WORKS* above, except for its seventh row) are said to be *null values* (*nulls*); all other ("normal") values are *not null*. Note that nulls are elements of a distinguished countable *NULLS* set that we may freely union with any attribute codomain, except for primary keys.

Although RDM, which is a purely "syntactic" model (i.e., deprived of any semantics), does not require it, it is best that tables correspond to object sets, be them actual (like *COMPOSERS* and *MUSICAL_WORKS*) or conceptual (like *TONALITIES*), their columns to properties of these sets, and their rows to elements (objects) of them, just like, for example, in Fig. 3.2.

Note that, just like in math, the order of rows in tables does not matter and that, unlike in math[30], the order of columns does not matter either: for example, in table *COMPOSERS* above, if we were swapping between them pairs of columns *FirstName* and *LastName*, as well as *BirthYear* and *PassedAwayYear*, both stored data and derivable information would be exactly the same.

Figure 3.3 summarizes relational constraint type equivalences between three formalisms notational conventions (also see Fig. 3.1).

Informal	*RDM*	*math*
(table) column concatenation	attribute list	function (Cartesian) product
column range	domain constraint	function codomain
mandatory (required) column	NOT NULL constraint	(implicit, by definition) totally defined function (whose codomain is disjoint from *NULLS*)
unique column	single key	one-to-one function
unique column concatenation (composition)	superkey	one-to-one function (Cartesian) product
unique composed key	composed key	minimally one-to-one function (Cartesian) product
foreign key column(s)	foreign key, referential integrity constraint, (typed) inclusion dependency	function images inclusion first order predicate
(table) check constraint	(relation) tuple (check) constraint	first order logic closed proposition

FIGURE 3.3   Equivalences between informal, RDM, and math relational constraint types terminology.

---

[30] where the Cartesian product is not commutative.

3.2   THE FIVE BASIC RELATIONAL CONSTRAINT TYPES

> *What we anticipate seldom occurs.*
> *What we least expected usually happens.*
> —Benjamin Disraeli

The following five relational constraint types are implemented by any RDBMS: (co)domain, not-null (mandatory, compulsory, required, totality), key (minimal uniqueness), foreign key (referential integrity, typed inclusion), and check (tuple).

3.2.1   DOMAIN (RANGE) CONSTRAINTS

> *The shallow consider liberty a release from all law, from every constraint.*
> *The wise man sees in it, on the contrary, the potent Law of Laws.*
> —Walt Whitman

Domain (*range*) *constraints* are those constraints that specify plausible data ranges for column values.

Programmers are acquainted with assigning data types to their variables. DBMSs also provide data types for table columns. Declaring such a type for a column is sometimes enough, for example: BOOLEAN, NUMBER(n), VARCHAR2(n).

But for most of the time it is not: even if, by definition, all computer data types are finite, they are generally huge and most of their values are implausible for actual db instances.

For example, if letting *EMPLOYEES.BirthDate* to store DATE values, depending on the DBMS version, users might store birth dates either from year 100 or in year 10,000, both of them being trivially implausible.

This is why, generally, column values should be restricted to plausible subsets of type [*minValue, maxValue*] $\subset$ *dataType*. For example, *EMPLOYEES.BirthDate* could be restricted to [1/1/1930, *SysDate*() — 365 * 18] $\subset$ DATE (as it is implausible to have, for example in 2014, employees older than 83 and younger than 18).

3.2.2 NOT-NULL (MANDATORY DATA) CONSTRAINTS

> *I have three kinds of friends:*
> *those who love me,*
> *those who pay no attention to me,*
> *and those who detest me.*
> —Nicolas de Chamfort

Very often, it is the case that some values are not applicable or, at least temporarily, are unknown. In table *MUSICAL_WORKS* of Fig. 3.2, for example, there are no values in some rows for the *No* columns: some musical works have both an opus and number (within that opus), like, for example, Chopin's Nocturne in G major opus 37 no. 2; others do not have a number (as there is only one musical work within that opus), as it is the case for all other musical works in Fig. 3.2—for which this number is not applicable. On the other hand, if we were, for example, to add to *COM-POSERS* Vladimir Cosma (born in 1940 and, thanks God, still alive at the time of finishing this volume), his passing away year, although applicable, remains unknown.

Conventionally, such values are called *nulls*. The ANSI report (ANSI/ X3/SPARC, 1975) distinguishes between 14 types of nulls; most of us consider, however, that there are only three basic types (the rest being either particular cases of these three or only due to implementation considerations): *nonexistent* (*inapplicable*), (*temporarily*) *unknown*, and *no-information* (the logical *or* between the first two: it is not known whether such a value exists or, if it were existing, nothing is known on its nature).

Examples for this last type do not abound; my favorite one is on Moldavian king Stephen the Great (1433–1504) and the Russian language: did he read Russian? As Russians existed at that time, even if it took another two centuries to become Moldavia's neighbors, it is possible that the answer would be affirmative; however, no proof exists for or against it, and the chances to find at least one in the foreseeable future are almost negligible.

Consequently, in dbs we have to accommodate with null values too. However, constraints should also be placed on their usage: what would be the meaning of a row having only nulls (even if except for its autonumber primary key)? Obviously, at least one non-surrogate key column per each table should not accept nulls: *not-null* (*mandatory, compulsory, required, totality*) *constraints* are used to specify such columns.

Please note that there is great misunderstanding on null values: some DBMSs (e.g., MS *SQL Server*) wrongly assume that there is only one null value, so columns accepting nulls are not allowed in unique constraints; others (IBM *DB2, Oracle,* and even MS *Access*) assume correctly that there is a countable infinite number of nulls (so they accept non not null columns into keys).

One final note on *dirty nulls*—that is not empty text strings made out of only non-printable/displayable characters (spaces, for example): obviously they should be banned too, but, unfortunately, very few DBMSs are doing it (and sometimes when they may be banned too, as, for example, in MS *Access,* this is not the default setting). Consequently, whenever a DBMS provides this facility too, we should always use it for all columns, regardless of the fact that they should accept nulls or not (what would be the sense to allow storing dirty nulls even in columns that accept nulls?); whenever a DBMS does not provide this facility, we should enforce dirty nulls banning through the software applications managing corresponding dbs.

3.2.3 KEY (MINIMAL UNIQUENESS) CONSTRAINTS

The truly important things in life –love, beauty, and one's own uniqueness–
are constantly being overlooked.
—Pablo Casals

In reality, there are both objects whose uniqueness is interesting and objects that we do not need to uniquely identify. For example, in the first category there are people (with their unique e-mail addresses, phone numbers, SSNs, etc.), while in the second are the chairs of a room/apartment/building.

In dbs we should always care for uniqueness, in each table, according to axiom *A-DA*04 from 5.1.1: for example, if the db should store detailed information on each chair, then they have to be uniquely labeled, for example with an inventory number, and these numbers stored for each chair in a *CHAIRS* table; if not, then we would probably abstract a *CHAIR_TYPES* table and for each of its rows we would store a unique type name (plus description, total number of chairs per type or, in other tables, per type and room/apartment/building, if needed).

A *key constraint* is a statement of the type "$C_1 \bullet \ldots \bullet C_n$ *key*", where $n > 0$ is a natural, C_i are columns of a table T, "$\bullet$" denotes concatenation[31] (which is, generally, omitted) of these columns and *key* means minimally unique[32]—that is it is unique and it does not include any other key (see, for example, Fig. 3.2). When $n = 1$, the key is called *simple*; when $n > 1$ it is called *concatenated*; if a unique column concatenation properly contains a key (i.e., the included key has smaller than n arity), then it is not a key (as it is not minimal), but a *superkey*; note that superkeys are of no actual, but only theoretical interest.

For example, in Fig. 3.2, *Tonality* is a simple key of *TONALITIES*, while *FirstName • LastName* and *Composer • Opus • No* are concatenated keys of *COMPOSERS* and *MUSICAL_WORKS*, respectively. *FirstName • LastName • BirthYear*, *FirstName • LastName • PassedAwayYear*, and *FirstName • LastName • BirthYear • PassedAwayYear* are all superkeys[33].

FirstName • LastName : *COMPOSERS* → ASCII(32) × ASCII(32) is, in this context, one-to-one (as all "old" composers had unique such pairs and today no composer would dare to take both first and last names of an "old" one and even if a composer were having, by chance, identical first and last names as another today's one, he/she would at least add a surname in order to differentiate him/herself) and also minimally one-to-one (as neither of its subproducts are one-to-one: *FirstName*, as there are several composers having same first name –for example Franz Schubert and Franz Liszt– and *LastName*, as there are several composers having same last name—for example Johann Strauss and Richard Strauss).

Similarly, *Composer • Opus • No* : *MUSICAL_WORKS* → [0, 99999999] × ASCII(16) × [1, 255] is one-to-one too (as, by definition, musical works are uniquely identified by their composer, opus (catalog) and number) and also minimally one-to-one (as neither of its subproducts are one-to-one: *Composer • Opus*, as there are several works by a same composer having same opus but different numbers, *Composer • No*, as there are several works by a same composer having same number but dif-

---

[31] Mathematically, this means mapping (Cartesian) product.

[32] That is minimally one-to-one (see Appendix).

[33] Recall that if $f : A \to B$ is one-to-one then, for any $g : A \to C$, $f \bullet g$ is one-to-one too (see Appendix).

ferent opuses, and *Opus • No*, as there are several works having same opus and number, but different composers).

Only keys need to be declared, not superkeys[34]: for example, if instead of *FirstName • LastName* you declare *FirstName • LastName • BirthYear* as a key, then implausible data might be stored (e.g., a second Antonio Vivaldi, born in 1927 or any other valid year than 1678, could be stored too); if *FirstName • LastName* and *FirstName • LastName • BirthYear* are both declared as keys, then no implausible data might be stored, but updates to *COMPOSERS* would be slower (and the db would need additional disk and memory space to store and process the superkey *FirstName • LastName • BirthYear*), as the system would enforce the superkey too.

3.2.3.1   Primary Surrogate (Syntactic) Keys

> *If we have data, let's look at data.*
> *If all we have are opinions, let's go with mine.*
> —Jim Barksdale

Although RDM does not require it, it is a best practice to declare for each table a *primary key*: a key that is not accepting nulls and which is the default referenced one by foreign keys[35]. Conventionally, its name is underlined.

Unfortunately, RDM allows users to declare any key as being the primary one; consequently, it is a widespread bad practice to declare as primary keys concatenated ones and/or keys containing columns that are non-numeric. For example, a vast majority of db "experts" would declare *FirstName • LastName* as *COMPOSERS'* primary key (and would not add *x* to this table); consequently, they would need *FirstName • LastName* as a foreign key in *MUSICAL_WORKS* (referencing *COMPOSERS*) and declare *FirstName • LastName • Opus • No* as its primary key (not adding *x* to this table either).

There are several disadvantages of this approach:

---

[34] Unfortunately, most RDBMSs (including, for example, *Oracle*, MS *SQL Server* and *Access*, etc.) allow for enforcement of both keys and superkeys!

[35] Moreover, for most RDBMS versions, table instances are stored by default in the order of their primary keys, if any.

✓ db scheme is more complicated and needs more space, having two more ASCII columns and a greater arity key (4 instead of 3) for table *MUSICAL_WORKS*;

✓ any updates to columns *FirstName* and *LastName* in *COMPOSERS* have to be immediately propagated to their homonym ones in *MUSICAL_WORKS*;

✓ any comparison/join between these two table columns takes much longer: instead of comparing two (small) numbers (which is done by the fastest CPU—the arithmetical-logic one), four strings of ASCII characters (each having 32 ones, which, in the worst case, means comparing 64 with 64 characters) have to be compared for each pair of rows;

✓ by using a systematic approach of this type, a table *CONCERTS_PROGRAMS*, which would store pairs of type $<c, w>$ for each time that a musical work w is programmed in a concert c, would need a foreign key of arity 4 referencing *MUSICAL_WORKS* and a primary key of arity at least 5 (assuming that the foreign key to *CONCERTS* is simple, which is generally not the case in such approaches); moreover, a table *PERFORMANCES* storing triples of type $<p, a, i>$ for each performance p of work w in concert c by artist a on instrument i would need a foreign key of arity at least 5 referencing *CONCERTS* and a primary key of arity at least 7 (assuming, again, the very improbable most favorable case when foreign keys referencing *ARTISTS* and *INSTRUMENTS* are simple); and so on and so forth…

To conclude with, it is always preferable to declare primary keys for all fundamental tables (derived and temporarily ones generally do not need keys or any other constraints), but they should be simple numeric (preferably auto generated) ones, deprived of any semantics (i.e., of the so-called *surrogate* or *syntactic* type), except for very special cases: for example, in rabbit/poultry/etc. farms, cages could be uniquely identified (and actually labeled) only by this surrogate primary key (auto generated) values.

Surrogate keys are generally denoted by "ID", used as such or as a prefix/suffix of the table name, or by a special character pre/suffix, like '#'; I am using x for them (see, for example, Fig. 3.2) for reasons that will become clear in Section 3.6 and especially in the second chapter of the second volume.

3.2.3.2 Semantic (Candidate) Keys

> *There are only two things in the world: nothing and semantics.*
> —Werner Erhard

Note that, unfortunately, there are lot of dbs whose fundamental tables do not have keys (and, generally, no other constraints either); others always have only the primary one (generally, of surrogate type). Any table should have all keys existing in the subuniverse modeled by the corresponding db, just like it should have any other existing constraint: a table lacking an existing constraint allows for implausible data storage in its instances.

All nonprimary keys are called *semantic* (*candidate*[36]) keys.

For example, let us consider table *COUNTRIES* from Fig. 3.4, where all columns except for *Population* are keys: *x*, the surrogate primary one, by definition; *Country*, as there may not be two countries having same names; *CountryCode*, as there may not be two countries having same codes[37]; finally, *Capital*, as there may not be two countries having same city as their capitals.

If we were not declaring, for example, *Capital* as a key too, then users might have entered 4 (instead of 5) in this column for the fifth row too, storing the highly implausible fact that Washington D.C. is not only U.S.A.'s capital, but also Romania's one.

Please note that uniqueness is not an absolute, but a relative property: in some contexts a column or concatenation of columns are unique, while in others (even in a same subuniverse) they are not.

For example, in the subuniverse modeled by Fig. 3.4, where only big cities are considered, there may not be two cities having same name in a same state; if we would need to store villages and small cities too, this product (*City • State*) would not anymore be unique, as there may be several such villages or even small cities of a same state having same name.

---

[36] Unfortunately (from the semantic point of view), in RDM *candidate* means that any of them might be chosen as the primary one.
[37] See ISO 3166 codes.

In this case, for example, in most of Europe (where states are called regions, lands, departments, cantons, etc., depending on countries) there are three rules:

 ➢ villages and small cities may be subordinated to bigger ones, generally called "communes" and there may not be two cities of a same commune having same name;
 ➢ there may not be two communes of a same state having same names;
 ➢ there may not be two identical zip codes in a same country (where small cities have one zip code, bigger cities none, but their street addresses have one too).

Figures 3.4 and 3.5 show the corresponding *CITIES* tables.

Another, even worse example (as the subuniverse is strictly the same), is computer file names: almost everybody knows that "there may not be two files having same name and extension in a folder", which means that the triple *FileName • FileExt • Folder* is a key; in fact, OSs enforce a supplementary constraint too: there may not be two files in a same folder having null (no) extensions and same name, which means that for the subset of files without extension *FileName • Folder* is a key (and *FileName • FileExt • Folder* is a superkey).

Consequently, first of all, semantic keys, just like any other type of fundamental (i.e., not derived) constraints may only be discovered and declared by humans: there may not ever be any tools, be them hardware, software, conceptual, etc., able to do such a job. Moreover, especially keys are not at all easy to discover: besides their relativeness, as the number of columns increases, the number of their products (theoretically very many of them being possible keys) increases exponentially. This is why Section 3.6 presents an algorithm for assisting their discovery.

Please note that, dually, again just like for any other constraint type, you should never assert keys that do not exist in the corresponding subuniverse: if you were doing it, you would aberrantly prevent users to store plausible data. For example, if in any table of Figs. 3.4 and 3.5 you were declaring *Population* as a key too, then no two corresponding objects (countries, states, communes, or/and cities) having same population figures could be stored simultaneously.

COUNTRIES (*x*, *Country*, *CountryCode*, *Capital*)

x	*Country*	*CountryCode*	*Capital*	*Population*
autonum-ber(3)	ASCII(128)	ASCII(8)	*Im(CITIES.x)*	[100000, 2000000000]
NOT NULL	NOT NULL	NOT NULL		
1	U.S.A.	USA	6	313,000,000
2	U.K.	UK	2	63,200,000
3	Germany	D	3	82,000,000
4	France	F	1	65,650,000
5	Romania	RO	4	19,050,000

STATES (*x*, *State* • *Country*, *Capital*)

x	*State*	*StateCode*	*Country*	*Capital*	*Population*
autonum-ber(8)	ASCII(128)	ASCII(8)	*Im(COUN-TRIES.x)*	*Im(CI-TIES.x)*	[100000, 50000000]
NOT NULL	NOT NULL		NOT NULL		
1	Ile-de-France	75	4	1	11,175,000
2	Greater London		2	2	8,300,000
3	Berlin	B	3	3	3,460,000
4	Bucharest	B	5	4	2,200,000
5	D.C.	DC	1	6	6,900,000
6	New York	NY	1	5	19,570,000

CITIES (*x*, *City* • *State*)

x	*City*	*State*	*Population*
autonumber(16)	ASCII(128)	*Im(STATES.x)*	[100000, 30000000]
NOT NULL	NOT NULL	NOT NULL	
1	Paris	1	2,250,000
2	London	2	2,860,000
3	Berlin	3	3,400,000
4	Bucharest	4	1,680,000
5	New York	6	8,400,000
6	Washington	5	618,000

FIGURE 3.4   A small geographical rdb for big cities.

3.2.3.3   Keys Including Computed Columns

Facts do not cease to exist because they are ignored.
—Aldous Huxley

Even if not that often, uniqueness sometimes involves columns fundamentally belonging to different tables, like in the zip codes example above for small cities only having one such code; another example is provided by some countries, including Romania, where businesses must have their headquarters in a city, but their names should be unique within the corresponding state (county).

Not only RDM can't cope with such keys by using only fundamental columns: not even math can, without computed mappings.

For example, in table *CITIES* of Fig. 3.5, in order to embed the rule that small cities only have one zip code, a *Zip* column has been added; in order to add the (more general) rule that there may not be two identical zip codes into a same country, a computed column *Country has been added too, plus the key *Country • Zip. Obviously, *Country is computable both mathematically[38] and in *SQL*[39].

3.2.3.4   Prime and Nonprime Mappings

You can use all the quantitative data you can get, but you still have to distrust it and use your own intelligence and judgment.
—Alvin Toffler

RDM also introduced the concepts of prime and nonprime (nonessential) attributes (columns): if it is part of at least one (unique) key, a column is said to be *prime* and otherwise *nonprime* (*nonessential*). Obviously, any single key column is prime.

---

[38] Mathematically, *Country = Country ° State ° Commune, where ° denotes *mapping composition* (see Appendix).
[39] By the statement:
```
SELECT CITIES.x, Country
FROM (STATES INNER JOIN COMMUNES
        ON STATES.x = COMMUNES.State) INNER JOIN CITIES
        ON CITIES.Commune = COMMUNES.x;
```
whereas, for example, *Country(4) (which is equal to 4) is computable by:
```
SELECT Country
FROM (STATES INNER JOIN COMMUNES
        ON STATES.x = COMMUNES.State) INNER JOIN CITIES
        ON CITIES.Commune = COMMUNES.x
WHERE CITIES.x = 4;
```

COMMUNES ($\underline{x}$, Commune • State)

$\underline{x}$	Commune	State	Population
autonumber(16)	ASCII(128)	Im(STATES.x)	[100000, 30000000]]
NOT NULL	NOT NULL	NOT NULL	
1	Paris	1	2,250,000
2	London	2	2,860,000
3	Berlin	3	3,400,000
4	Bucharest	4	1,680,000
5	New York	6	8,400,000
6	Washington	5	618,000

CITIES ($\underline{x}$, City • Commune, *Country • Zip)

$\underline{x}$	City	Commune	*Country	Zip	Population
autonumber(16)	ASCII(128)	Im(COMMUNES.x)	= Country(State (Commune))	ASCII (16)	[100000, 30000000]
NOT NULL	NOT NULL			NOT NULL	
1	Saint Mandé	1	4	94160	23,000
2	Suresnes	1	4	92150	40,000
3	Neuilly sur Seine	1	4	92200	60,000
4	Bezons	1	4	95870	26,300
5	Pantelimon	4	5	077145	23,300
6	Cernica	4	5	077035	2,800
7	Jilava	4	5	077120	12,000

FIGURE 3.5   Tables COMMUNES for big cities and CITIES for small ones.

For example, in Fig. 3.5, columns x from both tables, *Commune* and *State* from *COMMUNES*, as well as *City*, *Commune*, **Country*, and *Zip* from *CITIES* are prime, whereas columns *Population* from both tables are nonprime.

Just like RDM itself, defined as such, primeness and nonprimeness are purely *syntactic* concepts: they are applied only *after* the set of table keys are established.

As it will become clear why in (E)MDM (see the second chapter of the second volume), we need in fact slightly different *semantic* definitions for them, to be applied *before* the set of all keys for a table is established: a column is said to be *prime* in any given fixed subuniverse, if in that subuniverse it is either a key or might be part of a key and *nonprime* otherwise.

Consequently, in particular, this semantic definition of nonprimeness (the only interesting one, in fact, as primeness is somewhat the default one) of a column is: "*not only it is not a single key, but it cannot be part of any concatenated one either*" (in that particular subuniverse).

For example, generally, except for very few particular contexts (e.g., tables for time zones or colors), very many other numeric properties, like population, area, temperature, pressure, altitude and their dual (rivers, lakes, seas, oceans, etc.) depth, length, width, GDP, rating, probabilities, percentages, time zones, etc., some text ones like color, notes, even some names (like, for example, states/county ones in different countries), even some calendar dates (like, for example, discovery, national dates, etc.), etc., as well as most of the Boolean ones (e.g., *Married?*, *Divorced?*, *Widow?*, *Author?*, *Holiday?*, etc.) are nonprime.

Algorithm *A7/8–3* (see Section 3.6), which is assisting validation of the keys declared for any table and discovery of any other existing ones, proves that correctly declaring of all nonprime mappings is crucial in significantly reducing the otherwise huge number of possible cases to analyze by this exponential algorithm.

This is why we are using the above-introduced semantic concept of nonprimeness, instead of the primeness syntactic one.

3.2.3.5 Keys and Null Values

> *True genius resides in the capacity for evaluation of uncertain, hazardous, and conflicting information.*
> —Winston Churchill

As we have already seen, by definition, primary keys do not allow for null values. Unfortunately, some RDBMSs (e.g., MS *SQL Server*[40]) do not allow for null values in any other key (implicitly thus assuming that *NULLS* is not infinite, but only has one value).

Unfortunately, there are only two types of solutions for overcoming such aberrant restrictions:

---

[40] Which is quite amazing, as MS *Access* allows for them and there are decades already since Microsoft bought the *SQL Server* license from Sybase.

✓ (the "hard" one) enforce such keys through db management application code (i.e., trigger-type methods) or

✓ (the "soft" one) automatically generate explicit "unknown" rows, one per each involved table (e.g., if in table *COMMUNES* from Fig. 3.5 above *STATES* should also conceptually accept nulls, then an "unknown" state should be generated in corresponding table *STATES* for each country in *COUNTRIES* and these "unknown" states be used in column *COMMUNES.State* instead of nulls).

Dually, it is obvious that when nulls are accepted by RDBMSs for nonprimary keys, but they should not be accepted conceptually (like, for example, having two files with no extensions and same names in a same folder), there is only one solution: declare as key the corresponding (concatenation of) column(s) (in the above example: *Folder • FileName • FileExtension*) and enforce subkeys as keys (*Folder • FileName*, in the above example) for null values through db management application code (triggers or trigger-type methods).

3.2.4 FOREIGN KEY (REFERENTIAL INTEGRITY) CONSTRAINTS

> *Integrity without knowledge is weak and useless,*
> *and knowledge without integrity is dangerous and dreadful.*
> —Samuel Johnson

Links between tables, as well as those between a table and itself are done by *foreign keys*: pointer-type columns whose values should always be among the values of the corresponding referenced columns. For example, from Fig. 3.2, *Composer* and *Tonality* in *MUSICAL_WORKS*, from Fig. 3.4, *Capital* in both *COUNTRIES* and *STATES*, *Country* in *STATES*, *State* in *CITIES* and from Fig. 3.5, *State* in *COMMUNES* and *Commune* and **Country* in *CITIES* are all foreign keys; for example, by its definition, values of **Country* are always among the values of *STATES.Country*, which, in their turn, are always among the values of *COUNTRIES.x.*

This is why, on the second table header row, instead of a corresponding domain constraint, for all foreign keys is shown the so-called *referential integrity* constraint[41] (or (*typed*) *inclusion dependency*), stating that their values should always be among the values of the corresponding referenced column.

This does not mean that foreign keys do not have associated domain constraints: it only means that these constraints are derived ones, from the corresponding referential integrity constraints, so they do not need to be explicitly asserted. For example, in Fig. 3.4, from the fact that *COUNTRIES. Capital* should always take its values from within *CITIES.x* values, it follows that the former's domain constraint is the same as the one of the latter's (i.e., *autonumber(*16)). However, note that, unfortunately, RDBMSs do not automatically derive corresponding domain constraints; even worse, you are sometimes free to oversize foreign key domain constraints: they should only have same data type as corresponding referenced columns and be able to accommodate all of its values (i.e., at least, they cannot be undersized).

The referential integrity constraint between foreign key f and corresponding referenced column g it's generally denoted by $f \subseteq g$[42]. Obviously, this is a notational abuse[43]: in fact, its meaning is $Im(f) \subseteq Im(g)$, where *Im* is the *image* operator, which computes the set of values taken by its operand (i.e., the set of all of the values taken by the corresponding mapping). This is why, didactically, I prefer to denote such referential integrity by $Im(g)$ instead of, simply, g.

Note that failing to enforce this constraint[44] may result in storing the so-called *dangling pointers*: values that point to nonexisting values in the set of values of the referenced columns.

For example, if first data row from table *COUNTRIES* of Fig. 3.4 above were containing 7 (or any other number except those from 1 to 6) instead of 6 in column *Capital*, the corresponding referential integrity would be violated (as there is no row in the *CITIES* tables having 7 in its x column) and this 7 would consequently be a dangling pointer; in this case, the capital of the U.S. could not be actually computed from that rdb instance.

---

[41] Note that some RDBMSs use confusing terminology in this respect too: for example, MS *Access* table "design" mode considers referential integrity as being only half of this constraint type, the other half being enforceable through the "limit to list" property of foreign keys.

[42] This being the reason why RDM theory also refers to these constraints as being (*typed*) *inclusion dependencies*.

[43] Columns being mappings, it is senseless to talk about mapping inclusions.

[44] It may be done, for example, through the ALTER TABLE *schemeName.tableName* ADD CONSTRAINT *constraintName* FOREIGN KEY (*columnName*) REFERENCES *table-Name* (*colName*) statements, see Section 3.8 below.

Referential integrity constraints might be violated from both sides of the inclusion mathematical relation: as just seen above, from the left one by storing in a foreign key column a value which does not belong to the referenced column, but, dually, from the right one too, by deleting from a table a row which is referenced by at least one row.

For example, if the fifth *COUNTRIES*' row of Fig. 3.4 above (corresponding to Romania) were deleted, then table *STATES* would become inconsistent, as its fourth line (corresponding to Bucharest) would contain a dangling pointer value in column *Country*.

Also note that, as already discussed above, RDM allows for concatenated foreign keys and, moreover, for foreign keys referencing any columns, not necessarily being keys. Best practice, however, is to only use single foreign keys, always referencing primary keys (which should always be surrogate-type appropriately range restricted numeric ones—see Subsection 3.2.3.1).

As RDM, however, allows for defining of any foreign keys, provisions are made in *SQL* for easing referential integrity constraints: foreign keys definitions may also include the predicates "ON UPDATE | DELETE [RESTRICT | CASCADE | SET NULL | NO ACTION]".

If you only use single foreign keys referencing primary autonumber columns, you do not need any such "ON UPDATE" predicate: as autonumbers may not be updated, no such problem arises. On the contrary, when you use foreign keys referencing nonautonumber columns (which may also be the case of subset-type tables primary keys—see Subsection 3.2.4.3 below) you should decide whether not to allow referenced column's values modification (with "ON UPDATE RESTRICT", generally the default value), or to allow them and either ask the system to automatically also update corresponding foreign key values (with "ON UPDATE CASCADE"), or to update all these latter values by replacing them with nulls (with "ON UPDATE SET NULL"), but this should be possible (i.e., corresponding foreign keys should allow nulls). By choosing in such cases "ON UPDATE NO ACTION", dangling pointers might appear.

As the "ON DELETE" predicate has similar behavior for the above four parameters, cascading deletes is very risky and should therefore be very cautiously used: for example, let us consider a db storing details for all phone calls, which are detailed invoice lines for telecom invoices of worldwide customers, in which all corresponding foreign keys (from phone calling details to phone calls, from phone calls to invoices, from invoices to customers, from customers to their corresponding cities, from cities to states, from states to countries, and from countries to continents) have been defined with the predicate "ON DELETE CAS-CADE"; if the row corresponding to North America is deleted from the continents table, cascading will also automatically delete the three rows corresponding to Canada, U.S., and Mexico from the countries tables, then all some 100 rows of their corresponding states from the states table, then all probably some 1,000 rows of corresponding cities from the cities table, then probably some corresponding hundred thousand rows from the customers table, millions of rows from the invoices and phone calls ones, as well as some 1 billion rows from the phone call details one.

Consequently, not only for commercial dbs, "ON DELETE RESTRICT", which does not allow deleting referenced rows, is generally the best choice.

Note that various RDBMSs versions are generally offering only some of the above parameters and that even corresponding default values are varying.[45]

3.2.4.1   The Key Propagation Principle

When will the world know that peace and propagation
are the two most delightful things in it?
—Horace, 4th Earl of Orford Walpole

Foreign keys arouse from the following *Key Propagation Principle* (KPP), to be explained in what follows by an example (also see Sections 2.4 and 3.12): let us consider the tables *COUNTRIES* and *STATES* from Fig. 3.4, except for the foreign key column *Country* of *STATES*. In order to be able to link data from these tables, we should store for each state the country to which it belongs—that is the graph of mapping *Country* : *STATES* → *COUNTRIES* (as each state belongs to one and only one country).

As for any finite mapping $f : A \rightarrow B$, the first (brute force type) idea of storing its graph is a table with two columns: x (from A) and $f(x)$ (from

---

[45] Probably the most famous example is MS *Access*: when programmatically creating a foreign key, the default is "ON DELETE RESTRICT", whereas when interactively creating it, for example, with its *Lookup Wizard* (but not only), the default is "ON DELETE NO ACTION".

B). Table 3.2 shows this solution for the graph *STATES_COUNTRIES* of *Country* : *STATES* → *COUNTRIES*, where, for optimality reasons, *State* references *STATES.x* and *Country* references *COUNTRIES.x*.

Both from *A* and *B*, keys should be chosen for storing *f*'s graph (in order to uniquely identify any element, both from *A* and *B*). As all keys of a set are equivalent (see Exercise 3.42(*iv*)), for any mapping *f*, any key from *A* could be chosen as *x* and any key from *B* could be chosen as *f*(*x*). For optimality reasons, however, the best choice for them is always the two involved surrogate primary keys.

When we add this table to the rdb scheme in Fig. 3.4 it is almost impossible not to notice that both columns *x* and *State* of it are redundant, as they are duplicating values of column *x* from *STATES*. Consequently, an equivalent but much more elegant and optimal solution to this data modeling problem is simply adding the foreign key *Country* from *STATES_COUNTRIES* to table *STATES*, instead of adding the three columns of *STATES_COUNTRIES*, out of which two are redundant.

This process is called *key propagation* for obvious reasons: for any such mapping (i.e., functional relationships between two tables), according to KPP, instead of adding a new table, it is better to propagate in the domain table[46] a key from the codomain one[47]. This is also why propagated columns are called *foreign keys*: generally, they take values from keys of other (foreign) tables.

Country : *STATES* → *COUNTRIES*

Note that "foreign" in the "foreign key" syntagm is a false friend: when applying KPP to auto-mappings (i.e., to mappings defined on and taking values from a same set), corresponding foreign keys are referencing a key from the same table (not from a "foreign" one).

For example, applying KPP to *ReportsTo* : *EMPLOYEES* → *EMPLOYEES*, results in the "foreign" key *ReportsTo* of table *EMPLOYEES* that references its *x* column, as shown by Table 3.3, where, obviously, the first employee does not report to anybody, the second one reports to the first one, whereas the third one reports to the second one.

---

[46] Generally, the "many" one: in the above example, *STATES*. For the one-to-one ones, it is, generally, the one potentially having less rows than the other, like *COUNTRIES* (as compared to *CITIES*) for *Capital* (see Fig. 3.4).

[47] For those including one "many", it is the "one" one: in the above example, *COUNTRIES*. For the one-to-one ones, it is, generally, the one potentially having more rows than the other, like *CITIES* (as compared to *COUNTRIES*) for *Capital* (see Fig. 3.4).

TABLE 3.2   Brute Force Type Graph Storage Implementation of Mapping *Country :*
STATES → COUNTRIES

STATES_COUNTRIES (*x*, *State*)

	x	*f(x)*
x	*State*	*Country*
autonumber(8)	*Im(STATES.x)*	*Im(COUNTRIES.x)*
NOT NULL	NOT NULL	NOT NULL
1	1	4
2	2	2
3	3	3
4	4	5
5	5	1
6	6	1

TABLE 3.3   An Example of a Foreign Key Which is Actually Not "Foreign" At All

EMPLOYEES (*x, SSN*)

x	*FirstName*	*LastName*	*SSN*	*ReportsTo*
autonumber(4)	ASCII(64)	ASCII(64)	NAT(9)	*Im(x)*
NOT NULL	NOT NULL	NOT NULL	NOT NULL	
1	Backar	Omaba	111111111	
2	Hirally	Clitnon	222222222	1
3	Ramk	Tiggenstein	333333333	2

3.2.4.2   Orthogonality of Unique and Foreign Keys

In God we trust. All others must bring data.
—W. Edwards Deming

In fact, the syntagm "foreign key" is a double false friend: not only "foreign" is a false friend, but also "key" is one, as, generally, foreign keys are not also (unique) keys.

For example, let us consider table *COUNTRIES* from Fig. 3.4 augmented with a foreign key *Currency* (referencing a table *CURRENCIES*). Obviously, *Capital* is both a key and a foreign key, *Country* is a key but not a foreign key, *Currency* is a foreign key but not a key (as there are, for

TABLE 3.4   Proof of Orthogonality of Unique and Foreign Key Concepts

Column	Key?	Foreign key?
Population		
Currency		✓
Country	✓	
Capital	✓	✓

example, lot of countries having Euro as currency), whereas *Population* is neither a key nor a foreign key. Table 3.4 is summarizing these findings, which prove that the concepts of unique and foreign keys are orthogonal to each other.

3.2.4.3 Modeling Set Inclusions

Most of the fundamental ideas of science are essentially simple, and may, as a rule, be expressed in a language comprehensible to everyone.
—Albert Einstein

Let us consider *SUBORDINATED_COUNTRIES ⊆ COUNTRIES* that is the subset of countries and territories that are subordinated to other countries; if we need to store their corresponding subordinations together with their starting subordination years, the best solutions is to add a *SUBORDINATED_COUNTRIES* table as shown in Fig. 3.6.

We could have defined its surrogate primary key x as being of type autonumber too, but then we need an additional foreign key *Country* referencing *COUNTRIES* for storing subordinated countries. The solution from Fig. 3.6 is better, as it consolidates *Country* into x.

Generally, the best way to represent relationally an inclusion $T \subseteq S$ is by declaring primary surrogate key *T.x* as being a foreign key referencing primary surrogate key *S.x*. *T.x* may not be an autonumber: it should be of number type (and maximum cardinality equal to the one of *S.x*).

Mathematically, any set is a subset of its own (even if, just like the empty set, an "improper" one), but you should never try to define such inclusions, which would recursively include a table into itself: while, for example, MS *Access* is not letting you define such trivial inclusions ($T \subseteq T$),

COUNTRIES (x, Country, CountryCode, Capital)				
x	Country	CountryCode	Capital	Population
autonum-ber(3)	ASCII(255)	ASCII(8)	Im(CITIES.x)	[100000, 1500000000]
NOT NULL	NOT NULL	NOT NULL		
1	U.S.A.	USA	6	313,000,000
2	U.K.	UK	2	63,200,000
6	Canada	CA		34,500,000
7	Australia	AU		21,750,000
8	New Zealand	NZ		4,450,000
9	Guam	GU		186,000
10	U.S. Virgin Islands	VI		105,500

SUBORDINATED_COUNTRIES (x)

x	ColonialPower	SubordinationYear
Im(COUNTRIES.x)	Im(COUNTRIES.x)	[1400, CurrentYear()]
NOT NULL	NOT NULL	
6	2	1867
7	2	1901
8	2	1853
9	1	1898
10	1	1917

FIGURE 3.6   Example of relational modeling of a set inclusion.

even the current 12c version of *Oracle* doesn't have anything against it! It is true that the only harm done by enforcing such trivial inclusions is that it is wasting for nothing additional time for enforcing the corresponding referential integrity constraint for each row insertion into the corresponding table.

Moreover, pay attention to inclusion cycles too: recall from sets algebra that, for any sets $S_1, S_2, \ldots, S_n$, n natural, $S_1 \subseteq S_2 \subseteq \ldots \subseteq S_n \subseteq S_1 \Rightarrow S_1 = S_2 = \ldots = S_n$ (i.e., all sets involved in a cycle of inclusions are equal). First of all, equal sets should be relationally modeled by using only one table for all of them; but what is much, much worse is that most RDBMSs (including, for example, MS *Access* 2010 and *Oracle* 12c) are allowing you to define such cycles (even when they only involve two tables!), which results in not being then able to insert any line in any of the involved

tables. Consequently, we should always keep table inclusions *acyclic*—
that is we should never close cycles of included sets.

3.2.5   CHECK (TUPLE) CONSTRAINTS

> *The more constraints one imposes, the more one frees one's self.*
> —Igor Stravinsky

For *check constraints* there is no generally agreed definition; from most of
RDBMSs implementations, they are first order logic formulas that have to
be satisfied by all rows of a table and have (by notational abuse) a simplified, propositional logic type form: they are using parenthesis, the logical
not, *and*, and *or* operators connecting terms having the form $C \theta D$, where
all such C and D (from all terms) are columns of a same table, and θ is a
standard operator[48] (this being the reasons why these constraints are also
called *tuple constraints*, as they relate columns of a same table that have to
be satisfied by all tuples of that table).

For example, for table *COMPOSERS* of Fig. 3.2, check constraint
BirthYear + 3 < *PassedAwayYear* states that, for each row, corresponding
life duration should have been at least 3 years (assuming that no one could
ever compose before he/she is at least 3 years old). Both check constraints
of this table may be consolidated into a single one, by using either the
logical *and* (as (*BirthYear* + 3 < *PassedAwayYear*) *and* (*PassedAwayYear*
— *BirthYear* < 120)) or the logical *between … and* (as *PassedAwayYear* —
BirthYear between 3 *and* 120).

In fact, check/tuple constraints are first order logic formulas with only
one variable universally quantified, which, by notational abuse, is omitted.
For example, the above apparently propositional calculus formula stands
for ($\forall x \in COMPOSERS$)(*PassedAwayYear*(x) — *BirthYear*(x) *between* 3
and 120).

The limitation to only one table is essential: although syntactically
looking similar, no RDBMS is recognizing (and, consequently, enforcing),

---

[48] The set of *db standard operators* always include the corresponding math ones, plus some additional
ones, either derived from them, like *between … and*, or generally coming from regular expressions
manipulation, as *like* (which uses, just like in OSs, meta-characters too: '%' or '*' for any character
string, including empty ones, etc.).

for example, constraints like *COUNTRIES.Population ≥ STATES.Population* (see Fig. 3.4).

Note that you should not get confused because of theory and RDBMS terminologies: for example, not-null constraints are also considered by *Oracle* as being check constraints.

3.3   THE ALGORITHM FOR TRANSLATING E-R DATA MODELS INTO RELATIONAL SCHEMAS AND NON-RELATIONAL CONSTRAINT SETS (A1–7). AN RDM MODEL OF THE E-RDM

There are truths that you cannot justly understand without previously experiencing some errances.

—Lucian Blaga

In order to further refine especially restriction sets, but sometimes also even E-RDs, conceptual data modeling should not end only with establishing E-RDs and corresponding restriction sets, as it will be proved in the next volume.

However, when in a hurry, when accuracy of conceptual data modeling is not that big a concern, one may directly translate E-RDs and restriction sets into relational db schemas and non-relational db constraint sets by applying the algorithm presented in Subsection 3.3.1.

3.3.1   THE ALGORITHM FOR TRANSLATING E-R DATA MODELS INTO RELATIONAL SCHEMAS AND NON-RELATIONAL CONSTRAINT SETS (A1–7)

An algorithm must be seen to be believed.

—Donald Knuth

The algorithm is shown in Figs. 3.7–3.11[49]. Note that $A1$–7 assumes the following: the corresponding rdb exists and there are no duplicated object set names or mapping names defined on a same object set.

---

[49] whose complexity is linear in the sum of object sets collection, property and restriction sets cardinals, as proved in Section 3.12 (see Proposition 3.4).

Algorithm A1-7 (Translation of E-R data models into relational schemas and
     non-relational constraint sets)
Input: an E-R data model *M*
Output: the corresponding relational db scheme and an associated non-relational
    constraint set
Strategy:
 loop for all E-RDs *D* from *M*
  loop for all rectangles *R* in *D* (in bottom-up order, from only referenced, non-
    referencing object sets to non-referenced, only referencing ones)
   if R is not a computed set *then*
    createTable(R);
   else
    add a corresponding view *R* to the db scheme;
   end if;
  end loop (for all rectangles);
  loop for all diamonds *R* in *D* (in bottom-up order, from only referenced, non-
    referencing object sets to non-referenced, only referencing ones)
   createTable(R);
   loop for all R's underlying object set *S* occurences (having role *r*)
    addForeignKey(R, r, S);
    add to *R*'s scheme a *NOT NULL* constraint for foreign key *r*;
   end loop (for all R's underlying object set occurences);
  end loop (for all diamonds);
 end loop (for all E-RDs);
add all unused (non-relational) restrictions to the non-relational constraint set;
End Algorithm A1-7;

FIGURE 3.7   Algorithm A1–7 (Translation of E-R data models into corresponding
      relational schemas and non-relational constraint sets).

3.3.2 AN RDM MODEL OF THE E-RDM

Learning is a treasure that will follow its owner everywhere.
—(Chinese Proverb)

Section 3.4 is applying the algorithm *A*1–7 to the public library E-RD and
restriction set designed in Section 2.8.

Solved exercise 3.0 provides another example of its application (to a
subuniverse of invoice payments).

By applying this algorithm to the E-R data metamodel of the E-RDM
presented in Section 2.9 above, the RDM model of the E-RDM shown in

```
method addForeignKey(R, A, S)
add a numeric column A to table R;
if S is a table then
  declare A as foreign key having maximum value corresponding to S's
  cardinality restriction, and referencing primary key x of table S;
else code S' values numerically and add corresponding domain constraint to A;
end if;
end method addForeignKey;
```

FIGURE 3.8   Algorithm A1–7's *addForeignKey* method.

```
method createTable(R)
  if there is no table R and R has attributes or is referencing or it is not
    referencing, but it is not static and with small cardinality then create table R;
    if R is a subset of S then
      AddForeignKey(R, x, S); declare x as primary key;
      loop for all other sets T with R ⊆ T
        AddForeignKey(R, xT, T);
        add to R's scheme NOT NULL and UNIQUE constraints for xT;
      end loop;
    else add an autonumber primary key x, having maximum value corresponding
      to R's cardinality restriction;
    end if;
    completeScheme(R);
  end if;
end method createTable;
```

FIGURE 3.9   Algorithm A1–7's *createTable* method.

Fig. 3.12 is obtained (regardless of the fact that it is applied to the E-RD from Subsection 2.9.1 or to the one from Subsection 2.11.2).

Note that the unlabeled autofunctions of *OBJECT_SETS, DATA_ TYPES* and *RANGES*, which are modeling the partial ordering of the elements of these sets by inclusion, have all been labeled *Superset*, because their graphs store for each such element the corresponding smallest superset that includes them too.

For example, ASCII(16) ⊂ ASCII(32) ⊂ ASCII(64) ⊂ ASCII(128) ⊂ ASCII(255), [2, 16] ⊂ [1, 16] ⊂ [1, 255] ⊂ NAT(38) ⊂ INT(38) ⊂ RAT(38, 2), CURRENCY(38) ⊂ RAT(38, 2), [–2500, *Year(SysDate())*] ⊂ INT(38), [6/1/2011, *SysDate()*] ⊂ [6/1/2011, *SysDate()* + 300] ⊂ DATE, etc.

```
method addColumn(R, e)
  add column e to table R, having domain constraint (that is data type and range)
   corresponding to e's range restriction;
  if e is computed then add corresponding computation trigger;
end method addColumn;
```

FIGURE 3.10   Algorithm A1–7's completeScheme method.

```
method completeScheme(R)
  loop for all arrows A from R to S, except for set inclusion-type ones
     addForeignKey(R, A, S);
  end loop (for all arrows);
  loop for all ellipses e connected to R, except for the surrogate key x
     addColumn(R, e);
  end loop (for all ellipses);
  loop for all compulsory restrictions c associated to R
     add to R's scheme corresponding NOT NULL constraints;
  end loop (for all compulsory restrictions);
  loop for all uniqueness restrictions u associated to R
     add to R's scheme corresponding u key constraint;
  end loop (for all uniqueness restrictions);
  loop for all tuple-type restrictions t associated to R
     add to R's scheme corresponding tuple constraint t;
  end loop (for all other tuple-type restrictions);
end method completeScheme;
```

FIGURE 3.11   Algorithm A1–7's *addColumn* method.

This rdb may be used as the kernel of the metadata catalog (see Section 3.7) of a DBMS providing E-RDM to its users.

The example instance provided corresponds to the public library case study from the Section 2.8 (also see Section 3.4).

Note that there is no need (in this kernel metacatalog) for a table *ENTITY_TYPES*, as it would only have an *x* primary column, which would duplicate the one of *OBJECT_SETS*.

The associated non-relational constraint set is made out of the following two constraints:

OBJECT_SETS (x, Name)

X	Name	Description	MaxCard	Superset
auton(38)	ASCII(255)	ASCII(255)	NAT(38)	Im(x)
NOT NULL	NOT NULL	NOT NULL	NOT NULL	
1	PERSONS	The set of book (co-)authors and subscribers of interest to the library	10^6	
2	BOOKS	The set of written works of interest to the library	10^6	
3	CO-AUTHORS	The set of pairs <p, b> storing the fact that book b was also written by person p	$2*10^6$	
4	PUBLISHERS	The set of publishers of interest to the library	10^4	
5	EDITIONS	The set of book editions of interest to the library	10^6	
6	VOLUMES	The set of edition volumes of interest to the library	$2*10^6$	
7	VOLUME_CON-TENTS	The set of pairs <b, v> storing the fact that book b is (also) included in volume v	$4*10^6$	
8	COPIES	The set of volume copies that the library ever possessed	$32*10^6$	
9	BORROWS	The set of copy borrows by subscribers	10^{11}	
10	BORROW_LISTS	The set of pairs <b, c> storing the fact that volume copy c was (also) borrowed in borrow b	10^{12}	

NON_FUNCTIONAL_RELATIONSHIP_TYPES (x)

x	*Arity
Im(OBJECT_SETS.x)	NAT(2)
NOT NULL	
3	2
7	2
10	2

FIGURE 3.12   A RDM model of the E-RDM and a valid instance of it corresponding to the public library case study from Section 2.8.

— For each nonfunctional relationship-type object set there
should be at least two underlying object sets.   (RU5)

— The graph of *UNDERLYING_SETS* should be acyclic
(that is no non-functional relationship-type object set may

UNDERLYING_SETS ($\underline{x}$, Role • RelationshipType)

x	ObjectSet	RelationshipType	Role
auton(38)	Im(OBJECT_SETS.x)	Im(NON_FUNCTIONAL_RELATIONSHIP_TYPES.x)	ASCII(255)
NOT NULL	NOT NULL	NOT NULL	NOT NULL
1	1	3	Co-author
2	2	3	Book
3	6	7	Volume
4	2	7	Book
5	9	10	Borrow
6	8	10	Copy

DATA_TYPES ($\underline{x}$, Name)

x	Name	Superset
auton(38)	ASCII(255)	Im(x)
NOT NULL	NOT NULL	
1	ASCII(255)	
2	INT(38)	5
3	NAT(38)	2
4	CURRENCY(38)	5
5	RAT(38,2)	
6	DATE	

FIGURE 3.12   (Continued)

be defined on itself, neither directly, nor indirectly; equivalently: recursive definitions of non-functional relationship-type object sets are not allowed).   (RU6)

Note that constraints (RO6) and (RM7) need no explicit enforcement, as they are implicitly enforced by the closed world assumption and the structure of the rdb.

Also note that any software application built on top of this rdb should also automatically (re)compute (immediately after inserts, deletes or updates in column RelationshipType of table UNDERLYING_SETS) the column *Arity of table NON_FUNCTIONAL_RELATIONSHIP_TYPES.

RANGES ($\underline{x}$, *DataType* • *MinValue* • *MaxValue*)

$\underline{x}$	*DataType*	*MinValue*	*MaxValue*	*Superset*
auton(38)	*Im(DATA_TYPES.x)*	ASCII(255)	ASCII(255)	*Im(x)*
NOT NULL	NOT NULL	NOT NULL	NOT NULL	
1	1	0	255	
2	1	0	128	1
3	1	0	64	2
4	1	0	32	3
5	1	0	16	4
6	2	-2500	Year(SysDate())	
7	3	1	16	9
8	3	2	16	7
9	3	1	255	
10	4	1	8	
11	6	6/1/2011	SysDate()	12
12	6	6/1/2011	SysDate() + 300	

FIGURE 3.12   (Continued)

3.4   CASE STUDY: THE RELATIONAL SCHEME OF THE PUBLIC LIBRARY DATA MODEL

> *Science is what we understand well enough to explain to a computer.*
> *Art is everything else we do.*
> —Donald Knuth

As an example, by applying Algorithm *A*1–7 on the E-RD and its associated restriction set for a public library (see Sections 2.8.2 and 2.8.3), the following relational db scheme (see Section 3.4.2) and non-relational constraint set (see Section 3.4.1) are obtained:

3.4.1   PUBLIC LIBRARY NON-RELATIONAL CONSTRAINTS SET

> *A list is only as strong as its weakest link.*
> —Donald Knuth

MAPPINGS (*x*, *Name* • *DomainObjectSet*)

x	*Name*	*Domain-ObjectSet*	*Compul-sory?*	*OneTo-One?*	*Descrip-tion*
auton(38)	ASCII(255)	*Im(OBJECT SETS x)*	Boolean	Boolean	ASCII (255)
NOT NULL	NOT NULL	NOT NULL	NOT NULL	NOT NULL	
1	FName	1	F	F	
2	LName	1	T	F	
3	e-mail	1	F	F	
4	BTitle	2	T	F	
5	BYear	2	F	F	
6	FirstAuthor	2	T	F	
7	Co-author	3	T	F	
8	Book	3	T	F	
9	PosInList	3	T	F	
10	PubName	4	T	T	
11	Publisher	5	T	F	
12	ETitle	5	T	F	
13	EYear	5	F	F	
14	FirstBook	5	T	F	
15	Edition	6	T	F	
16	ISBN	6	F	T	
17	VNo	6	T	F	
18	VTitle	6	N	F	
19	Price	6	T	F	
20	Volume	7	T	F	
21	Book	7	T	F	
22	BookPos	7	T	F	
23	InvNo	8	T	T	
24	Volume	8	T	F	
25	Subscriber	9	T	F	
26	BorrowDate	9	T	F	
27	Borrow	10	T	F	
28	Copy	10	T	F	
29	DueReturnDate	10	T	F	
30	ActualReturnDate	10	F	F	

FIGURE 3.12   (Continued)

1. *PERSONS*

FName and *e-mail* should be compulsory for
subscribers   (RP7)

FUNCTIONAL_RELATIONSHIPS (x)

x	CodomainObjectSet
$Im(MAPPINGS.x)$	$Im(OBJECT\_SETS.x)$
NOT NULL	NOT NULL
6	1
7	1
8	2
11	4
14	2
15	5
20	6
21	2
24	6
25	1
27	9
28	8

ATTRIBUTES (x)

x	RangeRestriction
$Im(MAPPINGS.x)$	$Im(RANGES.x)$
NOT NULL	NOT NULL
1	2
2	3
3	1
4	1
5	6
9	8
10	2
12	1
13	6
16	5
17	9
18	1
19	10
22	7
23	4
26	11
29	12
30	11

FIGURE 3.12   (Continued)

2. ***VOLUMES_CONTENTS*** = (*BOOKS, VOLUMES*)

>For any edition, its first book should be the first one
>published in its first volume.   (RVC4)
>No edition may contain same book more than once.   (RVC5)

3. ***BORROWS_LISTS*** = (*BORROWS, COPIES*)

>No copy may be borrowed less than 0 days or more than
>300 days.   (RBL5)
>No copy may be simultaneously borrowed to more than
>one subscriber.   (RBL6)
>No copy may be returned before it was borrowed or after
>100 years since corresponding borrow date.   (RBL7)

(RP7) might seem as being the conjunction of two not-null constraints (one on *FName* and the other on *e-mail*), but, in fact, it is not, because it does not apply to all rows of *PERSONS*, but only to some of them (those corresponding to subscribers).

(RVC4) and (RVC5) are not relational as they involve columns from three tables: *x* and *FirstBook* (from *EDITIONS*), *x* and *Edition* (from *VOLUMES*), as well as *Volume*, *Book* and *BookPos* (from *VOLUME_CONTENTS*).

Similarly, (RBL5) involves columns *x* and *BorrowDate* (from *BORROWS*), as well as *Borrow* and *DueReturnDate* (from *BORROWS_LISTS*); (RBL6) involves *x*, *Subscriber* and *BorrowDate* (from *BORROWS*), as well as *Borrow*, *Copy* and *ActualReturnDate* (from *BORROWS_LISTS*); (RBL7) involves *x* and *BorrowDate* (from *BORROWS*), as well as *Borrow* and *ActualReturnDate* (from *BORROWS_LISTS*).

3.4.2 PUBLIC LIBRARY DB RELATIONAL SCHEME AND A PLAUSIBLE VALID INSTANCE OF IT

The hardest thing is to go to sleep at night, when there are so many urgent things needing to be done. A huge gap exists between what we know is possible with today's machines and what we have so far been able to finish.
—Donald Knuth

Figure 3.13 shows the corresponding public library rdb and a possible plausible instance of it.

PERSONS (<u>x</u>, *e-mail* • *FName* • *LName*)

x	*FName*	*LName*	*e-mail*	*Author?*
NAT(6)	ASCII(128)	ASCII(64)	ASCII(255)	BOOLEAN
NOT NULL		NOT NULL		
1		Homer		T
2	William	Shakespeare		T
3	Peter	Buneman	opb@inf.ed.ac.uk	T
4	Serge	Abiteboul		T
5	Dan	Suciu	suciu@cs.washington.edu	T

BOOKS (<u>x</u>, *FirstAuthor* • *BTitle* • *BYear*)

x	*FirstAuthor*	*BTitle*	*BYear*
NAT(6)	*Im(PERSONS.x)*	ASCII(255)	[-5000, Year(SysDate())]
NOT NULL	NOT NULL	NOT NULL	
1	1	Odyssey	-700
2	2	As You Like It	1600
3	4	Data on the Web: From Relations to Semi-structured Data and XML	1999

CO-AUTHORS (<u>x</u>, *Book* • *Co-author*)

x	*Book*	*Co-author*	*PosInList*
NAT(7)	*Im(BOOKS.x)*	*Im(PERSONS.x)*	[2, 16]
NOT NULL	NOT NULL	NOT NULL	NOT NULL
1	3	3	2
2	3	5	3

FIGURE 3.13   Public library rdb and a possible instance of it.

3.5   THE REVERSE ENGINEERING ALGORITHM FOR TRANSLATING RELATIONAL SCHEMAS INTO E-R DATA MODELS (*REA1–2*)

> *The softest things in the world overcome*
> *the hardest things in the world.*
> —Lao Tzu

Very often, probably much more often than having opportunities to design a new db, we are faced with a dual problem, of *reverse engineering* (RE)

PUBLISHERS (*x*, *PubName*)

x	*PubName*
NAT(4)	ASCII(128)
NOT NULL	NOT NULL
1	Apple Academic Press
2	Springer Verlag
3	Morgan Kaufmann
4	Penguin Books
5	Washington Square Press

EDITIONS (*x*, *Publisher* • *FirstBook* • *EYear*)

x	*Publisher*	*FirstBook*	*ETitle*	*EYear*
NAT (6)	*Im(PUBLI-SHERS.x)*	*Im(BOOKS .x)*	ASCII(255)	[-2500, Year(SysDate())]
NOT NULL	NOT NULL	NOT NULL	NOT NULL	
1	4	1	The Odyssey translated by Robert Fagles	2012
2	5	2	As You Like It	2011
3	3	3	Data on the Web: From Relations to Semi-structured Data and XML	1999

VOLUMES (*x*, *ISBN*)

x	*Edition*	*VNo*	*VTitle*	*ISBN*	*Price*
NAT (7)	*Im(EDI-TIONS. x)*	[1, 255]	ASCII(255)	ASCII(16)	[0, 100000]
NOT NULL	NOT NULL	NOT NULL			NOT NULL
1	1	1		0-670-82162-4	$12.95
2	2	1		978-1613821114	$9.99
3	3	1		978-1558606227	$74.95

FIGURE 3.13   (Continued)

VOLUMES_CONTENTS ($\underline{x}$, *Volume • Book*)

$\underline{x}$	*Volume*	*Book*	*BookPos*
NAT(7)	*Im(VOLUMES.x)*	*Im(BOOKS.x)*	[1, 16]
NOT NULL	NOT NULL	NOT NULL	NOT NULL
1	1	1	1
2	2	2	1
3	3	3	1

BORROWS ($\underline{x}$, *BorrowDate • Subscriber*)

$\underline{x}$	*Subscriber*	*BorrowDate*
NAT(11)	*Im(PERSONS.x)*	[6/1/2000, *SysDate*()]
NOT NULL	NOT NULL	NOT NULL
1	5	10/29/2012

COPIES ($\underline{x}$, *InvNo*)

$\underline{x}$	*InvNo*	*Volume*
NAT(8)	ASCII(32)	*Im(VOLUMES.x)*
NOT NULL	NOT NULL	NOT NULL
1	H-O-1	1
2	H-O-2	1
3	S-AYLI-1	2
4	ABS-DW-1	3
5	ABS-DW-2	3

BORROWS_LISTS ($\underline{x}$, *Borrow • Copy*)

$\underline{x}$	*Borrow*	*Copy*	*DueReturnDate*	*ActualReturnDate*
NAT (12)	*Im(BOR- ROWS.x)*	*Im(CO- PIES.x)*	[6/1/2000, *SysDate*() + 300]	[6/1/2000, *SysDate*()]
NOT NULL	NOT NULL	NOT NULL	NOT NULL	
1	1	1	11/29/2012	11/23/2012
2	1	3	12/29/2012	

FIGURE 3.13   (Continued)

type: we have to work with (and, so, understand) a db for which there are
no E-RDs and/or restriction set and/or description. This section presents,
in Fig. 3.14, algorithm *REA*1–2 that outputs these three items of an E-R
data model from any RDM scheme.

Algorithm REA1-2 (Reverse engineering of RDM schemas into E-R data models)
Input: an RDM scheme *S*
Output: a corresponding E-R data model
Strategy:
loop for all tables and views *T* of *S*
  if T is a table *then*
    add a rectangle labeled *T*;
  else
    add a dashed rectangle labeled *T*;
  end if;
  loop for all non-foreign key columns *c* of *T*
    if c is a computed column *then*
      add a dashed ellipse labeled *c* and attach it to rectangle *T*;
    else
      add an ellipse labeled *c* and attach it to rectangle *T*;
    end if;
  end loop (for all non-foreign keys columns);
  loop for all foreign keys *f* of *T* referencing table *U*
    if f is *T* 's primary key or a single not null key referencing a key of *U* then
      add an inclusion arrow from rectangle *T* to rectangle *U*;
    else
      add an arrow labeled *f* from rectangle *T* to rectangle *U*;
    end if;
  end loop (for all foreign keys);
  loop for all (unique) keys *k* of *T*
    if k is a concatenated key (i.e. a column product) *then*
      add corresponding uniqueness constraint to *T* 's restriction set;
    else
      double-edge the corresponding arrow (to rectangle *T*, but not tangent to
             it);
    end if;
  end loop (for all (unique) keys);
  loop for all other constraints of *T*
    add corresponding constraint to *T*'s restriction set, uniquely labeling it;
  end loop (for all other constraints);
end loop (for all tables and views);
describe the above obtained E-RDs and restriction set;
end algorithm REA1-2;

FIGURE 3.14 Algorithm REA1–2 for reverse engineering RDM schemas into
E-R data models.

In practice, before applying this algorithm, another RE algorithm has to be applied, in order to get the corresponding RDM scheme from an actual rdb (or, instead, apply directly the composition of these two algorithms: see, for example, Section 4.4).

Note that, as RDM is a purely syntactic formalism, the E-RDs obtained by reverse engineering may never contain relationship-type sets, but only entity-type ones; however, as proved in Section 2.10 of the second volume, these E-RDs are strictly equivalent with ones also having relationship-type sets.

Also note that if you do not need computed sets or when the number of views is high, you may want to apply the above algorithm only for tables and/or some views.

As an example, let's consider the RDM scheme from Fig. 3.15; by applying to it the algorithm $REA1$–2, the E-RD from Fig. 3.16, the associated restriction set from Fig. 3.17, and the description from Fig. 3.18 (obtained by E-RD and restrictions reverse engineering) are obtained.

Note that, in fact, *NEIGHBORS*, which stores pairs of countries <c, d> whenever c and d are neighbors, is a non-functional binary mathematical relation (i.e., a many-to-many relationship) defined on *COUNTRIES*, so, most probably, when applying algorithm $A0$ (see Section 2.9), one would rather model it as a diamond. Nothing differs, however, at the rdb level for this table: when applying algorithm $A1$–7 above to the *NEIGHBORS* diamond exactly the same table as in Fig. 3.15 is obtained (see Exercise 3.1).

In Volume 2, we will see that, generally, any relationship-type set may be equivalently modeled with an entity-type one, providing that all of its roles are included too as corresponding functional relationships. This means that the only fundamental mathematical relations needed in data modeling are the functions.[50]

3.6   THE ALGORITHM FOR ASSISTING KEYS DISCOVERY (A7/8–3)

---

[50] Moreover, this implies that E-RDM and functional data models are strictly equivalent as expressive powers.

CONTINENTS ($\underline{x}$, *Continent*)

$\underline{x}$	Continent
autonumber(2)	ASCII(64)
NOT NULL	NOT NULL
1	Africa
2	Asia
3	Europe
4	North America
5	South America

COUNTRIES ($\underline{x}$, *Country, CountryCode, Capital*)

$\underline{x}$	Country	Country Code	Capital	Continent
autonum-ber(3)	ASCII (255)	ASCII (8)	*Im(CI-TIES.x)*	Im(*CONTI-NENTS.x*)
NOT NULL	NOT NULL	NOT NULL		NOT NULL
1	U.S.A.	USA	4	4
2	U.K.	UK	2	3
3	Germany	DE	3	3
4	France	F	1	3
5	Romania	RO	5	3

STATES ($\underline{x}$, *State • Country, StateCode • Country*)

$\underline{x}$	State	State Code	Country
autonum-ber(8)	ASCII(128)	ASCII (8)	*Im(COUN-TRIES.x)*
NOT NULL	NOT NULL		NOT NULL
1	Ile-de-France	75	4
2	Greater London		2
3	Berlin	B	3
4	Bucharest	B	5
5	D.C.	DC	1

FIGURE 3.15 An rdb example.

NEIGHBORS (*x*, *Country* • *Neighbor*)

x	*Country*	*Neighbor*
autonumber(6)	*Im(COUNTRIES.x)*	*Im(COUNTRIES.x)*
NOT NULL	NOT NULL	NOT NULL
1	5	4

CITIES (*x*, *City* • *State*)

x	*City*	*State*
autonumber(16)	ASCII(128)	*Im(STATES.x)*
NOT NULL	NOT NULL	NOT NULL
1	Paris	1
2	London	2
3	Berlin	3
4	Washington	5
5	Bucharest	4

FIGURE 3.15   (Continued)

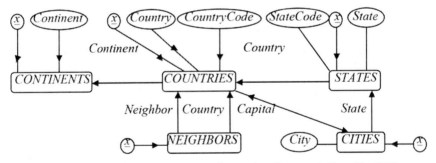

FIGURE 3.16   The E-RD corresponding to the rdb example from Fig. 3.15.

Do the difficult things while they are easy and do the great things while they are small. A journey of a thousand miles must begin with a single step.
—Lao Tzu

Generally, even db architects and designers are designing very few new dbs and are modifying/using much more other existing ones, which very rarely have all constraints (and, consequently, all keys as well) enforced.

1. ***CONTINENTS***
   a. Domain (range) restrictions:
      x: NAT(2)   (RD1)
      Continent: ASCII(64)   (RD2)
   b. Compulsory restrictions: x, *Continent*   (RC1)
   c. Uniqueness restrictions: x (primary key)   (RU1)
      Continent   (RU2)

2. ***COUNTRIES***
   a. Domain (range) restrictions:
      x: NAT(3)   (RD3)
      Country: ASCII(255)   (RD4)
      CountryCode: ASCII(8)   (RD5)
   b. Compulsory restrictions: x, *Country*, *CountryCode*,
                                  Continent   (RC2)
   c. Uniqueness restrictions: x (primary key)   (RU3)
      Country   (RU4)
      CountryCode   (RU5)
      Capital   (RU6)

3. ***STATES***
   a. Domain (range) restrictions:
      x: NAT(8)   (RD6)
      State: ASCII(128)   (RD7)
      StateCode: ASCII(8)   (RD8)
   b. Compulsory restrictions: x, *State*, *Country*   (RC3)
   c. Uniqueness restrictions: x (primary key)   (RU7)
      State • *Country*   (RU8)
      StateCode • *Country*   (RU9)

4. ***NEIGHBORS***
   a. Domain (range) restrictions:
      x: NAT(6)   (RD9)
   b. Compulsory restrictions: x, *Country*, *Neighbor*   (RC4)
   c. Uniqueness restrictions: x (primary key)   (RU10)
      Neighbor • *Country*   (RU11)

5. ***CITIES***
   a. Domain (range) restrictions:
      x: NAT(10)   (RD10)
      City: ASCII(128)   (RD11)
   b. Compulsory restrictions: x, *State*, *City*   (RC5)
   c. Uniqueness restrictions: x (primary key)   (RU12)
      State • *City*   (RU13)

FIGURE 3.17   The restriction set associated to the E-RD from Fig. 3.16.

There may be at most 100 continents, characterized only by their compulsory names, which may have at most 64 characters; there may not be two continents having same names.

At most 1,000,000 pairs of neighbor countries are also stored, only once.

There may be at most 1,000 countries, characterized by their compulsory names (at most 255 characters) and codes (at most 8 characters), each one of them belonging to only one continent (which is mandatory too); there may not be two countries having same names or codes in a same state. When known, country capitals are stored too; no city may simultaneously be the capital of two countries.

There may be at most 100,000,000 states, characterized by their compulsory names (at most 128 characters) and codes (at most 8 characters), each one of them belonging to only one country (which is mandatory too); there may not be two states having same names or codes in a same country.

There may be at most 10,000,000,000 cities, characterized by their compulsory names (at most 128 characters), each one of them belonging to only one state (which is mandatory too); there may not be two cities having same names or codes in a same state.

FIGURE 3.18   The description corresponding to the rdb example from Fig. 3.15.

As we will see in the second chapter of the second volume, keys should be designed earlier, on higher conceptual levels, not on the RDM one or, even lower, on the RDBMS managed rdbs one, but better later than never.

It is true that both this chapter, as well as the second chapter of the second volume are also providing reverse engineering algorithms, which, given any rdb scheme, are computing corresponding higher conceptual level schemas, so db designers might first apply them for obtaining an E-R data model, then apply $A1$ to translate this model into a(n) (E)MDM scheme, $A2$ to $A6$ (which includes $A3/3'$) to refine it and then $A7$, $A8$, and the needed algorithm from $AF8'$, in order to obtain the new rdb scheme with the keys designed according to $A3$ or $A3'$.

However, in order not to be obliged to apply all of these algorithms (especially for legacy rdbs, but not only), this section is introducing in Fig. 3.19 *A7/8–3*, an equivalent algorithm for assisting discovery of all keys in any table (where "//" are prefixing comments; [*x*], *x* real, denotes the *integer part* of *x* –for example, [5/2] = 2, [π] = 3; C(*n*, *j*) denotes the (natural) number of "*n choose j*", *n* and *j* naturals, *n* ≥ *j*,—that is *combinations of j elements taken from n ones*: C(*n*, *j*) = *n*!/(*j*! * (*n*—*j*)!), where *x*! = 1 * 2 * ... * (*x*–1) * *x* is the *factorial* of *x*, *x* natural; for example, C(4,2) = 4!/(2! * 2!) = 4 * 3 * 2/(2 * 2) = 6– see Appendix too; and statement numbers are shown for easing the understanding of the following example of applying *A7/8–3*).

Let us apply, for example, Algorithm *A7/8–3* to the Table 3.5 (also presented in Fig. 3.4, here augmented with two new columns for state phone prefixes and areas):

$T = STATES$, $m = 8$, $C_1 = x$, ..., $C_8 = Area$, $K = \{ x, Capital, State \bullet Country \}$, $k = 3$.

Note that, in order to simplify things, it is considered that each state has only one (national) telephone prefix, which is true only for small states/countries; for cases where there are several such prefixes per state (as in NY, for example), you can consider *TelPrefix* as storing the first one of them.

Results of running Algorithm *A7/8–3* for this input step by step are the following (only "actual" statements being numbered: *repeat for, else, end* ..., etc. are not):

01. $K' = \{ x, State \bullet Country, Capital \}$;
02. *x* is a key, by definition (being a surrogate column);
02. *Capital* is a key, as there may not be cities that are (simultaneously) capitals of two (or more) countries; as there are no superkeys of it, *K'* remains the same;
02. *State* • *Country* is a key, as there may not be two states of a same country having same names; as there are no superkeys of it, *K'* remains the same;
05. $SK = \{ x, State, Country, Capital \}$;
06. $T' = \{ StateCode, Population, TelPrefix, Area \}$;

Algorithm A7/8-3 (*table keys discovery assistance*)
  Input: a table T having m columns $C_1, \ldots, C_m$, $m > 0$ natural, and a set of
      associated keys K, card$(K) = k$, k natural.
  Output: K', card$(K') = l$, l natural, the set of all T keys.
  Strategy:
01. $K' = K$;
    // eliminate existing superkeys and wrong keys
    repeat for all keys u *in* K' (in bottom-up order)
02.   *if* u is not a key (in the given context) *then*
03. $K' = K' - \{ u \}$;
      else repeat for all superkeys s of u
04. $K' = K' - \{ s \}$;
        end repeat;
      end if;
    end repeat;
    // eliminate prime columns:
05. $SK = \{ C_i \mid C_i$ appears in K', $1 \leq i \leq m \}$;
06. $T' = T - SK$;
    // eliminate all other single keys, as well as all non-prime and not-
    // accepted[51] columns:
    repeat for all columns C *in* T'
07.   *if* C is unique (in the given context) *then*
08. $T' = T' - \{C\}$;   // eliminate C as well from further processing
09. $K' = K' \cup \{C\}$;   // add C to existing keys
      else
10.   *if* C is non-prime *or* not accepted *then* $T' = T' - \{C\}$ *end if*;
      end if;
    end repeat;
11. $T' = T' \cup \{$all non-key prime C_i, $1 \leq i \leq m \}$
12. $n = $ card(T');
13. *if* $n > 0$ *then*   // for all remaining columns, if any, look for other keys
14. $l = $ card(K');
15. $kmax = C(m, [m/2])$;   // the maximum possible numbers of T keys
16. $i = 2$;   // starting column products arity
17. $allSuperkeys = false$;   // initially, no superkeys possible for $i = 1$

FIGURE 3.19   Algorithm A7/8–3 (table keys discovery assistance).

---

[51] Depending on RDBMS versions, some data types, generally those needing "huge" storage sizes (for example memos, pictures, etc.) are not actually allowed to be declared as prime.

```
18.    while i ≤ n and l < kₘₐₓ and not allSuperkeys do
19.    allSuperkeys = true;      // all C(n, i) combinations might be superkeys
       repeat for all C(n, i) mapping products p made out of i elements
20.       if there is no x∈K' such that p is a subproduct of x then
21.          if p is not a superkey then     // at least one no superkey discovered
22.             allSuperkeys = false;        //    on the current level i
23.             if p is minimally unique (in the given context) then
24.                K' = K' ∪ { p };     // add newly found key
25.                l = l + 1;           // correspondingly increase K' cardinal
             end if;
          end if;                       // (of 21.: if p is not a superkey)
       end if;                          // (of 20.: if there is no x∈K' such that p ...)
       end repeat;
26.    i = i + 1;                       // increment level (column products arity)
       end while;
    end if;                             // (of 13.: if n > 0 ...)
27. if l = 0 then            // T does not even have a surrogate primary key!
28.    display "add a surrogate primary key to T, according to best practices!";
29. elseif l = 1 then
30.    if the only key of T is not a surrogate primary one then
31.       display "add a surrogate primary key to T, according to best practices!";
       else                  // the only key of T is the surrogate primary one
32.       if T does not correspond to a subset then
33.         display "is T corresponding to an object set of type poultry cages?";
34.         if answer is no then display     // T needs at least one semantic key
33.            "T needs at least one more column in order to have semantic keys too!";
          end if;
       end if;                          // (of 32.: if T does not correspond to a subset)
    end if;                             // (of 30.: if the only key of T is ...)
 end if;                                // (of 27.: if l = 0)
End Algorithm A7/8-3;
```

FIGURE 3.19   (Continued).

07. *StateCode* is not one-to-one (as there may be two states having same code, but from different countries), so neither T', nor K' are modified;

10. *StateCode* is prime (as there may not be two states having same code in a same country), so T' is not modified;

07. *Population* is not one-to-one (as there may be two states having same population, even in a same country), so neither T', nor K' are modified;

TABLE 3.5   Input Example Table for Applying Algorithm *A7/8–3*.

STATES (x, *State* • *Country, Capital*)

x	State	State Code	Country	Capital	Population	Tel Prefix	Area (km²)
auto-num-ber(8)	ASCII (128)	ASCII (8)	Im(COUN-TRIES.x)	Im(CI-TIES.x)	[100000, 500000000]	NAT (4)	[100, 3103200]
NOT NULL	NOT NULL		NOT NULL				
1	Ile-de-France	75	4	1	11,175,000	1	12012
2	Greater London		2	2	8,300,000	20	1572
3	Berlin	B	3	3	3,460,000	30	891
4	Bucha-rest	B	5	4	2,200,000	21	285
5	D.C.	DC	1	6	6,900,000	202	177
6	New York	NY	1	5	19,570,000	212	141300

10. *Population* is nonprime (as it cannot help, in any context, to uniquely identify states), so *T'* = { *State*, *StateCode*, *Country*, *TelPrefix*, *Area* };

07. *TelPrefix* is not one-to-one (as there may be two states having same telephone prefixes, but from different countries), so neither *T'*, nor *K'* are modified;

10. *TelPrefix* is prime (as there may not be two states having same telephone prefixes in a same country), so *T'* is not modified;

07. *Area* is not one-to-one (as there may be two states having same area, even in a same country), so neither *T'*, nor *K'* are modified;

10. *Area* is nonprime (as it cannot help, in any context, to uniquely identify states), so *T'* = { *StateCode*, *TelPrefix* };

11. *T'*= { *State*, *StateCode*, *Country*, *TelPrefix* };

12. $n = 4$;

13. as $4 > 1$:

14. $l = 3$;

15. $kmax = C(8, 4) = 8!/(4! * 4!) = 8 * 7 * 6 * 5 * 4 * 3 * 2/(4 * 3 * 2 * 4 * 3 * 2) = 8 * 7 * 6 * 5 /(4 * 3 * 2) = 7 * 2 * 5 = 70$;

16. $i = 2$;

17. *allSuperkeys* = *false*;

18. as $2 \leq 4$ *and* $3 < 70$ *and true* = *true*:

19. *allSuperkeys* = *true*;

20. *State* • *StateCode* is not already known to be a key, so:

21. *State* • *StateCode* is not a superkey, so:

22. *allSuperkeys* = *false*;

23. *State* • *StateCode* is not a key, as there may be two states having same names and codes in different countries

20. *State* • *Country* is already known to be a key, so it is skipped

20. *State* • *TelPrefix* is not already known to be a key, so:

21. *State* • *TelPrefix* is not a superkey, so:

22. *allSuperkeys* = *false*;

23. *State* • *TelPrefix* is not a key, as there may be two states having same names and telephone prefixes in different countries

20. *StateCode* • *Country* is not already known to be a key, so:

21. *StateCode • Country* is not a superkey, so:

22. *allSuperkeys = false*;

23. *StateCode • Country* is a key, as there may not be two states having same code in a same country

24. *K' = { x, State • Country, Capital, StateCode • Country }*;

25. *l = 4*;

20. *StateCode • TelPrefix* is not already known to be a key, so:

21. *StateCode • TelPrefix* is not a superkey, so:

22. *allSuperkeys = false*;

23. *StateCode • TelPrefix* is not a key, as there may be two states having same codes and telephone prefixes in different countries

20. *Country • TelPrefix* is not already known to be a key, so:

21. *Country • TelPrefix* is not a superkey, so:

22. *allSuperkeys = false*;

23. *Country • TelPrefix* is a key, as there may not be two states having same telephone prefixes in a same country

24. *K'={x, State • Country, Capital, StateCode • Country, Country • TelPrefix}*;

25. *l = 5*;

26. *i = 3*;

18. as $3 \leq 4$ *and* $5 < 70$ *and true = true*:

19. *allSuperkeys = true*;

20. *State • StateCode • Country* is not already known to be a key, so:

21. *State • StateCode • Country* is a superkey (as *State • Country* is a key)

20. *State • StateCode • TelPrefix* is not already known to be a key, so:

21. *State • StateCode • TelPrefix* is not a superkey, so:

22. *allSuperkeys = false*;

23. *State • StateCode • TelPrefix* is not a key, as there may be two states having same names, codes, and telephone prefixes in different countries

20. *State • Country • TelPrefix* is not already known to be a key, so:

21. *State • Country • TelPrefix* is superkey (as *State • Country* is a key)

20. *StateCode • Country • TelPrefix* is not already known to be a key, so:

21. *StateCode • Country • TelPrefix* is superkey (as *StateCode • Country* is a key)

26. *i* = 4;

18. as 4 ≤ 4 *and* 5 < 70 *and true* = *true*:

19. *allSuperkeys* = *true*;

20. *State • StateCode • Country • TelPrefix* is not already known to be a key, so:

21. *State • StateCode • Country • TelPrefix* is a superkey (as *State • Country* is a key)

26. *i* = 5;

18. as 5 ≤ 4 *and* 5 < 70 *and true* = *false* (because 5 > 4), the *while* loop is exited

27. 5 ≠ 0, so:

29. as 5 ≠ 1, execution of *A7/8–3* ends.

Corresponding output is $K' = \{\,x, State • Country, Capital, StateCode • Country, Country • TelPrefix\,\} \supseteq \{\,x, Capital, State • Country\,\} = K, l = 5 \geq 3 = k$. Table 3.6 shows the corresponding modified table scheme.

Note the following general considerations on Algorithm *A7/8–3*:

✓ In the best case (happening when *K* is empty and all *m* columns of *T* are keys: only the *m* iterations of the second *repeat* loop are needed, as, initially, *k* = 0, so the first *repeat* loop is not even entered and, at the end of the second one, *n* = 0, and the algorithm exits at statement 18), so *A7/8–3* only asks *m* questions: in this case, it only has to ask for each of the *m* columns, whether or not it is a key (as all answers are *yes* in this case).

✓ In the worst case (happening when *K* contains all possible combinations, so $k = 2^m - 1$, which are then all discarded and no key is discovered, so *l* = 0), all possible $2^m - 1 + 2^m + m - 1 = 2^{m+1} + m - 2$ combinations are considered (as there are, first, $2^m - 1$ questions on initially existing keys, all answered with *no*, then *m* questions on nonprimeness, all answered again with *no* and $2^m - 1$ questions on uniqueness, all answered with *no* too, where the only two ones that are not needed are corresponding to $C(m, 0)$, as there is no such thing as a mapping product made out of no

TABLE 3.6   Output Example Table for Applying Algorithm *A7/8–3* to Table 3.5

STATES (<u>x</u>, *State* • *Country, StateCode* • *Country, Country* • *TelPrefix,*
Capital)

<u>x</u>	State	State Code	Country	Capital	Population	Tel Prefix	Area (km²)
auto-num-ber(8) NOT NULL	ASCII (128) NOT NULL	ASCII (8)	Im(COUN -TRIES.x) NOT NULL	Im(CI-TIES.x)	[100000, 50000000]	NAT (4)	[100, 3103200]
1	Ile-de-France	75	4	1	11,175,000	1	12012
2	Greater London		2	2	8,300,000	20	1572
3	Berlin	B	3	3	3,460,000	30	891
4	Bucha-rest	B	5	4	2,200,000	21	285
5	D.C.	DC	1	6	6,900,000	202	177
6	New York	NY	1	5	19,570,000	212	141300

mappings), which means that the complexity of the algorithm is 2^{m+1} (i.e., exponential in the number of the table columns).

Consequently, in average, this algorithm needs some $2^m + m/2 - 1$ steps (i.e., half of the worst case one).

For example, in the worst case with $k = 0$, for $m = 4$ there are 19 combinations to consider, for $m = 5$ –36, for $m = 6$ –69 (still "bearable"!?), for $m = 7$ –134 (already "hard"), for $m = 8$ —263 (extremely "hard"), for $m = 9$ –520 (almost actually intractable), for $m = 10$ –1033 (already "hopeless").

Fortunately, for example, in my more than 35 years of professional career in this field, both in IT and University teaching, I have never encountered any example having $m > 7$ (out of thousands of them).

When $k > 0$, things are worsening, as in the first loop k additional questions have to be asked and answered, but they might get a little bit better then, if at least some initial keys are confirmed, as they will correspondingly reduce the number of asked questions on uniqueness and possibly also on nonprimeness (if single keys are reconfirmed).

Unfortunately, at least theoretically, they also might get really nasty if, initially, all or very many of the possible $2^m - 1$ combinations are wrongly declared as keys.

For example, in the extreme worse case, in which all possible $2^m - 1$ combinations are wrongly declared as keys, for $m = 4$ there are $19 + 16 - 1 = 34$ combinations to consider and for $m = 10$ there are $1033 + 1024 - 1 = 2056$ combinations to consider.

Please note that in step 0 of the algorithm, however, even if all of the $2^m - 1$ possible combinations were wrongly declared as keys and, consequently, $A7/8–3$ needs to deal with them all, in fact it will ask users only at most $C(m, [m/2])$ times whether or not some declared keys are actually keys, as it will automatically discard the remaining $2^m - 1 - C(m, [m/2])$ ones as superkeys.

For example, if, initially, $K = \{ x, Capital, State \bullet Country, State \bullet Capital, Country \bullet Capital, State \bullet Capital \bullet Country \}$, immediately after users are confirming that *Capital* is a key, the algorithm is automatically discarding superkeys *State* • *Capital*, *Country* • *Capital*, and *State* • *Capital* • *Country*.

Fortunately, first of all, although most RDBMSs stupidly allows for enforcing both keys and superkeys, the vast majority of currently existing dbs do not have (except, as always, for exceptions, but they are never systematical –that is including all possible ones–, but rather accidental) superkeys enforced[52].

Secondly, especially for semantic ones, current dbs rather lack of keys than having too much of them enforced; thirdly, most of the currently enforced keys are correct ones, as incorrect ones are noticed rather sooner than later at least by customers, so they need to be fixed almost immediately.

As such, from a practical point of view, K's cardinal and instance are not that much of an issue, as, generally, the former is small and the latter is almost valid.

✓ Consequently, however, especially for $m > 5$ and $card(SK) = 0$, it is crucial to minimize n by correctly identifying all nonprime columns.

✓ The algorithm's bottom-up strategy is the best possible, as it eliminates as soon as possible the maximum number of implausible combinations, namely all superkeys.

For example, supposing that *Capital* was not already declared as a key, then *STATES'* would have initially 7 elements (instead of 6), meaning that in the worst case there would have been $2^7 - 1 = 127$ cases, out of which 120 in the *while* loop; by discovering (in the first *repeat* loop) that *Capital* is a key, the number of cases drops to $2^6 = 64$, out of which only 57 in the *while* loop.

Generally, it is easy to see that each single key discovered (i.e., in the first *repeat* loop) is halving the initial number of nonsuperkeys combinations of the worst case.

Next, on level two, by discovering, for example, the key *StateCode • Country*, out of the initial $C(4, 3) + C(4, 4) = 4 + 1 = 5$ combinations to be analyzed on the following two levels, only $C(4, 3)/2 = 2$ remain.

Generally, by discovering a first key on any level i, $1 < i < n$, it can be shown that the number of combinations to be analyzed on remaining

---

[52] Is it because user common sense, which cannot even believe that such a stupidity exists and even if, exceptionally, discovers it, it does not use it, as it knows or at least feels that it would only aberrantly worsen their db throughput? Most probably yes, as it is hard to believe that, again, except for exceptions, they are much wiser than most of the RDBMS designers, who should have prevented any such stupidity.

levels (i.e., $i + 1, i + 2, \ldots$) are decreasing with increasing numbers as n increases, even if, as i increases, savings are less and less impressive, as they have less and less time to continue growing (e.g., when $i = 2$, with 1 for $n = 3$, from 1 to 0, with 3 for $n = 4$, from 5 to 2, with 12 for $n = 5$, from 16 to 4, with 15 for $n = 6$, from 42 to 27, with 30 for $n = 7$, from 98 to 68, etc.; when $i = 3$, with 1 for $n = 4$, from 1 to 0, with 4 for $n = 5$, from 6 to 2, with 6 for $n = 6$, from 22 to 16, with 14 for $n = 7$, from 63 to 49, etc.; when $i = 4$, with 1 for $n = 5$, from 1 to 0, with 3 for $n = 6$, from 7 to 4, with 14 for $n = 7$, from 29 to 49, etc.; when $i = 5$, with 1 for $n = 6$, from 1 to 0, with 3 for $n = 7$, from 8 to 5, etc.; when $i = 6$, with 1 for $n = 7$, from 1 to 0, etc.).

For example, discovering that the last level combination (which is made out of all n mappings) is a key does not save anything, whereas discovering a first key on the last but one level (i.e., $n - 1$) only saves one combination (the last one above).

Discovering subsequent keys on a same level is still decreasing the number of combinations to be analyzed on remaining levels, but not that much as the previous ones (as some combinations have been already discarded by these latter).

✓ *kmax* is computed according to an expression obtained through a theorem presented in section 3.12.

✓ Iteration in the *while* loop should end not only when the maximum level n combination was analyzed or when the maximum possible number of keys has been reached, but also when on the previous level all combinations where superkeys: trivially, on the current as well as on all subsequent ones, all combinations would be superkeys too (as they all include superkeys).

Consequently, Boolean variable *allSuperkeys* is used to detect when this latter condition holds.

✓ *A7/8–3* should be initially run once for each table; moreover, each time a table scheme changes (be it by adding new columns, dropping existing ones, or only changing their semantics), it should immediately be rerun for that table.

✓ All additional keys found by running *A7/8–3* should then be enforced; dually, all dropped ones should be removed (and, for both of them, the sooner, the better).

✓ *A7/8–3* cannot ever loop indefinitely (i.e., dually, it is always end-ing after a finite number of steps):

   (*i*) input m and k, as well as n, i, l, $kmax$, and $C(n, i)$ are always finite (being naturals);

   (*ii*) the first *repeat* loop is executed for at most $2^m - 1$ times (in the worst case, when $card(K) = 2^m - 1$, meaning that all possible keys and superkeys are initially enforced);

   (*iii*) the second *repeat* loop is executed for at most m times (in the worst case, when $card(SK) = 0$, meaning that no single keys are initially declared);

   (*iv*) the third *repeat* loop, the one inside *while*, is always executed $C(n, i)$ times;

   (*v*) finally, the *while* loop is executed at most $n - 1$ times, as i starts with 2 and is incremented by 1 in each such execution.

✓ *A7/8–3* always asks only legitimate questions[53] (i.e., nonprimeness ones only for single nonunique columns and uniqueness ones only for initially enforced keys and then for nonsuperkeys that are not known already to be keys).

Moreover, *A7/8–3* always asks all such possible questions[54] (i.e., dually, no possible question on nonprimeness or uniqueness is skipped).

✓ *A7/8–3* is only assisting its users: it is asking all and only legitimate questions, as it cannot respond to any of them (so it is only syntac-tically correct, being both sound and complete, the maximum that such an algorithm may ever reach).

Consequently, the semantic accuracy of its outputs is solely its corre-sponding user's responsibility.

✓ *A7/8–3* above is optimal relatively to its strategy: it asks the mini-mum number of questions in each possible case and it is doing its job with minimum number of statement executions.

However, it is not absolutely minimal: for example, the slight variation of it labeled *A3'* is bringing significant improvement in the case when the only semantic key of a table is the concatenation of all of its non-surrogate columns (see second volume of this book).

---

[53] Mathematically, such an algorithm is said to be *sound* (see Appendix).
[54] Mathematically, such an algorithm is said to be *complete* (see Appendix).

3.7 RDBMS METACATALOGS. RELATIONAL AND E-R DATA MODELS OF THE RDM

The purpose of models is not to fit the data, but to sharpen the questions.
—Samuel Karlin

3.7.1 RDBMS METACATALOGS

Everyday life is like programming, I guess.
If you love something you can put beauty into it.
—Donald Knuth

To begin with, this section is essentially presenting a relational model of the RDM: the schemas (and associated valid instances) of the main tables used by RDBMSs to store metadata, that is data on hosted dbs—the so called *metadata catalog* or *metacatalog*.

The considered tables store metadata on dbs, their tables, columns, and associated relational constraints. Actual metacatalogs have very many other columns for these kernel tables, as well as very many other tables (e.g., for users, groups of users, access right types, access rights of users and groups to db objects, views, triggers, stored functions and procedures, etc.).

Metacatalogs also store metadata on their schemas. Conventionally, we can consider that each RDBMS comes with a built-in system db (that we will call here *System*) storing metadata on its catalog, which is generally hidden to users, but accessible through read-only views. Besides it, for simplicity, we will consider in what follows a very simple user rdb called *BigCities*, made out of the two tables (*COUNTRIES* and *CITIES*) presented in Fig. 3.20.

Note that both tables have a primary key (denoted x), that *COUNTRIES* has two semantic keys (*Country*, as there may not be two countries having same name, and *Capital*, as no city may simultaneously be the capital of more than one country), whereas *CITIES* has only one such key (the product *City • Country*, as it is assumed that no country has several big cities with the same name).

COUNTRIES (_x_, *Country, Capital*)

_x_	Country	Capital
autonumber	ASCII(128)	Im(CITIES.x)
NOT NULL	NOT NULL	
1	U.S.A.	1
40	Romania	21

CITIES (_x_, *City* • *Country*)

_x_	Country	City
autonumber	Im(COUNTRIES.x)	ASCII (64)
NOT NULL	NOT NULL	NOT NULL
1	1	Washington
2	1	New York
21	40	Bucharest
24	40	Constanta

FIGURE 3.20   BigCities db.

Country (maximum 128 ASCII characters) and city (maximum 64 ASCII characters) names are mandatory, just as the country to which any city belongs. *Capital* is a foreign key referencing the *CITIES'* primary one and *Country* from *CITIES* is a foreign key referencing the *COUNTRIES'* primary one.

3.7.2 A RELATIONAL METAMODEL OF THE RDM

> *God is a challenge because there is no proof of his existence and therefore the search must continue.*
> —Donald Knuth

Table 3.7 shows the metacatalog table *DATABASES* and its instance for the two dbs above: *System* and *BigCities*. Note that db names (maximum 255 ASCII characters) are mandatory and unique (for any RDBMS).

This table actually has other columns too (e.g., *DateCreated, Password, Status*, etc.) and associated constraints.

TABLE 3.7   The *DATABASES* Metacatalog Table

DATABASES (<u>x</u>, *DBName*)

<u>x</u>	*DBName*
autonumber	ASCII(255)
NOT NULL	NOT NULL
0	System
1	BigCities

TABLE 3.8   The *RELATIONS* Metacatalog Table

RELATIONS (<u>x</u>, *RelationName* • *Database*, *PrimaryKey*)

<u>x</u>	*RelationName*	*PrimaryKey*	*Database*
autonumber	ASCII(255)	*Im(CONSTRAINTS. x)*	*Im(DATABASES. x)*
NOT NULL	NOT NULL		NOT NULL
0	DATABASES	0	0
1	RELATIONS	2	0
2	DOMAINS	7	0
3	ATTRIBUTES	5	0
4	CONSTRAINTS	9	0
5	KEYS_COLUMNS	11	0
6	FOREIGN_KEYS_COLS	14	0
7	COUNTRIES	27	1
8	CITIES	30	1

Table 3.8 presents the metacatalog table *RELATIONS* and its instance for the two dbs above. Note that table (relation) names (maximum 255 ASCII characters) are mandatory and unique within any db. As any column belongs to only one table and primary keys are columns, the *PrimaryKey* foreign key (referencing the *CONSTRAINTS*' primary one) is a semantic key too. As primary keys are optional, this column is not mandatory. The foreign key *Database* (referencing *DATABASES*' primary one) is mandatory, as any table belongs to a db. The first seven rows correspond to the *System* metacatalog tables, while the last two to the *BigCities* tables.

This table actually has many other columns too (e.g., *DateCreated*, *Tablespace*, *Status*, *Cardinal*, *AverageRowLength*, *Cached*, *Partitioned*, *Temporary*, *ReadOnly*, etc.) and associated constraints.

TABLE 3.9   The *DOMAINS* Metacatalog Table

DOMAINS (*x*, *DomainName*)

x	*DomainName*
autonumber	ASCII(32)
NOT NULL	NOT NULL
0	autonumber
1	BOOLE
2	INT
3	RAT
4	ASCII
5	UNICODE
6	DATE/TIME

Table 3.9 shows the metacatalog table *DOMAINS*, which stores both the RDBMS provided column data types and the users defined ones, as well as a partial valid instance for it. Note that domain names (maximum 32 ASCII characters) are mandatory and unique (for any RDBMS). This table has actually other columns too (e.g., *Owner*, *UpperBound*, *Length*, *CharacterSet*, etc.) and associated constraints, as well as very many other rows.

All *DOMAINS* seven rows are system ones, as *BigCities* does not contain any user-defined domain. Note that by INT, RAT, and DATE/TIME, respectively, it is understood corresponding OS/RDBMS representable subsets of the integers, rationals, and calendar dates, that autonumber is actually a synonym for INT (the only difference being that its values are automatically generated by the system and are unique for any primary key), and that BOOLE is the set {*True*, *False*}.

ASCII and UNICODE are subsets of the sets[55] of strings built upon these two character sets and having a maximum (RDBMS dependent) allowed length.

This table actually has other columns too (e.g., *minValue*, *maxValue*, etc.) and associated constraints as well.

Table 3.10 presents the metacatalog table *ATTRIBUTES* and its instance for the two dbs above (*System* and *BigCities*). Note that column (attribute)

---

[55] Mathematically, the freely generated monoids over the corresponding alphabets.

TABLE 3.10   The *ATTRIBUTES* Metacatalog Table

ATTRIBUTES (<u>*x*</u>, *AttributeName* • *Relation*)

<u>*x*</u>	*Attribute-Name*	*Total?*	*Relation*	*Domain*	*DConstr*
auto-number	ASCII(255)	BOOLE	*Im*(*RELATIONS.x*)	*Im*(*DOMAINS.x*)	ASCII(255)
NOT NULL	NOT NULL	NOT NULL	NOT NULL	NOT NULL	
0	x	yes	0	0	
1	DBName	yes	0	4	≤ 255
2	x	yes	1	0	
3	Relation-Name	yes	1	4	≤ 255
4	PrimaryKey	no	1	2	
5	Database	yes	1	2	
6	x	yes	3	0	
7	Attribute-Name	yes	3	4	≤ 255
8	Total?	yes	3	1	
9	Relation	yes	3	2	
10	Domain	yes	3	2	
11	DConstr	no	3	4	≤ 255
12	x	yes	2	0	
13	Domain-Name	yes	2	4	≤ 32
14	x	yes	4	0	
15	Constr-Name	yes	4	4	≤ 255
16	Constr-Relation	yes	4	2	
17	ConstrType	yes	4	4	'F', 'K', 'T'
18	*DB	no	4	2	
19	x	yes	5	0	
20	Position	yes	5	2	between 1–64
21	Key	yes	5	2	
22	Attribute	yes	5	2	

TABLE 3.10   (Continued).

x	Attribute-Name	Total?	Relation	Domain	DConstr
23	x	yes	6	0	
24	ForeignKey	yes	6	2	
25	FKAttribute	yes	6	2	
26	Ref-Attribute	yes	6	2	
27	Position	yes	6	2	between 1–64
28	x	yes	7	0	
29	Country	yes	7	4	≤ 128
30	Capital	no	7	2	
31	x	yes	8	0	
32	Country	yes	8	2	
33	City	yes	8	4	≤ 64

names (maximum 255 ASCII characters) are mandatory and unique within any table. The foreign key *Relation* (referencing *RELATIONS'* primary one) is mandatory, as any column belongs to a table.

The foreign key *Domain* (referencing *DOMAINS'* primary one) is mandatory too, as any column stores values of a desired data type. *Total?* is also mandatory, as for any column the system has to know whether or not to accept null values for it.

Note that columns *Domain*, *DConstr*, and *Total?* are storing domain (the first two) and not-null (the third) constraints. Note also that all primary keys are taking values into the autonumber subset of the integers, while foreign keys have integer values. The first twenty eight rows correspond to the *System* metacatalog tables, while the last six to the *BigCities* tables.

This table actually has very many other columns too (e.g., *Position, DefaultValue, LowestValue, HighestValue, DistinctValues, CharacterSet, AverageLength, Hidden, Computed*, etc.) and associated constraints.

Table 3.11 shows the metacatalog table *CONSTRAINTS*, which stores the relational constraints of the remaining three types ('K' for (unique) keys, 'F' for foreign keys, and 'T' for tuple/check ones), as well as its instance for the two dbs above (*System* and *BigCities*).

TABLE 3.11   The *CONSTRAINTS* Metacatalog Table
CONSTRAINTS (<u>*x*</u>, *ConstrName* • **DB*)

<u>*x*</u>	*ConstrName*	*ConstrRelation*	*Constr Type*	**DB*
autonumber	ASCII(255)	*Im*(*RELATIONS.x*)	{'F','K','T'}	*Im*(*DATABASES.x*)
NOT NULL	NOT NULL	NOT NULL	NOT NULL	
0	#DatabasesPK	0	K	0
1	DatabaseNameK	0	K	0
2	#RelationsPK	1	K	0
3	RelationNameK	1	K	0
4	PrimaryKeyK	1	K	0
5	#AttributesPK	3	K	0
6	AttributesKeyK	3	K	0
7	#DomainsPK	2	K	0
8	DomainNameK	2	K	0
9	#ConstraintsPK	4	K	0
10	ConstraintsK	4	K	0
11	#KeysColsPK	5	K	0
12	KeysCols-AttrKeyK	5	K	0
13	KeysColsPos-KeyK	5	K	0
14	#ForeignKeysPK	6	K	0
15	ForeignKeys-AttrK	6	K	0
16	Foreign-KeysPosK	6	K	0
17	Relations-DatabaseFK	1	F	0
18	Relations-PrimaryKeyFK	1	F	0
19	Attributes-RelationFK	3	F	0
20	Attributes-DomainFK	3	F	0
21	Constraints-RelationFK	4	F	0
22	KeysColsKeyFK	5	F	0

TABLE 3.11   (Continued).

x	ConstrName	ConstrRelation	Constr Type	*DB
23	KeysCols-AttributeFK	5	F	0
24	FKColsForeign-KeyFK	6	F	0
25	FKColsFKAt-tributeFK	6	F	0
26	FKColsRef-AttributeFK	6	F	0
27	#CountriesPK	7	K	1
28	Countries-CountryK	7	K	1
29	Countries-CapitalK	7	K	1
30	#CityPK	8	K	1
31	CitiesCitiesK	8	K	1
32	Countries-CapitalFK	7	F	1
33	CitiesCountryFK	8	F	1

Note that constraint names (maximum 255 ASCII characters) are mandatory and unique within any db. This is why the computed foreign key *DB (referencing *DATABASES*' primary one) is needed; its graph is computable[56].

Note that, except for the key, no constraints are needed on it. The foreign key *ConstrRelation* (referencing *RELATIONS*' primary one) is mandatory, as any constraint belongs to a table scheme. *ConstrType* is mandatory too, for partitioning the table's instance according to the three constraint types stored in it. The first twenty seven rows correspond to the *System* metacatalog tables, while the last seven to the *BigCities* tables.

This table actually has many other columns too (e.g., *Generated, Status, Deferrable, Deferred, Validated, Invalid*, etc.) and associated constraints.

---

[56] Mathematically, *DB = Database ° ConstrRelation*. Relationally:
```
SELECT CONSTRAINTS.x, Database
FROM CONSTRAINTS INNER JOIN RELATIONS
  ON CONSTRAINTS.ConstrRelation = RELATIONS.x
```

TABLE 3.12   The *KEYS_COLUMNS* Metacatalog Table
KEYS_COLUMNS (<u>x</u>, *Attribute • Key, Key • Position*)

x	*Key*	*Attribute*	*Position*
autonumber	*Im(CONSTRAINTS.x)*	*Im(ATTRIBUTES.x)*	[1, 64]
NOT NULL	NOT NULL	NOT NULL	NOT NULL
0	0	0	1
1	1	1	1
2	2	2	1
3	3	3	1
4	3	5	2
5	4	4	1
6	5	6	1
7	6	7	1
8	6	9	2
9	7	12	1
10	8	13	1
11	9	14	1
12	10	15	1
13	10	18	2
14	11	19	1
15	12	22	1
16	12	21	2
17	13	21	1
18	13	20	2
19	14	23	1
20	15	25	1
21	15	24	2
22	16	24	1
23	16	27	2
24	27	28	1
25	28	29	1
26	29	30	1
27	30	31	1
28	31	33	1
29	31	32	2

Table 3.12 presents the metacatalog table *KEYS_COLUMNS* and its instance for the two dbs above (*System* and *BigCities*). Note that column (attribute) positions (maximum 64 for IBM *DB2*, 32 for *Oracle*, 10 for MS *Access*, etc.) are mandatory within any key; moreover, in any position of any key, there may be only one column: hence, *Key • Position* is a semantic key of this table.

The foreign key '*Key*' (referencing *CONSTRAINTS*' primary one) is mandatory, as any key column belongs to a key constraint. The foreign key '*Attribute*' (referencing *ATTRIBUTES*' primary one) is mandatory too, as all columns of any key have to be known.

Note that no key should contain a same column more than once: hence, *Attribute • Key* is also a semantic key of this table. For example, its fourth and fifth rows ($x = 3, 4$) store the key *RelationName • Database* (constraint 3, named *RelationNameK*) of system table 1 (*RELATIONS*), which is made out of attributes 3 (*RelationName*) in position 1 and 5 (*Database*) in position 2. The first twenty four rows correspond to the *System* metacatalog keys, while the last six to the *BigCities* ones.

This table too may actually have other columns (e.g., *SortOrder*) and associated constraints.

Table 3.13 shows the metacatalog table *FOREIGN_KEYS_COLS* and its instance for the two dbs above (*System* and *BigCities*).

Note that all of its columns (attributes) are mandatory: the primary key *x*, by primary keys definition, and the foreign keys *ForeignKey* (referencing *CONSTRAINTS*' primary one), *FKAttribute*, and *RefAttribute* (both referencing *ATTRIBUTES*' primary one), as the foreign keys, their columns, and the corresponding referenced columns, respectively, have to be all known.

Also note that column (attribute) positions (maximum 64 for IBM *DB2*, 32 for *Oracle*, 10 for MS *Access*, etc.) are mandatory within any foreign key; moreover, in any position of any foreign key, there may be only one column: hence, *ForeignKey • Position* is a semantic key of this table.

Also note that, in fact, the RDM might not allow for concatenated foreign keys, as single ones are enough (and we will see in Subsection 3.11.1.3, best practice rule R-FK-1, that it is always better to use single numeric ones, always referencing single surrogate primary keys, in which case *Position* would not be needed anymore). This is why here, where all foreign keys are single ones, *Position* is always 1.

TABLE 3.13   The *FOREIGN_KEYS_COLUMNS* Metacatalog Table
FOREIGN_KEYS_COLS (*x*, *FKAttribute* • *ForeignKey*, *ForeignKey* • *Position*)

x	*ForeignKey*	*FKAttribute*	*Position*	*RefAttribute*
auto-number	*Im(CONSTRAINTS.x)*	*Im(ATTRIBUTES.x)*	[1, 64]	*Im(ATTRIBUTES.x)*
NOT NULL	NOT NULL	NOT NULL	NOT NULL	NOT NULL
0	17	5	1	0
1	18	4	1	14
2	19	9	1	2
3	20	10	1	12
4	21	16	1	2
5	22	21	1	14
6	23	22	1	6
7	24	24	1	14
8	25	25	1	6
9	26	26	1	6
10	32	30	1	31
11	33	32	1	28

No foreign key should contain a same column more than once: hence, *FKAttribute* • *ForeignKey* is also a semantic key of this table. For example, its first row (*x*=0) store the foreign key *Database* (*RelationsDatabaseFK*, constraint 17, attribute 5, position 1) of system table 1 (*RELATIONS*), which is referencing attribute 0 (*x*), the autonumber (domain 0) primary key (*#DatabasesPK*, constraint 0) of table 0 (*DATABASES*). The first ten rows correspond to the *System* metacatalog keys, while the last two to the *BigCities* ones.

This table too may actually have other columns and associated constraints as well.

Note that some actual RDBMS metacatalog tables are not even in BCNF (see Subsection 3.10.5). Moreover, surprisingly, some of them do not even correctly implement all of the RDM concepts.[57]

---

[57] For example, *Oracle*, MS *SQL Server*, and *Access* do not enforce either C_3 or C_{11}, that is minimal uniqueness (as they allow defining both keys and superkeys) or foreign keys acyclicity (as they allow attributes to reference themselves, both directly, except for MS *Access*, and indirectly).

Also note that RDBMS engines have to enforce, besides the relational constraints stored in *CONSTRAINTS* (see Table 3.11), many other non-relational constraints, including the ones shown in Fig. 3.21.

Note that for also storing tuple constraints you should either add at least a column to table *ATTRIBUTES* above or another kernel table (see Exercise 3.75).

3.7.3   AN E-R DATA MODEL OF THE RDM

There are truths so less striking that their discovery is almost creation.
—Lucian Blaga

By applying *REA*1–2 to the above RDBMS kernel metacatalog, the following E-R data model of the RDM is obtained, in which descriptions of the corresponding object sets and constraints, as well as table cardinality examples were added too for didactical reasons:

3.7.3.1   E-RD of the RDM

Friends share all things.
—Pythagoras

Figure 3.22 shows the E-RD of this fragment of the RDM.

3.7.3.2   Associated Restriction Set for the RDM

In fact what I would like to see is thousands of computer scientists let loose to do whatever they want. That's what really advances the field.
—Donald Knuth

1. ***DATABASES*** (The set of rdbs hosted by a RDBMS)
   a. *Cardinality*: $max(card(DATABASES)) = 10^4$ (*RDB0*)
   b. *Data ranges*: *DBName*: ASCII(255)   (*RDB1*)
   c. *Compulsory data*: *x*, *DBName*   (*RDB2*)

✓ C_0: *Relation surj* (any relation has at least one attribute)

✓ C_1: $(\forall x \in RELATIONS)(PrimaryKey(x) \notin NULLS \Rightarrow ConstrType(PrimaryKey(x)) =$ 'K') (primary keys are constraints of type *key*)

✓ C_2: $(\forall x \in RELATIONS)(\forall y \in KEYS\_COLUMNS)$ $(PrimaryKey(x) = Key(y) \Rightarrow Relation \circ Attribute(y) = x)$ (any attribute of the primary key of any relation has to be an attribute of that relation)

✓ C_3: $(\forall x \in CONSTRAINTS)(\forall y_1, ..., y_n \in KEYS\_COLUMNS,$ $n > 0,$ natural$)(ConstrType(x) =$ 'K' $\wedge Key(y_1) = ... = Key(y_n) =$ $x \Rightarrow Attribute(y_1) \bullet ... \bullet Attribute(y_n)$ is one-to-one) (keys are one-to-one)

✓ C_4: *ConstrRelation* $\circ$ *PrimaryKey* = $\mathbf{1}_{RELATIONS}$ (the primary key of any relation is a constraint of that relation)

✓ C_5: *ConstrRelation* $\circ$ *Key* = *Relation* $\circ$ *Attribute* (any attribute of a key should belong to the relation to which the corresponding key constraint is associated)

✓ C_6: $(\forall x \in KEYS\_COLUMNS)(PrimaryKey(Relation(Attribut e(x))) = Key(x) \Rightarrow Attribute(x)$ is part of the primary key of the relation $Relation(Attribute(x)))$ (primary key attributes are key attributes)

✓ C_7: $(\forall x \in FOREIGN\_KEYS\_COLS)$ $(ConstrType(ForeignKey(x)) =$ 'F') (referential integrities are constraints of type *foreign key*)

✓ C_8: *ConstrRelation* $\circ$ *ForeignKey* = *Relation* $\circ$ *FKAttribute* (any attribute of a foreign key should belong to the relation to which the corresponding referential integrity constraint is associated)

✓ C_9: $(\forall x, y \in FOREIGN\_KEYS\_COLS, x \neq y)$ $(ForeignKey(x)$ $= ForeignKey(y) \Rightarrow Relation \circ RefAttribute(x) = Relation \circ$ $RefAttribute(y))$ (all attributes referenced by a foreign key should belong to a same relation)

✓ C_{10}: $(\forall x \in FOREIGN\_KEYS\_COLS)(Im(FKAttribute(x)) \subseteq$ $Im(RefAttribute(x)))$ (referential integrity enforcement: values

FIGURE 3.21   Non-relational constraints associated to the above metacatalog.

of foreign keys should always be among those of the corresponding referenced columns)

✓ C_{11}: ($\forall x \in FOREIGN\_KEYS\_COLS$) (($FKAttribute \bullet RefAttribute$)($x$) $acyclic$) (no attribute should reference itself, neither directly, nor indirectly)

✓ C_{12}: ($\forall x \in CONSTRAINTS$)($\exists y \in FOREIGN\_KEYS\_COLS$) ($ConstrType(x) =$ 'F' $\Rightarrow ForeignKey(y)=x$) (any foreign key has at least one attribute)

✓ C_{13}: ($\forall x \in CONSTRAINTS$)($\exists y \in KEYS\_COLUMNS$)($ConstrType(x)$ = 'K' $\Rightarrow Key(y)=x$) (any key has at least one attribute)

FIGURE 3.21   (Continued).

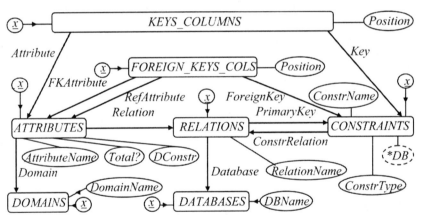

FIGURE 3.22   The E-RD of the RDM.

d. *Uniqueness*: x, DBName (there may not be two rdbs
   hosted on a same RDBMS and having same names)   (RDB3)

2. **DOMAINS** (The set of needed column data types)
   a. *Cardinality*: $max(card(DOMAINS)) = 10^2$ (RD0)
   b. *Data ranges*: DomainName: ASCII(32)   (RD1)
   c. *Compulsory data*: x, DomainName   (RD2)
   d. *Uniqueness*: x, DomainName (there may not be two
      domains managed by a same RDBMS and having
      same names)   (RD3)

3. **RELATIONS** (The set of tables of all the rdbs)

    a. *Cardinality*: *max*(*card*(*RELATIONS*)) = 10^8 (*RR0*)

    b. *Data ranges:RelationName*: ASCII(255)   (*RR1*)

    c. *Compulsory data*: x, *RelationName*   (*RR2*)

    d. *Uniqueness*: x, *RelationName • Database* (there may
       not be two tables of a same rdb having same names)   (*RR3*)

4. **ATTRIBUTES** (The set of columns of all tables)

    a. *Cardinality*: *max*(*card*(*ATTRIBUTES*)) = 10^{10} (*RA0*)

    b. *Data ranges*:

       AttributeName, DConstr: ASCII(255)   (*RA1*)

       Total?: Boolean   (*RA2*)

    c. *Compulsory data*: x, *AttributeName, Relation,*
       Domain, Total?   (*RA3*)

    d. *Uniqueness*: x, *AttributeName • Relation* (there may
       not be two columns of a same table having
       same names)   (*RA4*)

5. **CONSTRAINTS** (The set of constraints of all tables)

    a. *Cardinality*: *max*(*card*(*CONSTRAINTS*)) = 10^{10} (*RC0*)

    b. *Data ranges*:

       ConstrName: ASCII(255)   (*RC1*)

       ConstrType: {'F', 'K', 'T'}   (*RC2*)

    c. *Compulsory data*: x, *ConstrName, ConstrType,*
       ConstrRelation   (*RC3*)

    d. *Uniqueness*: x, *ConstrName • *DB* (there may not be
       two constraints in a same rdb having same names)   (*RC4*)

6. **FOREIGN_KEYS_COLS** (The set of pairs of type <*fk, c*>
  storing the fact that column *c* is a member of the foreign key
  constraint *fk*)

    a. *Cardinality*: *max*(*card*(*FOREIGN_KEYS_COLS*)) = 10^9 (*RFK0*)

    b. *Data ranges*: *Position*: [1, 64]   (*RFK1*)

    c. *Compulsory data*: x, *ForeignKey, FKAttribute,*
       RefAttribute, Position   (*RFK2*)

      d. Uniqueness: *x*,

      FKAttribute • ForeignKey (no attribute should take
      part more than once in a foreign key),   *(RFK3)*
      ForeignKey • Position (no position of a foreign key
      may be occupied by more than one attribute).   *(RFK4)*

7. **KEYS_COLUMNS** (The set of pairs of type $<k, c>$ storing the fact
    that column *c* is a member of the (unique) key constraint *k*)

      a. Cardinality: $max(card(KEYS\_COLUMNS)) = 10^9$ *(RK0)*
      b. Data ranges: *Position*: [1, 64]   *(RK1)*
      c. Compulsory data: *x, Key, Attribute, Position*   *(RK2)*
      d. Uniqueness: *x*,

      Attribute • Key (no attribute should take part more than
      once in a key),   *(RK3)*
      Key • Position (no position of a key may be occupied
      by more than one attribute).   *(RK4)*

All nonrelational constraints from Fig. 3.21 must be added here too.

3.7.3.3   Description of the RDM

> *There's ways to amuse yourself while doing things*
> *and that's how I look at efficiency.*
> —Donald Knuth

For example (for all the following maximum cardinalities), any RDBMS manages at most 10^4 rdbs, which are characterized by their compulsory unique name (at most 255 ASCII characters), and provides at most 10^2 data types, which are characterized by their compulsory unique name (at most 32 ASCII characters).

Any rdbs may include several relations, at most 10^8 in total, for all rdbs, which are characterized by their compulsory name (at most 255 ASCII characters), unique within their rdbs.

Any relation should include at least one attribute; attributes are characterized by their compulsory name (at most 255 ASCII characters), unique within their relation, NOT NULL constraint (does the corresponding column also accept null values or not?) and data type; optionally, a domain restriction (a propositional formula of at most 255 characters) may also be specified for attributes that should not accept all values of their domains, but only a subset of them. In total, for all rdbs, there may be at most 10^{10} attributes.

Any relation may have attached constraints, which are characterized by their compulsory name (at most 255 ASCII characters), unique within their rdbs, and type (one character, either 'F'—for foreign keys, 'K'—for (unique) keys, or 'T'—for tuple constraints). In total, for all rdbs, there may be at most 10^{10} constraints, out of which at most 10^9 foreign and unique keys.

Any foreign or unique key should include at least one and at most 64 attributes; all attributes of a key, be it foreign or unique, should belong to the relation to which the corresponding constraint belongs; no attribute may appear more than once in a key (be it foreign or unique).

For any key, the RDBMS should enforce its one-to-oneness —that is its values should uniquely identify the corresponding relation tuples, in any moment of time (or, dually, no duplicates should ever be allowed for the values of a key).

For any foreign key, the RDBMS should enforce its referential integrity: its values should always be among those of the corresponding referenced attributes.

Any attribute of a foreign key references another attribute of a same (or compatible) domain, acyclically (i.e., no attribute of a foreign key may recursively reference itself, neither directly, nor indirectly); all attributes of a foreign key should reference attributes belonging to a same relation.

Any relation may have at most one primary key, which must be a key composed of some of its attributes.

3.8   RELATIONAL SCHEMAS DEFINITION. *SQL DDL*

> *Morphological information has provided the greatest single source of data in the formulation and development of the theory of evolution and that even now, when the preponderance of work is experimental, the basis for interpretation in many areas of study remains the form and relationships of structures.*
> —Everett C. Olson

Let us consider in what follows the *DDL* of ANSI-92-type *SQL*. For creating a db named "Classical_Music" with the three tables from Fig.

```
CREATE DATABASE Classical_Music;
---------------------------------------------------
--  DDL for Table COMPOSERS
---------------------------------------------------
CREATE TABLE Classical_Music.COMPOSERS (
  x INT(8) GENERATED ALWAYS AS IDENTITY(1,1) PRIMARY,
  FirstName VARCHAR(32),
  LastName VARCHAR(32) NOT NULL,
  BirthYear INT(4) NOT NULL CHECK (BIRTH_YEAR BETWEEN
      1200 AND Year(SysDate)-3),
  PassedAwayYear INT(4) CHECK (PASSED_AWAY_YEAR
      BETWEEN 1200 AND Year(SysDate)),
CONSTRAINT COMPOSERS_Key UNIQUE (LastName,
FirstName),
CONSTRAINT COMPOSERS_LIFETIME_CHK CHECK
      (PASSED_AWAY_YEAR - BIRTH_YEAR BETWEEN 3 AND
120)
);
---------------------------------------------------
--  DDL for Table TONALITIES
---------------------------------------------------
CREATE TABLE Classical_Music.TONALITIES (
  x INT(2) GENERATED ALWAYS AS IDENTITY(1,1),
  Tonality VARCHAR(13) NOT NULL UNIQUE
);
---------------------------------------------------
--  DDL for Table MUSICAL_WORKS
---------------------------------------------------
CREATE TABLE Classical_Music.MUSICAL_WORKS (
  x INT(12) GENERATED ALWAYS AS IDENTITY(1,1) PRIMARY,
  Composer INT(8) NOT NULL,
  Opus VARCHAR(16) NOT NULL,
  No INT(3) CHECK (No BETWEEN 1 AND 255),
  Tonality INT(2),
  Title VARCHAR(255),
CONSTRAINT MUS_WORKS_Key UNIQUE (Composer, Opus, No),
CONSTRAINT MUS_WORKS_COMPOSER_FK FOREIGN KEY
      (Composer) REFERENCES Classical_Music.Composers,
CONSTRAINT MUS_WORKS_TONALITY_FK FOREIGN KEY
      (Tonality) REFERENCES Classical_Music.Tonalities
);
```

FIGURE 3.23 ANSI-92 SQL statements for creating the db scheme of Fig. 3.2.

3.2, the statements shown in Fig. 3.23 are needed (see algorithm *A*8 from Fig. 4.1).

Please note the following:

0. Actual RDBMS implementations syntax varies both as compared to the ANSI *SQL* standards and even between their versions. However, for most of them, ";" is ending statements.

1. Auto-generated numbers (*autonumbers*)[58] were introduced in the ANSI-2003 *SQL* standard.

2. "– –" are introducing (i.e., are followed by) comments.

3. As table names should be unique only within db schemas, full reference to any table should be of the type *db_name.table_name* (e.g., `Classical_Music.COMPOSERS`). When the context is unambiguous, the db names may be skipped.

4. Except for some *aggregation* ones (e.g., `COUNT, SUM, MAX, MIN, AVG`, obviously computing cardinals, total amounts, maximum, minimum, and arithmetic average values, respectively), there are no ANSI-92 *SQL* standard functions; however, *SysDate* (returning current system date) and *Year* (returning the year part of a calendar date value) above, as well as many others are provided by any RDBMS (possibly slightly renamed)[59].

5. `CREATE TABLE` statements are creating corresponding tables and their columns; for each column, its name is followed by its data type and, whenever needed, its maximum allowed length (e.g., `VARCHAR2(32)` means variable length strings of maximum 32 characters; `NUMBER(12,0)` means integer numbers (i.e., having 0 digits after the decimal point) using maximum 12 digits; `INT(12)` … `IDENTITY(1,1)` means an auto generated integer of maximum 12 digits, whose values start with and are incremented always by 1 (which is the default[60]). `PRIMARY` is declaring the corresponding column as being the corresponding table's primary key, while `NOT NULL` is declaring that the corresponding column is not accepting null values.

---

[58] As such, for example, *Oracle* continued not to provide them, to the despair of most of its users, up until December 2013, when version *12c* was released.

[59] Note that, for example, *Oracle* does not allow for *SysDate* either in table constraints or computed columns definitions; consequently, corresponding constraints may only be implemented through *PL/SQL* (see Subsection 4.3.2).

[60] Implementations are sometimes also requiring explicit start and increment values.

6. `ALTER TABLE` statements are declaring domain, check, unique (key) and referential integrity (foreign key) constraints. Note that, on one hand, these constraints might be declared in the `CREATE TABLE` statements too, using the `CONSTRAINT` clause and that, on the other, generally, `ALTER TABLE` statements may also add/modify columns, indexes, etc.

7. When needed, dbs, tables, columns, constraints, indexes, etc. may be removed by using the `DROP` statement; for example, for removing the constraint enforcing for column *No* of table *MUSICAL_WORKS* above values only between 1 and 255, you should execute the statement:

```
ALTER TABLE Classical_Music.MUSICAL_WORKS
DROP CONSTRAINT MUSICAL_WORKS_NO_CHK;
```

8. In any rdb scheme, all tables and constraints should have distinct names and within each table all columns should have distinct names.

9. *SQL DDL* also contains `CREATE`, `ALTER` and `DROP` statements for managing users and groups of users, as well as `GRANT` and `REVOKE` statements for granting to and revoking from them predefined access rights (e.g., create, alter, drop, read-only querying, insert, update, delete, etc.) to db objects (e.g., db and tables schemas, instances, etc.), as well as other object types (especially indexes, used for both enforcing unique constraints and speeding up query evaluation).

3.9 RELATIONAL INSTANCES MANIPULATION. *SQL DML.* RELATIONAL CALCULI AND ALGEBRA

> *Not everything that can be counted counts,*
> *and not everything that counts can be counted.*
> —Albert Einstein

Databases are not used only for data storage, but especially in order to retrieve and process it (e.g., by computing cardinals, totals, averages, minimums, maximums, etc.). This is why all computer data models include data manipulation languages. Querying languages are tools for formulating, syntax and semantic correctness checking, and computing answers to all possibly valid queries over any db. Data manipulation languages that include them are also providing statements for adding, modifying, and deleting rows in/from tables.

Although data manipulation languages are beyond the scope of this book, we have to review their basics, as they are needed even in db design, implementation and, especially, optimization.

3.9.1   INSERTING, UPDATING, AND DELETING DATA

Get the facts first and then you can distort them as much as you please.
—Mark Twain

For example, in order to populate the rdb scheme created by statements in Fig. 3.23 above with data presented in Fig. 3.2, you need the statements shown in Fig. 3.24.

Please note the following:

0.   Above form of the INSERT statement only inserts one row at a time, provided that valid data is specified for each column; for example, if in the last but one statement above the VALUES clause were (8,8,'83', null,1,'The 2nd piano concerto', null), its execution would fail as there is one more value than the number of fields in the *MUSI-CAL_WORKS* table; if it were (7,8,'83',null,1,'The 2nd piano concerto'), its execution would fail as *x* is a key, so no duplicates on this column are allowed; if it were (8,null,'83',null,1,'The 2nd piano concerto'), its execution would fail as column *Composer* does not accept nulls; if it were (8,8,'83',null,0,'The 2nd piano concerto'), its execution would fail as column *No* only accepts naturals between 1 and 255.

1. Several lines may be inserted at once in a table by using the alternate INSERT INTO … SELECT … statement.

2. For modifying existing data, *SQL* offers the UPDATE statement; for example, if we would like to add the fact that Don Giovanni starts in D minor, we could execute: UPDATE MUSICAL_WORKS SET Tonality=3 WHERE *x*=4; if we were trying, for example, to execute UPDATE MUSICAL_WORKS SET Tonality=7 WHERE *x*=4; instead, it would fail, as there is no tonality in table *TONALITIES* having *x* = 7.

```
-- INSERTING into CLASSICAL_MUSIC.COMPOSERS --
Insert into CLASSICAL_MUSIC.COMPOSERS (x,FirstName,
LastName,BirthYear,PassedAwayYear) values
(1,'Antonio','Vivaldi',1678,1741);
Insert into CLASSICAL_MUSIC.COMPOSERS (x, FirstName,
LastName, BirthYear, PassedAwayYear) values
(2,'Johann Sebastian','Bach',1685,1750);
Insert into CLASSICAL_MUSIC.COMPOSERS (x,FirstName,
LastName, BirthYear, PassedAwayYear) values
(3,'Joseph','Haydn',1732,1809);
Insert into CLASSICAL_MUSIC.COMPOSERS (x,FirstName,
LastName, BirthYear, PassedAwayYear) values
(4,'Wolfgang Amadeus','Mozart',1756,1791);
Insert into CLASSICAL_MUSIC.COMPOSERS (x,FirstName,
LastName, BirthYear, PassedAwayYear) values
(5,'Ludwig van','Beethoven',1770,1827);
Insert into CLASSICAL_MUSIC.COMPOSERS (x,FirstName,
LastName, BirthYear, PassedAwayYear) values
(6,'Franz','Schubert',1797,1828);
Insert into CLASSICAL_MUSIC.COMPOSERS (x,FirstName,
LastName, BirthYear, PassedAwayYear) values
(7,'Frédéric','Chopin',1810,1849);
Insert into CLASSICAL_MUSIC.COMPOSERS (x,FirstName,
LastName, BirthYear, PassedAwayYear) values
(8,'Johannes','Brahms',1833,1897);
-- INSERTING into CLASSICAL_MUSIC.TONALITIES--
Insert into CLASSICAL_MUSIC.TONALITIES (x,Tonality)
values (1,'B-flat major');
Insert into CLASSICAL_MUSIC.TONALITIES (x,Tonality)
values (2,'B minor');
Insert into CLASSICAL_MUSIC.TONALITIES (x,Tonality)
values (3,'D minor');
Insert into CLASSICAL_MUSIC.TONALITIES (x,Tonality)
values (4,'E major');
Insert into CLASSICAL_MUSIC.TONALITIES (x,Tonality)
values (5,'G major');
Insert into CLASSICAL_MUSIC.TONALITIES (x,Tonality)
values (6,'G minor');
-- INSERTING into CLASSICAL_MUSIC.MUSICAL_WORKS--
Insert into CLASSICAL_MUSIC.MUSICAL_WORKS
(x,Composer,Opus,No,Tonality,Title) values (1,1,'8 RV
269',null,4,'La primavera');
```

FIGURE 3.24   ANSI-92-type SQL statements for inserting instances of Fig. 3.2 into the rdb scheme presented in Fig. 3.23.

```
Insert into CLASSICAL_MUSIC.MUSICAL_WORKS
(x,Composer,Opus,No,Tonality,Title) values (2,2,'BWV
846-893',null,null,'The Well-Tempered Clavier');
Insert into CLASSICAL_MUSIC.MUSICAL_WORKS
(x,Composer,Opus,No,Tonality,Title) values (3,3,'H
1/94',null,5,'The "Surprise" symphony');
Insert into CLASSICAL_MUSIC.MUSICAL_WORKS
(x,Composer,Opus,No,Tonality,Title) values (4,4,'K
527',null,null,'Don Giovanni');
Insert into CLASSICAL_MUSIC.MUSICAL_WORKS
(x,Composer,Opus,No,Tonality,Title) values
(5,5,'125',null,3,'The 9th symphony');
Insert into CLASSICAL_MUSIC.MUSICAL_WORKS
(x,Composer,Opus,No,Tonality,Title) values (6,6,'D
759',null,2,'The "Unfinished" symphony');
Insert into CLASSICAL_MUSIC.MUSICAL_WORKS
(x,Composer,Opus,No,Tonality,Title) values
(7,7,'37',2,5,'Nocturne');
Insert into CLASSICAL_MUSIC.MUSICAL_WORKS
(x,Composer,Opus,No,Tonality,Title) values
(8,8,'83',null,1,'The 2nd piano concerto');
Insert into CLASSICAL_MUSIC.MUSICAL_WORKS
(x,Composer,Opus,No,Tonality,Title) values
(9,8,'115',null,2,'The clarinet quintet');
```

FIGURE 3.24   (Continued).

Caution: if you omit the optional WHERE clause of UPDATE then all corresponding rows will be updated. For example all rows of table *MUSICAL_WORKS* would contain 3 in its column *Tonality* after executing the statement: UPDATE MUSICAL_WORKS SET Tonality = 3;

3. For removing existing data, *SQL* offers the DELETE statement; for example, if we would like to remove Brahms' 2nd piano concerto, we could execute: DELETE FROM MUSICAL_WORKS WHERE $x=8$; if we were trying to delete any line in *TONALITIES*, we wouldn't be successful as all of them are referenced by at least one line in *MUSICAL_WORKS*.

Caution: if you omit the optional WHERE clause of DELETE then all corresponding rows will be deleted. For example all rows of table *MUSI-*

CAL_WORKS would be deleted, leaving the table empty, after executing the statement: `DELETE FROM MUSICAL_WORKS;`

3.9.2 QUERYING DATA

> *That is the essence of science: ask an impertinent question,*
> *and you are on the way to the pertinent answer.*
> —Jacob Bronowski

Any db scheme may be also viewed as a collection of queries, with its instance being the set of corresponding answers. For example, the db in Fig. 3.2 embeds the following three queries:

1. "Which are the first and last names, birth and passed away years of the composers of interest (i.e., those on whom data is stored?)?"
2. "Which are the composer, opus, number, tonality, and title of all of the musical works of interest?" and
3. "Which are the musical tonality names of interest?"

Especially in logical approaches of data modeling, but not only, it is considered that each table row is the answer to a query. For example, the first row of the *TONALITIES* table above is interpreted as "It is *true* that there is a tonality called 'B-flat major' whose db *x* code is 1", whereas the last one in table *COMPOSERS* as "It is *true* that, between 1833 and 1897, lived a composer named 'Johannes Brahms', whose db *x* code is 8".

Based on this embedded queries and answers, explicit queries may be built and their corresponding answers computed too. For example, considering the db in Fig. 3.2, the queries:

a. "Who are the composers having same first name?" (none in this case);
b. "Who were the composers contemporary with J.S. Bach?" (Antonio Vivaldi and Joseph Haydn);
c. "How many musical works were written in D-sharp minor?" (none in this case);
d. "Who are the composers who composed at least 2 musical works?" (Johannes Brahms in this case) etc.

a. SELECT COMPOSERS.FirstName, COMPOSERS.LastName,
   C.LastName AS otherLastName
   FROM COMPOSERS INNER JOIN COMPOSERS AS C
   ON COMPOSERS.FirstName = C.FirstName AND
      COMPOSERS.LastName <> C.LastName
   ORDER BY COMPOSERS.FirstName, COMPOSERS.LastName,
   C.LastName;
b. SELECT FirstName, LastName FROM COMPOSERS
   WHERE BirthYear <= 1750 AND PAssedAwayYear >= 1685
   AND x <> 2 ORDER BY FirstName, LastName;
c. SELECT COUNT(*) FROM MUSICAL_WORKS WHERE Tonality IN
   (SELECT x FROM TONALITIES WHERE Tonality = 'D-sharp
   minor');
d. SELECT FirstName, LastName FROM COMPOSERS WHERE x IN
   (SELECT Composer FROM MUSICAL_WORKS
    GROUP BY Composer HAVING COUNT(x)>1)
   ORDER BY FirstName, LastName;

may be written in ANSI-92-type *SQL* as follows, respectively (where "<>" means "not equal" and "COUNT (*)" means the number of all involved rows):

Please note the following:

0. SELECT is the only standard *SQL* querying statement (as there is no need for any other one).[61]
1. With only one tiny exception for some RDBMSs[62], its two first and only mandatory clauses are SELECT and FROM, in this order.
2. All expressions in the SELECT clause (separated between them by commas) will generate corresponding columns (which should have unique names) in the result, in corresponding order; these expressions may be constants, columns from the table instances occurring in the FROM clause, or arbitrary algebraic expressions based on these columns and constants, including some restricted forms of subqueries (i.e., queries embedded in parenthesis, just like, for example, in c. and d. above: see 18 to 25 below). Both column (in SELECT clauses) and table instances (in FROM ones) may be renamed (using the optional AS

---

[61] Some implementations include non-standard ones as well. For example, MS *Access* also provides TRANSFORM for computing crosstabulations.

[62] For example, in MS *Access*, SELECT @@IDENTITY; returns the latest auto-number value generated by the system.

operator); see, for example, *C* in a. above. Note that in some RDBSs the keyword AS is optional (e.g., in a. above, instead of COMPOSERS AS C, you could/should simply write COMPOSERS C).

3. Set quantifier predicates may be added to the SELECT clause in order to eliminate duplicates (DISTINCT, DISTINCTROW) or not (ALL, the default one). Note that DISTINCT removes duplicated in the end, only before final ordering (if any), while DISTINC-TROW removes them before evaluating the join operators (see notes 5 to 11). Other predicates may be added too (e.g., TOP/LIMIT n, requesting only for the first n rows of the answer, n natural).

4. The FROM clause specifies a table expression: an expression whose operands are table instances (possibly even several ones of a same table, like, for example, in a. above –where the second instance of *COMPOSERS* had to be renamed as *C*, in order to uniquely distinguish between them– or named subqueries) and whose operators are the Cartesian set product (denoted by *comma*) and/or the join operators.

5. The *join* operators are of two types: *inner* and *outer*. The *inner join* is a Cartesian product followed by a filtering WHERE-type condition.

6. The first optional clause is the WHERE one, which declares a condition (formalized as a Boolean expression) that all corresponding computed rows should satisfy. For example, in b. above, due to its WHERE clause, are selected only composers whose birth year is at most 1750 (J.S. Bach's passed away year) and whose passed away year is at least 1685 (J.S. Bach's birth year) and whose *x* is not 2 (i.e., J.S. Bach will not be selected in the result).

7. Most of the *Oracle* users do not use or even ignore the existence of inner joins (as well as the fact that comma in FROM means Cartesian product): they are explicitly using instead its definition given in 4 above. For example, they would write a. above as:

```
SELECT COMPOSERS.FirstName, COMPOSERS.LastName,
   C.LastName AS otherLastName
FROM COMPOSERS, COMPOSERS C
WHERE COMPOSERS.FirstName = C.FirstName AND
   COMPOSERS.LastName <> C.LastName
```

```
ORDER BY COMPOSERS.FirstName, COMPOSERS.LastName,
  C.LastName;
```

Note, however, that this is advisable in *Oracle*, *DB2*, and latest versions of *SQL Server* only because these RDBMS have internal optimizers that convert any such Cartesian products followed by WHERE filters into corresponding inner joins (for which computation best strategies are also chosen, depending on current data instances, indexes, etc.). Generally, this is not at all advisable for RDBMSs which do not have such optimizing capabilities (e.g., MS *Access* and *MySQL*), as they will first compute the Cartesian product and then will retain from it only the subset satisfying the WHERE filter (which is generally taking much, much more time than the corresponding inner join).

8. *Outer joins* are of three types: left, right, and full ones.

9. The *left* (*outer*) *join* unions the inner join and all other rows from its left operand, if any (padded with nulls for corresponding right operand's columns). For example, the statement:

```
SELECT FirstName, LastName, Opus, No,
  TONALITIES.Tonality, Title
FROM (COMPOSERS INNER JOIN MUSICAL_WORKS
  ON COMPOSERS.x = MUSICAL_WORKS.Composer) INNER JOIN
  TONALITIES ON MUSICAL_WORKS.Tonality = TONALITIES.x
ORDER BY FirstName, LastName, TONALITIES.Tonality,
  Opus, No, Title;
```

computes the result presented in Table 3.14, which does not contain either Bach's Well tempered clavier or Mozart's Don Giovanni, as they have no tonality; if we also need these two works, then we may use a left join with *TONALITIES* instead (Table 3.15 shows its result):

```
SELECT FirstName, LastName, Opus, No,
  TONALITIES.Tonality, Title
FROM (COMPOSERS INNER JOIN MUSICAL_WORKS
  ON COMPOSERS.x = MUSICAL_WORKS.Composer) LEFT JOIN
  TONALITIES ON MUSICAL_WORKS.Tonality = TONALITIES.x
ORDER BY FirstName, LastName, TONALITIES.Tonality,
  Opus, No, Title;
```

TABLE 3.14 Inner Join Result

FirstName	LastName	Opus	No	Tonality	Title
Antonio	Vivaldi	8 RV 269		E major	La primavera
Franz	Schubert	D 759		B minor	The "Unfinished" symphony
Frédéric	Chopin	37	2	G major	Nocturne
Johannes	Brahms	115		B minor	The clarinet quintet
Johannes	Brahms	83		B-flat major	The 2nd piano concerto
Joseph	Haydn	H 1/94		G major	The "Surprise" symphony
Ludwig van	Beethoven	125		D minor	The 9th symphony

TABLE 3.15 Left Join Result

FirstName	LastName	Opus	No	Tonality	Title
Antonio	Vivaldi	8 RV 269		E major	La primavera
Franz	Schubert	D 759		B minor	The "Unfinished" symphony
Frédéric	Chopin	37	2	G major	Nocturne
Johann Sebastian	Bach	BWV 846–893			The Well-Tempered Clavier
Johannes	Brahms	115		B minor	The clarinet quintet
Johannes	Brahms	83		B-flat major	The 2nd piano concerto
Joseph	Haydn	H 1/94		G major	The "Surprise" symphony
Ludwig van	Beethoven	125		D minor	The 9th symphony
Wolfgang Amadeus	Mozart	K 527			Don Giovanni

10. The *right* (*outer*) *join* is the dual of the left one: it unions the inner join and all other rows from its right operand, if any (padded with nulls for corresponding left operand's columns). For example, the statement:

```
SELECT FirstName, LastName, Opus, No,
  TONALITIES.Tonality, Title
FROM TONALITIES RIGHT JOIN (COMPOSERS INNER JOIN
  MUSICAL_WORKS
  ON COMPOSERS.x = MUSICAL_WORKS.Composer)
  ON MUSICAL_WORKS.Tonality = TONALITIES.x
ORDER BY FirstName, LastName, TONALITIES.Tonality,
  Opus, No, Title;
```

is strictly equivalent to the one above (and it also computes the result shown in Table 3.15).

11. The *full outer join* is the union of the left and right ones. For example, the statement:

```
SELECT FirstName, LastName, Opus, No,
  TONALITIES.Tonality, Title
FROM (COMPOSERS INNER JOIN MUSICAL_WORKS
  ON COMPOSERS.x = MUSICAL_WORKS.Composer) FULL JOIN
  TONALITIES ON MUSICAL_WORKS.Tonality = TONALITIES.x
ORDER BY FirstName, LastName, TONALITIES.Tonality
  Opus, No, Title;
```

computes the result shown by Table 3.16.

12. Sometimes we need to compute aggregations (e.g., cardinals, sums, minimums, maximums, averages, etc.) on all rows of a table or query, like, for example, in c. above. Other times, we need to compute them by *groups* (*partitions*[63]) of rows, like, for example, in d. above (where we need to compute the number of musical works per composer). Partitioning is achieved in SELECT statements by the second optional clause named GROUP BY.

---

[63] For example, the set of *MUSICAL_WORKS* rows may be partitioned by *Tonality* values into six partitions: one for those that do not have tonality values (here only "The well-tempered clavier" and "Don Giovanni") and five for those that have, namely one for each of the five distinct tonalities (three of them made out of only one row in this example, while the one corresponding to G major, which comprises Haydn's "Surprise" symphony and Chopin's Nocturne and the one corresponding to B minor, including Schubert's "Unfinished" symphony and Brahms' clarinet quintet having two rows each). Generally, partitions are the classes of an equivalent relation. In this case, the equivalent relation is "*has same tonality as*".

TABLE 3.16 Full Join Result

FirstName	LastName	Opus	No	Tonality	Title
Antonio	Vivaldi	8 RV 269		E major	La primavera
Franz	Schubert	D 759		B minor	The "Unfinished" symphony
Frédéric	Chopin	37	2	G major	Nocturne
Johann Sebastian	Bach	BWV 846–893			The Well-Tempered Clavier
Johannes	Brahms	115		B minor	The clarinet quintet
Johannes	Brahms	83		B-flat major	The 2nd piano concerto
Joseph	Haydn	H 1/94		G major	The "Surprise" symphony
Ludwig van	Beethoven	125		D minor	The 9th symphony
Wolfgang Amadeus	Mozart	K 527		G minor	Don Giovanni

13. The GROUP BY clause may contain several columns from the table expression of the FROM clause (separated by commas), which are defining partition hierarchies. For example, if we were to compute how many works were composed by composers in each tonality, the following statement partitions each partition corresponding to a composer into as many subpartitions as the number of tonalities that he/she used (in this example, all partitions are subpartitioned in only one subpartition, as there are only one work per composer, except for the one corresponding to Brahms, which is subpartitioned in two, as there are two works by him, each of which was written in a distinct tonality):

```
SELECT FirstName, LastName, TONALITIES.Tonality,
  Count(MUSICAL_WORKS.x) AS WorksNoByTonality
FROM (COMPOSERS INNER JOIN MUSICAL_WORKS
  ON COMPOSERS.x = MUSICAL_WORKS.Composer) LEFT JOIN
  TONALITIES ON MUSICAL_WORKS.Tonality = TONALITIES.x
GROUP BY FirstName, LastName, TONALITIES.Tonality
ORDER BY FirstName, LastName, TONALITIES.Tonality;
```

The result of this query is shown in Table 3.17:

TABLE 3.17   Multiple Grouping Result

FirstName	LastName	Tonality	WorksNoByTonality
Antonio	Vivaldi	E major	1
Franz	Schubert	B minor	1
Frédéric	Chopin	G major	1
Johann Sebastian	Bach		1
Johannes	Brahms	B minor	1
Johannes	Brahms	B-flat major	1
Joseph	Haydn	G major	1
Ludwig van	Beethoven	D minor	1
Wolfgang Amadeus	Mozart		1

14. The GROUP BY clause has an associated "*golden rule*": whenever a SELECT statement contains such a clause, its SELECT clause may contain columns as such only from the GROUP BY clause; any other column that might appear in its SELECT clause should be an argument of an aggregation function. For example, in the statement:

```
SELECT Tonality, Count(x) AS WorksNoByTonality
FROM MUSICAL_WORKS
GROUP BY Tonality
ORDER BY Tonality;
```

x (or any other column from *MUSICAL_WORKS* except for *Tonality*) may not appear as such in its SELECT clause. Indeed, suppose that we would forget to apply *Count* to x (or try to convince ourselves that it is not possible not to apply to x an aggregate function); any RDBMS would reject it syntactically, as senseless: for example, from the groups containing two rows, which x value should be selected and why?

15. The third optional clause, HAVING, is similar to the WHERE one, except for the moment when it is applied: WHERE is applied *before*, while HAVING is applied *after* grouping. Consequently, HAVING could do the job of WHERE too, but WHERE cannot do HAVING's job: in d. above, for example, if we were trying to replace the subquery with SELECT Composer FROM MUSICAL_WORKS

> WHERE COUNT(x) > 1 GROUP BY Composer, the query would be rejected.

16. Although WHERE might generally not be used at all, its usage saves enormous amount of time (and temporary disk and memory space as well); for example, let us consider a slight variation of d. above, when only composers that wrote (no matter how many) musical works in D minor are of interest; the following statement would compute this result correctly:

```
SELECT FirstName, LastName FROM COMPOSERS WHERE x IN
   (SELECT Composer FROM MUSICAL_WORKS
   GROUP BY Composer, Tonality
   HAVING Tonality = 3)
 ORDER BY FirstName, LastName;
```

However, this is not an optimal solution: when evaluating the subquery, the RDBMS will partition the whole *MUSICAL_WORKS* instance by *Composer* and then by *Tonality* –obtaining 9 partitions–, and will eliminate all of them, but one—as only Beethoven's 9th symphony is in D minor.

The equivalent subquery:

```
SELECT DISTINCT Composer FROM MUSICAL_WORKS
WHERE Tonality = 3
```

runs much faster (and with much fewer memory needed) to produce same results: from the *MUSICAL_WORKS* instance, the RDBMS will select only the composers of those written in D minor (in this case, only one Beethoven's occurrence) and will eliminate duplicates, if any, without even needing grouping!

Consequently, you should always use WHERE for all conditions that may be evaluated before grouping (i.e., all that do not contain aggregate functions) and HAVING only for those that may not be evaluated before grouping, but only after it (i.e., only those that do contain aggregate functions on groups).

For example, in order to compute the set of composers that lived in the nineteenth century and who composed at least two musical works (together with the total number of their musical works), the following statement is one of the best *SQL* solutions possible (its result is, obviously, <"Johannes", "Brahms", 2>):

```
SELECT FirstName, LastName, Count(MUSICAL_WORKS.x) AS
   TotalWorksNo
FROM COMPOSERS INNER JOIN MUSICAL_WORKS
   ON COMPOSERS.x = MUSICAL_WORKS.Composer
WHERE BirthAwayYear Between 1800 AND 1899 OR
   PassedAwayYear Between 1800 AND 1899
GROUP BY FirstName, LastName
HAVING Count(MUSICAL_WORKS.x) > 1
ORDER BY Count(MUSICAL_WORKS.x) DESC, FirstName,
   LastName;
```

17. The fourth optional clause, ORDER BY, accepts a comma separated set of pairs <*e, so*>, where *e* are algebraic expressions based on the columns of the FROM table expression and *so* are the sort order predicates ASC(ENDING), the default one (which may consequently be omitted) and DESC(ENDING). For example, if we would like to get the results of b. above first in descending order of their passed away year and only then in the ascending one of their first and then last names, the following statement should be run instead:

```
SELECT FirstName, LastName FROM COMPOSERS
WHERE BirthYear <= 1750 AND PAssedAwayYear >= 1685
   AND x <> 2
ORDER BY PAssedAwayYear DESC, FirstName,
   LastName;
```

18. *Subqueries* are SELECT statements embedded (syntactically, between parenthesis) into other SELECT, INSERT, UPDATE, and DELETE statements (see, for example, c. and d. above); they may not contain the ORDER BY clause. There are four types of subquery usages:

- (*SELECTstatement*)
- *comparison* [ANY | ALL | SOME] (*SELECTstatement*)
- *expression* [NOT] IN (*SELECTstatement*)
- [NOT] EXISTS (*SELECTstatement*)

    where *SELECTstatement* is a subquery, *comparison* is an expression followed by a comparison operator (comparing the expression result to the subquery one) and *expression* is an expression whose values are to be searched among the ones of the corresponding subquery evaluation result.

19. First above usage type is allowed only in SELECT (provided their result is only one value), FROM and GROUP BY clauses; the other three ones may be used only in WHERE and HAVING clauses. Columns defined by subqueries in the SELECT and FROM clauses may not be used in the host SELECT statement GROUP BY clause. Subqueries used in FROM clauses must be named by using the AS operator (in order to use them in ON subclauses, as well as to unambiguously refer to their computed columns).

20. When subqueries of the second and third type above are returning multicolumns row sets, the corresponding expressions should contain same arity and data type rows (syntactically enclosed in parenthesis and comma separated).

21. If the second type above is used without predicates (ANY, ALL or SOME) –that is only with =, <>, <, <=, >, or >= –, then the subquery should return only one value; if a predicate is used too, then the subquery may return any number of columns and rows.

22. ANY and SOME are both equivalent to the existential quantifier, whereas ALL is equivalent to the universal quantifier. Note that < ANY means "less than maximum"; > ANY, "more than minimum"; < ALL, "less than minimum"; and > ALL, "more than maximum".

23. IN stands for "included in" (also expressible as = ANY), whereas NOT IN for "not included in" (equivalent to <> ANY).

24. [NOT] EXISTS are comparisons of the subquery evaluation result to the empty set.

25. Subqueries using synonyms of table instances from the host query are *dynamic* and need reevaluation for each row in the host query instance. The rest are *static* and need evaluation only once, before evaluating the host query.

26. Unions may be computed by using the UNION operator that can merge both table instances and/or query results. Note that for unions, on one hand, result column names are given by their first operand, while ordering can only be done at the last one. Also note

that, by default, UNION eliminates duplicates; you can ask for preserving them by using UNION ALL instead.

27. Not all query results are updatable: for example, those obtained through DISTINCT, GROUP BY, and UNION are not.

28. One might also ask queries whose answer cannot be computed only based on a db because either there is no such stored data or queries are incorrectly or ambiguously formulated.

For example, considering Fig. 3.2, the answer to the query "Which is the tonality in which Mozart wrote his Requiem?" (SELECT Tonality FROM MUSICAL_WORKS WHERE Composer = 4 and Title = 'Requiem';) is the empty set (to be interpreted as "There is no data on a work by Mozart called Requiem"); the answer to the query "In which year was first publicly performed Brahms' 2nd piano concerto?" (SELECT FirstPublicPerformanceYear FROM MUSICAL_WORKS WHERE Composer = 8 and Title = 'The 2nd piano concerto';) is a syntactic error (to be interpreted as "this db does not store musical works first public performance year"); the answer to the query "Who are the composers whose birth year is greater than their last name?" (SELECT FirstName, LastName FROM COMPOSERS WHERE BirthYear > LastName;) is a semantic error (to be interpreted as "there is no meaning in comparing birth years, which are naturals, with last names, which are text strings").

Note that the following two subsections closing 3.9.2, devoted to relational calculus and algebra, may be skipped by all those not interested in the theoretical foundations of querying. However, especially the second one (on the relational algebra) could be interesting in order to understand how the SELECT *SQL* statement is internally evaluated by RDBMSes. Moreover, this second subsection is also providing intuition on the math formalization (presented in section 3.12) of the GROUP BY clause.

3.9.2.1   Relational Calculus

Programming is one of the most difficult branches of applied mathematics;
the poorer mathematicians had better remain pure mathematicians.

—Edsger W. Dijkstra

SQL is based on *relational calculus* (*RC*): a family of languages based on first-order predicate calculus. As their operands are tables, which are bi-dimensional objects, there are two such subfamilies: *tuple RC* (*TRC*), from which *SQL* stems[64], whose variables are rows (tuples), and *domain RC* (*DRC*), from which *Query-by-Example* (*QBE*)[65] stems, whose variables are columns (domains).

The first RC major simplification as compared to first order logic is lack of mapping symbols, except for the standard Boolean ones and a handful of so-called *aggregate* ones (whose set depend on the DBMSs implementations, but generally include COUNT, for cardinal, SUM, MAX, MIN, and AVG, for arithmetic average). Moreover, except for standard math operators (as, for example, the comparison ones: $=, \neq, <, \leq, >, \geq$), all other predicate symbols (those that in 1OL correspond to math relations) are interpreted as db table instances.

Second major simplification is that there are no propositions[66].

Third difference, a minor syntactic one, lies in notations (which does not affect at all semantics) that are slightly modified because RDM, as we have already seen, replaces for relations' underlying sets the math positional (not necessarily distinct) name lists with unordered collections of unique names ones.

Being subsets of 1OL, RC languages only contain declarative (non imperative), nonprocedural expressions, stating results' properties (without any indication whatsoever on how these results are to be computed). This is their most important source of power, compactness, and elegance: for example, a *SQL* SELECT statement of a dozen of lines might be equivalent to a *Java* method of some 120 statements or, sometimes, even much more.

---

[64] *SQL* is the most frequently used, but there are other equivalent languages too as, for example, *Quel*, *ISBL*, *UnQL*, *YQL*, as well as subsets and/or extensions of them designed and used for particular domains, as *DMX* (for data mining), *MQL* and *SMARTS* (for chemistry), *MDX* (for *OLAP* dbs), *TMQL* (for topic maps), *XQuery* (for *XML* files), etc.

[65] *QBE* is the only successful *DRC* language, which was imitated by nearly all RDBMSs (e.g., MS calls it *Query Design* mode in *Access* and *Query Designer* in *SQL Server*, pretending that they are *Visual QBEs*). Its success is due to its graphical nature, simplicity, and intuitiveness.

[66] In dbs, constraints are propositions, while queries are open formulas. The db instance subsets for which queries are *true* are considered to be their interpretation (meaning).

CRT and CRD are equivalent as expressive powers[67], with CRT formulas being slightly simpler. This allows, for example, translating *QBE* in *SQL* (and vice-versa too).

3.9.2.2   Relational Algebra

> *Arithmetic! Algebra! Geometry! Grandiose trinity! Luminous triangle!*
> *Whoever has not know you is senseless.*
> —Comte de Lautréamont

The *relational algebra* (*RA*) is an extension of the algebra of sets (*SA*) with five fundamental and several derived operators; RA operands are table instances.

Recall that fundamental algebra of sets operators are the *union*, *intersection*, *complementation*, and *Cartesian product* (denoted $\cup$, $\cap$, $^\complement$, $\times$, respectively). Derived ones include, for example, *difference* (or *relative complement*: \, —) and *direct sum* ($\oplus$). Moreover, besides *equality*, *inclusions* ($\subseteq$, $\subset$, $\supset$, $\supseteq$, $\not\subset$) play a significant role.

Fundamental RA proper initial (i.e., added from the first RDM specification) operators are selection, projection and renaming.

Selection is a "horizontal splitting" operator: it is selecting from a RA expression *E* instance only those rows that satisfy a logic formula *F* (notation: $\sigma_F(E)$). *SQL* implements it in its clauses `WHERE` and `HAVING`. Trivially, if all *E*'s rows satisfy *F*, then no selection is actually performed, the result being equal to *E*'s instance; dually, if no *E*'s row is satisfying *F*, then the result of the selection is the empty instance (of the *E*'s scheme).

Note that selection compositions may be consolidated into only one selection, by using the logical *and* operator: $\sigma_F(\sigma_G(E)) = \sigma_{F \wedge G}(E)$.

Projection is a dual "vertical splitting" operator: it is retaining from a RA expression *E* corresponding instance only desired explicitly stated columns included in its not empty selection set (also called *target list*) *S* (notation: $\pi_S(E)$). *SQL* implements it in its clause `SELECT`. Trivially, if *S* = *E*, no projection is actually performed: $\pi_E(E) = E$; in order to simplify writing in this case (i.e., not being obliged to mention all *E*'s columns

---

[67] In the rdbs querying context, a language *expressive power* means the class of queries that the language can express.

C_1, ..., C_n if we want them in this db order), *SQL* offers a corresponding abbreviation meta-character, generally '*': instead of SELECT C_1, ..., C_n FROM E, we can simply write SELECT * FROM E.[68] Dually, if $S = \emptyset$ (which is only theoretically possible: RDBMSs only accept target lists with at least one column), then the result is the empty set (both as scheme and instance).

Note that, theoretically, π affects not only E's scheme (by dropping some of its columns), but also its instance, by dropping, after projection, all duplicates (as instances should be sets). Obviously, if E contains a key, then no duplicates are possible. In order for users not to lose data by lack of knowledge or enough care, in fact, RDBMSs are not automatically removing duplicates after projection: it is to the user to explicitly ask for it by specifying the predicates DISTINCT or DISTINCTROW in the SELECT clause (see note 3 in Subsection 3.9.2).

Also note that projection compositions may be consolidated into only one projection, by considering only the last target list: $p_{S2}(p_{S1}(E)) = p_{S2}(E)$.

Finally, as for some queries we might need either several instances of a same expression, or to name calculated columns or subquery results, or to link dynamic subqueries with their host queries, or to rename result columns, RA also includes a *renaming* operator, generally denoted by $\rho_{B \leftarrow A}(E)$, meaning that all occurrences of A in E are renamed as B. *SQL* implements renaming with the *AS* operator (that can be used both in SELECT, FROM, WHERE and HAVING clauses) and refers to new names as *aliases* of the old corresponding ones (which might also be implicitly generated by the system). As renaming is associative, $\rho_{B1 \leftarrow A1}(...\rho_{Bn \leftarrow An}(E)...)$ might be simply written as $\rho_{B1, ..., Bn \leftarrow A1, ..., An}(E)$. Note that *SQL* programmers are also declaring corresponding *AS* defined aliases simply for using shorter table and/or column names (than the dbs ones).

Note that projection and renaming are performed by RDBMSs in virtual no time, in internal memory, together with selections (that are anyhow taking, in average, only some 5% of the total evaluation time of typical queries, especially when indexes on the involved columns are present— see the 2nd volume of this book).

---

[68] Unfortunately, perhaps the most frequently used statement in *MySQL* (be it under *PHP* or whatever other high level programming platform), as its users do not generally care of projecting desired expressions on only actually needed columns.

Two other proper relational fundamental operators have been added lately (when *SQL* was designed): ordering and grouping. Note that as their notation is not "standardized", we will use in what follows the *SQL* corresponding ones.

Users are generally not happy at all with unordered results, so that an *ORDER BY* operator has been added to RA as well, implemented by the *SQL* ORDER BY clause (see note 17 above).

Note that ordering is generally an expensive operator (even in the presence of appropriate indexes): consequently, *SQL* does not allow, for example, ordering of subqueries or union operators results; for us, as programmers, it is also important to keep in mind to almost always order results (i.e., to left them unordered only when ordering is not a must), but only as much as needed (i.e., not also ordering on columns that need not to be ordered).

Finally, as very often aggregate functions need to be applied for partitions rather than on the whole instance of an RA expression, the *kernel* and *quotient set* operators have been considered too (and implemented by the GROUP BY *SQL* clause—see notes 12 to 14 above), in order to compute desired partitions.

Note that SELECT S FROM E WHERE F GROUP BY C_1, ..., C_n HAVING G ORDER BY L is first computing *E*'s scheme and instance, then applies *F*, then partitions this subinstance into the equivalence classes of the equivalence relation $ker(C_1 \cdot ... \cdot C_n)$[69], applies *G* on this quotient set[70], and, finally, orders the result according to *L*, so that it is translated into the following equivalent RA expression:

$$ORDER\_BY_L(\pi_S(\sigma_G(\sigma_F(E)/_{ker(C1 \cdot ... \cdot Cn)}))).$$

For example, the *SQL* statement from 16 above:

```
SELECT FirstName, LastName FROM COMPOSERS WHERE x IN
  (SELECT Composer FROM MUSICAL_WORKS
  GROUP BY Composer, Tonality
  HAVING Tonality = 3)
ORDER BY FirstName, LastName;
```

---

[69] *ker* is the *kernel* (*nucleus*) operator on mappings (see Appendix), which places in a same equivalence class all elements of the operand mapping's domain for which the mapping computes a same value.
[70] A *quotient set* is the set of partitions generated by an equivalence relation.

is automatically translated by RDBMSs into the following equivalent RA expressions (the first one evaluating the subquery and the second one evaluating the final result):

$$sq: \pi_{Composer}(\sigma_{Tonality\,=\,3}(MUSICAL\_WORKS/_{ker(Composer\,\bullet\,Tonality)}));$$
$$ORDER\_BY_{FirstName,\,LastName}(\pi_{FirstName,\,LastName}(\sigma_{x\,\in\,sq}(COMPOSERS))).$$

Generally, SELECT S FROM E WHERE F GROUP BY G HAVING H is translated by RDBMSs into the RA expression $\pi_S(\sigma_H(\sigma_F(E)/_{ker(G)}))$.

Note that the grouping operator is also an expensive one (even in the presence of appropriate indexes): consequently, we should not ever group on columns that are not needed and we should always filter out all possible rows before grouping, by fully using the power of the WHERE clauses.

As we've seen in notes 5 to 11 above, a family of seven *join* derived RA operators are included too: the *inner* ones because not all RDBMSs have optimizers able to replace composition of selections and Cartesian products by corresponding inner joins (which are then evaluated much faster, depending on the context, either by processing them in main memory—for small instances—or by exploiting existing indexes, or even by building temporary ones, etc.) and the *outer* ones for simplifying both *SQL* programming and internal evaluations.

For example, the a. *SQL* statement above:

```
SELECT COMPOSERS.FirstName, COMPOSERS.LastName,
    C.LastName AS otherLastName
FROM COMPOSERS INNER JOIN COMPOSERS C
    ON COMPOSERS.FirstName = C.FirstName AND
    COMPOSERS.LastName <> C.LastName
ORDER BY COMPOSERS.FirstName, COMPOSERS.LastName,
    C.LastName;
```

is translated by any RDBMS into the following equivalent RA expression (note that, generally, inner joins are denoted in RA by the "bowtie" sign $|\!><\!|$):

$$ORDER\_BY_{FirstName,\,LastName,\,otherLastName}(\pi_{FirstName,\,LastName,\,otherLastName}(\rho_{FirstName,\,LastName,\,otherLastName\,\leftarrow\,COMPOSERS.FirstName,\,COMPOSERS.LastName,\,C.LastName}(COMPOSERS$$
$$|\!><\!|_{COMPOSERS.FirstName\,=\,C.FirstName\,\wedge\,COMPOSERS.LastName\,\neq\,C.LastName}\rho_{C\,\leftarrow\,COMPOSERS}(COMPOSERS)))).$$

However, the *SQL* statement from 7 above:

```
SELECT COMPOSERS.FirstName, COMPOSERS.LastName,
  C.LastName AS otherLastName
FROM COMPOSERS, COMPOSERS C
WHERE COMPOSERS.FirstName = C.FirstName AND
  COMPOSERS.LastName <> C.LastName
ORDER BY COMPOSERS.FirstName,
  COMPOSERS.LastName, C.LastName;
```

is translated by any RDBMS having an optimizer that translates $\sigma_F(E_1 \times E_2)$ into the equivalent $E_1 \bowtie|_F E_2$ exactly like above, while those RDBMS not having such a facility are translated it into the following equivalent RA expression (which is taking much more time and space to evaluate, both of them increasing to the square of increases in the *COMPOSERS* instance cardinal):

$$\text{ORDER\_BY}_{FirstName, LastName, otherLastName}(\pi_{FirstName, LastName, otherLastName}(\rho_{FirstName,}$$
$$_{LastName, otherLastName \leftarrow COMPOSERS.FirstName, COMPOSERS.LastName, C.LastName}(\sigma_{COMPOSERS.}$$
$$_{FirstName = C.FirstName \wedge COMPOSERS.LastName \neq C.LastName}(COMPOSERS \times \rho_{C \leftarrow}$$
$$_{COMPOSERS}(COMPOSERS)))).$$

If $F = E_1.A = E_2.A$, then the join is called *natural* (and F may be omitted); in our opinion, you should always avoid it at least from the following two reasons:

✓ the most important is that this unfortunate practice of naming identically two join columns from two tables is firstly obscuring semantics of such foreign keys; for example, if $E_1.A = CITIES.CityID$ and $E_2.A = PEOPLE.CityID$ how could anyone guess what stores *PEOPLE.CityID*? Are they birth places or home towns or etc.? Secondly, when two foreign keys both referencing same table are needed in a same table, only one join may remain natural and semantic confusion increases; for example, if in the above example both birth places and home towns have to be stored for people, which one of *PEOPLE.CityID* and *PEOPLE.CityID1* is storing what? Isn't it much better to ignore natural joins and call them instead *PEOPLE.BirthPlace* and *PEOPLE.HomeTown*?

✓ moreover, if two tables have more than one pair of identically named columns, most *QBE* implementations will automatically generate natural joins for all of them: if you don't actually need all of them and if you do not notice or forget do delete the unwanted ones, your query will be wrong; for example, if E_1 and E_2 both have columns A and B, then the automatically generated join condition is $F = E_1.A = E_2.A \wedge E_1.B = E_2.B$.

If $F = E_1.A = E_2.B$, then the join is called *equijoin*; trivially, natural joins are particular cases of equijoins (when $B = A$).

If $F = E_1.A \ \theta \ E_2.B$, where $\theta \in \{\neq, <, \leq, >, \geq\}$, then the join is called *θ-join*.

All RDBMSs are conceptually[71] translating the *SQL* left join statement from 9 above:

```
SELECT FirstName, LastName, Opus, No,
    TONALITIES.Tonality, Title
FROM (COMPOSERS INNER JOIN MUSICAL_WORKS
    ON COMPOSERS.x = MUSICAL_WORKS.Composer) LEFT JOIN
    TONALITIES ON MUSICAL_WORKS.Tonality = TONALITIES.x
ORDER BY FirstName, LastName, TONALITIES.Tonality,
    Opus, No, Title;
```

into the following equivalent RA expression:

$$\text{ORDER\_BY}_{FirstName, LastName, TONALITIES.Tonality, Opus, No, Title} (\pi_{FirstName, LastName, TONALITIES.Tonality, Opus, No, Title}((COMPOSERS \bowtie_{COMPOSERS.x = MUSICAL\_WORKS.Composer} MUSICAL\_WORKS) \bowtie_{TONALITIES.x = MUSICAL\_WORKS.Tonality} TONALITIES) \cup$$
$$\pi_{FirstName, LastName, \tau Tonality \leftarrow NULL, Opus, No, Title} (\sigma_{Tonality \notin Im(TONALITIES.x)} (COMPOSERS \bowtie_{COMPOSERS.x = MUSICAL\_WORKS.Composer} MUSICAL\_WORKS))).$$

Note that many books and even research papers are wrongly presenting inner joins as being fundamental operators, whereas they are in fact derived ones: $E_1 \bowtie_F E_2 = \sigma_F(E_1 \times E_2)$, for any RA expressions E_1, E_2 and any filter F.

A row of an inner join operand which is taking part in the result is called *joinable* (with respect to that join—that is with its filter F). If all rows of both operands are joinable, then the corresponding inner join is called a *complete* one. For example, the join from the a. *SQL* statement above is a complete one, whereas the first *SQL* statement from 9 above is not, as both the fourth row of *MUSICAL_WORKS* (because Don Giovanni does not have an associated tonality) and the sixth one from table *TONAL-ITIES* (as there is no musical work written in G minor) are not joinable.

---

[71] Actually, the best of them will optimize it (as union is the most expensive operator) by embedding the union within the inner join.

There is also a dual notion to the full join, which is also important in higher RDM so-called normal forms: a table is said to have a *loss-less(–join) decomposition* (*LD*) with respect to some of its projections if its instance can be exactly reconstructed from these projections (i.e., if $\pi_{X1}(r) \mathrel{|><|} \ldots \mathrel{|><|} \pi_{Xn}(r) = r$, where r is the instance of a table R and $X_1, \ldots,$ X_n are (concatenations of) its columns).[72]

Joins are dual to projections, as they are "vertical composition" operators (building up tables with columns from two tables, through the embedded Cartesian product).

The nature of this duality is much more profound than graphical appearances; it can be shown, for inner joins, that:
- ✓ the instance of the projection of a join between n RA expressions over the scheme of any of its operands is a subset of that operand's instance;
- ✓ above inclusions are equalities when joins are full;
- ✓ the composed operator defined above is idempotent (i.e., repeatedly applying same projection of a join over the scheme of one of its operands yields same result as applying it only once) and, dually, that:
- ✓ any RA expression instance is a subset of the instance of the join of any of its scheme full projections (i.e., projections whose union is the expression's scheme);
- ✓ for any full projections of a RA expression the composition between the corresponding join and projections is idempotent too.

When F is missing, joins are equal to the corresponding Cartesian products: $E_1 \mathrel{|><|_\varnothing} E_2 = E_1 \times E_2$.

Joins of same type are associative (consequently, they may also be considered as n-ary operators, n natural) and inner as well as the full outer ones are commutative too.

However, composition of several types of joins is sometimes ambiguous, so they are not allowed: for example, $E_1 \mathrel{^{L}|><|_F} E_2 \mathrel{|><|_G} E_3 \mathrel{^{L}|><|_H} E_4$, where $^{L}|><|$ denotes a left join, is ambiguous as it is not associative—$(E_1 \mathrel{^{L}|><|_F} E_2) \mathrel{|><|_G} (E_3 \mathrel{^{L}|><|_H} E_4) \neq E_1 \mathrel{^{L}|><|_F} (E_2 \mathrel{|><|_G} (E_3 \mathrel{^{L}|><|_H} E_4))$.

---

[72] Note that, in this context, "lossless" is a false-friend: lossy joins are not "loosing" stored data, but, on the contrary, computing additional, not stored data.

Also note that joins (even if heavily optimized) are the third more time consuming (after unions and Cartesian products) operators (taking some 95% of the average time for typical queries not having groupings and orderings, and, as approximate orders of magnitude, some 66% of the ones having groupings and orderings, which are spending only some 10% for grouping, 5% for filtering before grouping, 1% after and 18% for ordering). Outer joins (and especially the full ones) are more expensive than inner ones.

As, most frequently, RA expressions only use the select, project, and join operators (and much rarely renaming, Cartesian product, union, etc.) they are also called *SPJ expressions*.

TABLE 3.24   Equivalences between RA and *SQL*

RA	SQL
$R \cup S$	SELECT * FROM R UNION SELECT * FROM S[73]
$R \cap S$	SELECT * FROM R INNER JOIN S[74] or: SELECT * FROM R WHERE R.* IN (SELECT * FROM S)
$R - S$	SELECT R.* FROM R LEFT JOIN S ON R.X = S.Y WHERE S.Y IS NULL or: SELECT * FROM R WHERE R.* NOT IN (SELECT * FROM S)
$R \times S$	SELECT R.*, S.* FROM R, S
$\rho_{y \leftarrow x}(R)$	SELECT x AS y FROM R
$\pi_{x1, ..., xn}(R)$	SELECT $x1$, ..., xn FROM R
$\sigma_F(R)$	SELECT * FROM R WHERE F
$R \bowtie_F S$	SELECT R.*, S.* FROM R INNER JOIN S ON F or: SELECT R.*, S.* FROM R, S WHERE F[75]
$\pi_S(\sigma_H(\sigma_F(E)/$ $_{ker(G)}))$	SELECT S FROM E WHERE F GROUP BY G HAVING H

[73] R and S must have same number of columns, pairwise compatible.

[74] R and S have to be joinable on at least one column.

[75] Beware that, if the corresponding RDBMS does not have a query optimizer able to detect and replace the Cartesian product followed by a selection with a corresponding join, then this alternative solution is very inneficiently computed and may take hours or even days to complete for large instances.

Other derived, less frequently used RA operators are *left* and *right semijoins* (joins projected on their corresponding operands: the left one is $T \mathbin{|\!\!>\!\!<} U = \pi_T(T \mathbin{|\!\!>\!\!<} U)$, whereas the right one is $T \mathbin{>\!\!<\!\!|} U = \pi_U(T \mathbin{|\!\!>\!\!<} U)$) and *left* and *right antijoins* (or *antisemijoins*: $T \mathbin{|\!\!>} U = T—T \mathbin{|\!\!>\!\!<} U$, $T \mathbin{<\!\!|} U = U—T \mathbin{>\!\!<\!\!|} U$), as well as *division* ($T \div U = \pi_{C1, \ldots, Cn}(T)—\pi_{C1, \ldots, Cn}(\pi_{C1, \ldots, Cn}(T) \times U—T)$, where $C_1, \ldots, C_n$ are T's columns which are not joined to U's ones).

Note that RA's expressive power is considered the minimum acceptable standard for any relational querying language: any language that has at least its expressive power is called *complete*. Also note that RA and RCs are incomparable from this viewpoint: while RCs are, generally, much more expressive, they cannot, however, express unions. In order for them to have at least RA's expressive power (and thus becoming comparable) the union operator has to be added to RCs too. Also note that RA expressions can compute all possible table instances only containing data stored in corresponding rdb instances.

Table 3.24 shows the equivalences between RA and *SQL*.

3.9.3 VIEWS, STORED PROCEDURES, FUNCTIONS AND TRIGGERS. TRANSACTIONS AND ACIDITY

> *Make everything as simple as possible, but not simpler.*
> —Albert Einstein

Most RDBMSs provide saving queries (under unique names per db) as *views*, which are considered being computed tables; the rationale is that their parsing, compiling, and optimization is done only once and, each time that their results are needed, they may be referred to in any FROM clause just like an ordinary table. Note that, generally, views do not accept parameters[76] and that their names should be distinct not only from any other view, but also from any other table of the db. Often, hierarchies of views are declared, saved, and used later too.

Similarly, any parameterized sequences of extended *SQL* manipulation statements may be saved too (under unique names per db) as *stored*

---

[76] However, for example, MS *Access* ones (called *queries*) do accept them, as any symbol which is not a constant, a table, or column name is considered as being a parameter.

procedures, *functions*, and *triggers*: stored procedures are the equivalent of subroutines, while triggers are event-driven methods (mainly for enforcing non-relational constraints, but also for any other event-driven tasks, like, for example, generating unique numbers for surrogate keys). Procedures and functions may be consolidated into *packages*, the equivalent of high-level PLs libraries.

Both *DDL* and *DML* statements are executed ACIDly, where *ACID* is the acronym for *A*tomic, *C*onsistent, *I*solated, and *D*urable, that is their execution is reliable because it is:

   ✓ *A*tomic—that is indivisible, "all of nothing": either all concerned data table rows or none are processed;

   ✓ *C*onsistent—that is preserving corresponding db consistency (dually, modifications made to the scheme or instance must not violate any db constraint);

   ✓ *I*solated—that is independent of any other concurrent statements which might be executed simultaneously (more precisely, *serialized*: as if concurrent statements would have been executed serially, in one of the possible *serialization* orders);

   ✓ *D*urable—that is their scheme or instance modifications are saved on nonvolatile memory and remain stored there (even if the system/db crash, power fails, etc. immediately after their execution terminates).

Very often, a sequence of *SQL* statements (be it in-line or stored in a procedure, function or trigger) should also be executed ACIDly: for example, withdrawing money from a bank account in order to credit another one should execute ACIDly. To achieve it, such sequences are embedded into *explicit transactions*: *SQL* code enclosed into a `START (BEGIN) TRAN(SACTION)` and a corresponding `COMMIT` or `ROLLBACK` dedicated *SQL* statements.

Any transaction should end with both a `COMMIT` (for making all transaction modifications durable in case of success) and a `ROLLBACK` (for discarding all transaction modifications in case of failure).

Note that transactions may be nested on any number of levels and that *DDL* statements are *implicit transactions* (i.e., they do not need explicit `COMMIT`).

3.9.4   RELATIONAL LANGUAGES LIMITATIONS. EMBEDDED AND EXTENDED *SQL*

Raise your quality standards as high as you can live with,
avoid wasting your time on routine problems, and always try to
work as closely as possible at the boundary of your abilities.
Do this because it is the only way of discovering how that boundary
should be moved forward.
—Edsger Dijkstra

First relational languages (*RL*s) limitation, a minor one, is the lack of function symbols, except for the aggregate ones. Practically, RDBMSes provide lot of library functions that empower their *SQL*s.

Second RLs limitation, a major one, is their incapacity to compute, for example, *transitive closures* of binary homogenous relations (i.e., for example, the set of all ancestors and/or descendants of somebody, or the set of all files belonging to an OS logical drive, or the set of all employees subordinated, directly or indirectly, to any given employee, etc.)[77].

Consequently, two types of solutions were devised: *embedded* and *extended SQL*.

The vast majority of high-level PLs are embedding *SQL*, that is include at least one statement that accepts as parameters text strings containing *SQL* statements and/or clauses or views and/or stored procedure names and their actual parameters that are passed (generally, through middlewares like *ADO*[78]) to *SQL* engines; if they are queries, their results are then passed back from these engines to the host high-level PL in accordingly structured memory buffers called *data* (or *record*) *sets*.

The dual approach consists of extending *SQL* with high-level PL constructs (variable declarations, *if*, *while*, etc.[79]): IBM *DB2 SQL PL*, *Oracle PL/SQL*, Sybase and MS *T-SQL*, etc. are examples of extended *SQL*s.

---

[77] In fact, RLs limitations are much more severe: they cannot express any unbounded computation (i.e., computations on unbounded data), as they are not *Turing complete* (or *computationally universal*) *languages*.

[78] The acronym for *ActiveX Data Objects*, a MS COM-type middleware between PLs and *OLE DB* (the acronym for *Object Linking and Embedding DataBase*, a MS application programming interface (API) providing uniform manner access to data stored in a variety of formats).

[79] Moreover, in order to also compute transitive closures (more generally, in fact, basic linear and even nonlinear and mutual recursion), *SQL* was extended by powerful RDBMSs with a "WITH RECURSIVE" statement (see first Chapter of the 2nd volume of this book).

Note that, unfortunately, there is no standard extended *SQL*; dually, fortunately, middlewares like *ADO* are generally successful in automatically performing most of the needed translations between different *SQL* idioms.

Explicit embedded *SQL* is executed slower, just like dynamic *SQL* (see next subsection); only embedded calls to views and/or stored procedures are executed almost as fast as corresponding static *SQL* (see next subsection), as *ADO* is very fast, adding almost negligible overhead.

Extended *SQLs* generally do not provide the full power of high-level PLs: there are no graphic user interfaces (*GUI*), networking, web, etc. capabilities. Fortunately, their performances are, generally, the best possible.

Consequently, the majority of well-architectured, designed and developed db (software) applications use a mixture of both embedded and extended *SQL*. A major challenge in this area is to optimally divide their intermediate business logic (*BL*) levels between extended *SQL* and host high-level PL code: the best approach is to maximize application speed (which is not always easy to achieve).

3.9.5   STATIC VERSUS DYNAMIC *SQL*

> *Simplicity is a great virtue but it requires hard work to achieve it and*
> *education to appreciate it.*
> *And to make matters worse: complexity sells better.*
> —Edsger Dijkstra

Static SQL is code created before execution and which is not changing during execution, be it inline or as views and/or parameterized stored procedures.

Dynamic SQL is *SQL* code created dynamically (be it from scratch or using static *SQL* that is modified), during its execution, either inline or in stored procedures, be it in host high-level PLs embedding *SQL* or in extended *SQLs*.

Static *SQL* has no disadvantage and the following speed advantages:

✓   Only at creation/replacement time, its code is parsed and, if ok, then translated into internal RA expressions, which are then optimized, and for which, finally, the system builds and saves best execution plans possible, depending on current instance, indexes and, optionally, on DBA directions too. Note that optimization is done not only locally, for each statement, but globally too, for each views hierarchy, transaction, and stored procedure.

✓ When executed (which is generally very frequent when compared to creation/replacement), there is no need for any of the above steps to be redone: the system just executes directly the corresponding execution plan (possibly parameterized with actual parameter values).

Dynamic *SQL* has no significant advantage, except for increasing developers' productivity, and the following speed disadvantage: for each execution, each dynamic statement has to be parsed, translated into internal RA expressions, which are then optimized, and for which, finally, the system builds (but does not save) and executes best execution plans possible. Optimization may be done only locally, for each statement, but not globally too, either for inline code or for stored procedures.

Any dynamic *SQL* code may be equivalently rewritten statically.

It is true that, sometimes, dynamic *SQL* is easier to develop than its equivalent static one; as such, it could be a temporary solution when deadlines are tight. However, keep in mind that, generally, there are no other more permanent solutions than the temporary ones! Normally, once deadlines are met successfully, dynamic *SQL* should be replaced by equivalent static one.

Unfortunately, there are always new deadlines, almost all the time tighter, plus fatigue, which are not at all encouraging rewriting *SQL* code. This leads to a *PHP+MySQL*-type architecture, which is extremely slow, not using RDBMS's powerful views and static parameterized stored procedure facilities.

Here is an example of equivalent static and dynamic *SQL*, for querying some data for which end-users might ask for any combination of n conditions $c_1, \ldots, c_n, n > 0$ natural; obviously, there are 2^n combinations:

```
CASE c₁ IS NULL AND ... AND cₙ IS NULL:
   SELECT t1 FROM te;
CASE c₁ IS NOT NULL AND c₂ IS NULL AND ... AND cₙ IS NULL:
   SELECT t1 FROM te WHERE c₁;
```

```
...
CASE c₁ IS NOT NULL AND ... AND cₙ IS NOT NULL:
    SELECT tl FROM te WHERE c₁ AND ... AND cₙ;
```

In order to reduce complexity from exponential (2^n) to linear (n), most programmers, unfortunately, write the following type of dynamic embedded *SQL* code:

> sqlStr:= "SELECT tl FROM te";
> where:= *false*;
> *for* i:= 1, *n begin*
> > *if not IsNull(c_i) then begin*
> > > *if not* where *then begin*
> > > > sqlStr:= sqlStr & " WHERE ";
> > > > where:= *true*;
> > >
> > > *end*;
> > > sqlStr:= sqlStr & " AND " & c_i;
> >
> > *end*;
>
> *end*;
> *execute* sqlStr;

The following static mixed extended and embedded solution is equivalent and has same linear complexity:

```
CREATE OR REPLACE PROCEDURE p (
    c₁ VARCHAR DEFAULT NULL, ..., cₙ VARCHAR DEFAULT
NULL) IS
    BEGIN
        SELECT tl FROM te WHERE
            ((c₁ IS NOT NULL AND c₁) OR c₁ IS NULL) AND ... AND
            ((cₙ IS NOT NULL AND cₙ) OR cₙ IS NULL);
    END;
```

This is executed only once, for creating parameterized stored procedure *p*; the only necessary embedded *SQL* statement for calling it is of the type:

> *execute* "p(" & c_1 & " , " & ... & " , " & c_n & ");";

Whenever a condition c_i is null, corresponding WHERE line reduces to "AND true", which is neither applying c_i, nor tampering with the rest of the conditions; whenever condition c_i is not null, corresponding WHERE line reduces to "AND c_i", which is applying c_i, so that exactly same desired functionality is achieved also in only n steps.

3.10   HIGHER AND THE HIGHEST RDM NORMAL FORMS

> *All forms of distorted thinking must be corrected.*
> —John Bradshaw

Readers who are not interested in why db design should not be done in RDM could skip this subsection, except for 3.10.1, without losing anything.

There is a hierarchy of relational normal forms (*NF*s), each of which is essentially trying to eliminate particular cases of the so-called *data manipulation anomalies*: insertion, update, and deletion ones.

3.10.1   DATA MANIPULATION ANOMALIES

> *Experience is the teacher of all things.*
> —Julius Caesar

A table presents an *insert anomaly* if one cannot insert in it a line, except for adding nulls to some of its columns. For example, Table 3.18 has such an anomaly: no customer data may be stored unless that customer places at least one order.

A table presents an *update anomaly* if one should update same data on several rows. For example, Table 3.18 has several such anomalies too: for example, customer having id 1 has different names and its order number 1 has different dates (even for a same order, for both of them!).

A table presents a *delete anomaly* if, by deleting data on some facts, data on some other ones is deleted too. For example, Table 3.18 has such an anomaly too: by deleting its last line, the only one storing data on orders of customer "Some Customer", data on this customer is lost too.

TABLE 3.18   A Table Having an Insertion, a Deletion, and Several Update Anomalies

Customer-ID	Customer-Name	OrderNo	OrderDate	OrderedItems
NAT(4)	ASCII(64)	NAT(4)	[1/1/2010, SysDate()]	Im(PRODUCTS.x)
NOT NULL	NOT NULL	NOT NULL	NOT NULL	NOT NULL
1	Some Cola	1	3/15/2013	1
1	Someother Cola	1	3/16/2013	2
2	Some Customer	2	3/16/2013	1

3.10.2   FUNCTIONAL DEPENDENCIES

> *Nothing is a waste of time if you use the experience wisely.*
> —Auguste Rodin

Among other NFs, the 2nd, the 3rd, and the Boyce-Codd one, situated at the base of the relational NFs hierarchy are based on a constraint called *functional dependency.*

Let A and B be any two columns of a table T; B is said to be *functionally dependent on A* (or, equivalently A is said to *functionally determine B*) if and only if, by definition, to any value of A there is only one corresponding value of B. Symbolically, any such assertion, which is called a *functional dependency (FD)* or, abbreviated, *dependency*, is denoted by $A \rightarrow B$ (or, sometimes, dually: $B \leftarrow A$). A is called the *left hand* and B the *right hand* of $A \rightarrow B$.

For example, in Table 3.6, x and *Capital* are each functionally determining any other column (including themselves): for example, both $x \rightarrow$ *Capital* and *Capital* $\rightarrow x$, as well as *Capital* $\rightarrow$ *Population* hold.

These notions are immediately extendable to column concatenations: let X and Y be any such column products of a table T; Y is said to be *functional* dependent *on X* if for any rows having same values for X, corresponding values for Y are the same.

For example, in Table 3.6, *Country • State* $\rightarrow$ *Population • Area.*

Obviously, $X \to Y$ where $Y \subseteq X$ is trivial, as it holds in all instances of any table T including X and Y, from any rdb[80]. For example, *Country • State $\to$ Country*, *Country • State $\to$ State*, and *Country • State $\to$ Country • State*, are all trivial.

Note that, generally, if $X \bullet Y = T$, then $X \to Y$ holds in T if and only if X is a key of T (so, *key dependencies* are a particular case of the functional ones).

Also note that the implication problem for FDs may be solved linearly.

3.10.3   THE 2ND NORMAL FORM

> *Experience is a good school, but the fees are high.*
> —Heinrich Heine

A FD $X \to A$ is said to be *partial* if X is part of a key and A is non-prime (i.e., it is not part of a key). A table T is in *2nd NF* (*2NF*) if it is in 1NF and does not have any partial functional dependency.

For example, all above tables are in 2NF, while Table 3.18 is in 1NF, but not in 2NF, because of the partial FD *OrderNo $\to$ OrderDate*.

3.10.4   THE 3RD NORMAL FORM

> *Experience is simply the name we give to our mistakes.*
> —Oscar Wilde

Z is said to be *transitively dependent* on X if it functionally depends on X both directly and indirectly, that is there are at least three FDs of the type $X \to Y$, $Y \to Z$, and $X \to Z$, with Y not being unique. Moreover, $X \to Z$ is called *transitive*. A table T is in *3rd NF* (*3NF*) if it is in 2NF and does not have any transitive functional dependency.

For example, all above tables, except for 3.18 are in 3NF, while Table 3.19 is in 2NF, but not in 3NF, because of the transitive FD *Country $\to$*

---

[80] Moreover, trivially, $\emptyset \to \emptyset$ and $X \to \emptyset$ are trivial FD*s* too, completely uninteresting consequently, whereas FD*s* of type $\emptyset \to Y$, which implies that Y is a constant function, should never be asserted: on the contrary, such completely uninteresting constant Ys should be eliminated from all schemes.

TABLE 3.19   An Example of a Table in 2NF, but Not in 3NF

COUNTRIES ($\underline{x}$, *Country*)

$\underline{x}$	*Country*	*Currency*	*CurrencySymbol*
autonumber(3)	ASCII(255)	*Im(CURRENCIES.x)*	ASCII(4)
NOT NULL	NOT NULL	NOT NULL	NOT NULL
1	U.S.A.	1	USD
2	U.K.	2	BP
3	Germany	3	Euro
4	France	3	Euro

CurrencySymbol (as there also hold *Country* → *Currency* and *Currency* → *CurrencySymbol*).

Note that, theoretically, experts and, for example, *Oracle* too considers 3NF as the minimum acceptable form of an rdb.

3.10.5 THE BOYCE-CODD NORMAL FORM

A pessimist sees the difficulty in every opportunity; an optimist sees the opportunity in every difficulty.
—Winston Churchill

A table *T* is in *Boyce-Codd NF* (*BCNF*) (or *3.5NF*) if all of its nontrivial functional dependencies contain a key in their left hands.

For example, all above tables are in BCNF, while Table 3.20 is in 3NF, but not in BCNF, because of the FD *City* → *ZipCode*.

TABLE 3.20   An Example of a Table in 3NF, but Not in BCNF

SMALL_CITIES_ADDRESSES ($\underline{x}$, *ZipCode*)

$\underline{x}$	*City*	*Street*	*ZipCode*
autonumber(8)	*Im(CITIES.x)*	ASCII(128)	ASCII(16)
NOT NULL	NOT NULL	NOT NULL	NOT NULL
1	4	Edouard Vaillant	95870
2	4	Emile Zola	95870
3	4	Henri Barbusse	95870

Any BCNF table is also in 3NF, but BCNF is not always achievable: for example, the set of FDs $\{A \bullet B \to C, C \to B\}$ cannot be represented by a BCNF scheme.

3.10.6 MULTIVALUED DEPENDENCIES AND THE 4TH NORMAL FORM

> *Knowledge rests not upon truth alone, but upon error also.*
> —Carl Jung

Table 3.21 is in BCNF (as both of its FDs, *Student • Course • Club → x* and *x → Student • Course • Club*, have keys as their left hand sides), but exhibits, however, bad design issues.

No FDs are governing this table; unfortunately, two unrelated between them enrollment types (courses and clubs) are each related, however, to students. RDM calls this a *multivalued dependency* (*MVD*) and denotes it by *Student →→ Course • Club*.

Generally, a *multivalued dependency* $X \to\to Y \bullet Z$, where X, Y, and Z are (concatenations of) columns of a table T, is defined as a constraint (of "tuple generating" type) asserting that whenever two rows of T have same values a for X, for which $Y \bullet Z$'s values are $<b, c>$ and $<d, e>$ respectively, then there should also exist other two rows in the same T's instance having a values for X, but $<b, e>$ and $<d, c>$ values for $Y \bullet Z$ (i.e., all combinations of Y and Z values should exist for each common X value). Note that a MVD $X \to\to U$ of T is trivial if $U \subseteq X$ or if $T = X \bullet U$.

A table is said to be in the *4th NF* (*4NF*) if any nontrivial MVD has a superkey as its left-hand side.

Table 3.21 is in BCNF (as it has no FDs), but not in 4NF (as *Student* is not a key); this is why it stores that much redundant data for each student in both *Course* and *Club*.

Note that:

✓ any table in 4NF is in BCNF too;
✓ tables that are not in 4NF have their semantics overloaded with 2 distinct related ones;

TABLE 3.21   An Example of a Table in BCNF, but Not in 4NF

STUDENTS_COURSES_AND_CLUBS_ENROLLMENTS (<u>x</u>, Student • Course • Club)

x	*Student*	*Course*	*Club*
autonumber(8)	Im(STUDENTS.x)	Im(COURSES.x)	Im(CLUBS.x)
NOT NULL	NOT NULL	NOT NULL	NOT NULL
1	1	1	1
2	1	1	2
3	1	2	1
4	1	2	2

✓   FDs are particular cases of MVDs; dually, if $X \rightarrow Y$ holds, then $X \rightarrow\rightarrow Y$ holds too (but the reverse is not true, which means that FDs are stronger than MVDs);

✓   MVDs have the *complementary property*: if $X \rightarrow\rightarrow U$ holds, then $X \rightarrow\rightarrow T - U$ holds too (i.e., MVDs always exist in complementary pairs);

✓   a table in 4NF cannot have more than one pair of complementary nontrivial MVDs;

✓   generally, a table scheme is in BCNF or 4NF, respectively, if any of its FDs or MVDs *d*, respectively, is implied by at least one of its keys;

✓   the implication problem for FDs and MVDs is a hard one.

3.10.7 JOIN DEPENDENCIES, THE 5TH, AND THE PROJECTION-JOIN NORMAL FORMS

> *The words of truth are always paradoxical.*
> —Lao Tzu

In fact, MVDs are only the simplest particular case (for $n = 2$) of a yet weaker constraint type called *join dependency* (*JD*), denoted $|><| [X_1, X_2, ..., X_n]$, where $n > 1$ natural, $X_1, X_2, ..., X_n \subset T$ and $X_1X_2...X_n = T$ (i.e., all

X_i are (concatenations of) columns of a table T and any column of T is belonging to at least one X_i): T is said to *satisfy* JD $|><|$ $[X_1, X_2, ..., X_n]$ if it has a lossless decomposition over $X_1, X_2, ..., X_n$. A JD is trivial if there is an i, $1 \leq i \leq n$, such that $X_i = T$. Note that $X \rightarrow\rightarrow Y$ holds in T if and only if JD $|><|$ $[X \bullet Y, X \bullet (T — Y)]$ holds too (which proves that MVDs are just particular cases of JDs).

A table T is said to be in the *5th (5NF) normal form* if, for all of its non-trivial JDs, each corresponding projection includes a key of the original table. Any T in 5NF is also in 4NF.

A table T is said to be in the *projection-join normal form* (*PJNF*) if all of its nontrivial JDs are implied by the set of its keys (i.e., for any JD $|><|$ $[X_1, X_2, ..., X_n]$, $n > 1$ natural, any X_i, $1 \leq i \leq n$, is a superkey). Any T in PJNF is also in 5NF. Dually, any 3NF table only having single keys is also in PJNF.

A table in 4NF is also in 5NF if it cannot be losslessly decomposed into three or more smaller relations: for example, Table 3.22 is in 4NF, but not in 5NF, as it should satisfy JD $|><|$ $[x \bullet SSN, Job, Skill]$, with neither *Job*, nor *Skill* being keys, whereas all its four projections on $x \bullet$ *SSN* (*EMPLOYEES*), $x \bullet$ *Job* (*EMPLOYEES_JOBS*), $x \bullet$ *Skill* (*EMPLOY-EES_SKILLS*), and *Job* $\bullet$ *Skill* (*JOBS_SKILLS*) are in 5NF.

Note that, generally, all tables that are not in 5NF have their semantics overloaded with n distinct related ones, $n > 2$ natural being the arity of their longest JD.

Also note that the implication problem for JDs is a hard one too.

TABLE 3.22   An Example of a Table in 4NF, but Not in 5NF
EMPLOYEES (<u>*x*</u>, *SSN*)

<u>*x*</u>	*SSN*	*Job*	*Skill*
autonumber(4)	NAT(9)	*Im(JOB_TYPES.x)*	*Im(SKILL_TYPES.x)*
NOT NULL	NOT NULL	NOT NULL	NOT NULL
1	123456789	1	2
2	987654321	2	1

3.10.8 THE DOMAIN-KEY NORMAL FORM

> *Experience teaches only the teachable.*
> —Aldous Huxley

Before being able to define the so-called "ultimate" RDM NF, the *Domain-Key* one (*DKNF*), as it has been shown that any table in DKNF is in 5NF too, we need to investigate interactions between domain and other type of constraints and also formally define compatibility between rows and instances and their associated anomalies.

Trivially, empty domains only allow for corresponding empty instances, which trivially satisfy any FD, MVD, and JD. If the domain of a column C of a table T only has one value, then any FD $X \rightarrow C$ holds, for any (concatenation of) column(s) X of T. Moreover, no JD having n components, $n > 2$ natural, may hold in any table not having at least n values for each of the involved columns. This is why, theoretically, RDM considers infinite domains.

It can be shown that, for any table, if all of its involved columns may take at least two values, then any of its FDs and MVDs is implied only by FDs and MVDs (not by its domain constraints) and that if they may take at least n values, $n > 2$ natural, then any of its JDs with n components is implied only by its keys.

Domain and key constraints are called *primitive constraints*.

By definition, a table T having an associated constraint set CS is said to be in *DKNF* if any constraint c of CS is implied by the primitive constraints belonging to CS^+.

Row t is said to be *compatible* with table T if, together with T's instance, it satisfies T's primitive constraints (i.e., all of its values satisfy corresponding domain constraints and none of them is duplicating values of T's keys).

A table T is said to have an *insertion anomaly* whenever there is a consistent instance of and a compatible row with it, but the union of that instance and that compatible row is not a consistent instance of T; dually, T is said to have a *deletion anomaly* whenever there is a consistent instance and a row of it such that when removing that row its remaining instance is inconsistent.

DKNF is characterized as being *anomaly free* (i.e., *T* is in DKNF if and only if it has no insertion or deletion anomalies).

For example, Table 3.23 (also see Fig. 3.4), with $CS = PCs \cup FDs \cup RI$, where $PCs = \{x \in [0, 10^3]$, *Country* $\in$ ASCII(128), *CountryCode* $\in$ ASCII(8), *Capital* $\in [0, 10^{16}]$, *Population* $\in [10^5, 2 \times 10^9]$, *x* key, *Country* key, *CountryCode* key, *Capital* key} contains its primitive constraints, $FDs = \{x \rightarrow$ *Country* • *CountryCode* • *Capital* • *Population*, *Country* $\rightarrow$ *x* • *CountryCode* • *Capital* • *Population*, *CountryCode* $\rightarrow$ *x* • *Country* • *Capital* • *Population*, and *Capital* $\rightarrow$ *x* • *Country* • *CountryCode* • *Population*} its functional dependencies, and $RI = \{$*Capital* $\subseteq$ *CITIES.x*$\}$ its referential integrity constraint is in 5NF (as there are no MVDs or JDs and all four FDs have keys as their left hand side, so it is in BCNF), but not in DKNF, as its referential integrity constraint is not implied by its primitive ones: for example, tuple <7, "Japan", "JP", 7, 128,000,000> is compatible with this table, but it cannot be added to it, because it would violate this referential integrity constraint, as 7 does not belong to *Im*(*CITIES.x*) (see Fig. 3.4).

Note that most of the published examples of 5NF tables not in DKNF are in 5NF, in fact, only because some FD (or MVD, very rarely JD) is purposely "hidden"[81]—that is, although it is obvious and, consequently, should have been asserted in the first place, it is initially omitted; when added, generally, the corresponding scheme is not even in 2NF. Such "examples" are not only unconvincing at all, but, in fact, counterexamples of how not to give wrong examples: not initially including a constraint in a table's associated constraint set is not contradicting DKNF definition! Consequently, such tables are, in fact, according to their initial constraint set in DKNF too: only after adding the "hidden" constraint they are not in DKNF anymore, but not because they violate its definition, but simply because they are not even in 2/3/4NF.

Unfortunately, counterexamples like the above ones (i.e., the ones obtained by also considering inclusion dependencies associated to foreign keys) are dismissed by some, as they wrongly consider referential integrity as being domain constraints instead: for them, *RI* is always empty and, for example, *Capital* $\subseteq$ *CITIES.x* above is replaced by the

---

[81] Unfortunately, published true counter-examples of tables in 5NF but not in DKNF are very, very rare; a fortunate exception is the one given by *Wikipedia*.

TABLE 3.23   An Example of a Table in 5NF, but Not in DKNF
COUNTRIES (*x*, *Country*, *CountryCode*, *Capital*)

x	*Country*	*CountryCode*	*Capital*	*Population*
autonumber(3)	ASCII(128)	ASCII(8)	*Im(CITIES.x)*	$[10^3, 2*10^9]$
NOT NULL	NOT NULL	NOT NULL		
1	U.S.A.	USA	6	313,000,000
2	U.K.	UK	2	63,200,000
3	Germany	D	3	82,000,000
4	France	F	1	65,650,000
5	Romania	RO	4	19,050,000
6	Canada	C		

"domain constraint" *Capital* $\in$ [1, 6], so Table 3.23 would be in DKNF in their opinion.

Obviously, this is not true, as inclusion dependencies are not particular types of domain constraints (e.g., *Capital* $\in$ [1, 6] may be equivalent to *Capital* $\subseteq$ *CITIES.x* only for some periods of time, but extremely rarely forever): on the contrary, they could rather be thought of as dynamic extensions of the domain constraints, which are static (as both involved mapping images may and generally are changing in time, most of the times even very frequently).

In other words, as we've already seen in Subsection 3.2.4 above, domain constraints for foreign keys are derived from the corresponding referential integrity ones, not vice-versa.

Please also note that DKNF does not mention FDs, MVDs, JDs or any other constraint types, but only domain and key ones, so its associated implication problem is relatively simple to solve, as keys are particular types of FDs and domain constraints do not contribute to implication.

Unfortunately, DKNF neglection of referential integrities too makes it unacceptably vulnerable, as it does not ban dangling pointers. Consequently, we defined a higher NF, the *Domain-Key Referential Integrity* one (*DKRINF*), which is presented in the second chapter of the second volume of this book and which adds to DKNF referential integrities too.

3.11   BEST PRACTICE RULES

The U.S. states that allow for citizens' initiatives tend to have fewer laws and lower taxes that the ones that don't. But the beauty of the system is that it encourages the spread of best practices.
—Daniel Hannan

At the RDM level, to the db axioms presented in section 5.1 bellow correspond the following best practice rules (which are divided into two categories, corresponding to the *DDL* and *DML*, respectively).

3.11.1   RELATIONAL SCHEMAS

Knowledge and error flow from the same mental sources, only success can tell the one from the other.
—Ernst Mach

3.11.1.1   Tables

Einstein's results again turned the tables and now very few philosophers or scientists still think that scientific knowledge is, or can be, proven knowledge.
—Imre Lakatos

R-T-1. (Table creation, update, and drop)
1. For any dynamic fundamental data object set having at least one dynamic property a corresponding table should be added in the corresponding relational scheme. For example, even for a set of configuration data always having only one element, a corresponding table (which will always have only one line and one column) should be added.
2. On the contrary, for (not only small instances) static sets (i.e., sets whose elements should never be updated; for example, people title names, rainbow colors, payment methods, etc.), corresponding static enumerated sets or ranges may be used instead in the corresponding "foreign-like" keys.

3. Sets having only one calendar date property should be relationally implemented as columns (e.g., instead of a table *BILLING_DATES* and possibly several foreign keys referencing it, no table should be added, but the corresponding foreign keys have to be downgraded to normal, date type, *BillingDate* columns).

4. When updating table schemas having nonvoid instances, adding new compulsory columns (or adding NOT NULL constraints to existing ones) will not be accepted by some RDBMS (including *Oracle*); obvious standard solution to this problem is saving corresponding instance into a temporary table, deleting it, adding(/modifying) column, reimporting instance back (also providing values for the altered column whenever needed), and finally dropping the temporary table. Always remember in such cases (and, generally, in all cases having a potential risk of data loss) Murphy's Law "If something may go wrong, it will!": backup involved data before deletion/update and embed corresponding *SQL* statements in a transaction that should commit only when things went ok and rollback otherwise.

R-T-2. (Relational set inclusion implementation)

1. Any first pair <subset, set> of objects (let them be denoted as $R \subseteq S$, and having primary keys #R and #S, respectively) should be relationally implemented with a pair of corresponding tables as follows: #R should also be declared as a foreign key referencing #S (note that #R cannot be declared as an autonumber!); if S is not a subset, then primary key #S should be declared as an autonumber; otherwise, it should be declared only as an integer.

2. Any other subsequent pair <subset, other_set> (let them be denoted as $R \subseteq T$, and having primary keys #R and #T, respectively) should be relationally implemented with a pair of corresponding tables as follows: an one-to-one not null column labeled T should be added to table R as a foreign key referencing #T; if T is not a subset, then primary key #T should be declared as an autonumber; otherwise, it should be declared only as an integer.

R-T-3. (DKRINF—Domain-Key Referential Integrity Normal Form)
Any fundamental db table scheme should be at least in the DKRINF.

Please note that RDM does not provide any mean to achieve DKRINF, but there are algorithmically approaches that translate semantic NF schemas of higher level data models (e.g., (E)MDM) into DKRINF ones (see Algorithm *A*7, in the second volume of this book).

3.11.1.2   Keys

> *Angels are like diamonds. They can't be made, you have to find them.*
> *Each one is unique.*
> —Jaclyn Smith

R-K-1.(Primary keys)
Recall that, by definition, primary keys are unique and do not accept NULLs.

1. Any fundamental table should have a primary key (implementing the corresponding surrogate key).
2. Any primary key should be of integer type (range restricted according to the maximum possible corresponding instances cardinality—see R-DA-12), and except for subsets (see R-T-2) they should be of the autonumber type.
3. By exception, not referenced tables might have concatenated primary keys instead, provided that all of their columns are of type integer. You should never make such exceptions for referenced tables: as soon as such a table is referenced, the former primary key should be downgraded to a non primary (semantic) one, and a sole surrogate primary key should be added to it.
4. Derived/computed tables may have no keys (be them primary or not).

R-K-2.(Semantic keys)

1. Any fundamental table should have all of the corresponding semantic keys. With extremely rare exceptions (see the rabbit cages example in R-DA-15 above), any nonsubset table should consequently have at least one semantic key: any of its lines should differ in at least one non primary key column (note that there is an infinite number of NULLs, all of them distinct!).

2. Only subset tables (see R-T-2 above), poultry/rabbit cage-type tables, and derived/computed ones might not have semantic keys (but they may have such keys too, as well).

3.11.1.3 Foreign Keys

All of life is a foreign country.
—Jack Kerouac

R-FK-1.(Foreign keys)

1. Any foreign key should always reference the primary key of the corresponding table. Hence, it should be a sole integer column: dually, you should never use concatenated columns or non integer columns as foreign keys.
2. Moreover, their definitions should match exactly the ones of the corresponding referenced primary keys (see R-K-1 above).

This rule is not simply about minimum db space, but mainly for processing speed:

➤ numbers are processed by the fastest logic subprocessors, the arithmetic-logic ones (and the smaller the number, the fastest the speed: for not huge numbers, only one simple CPU instruction is needed, for example, in comparing two such numbers for, let's say, a join), whereas:

➤ character strings need the slowest processors (and a loop whose number of steps is directly proportional to the string lengths);

➤ moreover, nearly a thousand natural numbers, each of them between 0 and nearly 4.3 billion, are read from the hard disk (the slowest common storage device) with only one read operation (from a typical index file),

➤ while reading a thousand strings of, let's say, 200 ASCII characters needs 6 such read operations…

3.11.2 *DATA MANIPULATION*

An expert is a person who has made all the mistakes that can be made in a very narrow field.
—Niels Bohr

3.11.2.1 General Rules

You have to learn the rules of the game. And then you have to play better than anyone else.
—Albert Einstein

R-G-0.(Think in sets of elements, not elements of sets)
Whenever you design queries, think in terms of sets of elements, not elements of sets.

For example, by using either extended or embedded *SQL*, you can always process rows of tables one by one, just like in ordinary sequential programming; however, especially as you are expected to always design the fastest possible solutions, do not ever forget that *SQL* is optimized for performing in parallel any (parameterized) data processing tasks on any number of rows and not just one.

Consequently, for example, whenever not compulsory (which is extremely rare), use pure *SQL* and forget about cursors.

R-G-1.(Use parallelism)
Whenever possible, be it for querying and/or updating, use parallel programming. For example:

➢ whenever available, partition large tables instances (e.g., by months/years/etc.) and use parallel queries on several such partitions;

➢ whenever available, use parallel-enabled functions (including pipelined table ones), which allows them to be used safely in slave sessions of parallel *DML* evaluations;

➢ process several queries in parallel by declaring and opening multiple explicit cursors, especially when corresponding data is stored on different disks and the usage environment is a low-concurrency one.

R-G-2.(Minimize I/Os)
Always factorize I/O operations, for keeping them to the minimum possible. For example:

➢ always use only one UPDATE *SQL* statement to update several columns simultaneously, instead of several such statements—one per column; for example, replace:

```
UPDATE T SET A = v;
UPDATE T SET B = w;
```
by:
```
UPDATE T SET A = v, B = w;
```
➢ always update only needed rows; for example, replace:
```
UPDATE T SET A = TRUE WHERE B = w;
```
by:
```
UPDATE T SET A = TRUE
WHERE B = w AND A = FALSE;
```
➢ when selecting data, always select only needed columns, from only needed tables, and eliminate all unneeded rows as soon as possible; for example, if, in the end, you only need male person names, (even if you are for the time being, for example, only a *PHP + MySQL* developer) replace:
```
SELECT * FROM PERSONS;
```
by:
```
SELECT Name FROM PERSONS WHERE Sex = 'M';
```
➢ use cache and, generally, internal memory always when available; for example, small static tables, functions, and queries should be cached whenever they are frequently used.

R-G-3.(Reuse db connections)
1. Avoid continually creating and releasing db connections.
2. In web-based or multi-tiered applications where application servers multiplex user connections, connection pooling should be used to ensure that db connections are not reestablished for each request.
3. When using connection pooling always set a connection wait time-out to prevent frequent attempts to reconnect to the db if there are no connections available in the pool.
4. Provide applications with connect-time failover for high-availability and client load balancing.

R-G-4.(Avoid dynamic SQL)
Whenever possible, avoid dynamic *SQL* and use static one instead; when it is absolutely needed, prefer the native dynamic one and keep dynamicity to the minimum possible.

Besides poor execution speed, note that successfully executing a dynamic *SQL* statement also involves an automatic commit performed by the system, which may have undesired side effects on concurrency and data integrity.

R-G-5.*(Minimize undo segments querying)*
From the application design phase, always keep querying in parallel with updating same data to the absolute minimum possible (in order to minimize undo segments querying).

R-G-6.*(Unit testing)*
Do not ever deliver any query, view, stored procedure, function, or trigger without, at least, completely and thoroughly unit testing it.

All subsequent tests should be run only with nearly perfectly functioning units.

3.11.2.2   Data Modifications

> *Our creation is the modification of relationship.*
> —Rabindranath Tagore

R-M-0.*(Backup)*
Before modifying data values, especially for massive updates and deletes, always backup the corresponding table instances (so that if anything goes wrong you may restore previous values).

Alternatively, you can disable automatic committing and issue explicit COMMITs only after checking that everything went as desired.

Moreover, regular backup (either full or incremental) of any db is crucial to:
  ➢ avoid data loss and
  ➢ minimize db recovery time.

R-M-1.*(Test SELECTs first)*
Before modifying data values by using SELECT subqueries/clauses, especially for massive modifications, firstly test corresponding SELECT statement results, for checking whether or not they correctly compute the set of desired rows (be it for insertion, update, or deletion).

For example, before executing statement `DELETE FROM T WHERE` x `IN (SELECT ...);`, first execute its subquery and make sure that it returns all and only the rows that you want to delete from table T.

R-M-2.(Use rollback transactions)
Even when "nothing can go wrong", embed data modification statements in transactions having not only `COMMIT`s, but `ROLLBACK`s too: not even RDBMSs are perfect, communications may fail, power outages may occur, etc.!

R-M-3.(Caution with the ON DELETE CASCADE "atomic bombs")
Use the `ON DELETE CASCADE` predicate with same caution as when manipulating atomic bombs: it is a potential data "mass-extinction" weapon!

R-M-4.(Rule out dirty nulls)
Whenever available, instruct DBMSs to ban dirty nulls (not null, but unprintable characters); when such a feature is not available, add it to your dbs through triggers or trigger-type methods.

R-M-5.(Immediately purge unneeded old data)
Do not forget that, generally, when executing *SQL* `DELETE` statements, RDBMSs do not physically, but only logically delete (in fact, they mark for deletion) corresponding deleted data: immediately when it becomes obsolete, purge it explicitly.

3.11.2.3   Querying

Sometimes it is not enough to do our best; we must do what is required.
—Winston Churchill

R-Q-0.(Data and processing minimality)
Always use only needed data and operations: for example,
- do not select more data than needed;
- do not use more table instances than needed;

- filter data as soon as possible;
- do not group data when not needed;
- do not order data until the end and never beyond a key.

Moreover,

- decompose your complex queries into the smallest possible sub-queries,
- program each of them only using primary surrogate keys and corresponding foreign keys, and
- get semantic needed data only in the last step.

R-Q-1.(WHERE-HAVING rule)

Always use the WHERE clause to eliminate all unneeded rows before grouping (if any); always use the HAVING clause only for conditions involving application of aggregation functions on groups.

Dually, never place on HAVING a filter that can be placed on WHERE.

R-Q-2.(WHERE rules)

a. In any adjacent conditions pair, the first condition should eliminate more rows than the second one.
b. Do first inner and then outer joins.
c. Join tables in their increasing number of rows order.
d. Do not prevent index usage by applying unneeded functions to the corresponding columns; take also care of implicit conversions.
e. Not only in *Oracle*, if you use filtered Cartesian products instead of explicit joins, always place non join filters before join conditions.

R-Q-3.(Cautiously use static subqueries)

Subqueries allow for much more elegant and close to mathematical logic queries, but, unfortunately, due to lack of adequate RDBMSs optimizations, even the static (uncorrelated) ones are generally less efficient than corresponding equivalent subqueryless queries: consequently, use them cautiously.

R-Q-4.(Avoid dynamic subqueries)

Use dynamic (correlated) subqueries if and only if there is no other equivalent solution or speed is not critical (recall that, unlike for static (uncorrelated) ones, for which subqueries are evaluated only once, before evaluating the corresponding query, the dynamic ones need to be reevaluated for each row in the corresponding query).

R-Q-5.(Always sort intelligently)

Never present unordered results (except for cases when explicitly asked to do so). Always order intelligently, in the best possible meaningful sorting order. Do not use column positions for sorting, but only column names: code maintenance and debugging are much easier when using names.

3.11.2.4   Extended SQL

> *Once you have a firefighter in your family,*
> *your family and the families from his crew become one big extended family.*
> —David Leary

R-E-1.(Use named constants)

On one hand, do not use hardcoded, but named constants: your code becomes much easier to understand and maintain. On the other, whenever possible, do not use constant functions instead of constants: your code becomes slower.

For example, instead of:

if obType = 1 *then ...*   // 1 is a hardcoded constant

first declare, for example, the named integer constant *Orders*:

```
Orders constant int := 1;
```

and then use it:

if obType = Orders *then ...*

Others, unfortunately always use constant functions instead:

```
FUNCTION Get_Orders_ID(b boolean default true)
RETURN int;
```

...

```
FUNCTION Get_Orders_ID(b boolean default true)
RETURN int is
  BEGIN
     return 1;
```

END Get_Orders_ID;

...

if obType = Get_Orders_ID *then* ...

It is true that, for example, unfortunately, in *Oracle*, you can use named constants only in *PL/SQL*, but not also in *SQL* (like, for example, DDL statements, views, triggers, etc.), which needs constant functions instead; however, this is not a reason not to use constants in *PL/SQL*.

3.12   THE MATH BEHIND RDM

> *Difficulties mastered are opportunities won.*
> —Winston Churchill

Definition 3.0 The *scheme* of a rdb is a triple $< S, \mathcal{F}, C >$, where:

➤ S is a finite nonempty collection of atomic sets (i.e., they are not Cartesian products or power sets, etc.) having unique names (across the rdb);

  ✓ $(S, \subseteq)$ is a poset;

  ✓ $S = \mathcal{R} \oplus \mathcal{V} \oplus CS \oplus \{\text{NULLS}, \emptyset\}$, where:

   • $\mathcal{R}$ is a finite nonempty collection of finite (fundamental object) sets called *relations*;

   • $\mathcal{V}$ is a finite nonempty collection of (value) sets called *domains*;

   • CS is a finite collection of finite (object) sets computed from those in $\mathcal{R}$ (by using RA set operators);

   • NULLS is a distinguished countable set whose elements are called *nulls*;

➤ $\mathcal{F} \subseteq (\mathcal{R} \oplus \mathcal{V} \oplus CS \oplus \{\text{NULLS}\})^{\mathcal{R} \oplus CS}$ is a finite nonempty set of functions called *(relation) attributes*;

  ✓ $\forall R \in \mathcal{R} \oplus CS$, let $f_1, ..., f_k, k > 0$ natural, be all $\mathcal{F}$ functions defined on R (note that they all must have distinct names);

   • the product $f_1 \bullet ... \bullet f_k : R \rightarrow codom(f_1) \times ... \times codom(f_k)$ is called a *relation scheme* (of R);

   • its graph is called the corresponding *relation instance (content)* or sometimes, by notational abuse, simply *relation*;

- the union of all relation schemas is called the *db scheme*;
- the union of all relation instances is called the *db instance* (*content*) or sometimes, by notational abuse, simply *database*;

✓ $\mathcal{F} = \mathcal{A} \oplus \mathcal{M} \oplus C\mathcal{F}$, where:

- $\mathcal{A} \subseteq (\mathcal{V} \oplus \{\text{NULLS}\})^{\mathcal{R} \oplus CS}$ is a finite nonempty set of (fundamental) functions;
- $\mathcal{M} \subseteq (\mathcal{R} \oplus CS \oplus \{\text{NULLS}\})^{\mathcal{R} \oplus CS}$ is a finite set of (*structural* fundamental) functions; $\forall f \in \mathcal{M}, f : R \to R'(R, R' \in \mathcal{R} \oplus CS)$, and $\forall R'$ attribute $g : R' \to codom(g)$, there is a mapping $fk = g \circ f : R \to codom(g)$, called a *foreign key* (of R, *referencing* g in R'); trivially, $Im(fk) \subseteq Im(g)$, which in RDM is called a *referential integrity constraint* (or a *typed inclusion dependency*) and is abusively denoted by $fk \subseteq g$; note that both g and fk may be function products too and that R and R' need not be distinct; also note that, unfortunately[82], all functions in $\mathcal{M}$ are implemented in RDM relations by a corresponding foreign key;
- $C\mathcal{F} \subseteq (\mathcal{R} \oplus \mathcal{V} \oplus CS \oplus \{\text{NULLS}\})^{\mathcal{R} \oplus CS}$ is a finite set of computed functions from those of $\mathcal{A} \oplus \mathcal{M}$ (by using function algebra and RA function operators);

➢ C is a finite nonempty set of closed first order logical formulas called *constraints* (or *dependencies*) whose predicates are either standard mathematical or existence ones associated to the sets in $\mathcal{R}$ and whose names must be unique (across the rdb);

✓ $C = \mathcal{D} \oplus \mathcal{E} \oplus \mathcal{K} \oplus \mathcal{T}$, where:

- $\mathcal{D}$ is a finite nonempty set of *domain constraints* of the type $codom(f) = V \in \mathcal{V} \oplus \{\text{NULLS}\}, \forall f \in \mathcal{F}$, generally denoted $f \subseteq V$; note that, unfortunately, domain constraints have to be explicitly asserted for foreign keys too (although they could have been automatically inferred from the corresponding referential integrity constraints) and, even worse, they may be oversized (e.g., $codom(fk) \supset codom(g)$, see foreign keys definition above); also note

---

[82] Unfortunately especially as g needs not be one-to-one, which would have always guaranteed that fk is not only existing, but is also unique.

that, unfortunately, V needs not to be finite[83], which is not guaranteeing data plausibility;

- $\mathcal{E}$ is a finite set of *existence constraints* denoted $f \mid\!\!- g$ (where f and $g \in \mathcal{F}$, are distinct of each other and have a same domain R), which means: $(\forall x \in R)(f(x) \notin \text{NULLS} \Rightarrow g(x) \notin \text{NULLS})$. Note that, in fact, no RDBMS is providing such constraint type, but only its particularization $\emptyset \mid\!\!- f$, meaning $codom(f) \cap \text{NULLS} = \emptyset$, which are called *NOT NULL* (or *required values*) *constraints*. Also note that, unfortunately, both RDM and RDBMSs allow for empty sets of NOT NULL constraints, thus allowing for empty rows in tables, which are, trivially, senseless.

- $\mathcal{K}$ is a finite set of *key constraints*, denoted generally $f\,key$, $f \in \mathcal{F}$, and meaning that f is minimally one-to-one. Note that f may be a function product and that, unfortunately, this set may also be empty (both in RDM and RDBMSs), thus allowing for table instances containing from redundant and confusing duplicates to only a same row duplicated billions of times, and that (for function products), although RDM makes the difference between one-to-one (called *superkeys*) and minimally one-to-one (*keys*) ones, RDBMSs do not: you can freely declare any combination of keys and superkeys (which is trivially absurd, as, on one hand, they could easily prevent superkeys declarations and, on the other, they are losing both disk space and data manipulation speed with every enforced superkey). A mapping which is either one-to-one or is a member of at least one minimally one-to-one mapping product is said to be *prime* (and otherwise *nonprime*).

- $\mathcal{T}$ is a finite set of *tuple* (or *check*) *constraints*, for which there is neither a RDM commonly agreed precise definition, nor a RDBMS industry standard. Generally, they are restricted

---

[83] Even if, in actual RDBMS implementations, all value sets (generally called, just like in computer programming, *data types*) are finite, they are too large: for example, *Oracle* allows calendar dates starting with 4712 BC and up to 9999, while *MS Access*, between years 100 and 9999; if you only declare that employees birth date is a calendar date (which is, unfortunately, not only possible, but the *de facto* standard practice), then end-users might store data about employees being born some nearly 7,000 years ago or to be born in some 7,000 years from now on!

forms of first order logic propositions, denoted as zero-th order logic ones: they only mention functions defined on a same relation and use only one universally quantified variable (which allows for discarding it: for example, instead of $((\forall x \in PEOPLE)(BirthDay(x) \leq DeathDay(x)))$, $BirthDay \leq DeathDay$ is written instead). Note that, except for referential integrities, there are no other interrelational, but only intrarelational constraints (which is natural for the domain, existence, and key types ones, but constitutes a severe limitation for the tuple ones: for example, $BirthDay \leq MarriageDay \leq DivorceDay \leq DeathDay$, that is, for example, $(\forall x \in PEOPLE)$ $(\forall z \in Marriages)$ $(Husband(z) = x \Rightarrow BirthDay(x) \leq MarriageDay(z) \leq DivorceDay(z) \leq DeathDay(x))$, is not assertable either in RDM or in RDBMSs, as it needs two universally quantified variables and functions defined on two sets).

Definition 3.1 (*First Normal Form*) A relation scheme for which all function members are atomic, that is they are not either function products or taking values from Cartesian products or power sets, etc., is said to be in the *First Normal Form* (*1NF*). An rdb scheme whose relations are all in 1NF is in 1NF too.

Definition 3.2 (*function product chains*) A *function product chain* (or simply *chain*) of length n, $n > 1$, natural, made up of subsets of a function product $\Pi^{n}_{j=1}S_j$ is a sequence $M_1 \subset M_2 \subset \ldots \subset M_{n-1} \subset \Pi^{n}_{j=1}S_j$ such that $|M_i| = i$, for $i = 1, 2, \ldots, n - 1$.

Note that, trivially, the number of chains of length n that can be constructed in $\Pi^{n}_{j=1}S_j$, $n > 0$, natural, is $n!$

Proposition 3.0 (*Maximum number of keys*) For any relation scheme $\hat{o} = f_1 \bullet \ldots \bullet f_n : O \rightarrow codom(f_1) \times \ldots \times codom(f_n)$, $n > 0$ natural, $\forall O \in \mathcal{R} \oplus CS$, the maximum number of keys is $C(n, [n/2])$.

Proof:

Let $X \subset \hat{o}$, $|X| = r$, $r < n$; the number of chains of length n that include X and have the form $M_1 \subset \ldots \subset M_{r-1} \subset X \subset M_{r+1} \subset \ldots \subset \hat{o}$ is equal to $r!(n-r)!$.

Let $1 \leq i, j \leq n$, $i \neq j$; by definition, X_i and X_j might simultaneously be keys *if* $X_i \not\subseteq X_j$ and $X_j \not\subseteq X_i$. Obviously, if X_i and X_j are not $\subseteq$-comparable, then any chain containing X_i is different of any chain containing X_j.

Denoting by p the number of possible simultaneous keys and by $|M_i| = n_i$,

it follows that $\displaystyle\sum_{i=1}^{p} n_i!(n - n_i)! \leq n!$; as $\displaystyle\max_{1 \leq n_i \leq n} (C(n, n_i)) = C(n, [n/2])$, this

implies that: $\displaystyle\frac{p}{C_n^{\left[\frac{n}{2}\right]}} \leq \sum_{i=1}^{p} \frac{1}{C_n^{n_i}} \leq 1 \Rightarrow \max(p) \leq C(n, [n/2])$.

Conversely, by considering ô subsets family having $m = [n/2]$ elements, it can be easily proved that $\max(p) \geq C(n, [n/2])$. Consequently, it follows that $\max(p) = C(n, [n/2])$.   *Q.E.D.*

Proposition 3.1 (*Algorithm A7/8–3 characterization*) A7/8–3 has the following properties:

(*i*) its complexity is $O(2^m)$;

(*ii*) it is complete (i.e., it is generating all possible keys);

(*iii*) it is sound (i.e., it is generating and outputting only possible keys);

(*iv*) it is relatively optimal (i.e., it generates the minimum possible number of questions relative to its strategy).

Proof:

(*i*) (*Complexity*) Obviously, as it has three embedded loops, all of them finite (hence, it never loops infinitely) and the first one may take at most $2^m - 1$ steps (as all superkeys may have been wrongly declared as keys), the second one may take at most $C(m, [m/2])$ steps (as the maximum possible number of wrong keys may have been previously declared), while the third one may take at most $2^m - 1 - m$ steps (when no keys are found). Consequently, in the worst case scenario, the total number of steps is $2^m - 1 + C(m, [m/2]) + 2^m - 1 - m = 2^{m+1} + C(m, [m/2]) - m - 2$.

(*ii*) (*Completeness*) All possible keys are considered, as the algorithm is generating in its third loop all $2^m - 1$ non-void combinations of any set of mappings having cardinal m or all of the maximum $C(m, [m/2])$ possible simultaneous keys.

(*iii*) (*Soundness*) Moreover, as each one of them is first checked to ensure that it is not a superkey and only then db designer is asked whether it is a key or not, it follows that the algorithm is computing only possible remaining keys.

(iv) (*Optimality*) In all three loops, each key is eliminating the maximum possible number of further questions to ask db designers and no possible key is processed more than once.   Q.E.D.

Proposition 3.2 (*Key Propagation Principle formalization*) Let R and R' be any (object) sets (not necessarily distinct), $f : R \rightarrow R'$ any function between them and $a : R' \rightarrow A$ be any one-to-one function or function (Cartesian) product of R';

(i) $\exists! fk : R \rightarrow A$ such that $fk = a \circ f$ (with fk called a *foreign key* of R referencing a of R' and being considered as a key "propagated" from R' to R):

$$R \xrightarrow{\ f\ } R'$$
$$\exists! fk \searrow \circlearrowleft \nearrow a$$
$$A$$

(ii) $Im(fk) \subseteq Im(a)$ (called the associated *referential integrity* constraint)
(iii) $card(Im(fk)) = card(Im(f))$
(iv) if f is one-to-one, then fk is one-to-one too.

Proof:
(i) Let $fk : R \rightarrow A$ such that, $\forall x \in R, fk(x) = a(f(x))$; trivially, fk is totally defined: $\forall x \in R, \exists y \in Im(f) \subseteq R'$ such that $y = f(x)$ and $\exists z \in Im(a) \subseteq A$, such that $z = a(y) = a(f(x)) = fk(x)$; moreover, fk is well-defined: $\forall x, y \in R$ such that $x = y$, according to $a \circ f$ definition, $a(f(x)) = a(f(y)) \Rightarrow fk(x) = fk(y)$.
Obviously, $fk = a \circ f$: $\forall x \in R$, let $y = a(f(x)) \in Im(a) \subseteq A$; according to fk definition, $fk(x) = y \Rightarrow a(f(x)) = fk(x) \Rightarrow a \circ f = fk$.
Moreover, fk is the only function having these properties: let us suppose that $\exists fk' : R \rightarrow A$ such that $a \circ f = fk' \Rightarrow \forall x \in R, a(f(x)) = fk'(x)$; according to fk definition, $fk(x) = a(f(x)) \Rightarrow fk(x) = fk'(x) \Rightarrow fk \equiv fk'$.
(ii) Trivially, $\forall y \in Im(fk), \exists x \in R$ such that $y = fk(x)$; as f is totally defined, it follows that $\exists z \in R'$ such that $z = f(x)$; as a is totally defined, it follows that $\exists u \in Im(a)$ such that $u = a(z) = a(f(x)) = fk(x) = y$.

(*iii*) Trivially, $\forall y \in R' \setminus Im(f)$, $\neg(\exists x \in R$ such that $y = f(x)) \Rightarrow \neg(\exists z \in A$ such that $a(y) = z \in Im(a°f)) \Rightarrow card(Im(fk)) = card(Im(a°f)) \leq card(Im(f))$; conversely, let us suppose that $card(Im(fk)) < card(Im(f))$; it follows that $\exists x, y \in R$ such that $x \neq y$ and $fk(x) = fk(y)$, but $f(x) \neq f(y)$; as a is one-to-one, it follows that $a(f(x)) \neq a(f(y)) \Rightarrow fk(x) \neq fk(y) \Rightarrow \neg(card(Im(fk)) < card(Im(f))) \Rightarrow card(Im(fk)) = card(Im(f))$.

(*iv*) Let f and a be one-to-one and suppose that fk it is not $\Rightarrow \exists x, y \in R$ such that $x \neq y$, but $fk(x) = fk(y)$; as f is one-to-one, $f(x) \neq f(y)$; as a is one-to-one too, $fk(x) = a(f(x)) \neq a(f(y)) = fk(y) \Rightarrow fk(x) \neq fk(y)$.

Q.E.D.

Note that, trivially, even if a is not one-to-one, then fk still exists, it is the only function satisfying 3.2(*i*) above, and its associated referential integrity constraint still holds (as a's one-to-oneness is not needed in their proofs).

Unfortunately, this allows RDM to define foreign keys referencing any columns or column products, be them one-to-one or not.

Obviously, referencing a column or column product which is not one-to-one has two disadvantages (also see Exercise 3.43) and no advantage; let us consider, for example, $a = CityName : CITIES \rightarrow ASCII(256)$, which is not generally one-to-one, as there may be cities having same names even inside a same country; suppose, for example, that $CityName(1) = CityName(2)$; trivially, any foreign key fk referencing $CityName$ has the following two disadvantages:

- $card(Im(fk)) \leq card(Im(f))$, that is fk is not always unambiguously identifying the elements of R' computed by f; for example, if $f = BirthPlace : PEOPLE \rightarrow CITIES$, $BirthPlace(1) = 1$, $BirthPlace(2) = 2$, then, although, in fact, people 1 and 2 were born in different cities, the corresponding foreign key $fk = CityName °BirthPlace$ would mislead us by indicating that both of them were born in a same city (as, in fact, it does not compute unique cities, but their names, which are the same in this case)
- fk does not preserve anymore f's one-to-oneness; for example, if $f = Capital : COUNTRIES \rightarrow CITIES$, which is obviously one-to-one (as no city may simultaneously be a capital of two or more countries), $Capital(1) = 1$, $Capital(2) = 2$, then, although, in fact,

countries 1 and 2 have as capitals different cities, the corresponding foreign key $fk = CityName \circ Capital$ would mislead us by indicating that both of them have as capital a same city (for reasons similar to the above ones).

Consequently, you should never use foreign keys referencing not one-to-one columns or column products; moreover, as all keys of any set are equivalent (see Exercise 3.42(iv)), for minimality reasons (both as needed disk storing space and, especially, querying speed when joining R and R' for finding all $x \in R$ such that either $fk(x) = a(f(x))$ or $fk(x) \neq a(f(x))$)) the best solution is to always reference minimal single keys of R'; trivially, when not erroneously oversized, always the best such choice is the surrogate (primary) key of R'.

Moreover, (ii) above proves that referential integrities are implied by the KPP, so that it is not necessary to assert them explicitly.

Definition 3.3 (Functional dependencies) Let $f : S \to A$ and $g : S \to B$ be any two attributes of S; f is said to *functionally determine* g (or g is said to be *functionally dependent* on f, or that there is a *functional dependency* between f and g, denoted $f \to g$) *iff* $ker(f) \subseteq ker(g)$.

Proposition 3.3 (Functional dependencies characterization) $f \to g \Leftrightarrow \exists! \, fd : Im(f) \to Im(g)$ such that $g = fd \circ f$:

$$
\begin{array}{c}
\exists! \, fd \\
Im(f) \longrightarrow Im(g) \\
f \nwarrow \quad \nearrow g \\
S
\end{array}
$$

Proof:

($\Rightarrow$) Let us define $fd : Im(f) \to Im(g)$ such that $\forall y \in Im(f)$, $fd(y) = z$, where $y = f(x)$ and $z = g(x)$, and firstly prove that fd is well-defined:

(*fd* totally defined) According to mapping image definition, $\forall y \in Im(f)$, $\exists x \in S$, $y = f(x)$; as g is totally defined, $\forall x \in S$, $\exists z \in Im(g)$, $z = g(x) \Rightarrow \forall y \in Im(f)$, $fd(y) = z$.

(*fd* functional) Let t, $u \in Im(f)$, $t = u$; by fd definition, $\forall x, y \in S$ such that $fd(t) = v$, where $t = f(x)$, $v = g(x)$, and $fd(u) = w$, where $u = f(y)$ and $w = g(y)$; $t = u \Rightarrow f(x) = f(y) \Rightarrow (x, y) \in ker(f)$; as $ker(f) \subseteq ker(g)$, it follows that $(x, y) \in ker(g) \Rightarrow g(x) = g(y) \Rightarrow v = w \Rightarrow fd(t) = fd(u)$.

Let us also show that $g = fd \circ f$; $\forall x \in S$, denote $z = g(x) \in Im(g)$ and $y = f(x)$ $\in Im(f)$; according to fd definition, $fd(y) = z \Rightarrow fd(f(x)) = g(x) \Rightarrow fd \circ f = g$.

Finally, we prove that fd is unique: assume that $\exists h : Im(f) \to Im(g)$ such that $h \circ f = g$; it follows that, $\forall x \in Im(f), \exists y \in S, x = f(y)$ such that $h(f(y)) = g(y)$; according to fd definition, $fd(x) = g(y) \Rightarrow fd(x) = h(x) \Rightarrow fd \equiv h$.

($\Leftarrow$) Assume $\neg(f \to g)$; $\neg(ker(f) \subseteq ker(g)) \Rightarrow \exists (x, y) \in ker(f)$ such that $(x, y) \notin ker(g)$, that is $\exists (x, y) \in S^2, f(x) = f(y)$ and $g(x) \neq g(y)$; as fd functional, $f(x) = f(y) \Rightarrow fd(f(x)) = fd(f(y))$, which, by fd definition, implies that $g(x) = g(y)$; this trivially means that the assumption is false and, consequently, $f \to g$.   Q.E.D.

Note that the functional dependency relation is a (partial) preorder (see Exercise 3.2) and that it is neither symmetric, nor antisymmetric.

Proposition 3.4 (*Algorithm A1–7 complexity*) Let the total number of distinct fundamental rectangles (entity-type sets) be e and the corresponding ones for fundamental diamonds (relationship-type sets) be d, for roles (canonical Cartesian projections) be r, for arrows (functional relationships) be f, for ellipsis (attributes) be a, for restrictions (constraints) be c, and for computed sets (views) be v (e, d, r, f, a, c, v naturals); $A1–7$ has complexity $O(e + d + r + f + a + c + v)$.

Proof:

Obviously, $A1–7$ only processes once any rectangle, diamond, role, arrow, ellipsis, and constraint; its main method (Fig. 3.7) has three loops: the first one is executed e times, the third one v times, and the second one is executed $d + r$ times, with each of the $e + d$ times calling method *CreateTable* once and with each of the r times calling method *addForeignKey* once; method *CreateTable* (Fig. 3.8) also calls method *addForeignKey* in a loop for a total number of i times, where i is the total number of canonical inclusions (i natural, $i \leq f$); if there is no surrogate key for the current object set, then *CreateTable* executes a supplementary step for it, in order to automatically add a primary key to the corresponding table; methods *addForeignKey* (Fig. 3.9) and *addColumn* (Fig. 3.11) always perform only one step; finally, method *completeScheme* (Fig.3.10) has six loops: the first one is performed in total a times; the second one for $f - i$ times; the last four ones, together, for c times; trivially, all of the above loops are finite (hence, $A1$–7 never loops infinitely). Consequently, in the worst case scenario (when no surrogate key is declared in the input E-R data model), the total number of steps is $2*(e + d) + r + f + a + c + v$.   Q.E.D.

Proposition 3.5 (*Algorithm REA1–2 complexity*) Let t be the total number of tables and views of S and the corresponding ones for foreign keys be f, for non foreign keys be a, for keys be k, for NOT NULL constraints be n, and for tuple (check) constraints be c (t, f, a, k, n, c naturals); $REA1$–2 has complexity $O(t + 2*a + f + k + n + c)$.

Proof:

Obviously, $REA1$-2 only processes once any table, view, column (be it foreign key or not), and constraint (be it domain, key, NOT NULL, or tuple/check); it has one outer loop, which is executed t times, and which embeds other four ones: in total, the first one is executed a times (for creating the ellipses), the second one f times (for the arrows), the third one k times (for the keys), and the last one $a + n + c$ times (for adding the corresponding domain, NOT NULL, and tuple/check constraints, respectively); trivially, all of the above loops are finite (hence, $REA1$-2 never loops infinitely). Consequently, the total number of steps is $t + 2*a + f + k + n + c$. *Q.E.D.*

3.13   CONCLUSION

> *I think and think for months and years. Ninety-nine times,*
> *the conclusion is false.*
> *The hundredth time I am right.*
> — Albert Einstein

3.13.1   RDM PROS

> *We chose it because we deal with huge amounts of data.*
> *Besides, it sounds really cool.*
> — Larry Page

RDM introduction had as a declared goal guaranteeing data independency from the physical, implementation level and outcoming restrictions of previous data models.

Its most frequently used predecessors, the hierarchical and network data models, include explicit references to implementation characteristics, like pointers, which are almost excluding any potential user who is not a computer programmer. Moreover, these models lack formal definitions, hence precision, consistency, and rigor.

However, RDM's success is not based only on math formalization and the high data independency level from any possible physical implementation, but also on the simplicity of the sole used data structure (the relation), on its natural and expressive representation as a table, and, above all, on the power and elegance of the relational query languages.

RDM is a cornerstone in conceptual data modeling and querying at least for the following four reasons:

1. The vast majority of both commercial and open source existing DBMS are based on it for already some 35 years and even today, with the advent of *NoSQL* ones, RDM remains the best for storing and high speed processing of homogeneous data, even for the fore-seeable future.

2. Any higher-level conceptual homogeneous data model has to include an algorithm for translating its schemas into RDM ones, as the most natural way to store discrete homogeneous data is table-like.

3. The vast majority of data query languages, be them relational or not, are based (externally) on the relational calculus (i.e., are *SQL*-like and/or *QBE*-like) and (internally) on the relational algebra, *SQL* being a de facto db standard.

4. The RDM mathematic theory is still open and fundaments or at least inspires the majority of research endeavors in other data and knowledge models, as well as the general discrete models theory, logic, and even imperative procedural programming.

RDM theory has two major parts: the constraint (dependency) and the query ones. Although it can almost entirely be considered as a (finite) specialization of mathematical logic (with its two parts corresponding to the closed and, respectively, open first order formulae), its roots come also from the seminaïve algebraic theory of sets, relations and functions, as well as from the computational complexity, computer programming, formal languages, etc.

Besides the five basic type ones implemented by nearly all RDBMSs, dependency theory formalized a plethora of integrity constraints, out of which only the most "famous" ones are presented in Section 3.10 (some others are briefly discussed in the last section of this chapter).

Relational languages used by programmers are of logic type, being declarative and nonprocedural; internally, they are translated by RDBMSs into optimized relational algebra expressions. Their main limitation (not being able to compute transitive closures) is overcome by recursive extensions that are presented in the first chapter of the second volume of this book.

RDM is also important because of its very interesting interactions with logic and object-oriented programming, as well as expert systems and AI.

Moreover, I consider that it is a must to include RDM in any db book, even if the literature in this field is that abundant, at least for stating your point of view on its most important results, its limits, and how to overcome them.

Hopefully, besides this point of view, this chapter also provides a rich collection of correct relational solutions for interesting subuniverses of discourse, as well as some of our results in the field (especially the algorithms $A1–7$, $A7/8–3$ and $REA1–2$, the best practice rules, the formalization of the SQL GROUP BY clause with quotient sets computed by the kernel of mapping products equivalence relations, the sections 3.7 –including its RDM and E-R data models of the RDM–, and 3.12 –mainly including the formalization of the KPP and the proof that any table having n columns may have at most $C(n, [n/2])$ keys–, and the subsections 3.2.3.3, 3.2.3.4, 3.2.4.2, 3.2.4.3, and 3.3.2—the relational model of the E-RDM); the *Domain-Key-Referential-Integrity Normal Form* (*DKRINF*) is characterized in the second volume of this book.

3.13.2   RDM CONS

> *Truth is not defined by how many people believe something.*
> *Ask. Question. Think. Decide for yourself.*
> —Jonathan Lockwood Huie

The main disadvantage of RDM is that it is purely syntactic, while, for example, both domain and key constraints, the only two constraint types needed for defining DKNF, its highest normal form, as well as the existence (hence not null) and tuple ones are intrinsically semantic.

Consequently, no algorithms could ever design domain or key or existence or tuple constraints: the best things that may be designed for discovering and fine tuning them are assistance algorithms, like, for example, $A0$ (for domain, existence, and tuple ones) and $A7/8–3$ (for keys).

Moreover, all other RDM constraints are semantic, just like any other possible constraints, in any other context, except for the trivial and implied ones (like, for example, the referential integrities, which are not at all interesting for db design, but, on the contrary, have to be eliminated): for example, no algorithm could ever decide whether or not a column should accept nulls in a given context (mandatory type constraints also being relative: for example, *Capital* may accept nulls in commercial dbs, but not in U.N. or geographical ones) or whether or not net salaries should always be less than or equal to corresponding brut ones.

Even much more convincing is that no algorithm may ever decide that people birth places are cities, or that folders are files, or that "reports to" are employees (or anything else), etc. Same goes for FDs, MVDs, JDs, etc., too.

Second argument is that there are two RDM dual approaches for db design, both essentially unsuccessful: *synthesis* (bottom-up) and *decomposition* (top-down). Both start with a constraint set (generally made up of FDs, so keys too, as a particular case, MVDs, JDs, etc.) and:

  ➤ *synthesis*—with the set of all needed columns (i.e., unary tables or vectors), while
  ➤ *decomposition*—with a single "universal" table, made out of all needed columns,

both of them trying to obtain corresponding table schemas, in the desired NF, preferably also being lossless and *preserving dependencies* (i.e., all initial dependencies being retrievable in the final db scheme, no matter in which table).

Currently, decomposition is the main strategy used for achieving desired NFs. For example, column *CurrencySymbol* from Table 3.19 should be moved to a table *CURRENCIES*; column *ZipCode* from Table

3.20 should be moved to a table *SMALL_CITIES*; Table 3.21 should be decomposed into tables *STUDENTS_COURSES_ENROLLMENTS* having columns *x*, *Student*, *Course* and keys *x* and *Student • Course* and *STUDENTS_CLUBS_ENROLLMENTS* having columns *x*, *Student*, *Club* and keys *x* and *Student • Club*; Table 3.22 should be decomposed into tables *EMPLOYEES* having columns *x*, *SSN* and keys *x* and *SSN*, *JOBS_HISTORY* having columns *x*, *Employee*, *Job* and keys *x* and *Employee • Job*, *EMPLOYEES_SKILLS* having columns *x*, *Employee*, *Skill* and keys *x* and *Employee • Skill*, and *JOBS_SKILLS* having columns *x*, *Job*, *Skill* and keys *x* and *Job • Skill*.

It has been shown that, unfortunately, for both these strategies, obtainable results are not at all encouraging, as:

- ✓ it is not always possible to obtain lossless and dependency preserving results;
- ✓ even when this is achievable, sometimes it is *nondeterministic* (i.e., there are several equivalent solutions, out of which the rdb designers should then choose the "best semantically fitted" one; moreover, being purely syntactic, RDM cannot provide any guidelines to do it);
- ✓ the lossless decomposition property does not generally prevent storing aberrant data;[84]
- ✓ generally, given an arbitrary set of RDM dependencies (constraints), one cannot always obtain any desired RDM NF;
- ✓ even when obtaining of a desired NF is possible, sometimes it takes much too long to obtain it[85], too much to be practically reasonable to wait for[86] and, to further complicate things, sometimes this process is also nondeterministic.
- ✓ there is no algorithm for obtaining DKNF, the highest relational normal form.

Thirdly, both bottom-up and top-down rdb design strategies are essentially trying to compute object sets (i.e., tables, in this paradigm) from

---

[84] Fact that, paradoxically, was and perhaps is still considered by some as an advantage.

[85] As some of such algorithms, like, for example, finding tables of an rdb that violate BCNF, are NP-complete.

[86] For example, assuming that an algorithm only needs 2^n seconds to decompose a $n = 3,000$ columns universal relation (which is only an average actual complex one), this means an unconceivable period of time: for example, if $n = 30$, this means some 34.5 years; when powering 34.5 with 100, you have to start counting in light years already!

mappings' domains and math relations' underlying sets, which is trivially absurd: for centuries already and for all foreseeable centuries to come, both relations and mappings (their particular case) are defined on such sets (and taking values from such sets) and not vice-versa. For example, from the *Altitude* mapping values (see Table 3.1), one cannot ever infer that its domain is the Carpathians, or the Rockies, or the Earth's, or the Moon's mountains, or, even worse, any other object type which is measurable in meters …

Fourthly, as DKRINF is proving it in the second volume of this book, DKNF, the "ultimate" NF, is too weak even in relational terms, as it does not take also into account referential integrity ((typed) inclusion dependencies (*IND*s)). Moreover, not even DKRINF, even if it also takes into account INDs, is not satisfactory either, as it also allows for storing implausible data when non-relational type constraints are also considered (which is compulsory).

For example, if to the db in Fig. 3.4 we associate the constraint set $CS = PCs \cup FDs \cup NRC$, where:

✓ *PCs* = {*COUNTRIES.x* $\in$ [0, 999], *COUNTRIES.Country* $\in$ ASCII(128), *CountryCode* $\in$ ASCII(8), *COUNTRIES.Capital* $\in$ [0, 10^{16}], *COUNTRIES.Population* $\in$ [10^5, 2×10^9], *COUNTRIES.x* key, *COUNTRIES.Country* key, *CountryCode* key, *COUNTRIES.Capital* key, *COUNTRIES.Capital* $\subseteq$ *CITIES.x*, *STATES.x* $\in$ [0, 10^8], *STATES.State* $\in$ ASCII(128), *StateCode* $\in$ ASCII(8), *STATES.Country* $\in$ [0, 999], *STATES.Capital* $\in$ [0, 10^{16}], *STATES.Population* $\in$ [10^5, 5×10^7], *STATES.x* key, *STATES.Country* • *StateCode* key, *STATES.Capital* key, *STATES.Capital* $\subseteq$ *CITIES.x*, *CITIES.x* $\in$ [0, 10^{16}], *CITIES.State* $\in$ [0, 10^8], *City* $\in$ ASCII(128), *CITIES.Population* $\in$ [10^5, 3×10^7], *CITIES.x* key, *CITIES.State* • *City* key, *CITIES.State* $\subseteq$ *STATES.x*} contains its extended primitive constraints (i.e., also including INDs),

✓ *FDs* = {*COUNTRIES.x* $\rightarrow$ *COUNTRIES.Country* • *CountryCode* • *COUNTRIES.Capital* • *COUNTRIES.Population*,
COUNTRIES.Country $\rightarrow$ *COUNTRIES.x* • *CountryCode* • *COUNTRIES.Capital* • *COUNTRIES.Population*,
CountryCode $\rightarrow$ *COUNTRIES.x* • *COUNTRIES.Country* • *COUNTRIES.Capital* • *COUNTRIES.Population*,

> $COUNTRIES.Capital \rightarrow COUNTRIES.x \bullet COUNTRIES.Country \bullet$
> $CountryCode \bullet COUNTRIES.Population,$
> $STATES.x \rightarrow STATES.State \bullet STATES.Country \bullet StateCode \bullet$
> $STATES.Capital \bullet STATES.Population,$
> $STATES.Country \bullet STATES.State \rightarrow STATES.x \bullet StateCode \bullet$
> $STATES.Capital \bullet STATES.Population,$
> $STATES.Capital \rightarrow STATES.x \bullet STATES.State \bullet STATES.Country \bullet$
> $StateCode \bullet STATES.Population,$
> $CITIES.x \rightarrow CITIES.State \bullet CITIES.City \bullet CITIES.Population,$
> $CITIES.City \bullet CITIES.State \rightarrow CITIES.x \bullet CITIES.Population\}$
> its functional dependencies, and
> ✓ $NRC = \{STATES.Country \circ CITIES.State \circ COUNTRIES.Capital = 1_{COUNTRIES}, CITIES.State \circ STATES.Capital = 1_{STATES}\}$ includes only the non-relational constraints "no country may have as its capital a city of another country" (or, dually, "no city may be the capital of a country to which it does not belong") and "no state may have as its capital a city of another state" (or, dually, "no city may be the capital of a state to which it does not belong"), this db is not anymore in DKRINF, as its nonrelational constraints are not implied by its extended primitive ones.

For example, row <7, "Japan", "JP", 5, 128,000,000> is compatible with table *COUNTRIES*, but it cannot be added to it, because it would violate the first non-relational constraint, as 5 in *CITIES* (see Fig. 3.4) is New York, which does not belong to Japan, but to the U.S.; similarly, even if, previously, in table *CITIES* a row <7, "Buffalo", 6, 262,000> were inserted, row <7, "Constanta", "CT", 5, 630,000>, although compatible with *STATES*, could not be inserted into this table, as the second non-relational constraint would be violated (trivially, Buffalo, a city of the U.S. New York State, cannot be the capital of the Romanian county Constanta).

Finally (fifthly), as the example above proves it and as this will hopefully become crystal clear in the second volume of this book, RDM is much too poor as constraint types, allowing storage of implausible data even for very simple DKRINF and DKNF dbs: much higher, semantic type data models, providing much more powerful constraint types than

RDM ones are needed for db design, in order to fully guarantee data plausibility.

For (just another) example, RDM is not able to accept even very common constraints such as "population of a country should be greater or equal than the sum of populations of its states" and "population of a state should be greater or equal than the sum of populations of its cities", which are very similar to check (tuple) ones, except for the fact that they involve columns from different tables.

Consequently, db design should always be done at a higher, semantic level and RDM should only be used as an abstraction layer above RDBMS versions.

On the contrary, querying should always be done relationally, as it is not only powerful, simple, and elegant, but also benefits from very fast processing.

3.14 EXERCISES

Exercise to stimulate, not to annihilate. The world hasn't formed in a day, and neither were we.
Set small goals and build upon them.
—Lee Haney

3.14.1 SOLVED EXERCISES

Problems are only opportunities in work clothes.
–Henry J. Kaiser

3.0. (a) Apply algorithm $A1$–7 to the E-R data model of Exercise 2.0 above and provide a valid instance of at least two rows per table.

(b) Apply algorithm $A7/8$-3 to the above obtained relational scheme in order to discover and add to it any other existing keys (*hint*: skip verifying existing keys as we just correctly identify all of them).

> *SubsidiaryOf* should be acyclic. (*P6*)
> All and only the subsidiaries of a bank should be
> of bank type. (*P7*)
> No payment for an invoice can be done before the
> invoice is established or after more than 100 years
> after corresponding establishing date. (*PD6*)
> Whenever *Type* is 'c' (cash), *To, From*, and *Card* should
> be null. (*PD7*)
> Whenever *Type* is 'C' (card), *From* should be null,
> whereas *To* and *Card* should not be null. (*PD8*)
> Whenever *Type* is 'w' (bank wire), *Card* should be null
> and *To* and *From* should not be null. (*PD9*)
> Sum of all payments for a same invoice should not be
> greater than the corresponding invoice total amount. (*PM4*)
> Bank subsidiaries should always be bank type partners. (*A5*)
> No card payment may be done with an expired card. (*C8*)

FIGURE 3.25 Non-relational constraint set from Exercise 2.0.

Solution:
(a) The associated non-relational constraint set (that will have to be enforced either through db triggers or by db applications) is presented in Fig.3.25.

Obviously, acyclicity (*P6*) is not a relational type constraint.

Each one of (*PD6*), (*PD7*), (*PM4*), and (*C8*) involve columns from more than one table.

(*P7*), (*PD8*), (*PD9*), and (*A5*) apply to only some rows of the corresponding tables.

Corresponding relational scheme, together with a test plausible and valid instance are presented in Fig. 3.26.

(b)
(*i*) Table *COUNTRIES*
$m = 2$; $K' = K = \{x, Country\}$; $T' = \emptyset$; $n = 0$; $l = k = 2$;

(*ii*) Table *CITIES*
$m = 3$; $K' = K = \{x, Country \bullet ZipCode\}$; $T' = \{City, Country, ZipCode\}$; $n = 3$; $l = k = 2$; $k_{max} = C(3, 1) = 3$;

PARTNERS (*x*, *Name* • *Location* • *Address*)

x	*Name*	*Location*	*Address*	*Subsidiary Of*	*Bank?*
auto-num(5)	ASCII (128)	*Im(CI-TIES.x)*	ASCII(255)	*Im(PART-NERS.x)*	BOOLE
NOT NULL	NOT NULL	NOT NULL	NOT NULL		NOT NULL
1	Champagne Nicolas Feuillate	1	CD 40 A «Plumecoq», 51530		F
2	ING Commercial Banking	2	Amsterdamse Poort Building Bijlmerplein 888 1102 MG		T
3	ING Bank N.V. Amsterdam Romania	3	Bd. Iancu de Hunedoara nr. 48, Sector 1, 011745	2	T

INVOICES (*x*, *No.* • *Date* • *Supplier*) *Supplier* ≠ *Customer*

x	*Supplier*	*Customer*	*No.*	*Date*	**TotAmount*
auto-num (7)	*Im(PART-NERS.x)*	*Im(PART-NERS.x)*	ASCII (16)	[1/1/2000, SysDate()]	CURREN-CY(8)
NOT NULL	NOT NULL	NOT NULL	NOT NULL	NOT NULL	NOT NULL
1	1	2	12	11/21/2012	1,000
2	1	3	13	11/21/2012	200
3	1	3	123	12/21/2012	300

FIGURE 3.26   Relational Scheme from Exercise 2.0 and a Valid Test Instance.

$i = 1$ (C(3, 1) = 3)

1. *City* is not a key, as there may be cities having same names but different zip codes in a same country or even same zip codes too but from different countries;

2. *ZipCode* is not a key, as there may be cities having same zip codes but different names in a same country or even same names too but in different countries;

PAYMENTS (_x_, Invoice • PaymDoc)

_x_	PaymDoc	Invoice	Amount
auto-num(8)	Im(PAYM_DOCS.x)	Im(INVOI-CES.x)	CURREN-CY(6)
NOT NULL	NOT NULL	NOT NULL	NOT NULL
1	1	1	500
2	2	1	500
3	3	2	200
4	3	3	300

ACCOUNTS (_x_, BankSubsidiary • No.)

_x_	BankSub-sidiary	No.	Holder	Amount
auto-num (6)	Im(PART-NERS.x)	ASCII(32)	Im(PART-NERS.x)	CURREN-CY(32)
NOT NULL	NOT NULL	NOT NULL	NOT NULL	NOT NULL
1	2	HO01INGB0000 999900111111	1	10,000,000
2	2	HO01INGB0000 999900000000	2	99,999,999
3	3	RO81INGB0000 999900123456	3	100,000

PAYM_DOCS (_x_. No. • Date • Debtor • Type) From ≠ To

_x_	Debtor	From	To	No.	Date	Type	Card	*TotAmount
auto-num (8)	Im(PART-NERS.x)	Im(AC-COUNTS.x)	Im(AC-COUNTS.x)	ASCII (16)	[1/1/2000, SysDate()]	{'c', 'C', 'w'}	Im (CARDS.x)	CURRENCY (8)
NOT NULL	NOT NULL			NOT NULL	NOT NULL	NOT NULL		NOT NULL
1	2	2	1	210	11/30/2012	w		500
2	2	2	1	320	12/30/2012	w		500
3	3	3	1	101	12/30/2012	w		500

FIGURE 3.26   (Continued).

3. _Country_ is not a key, as there may be several cities of a same country, but having different zip codes and/or names;

$i = 2$ (C(3, 2) = 3)

1. _Country_ • _ZipCode_ is already known to be a key;

2. _City_ • _ZipCode_ is not a key, as there may be cities having same name and zip codes, but from different countries;

CARDS (x, *Type* • *No.* • *Year*)

x	Type	No.	Month	Year	Security Code	D/C	Account
auto-num (6)	$Im(CARD$ $\_$ $TYPES.x)$	$[10^{15};$ 10^{16}-1]	[1, 12]	$[Year(SysD$ $ate()) - 2;$ $Year(SysDa$ $te()) + 2]$	[000; 999]	{'D', 'C'}	$Im(AC$-$COUNT$ $S.x)$
NOT NULL	NOT NULL	NOT NULL	NOT NULL	NOT NULL	NOT NULL	NOT NULL	NOT NULL
1	3	1234 5678 9876 5432	1	2015	123	C	1

CARD_TYPES (x, *CardType*)

x	CardType
autonum(2)	ASCII(16)
NOT NULL	NOT NULL
1	American Express
2	Visa
3	Visa Electron
4	Maestro

CITIES (x, *Country* • *ZipCode*)

x	City	ZipCode	Country
autonum(5)	ASCII(64)	ASCII(16)	$Im(COUNTRIES.x)$
NOT NULL	NOT NULL	NOT NULL	NOT NULL
1	Chouilly	51530	1
2	Amsterdam	10	3
3	Bucharest	0	2

COUNTRIES (x, *Country*)

x	Country
autonum(3)	ASCII(128)
NOT NULL	NOT NULL
1	France
2	Romania
3	Holland

FIGURE 3.26   (Continued).

3. *City* • *Country* is not a key, as there may be cities from a same country having same name, but having different zip codes;

$i = 3$ (C(3, 3) = 1)

City • *Country* • *ZipCode* is a superkey;
To conclude with, $K' = K = \{x, Country • ZipCode\}$; $l = 2$.

(*iii*) Table *CARD_TYPES*
$m = 2$; $K' = K = \{\underline{x}, CardType\}$; $T' = \emptyset$; $n = 0$; $l = k = 2$;

(*iv*) Table *PARTNERS*
$m = 6$; $K' = K = \{x, Name • Location • Address\}$; $SK = \{x\}$; $T' = \{Name,$
Location, Address, SubsidiaryOf, Bank?$\}$; $n = 5$; $l = 2$; $k_{max} = C(5, 2) = 10$;
$i = 1$ (C(5, 1) = 5)

1. *Name* is not a key, as there may be several partners having same names, but different types and/or being located in different cities and/or addresses and/or being subsidiaries of different partners;
2. *Location* is not a key, as there may be several partners being located in same cities, but having different names and/or types and/or being located at different address in that city and/or being subsidiaries of different partners;
3. *Address* is not a key, as there may be several partners having same address within their cities, but different names and/or types and/or being located in different cities and/or being subsidiaries of different partners;
4. *SubsidiaryOf* is not a key, as there may be several partners being subsidiaries of a same partner or not being subsidiaries, but having different names and/or types and/or being located in different cities and/or addresses;
5. *Bank?* is not a key, as there may be several partners having same type, but different names and/or being located in different cities and/or addresses and/or being subsidiaries of different partners;
   $i = 2$ (C(5, 2) = 10)
1-3. *Name* • *Location*, *Name* • *Address*, and *Location* • *Address* are subproducts of *Name* • *Location* • *Address*, so they cannot be keys;
4. *Name* • *SubsidiaryOf* is not a key, as there may be partners having same names and being subsidiaries of a same partner or not being subsidiaries, but being located in different cities and/or different addresses and/or having different types;

5. *Name • Bank?* is not a key, as there may be partners having same names and type, but being located in different cities and/or different addresses and/or being subsidiaries of other partners;

6. *Location • SubsidiaryOf* is not a key, as there may be partners located in same cities and being subsidiaries of a same partner or not being subsidiaries, but being located at different addresses and/or having different names and/or types;

7. *Location • Bank?* is not a key, as there may be partners located in same cities and having same type, but being located at different addresses and/or having different names and/or being subsidiaries of other partners;

8. *Address • SubsidiaryOf* is not a key, as there may be partners located at same addresses within their cities and being subsidiaries of a same partner or not being subsidiaries, but being located in different cities and/or having different names and/or types;

9. *Address • Bank?* is not a key, as there may be partners located at same addresses within their cities and having same type, but being located in different cities and/or having different names and/or being subsidiaries of other partners;

10. *SubsidiaryOf • Bank?* is not a key, as there may be partners having same type and being subsidiaries of a same partner or not being subsidiaries, but being located in different cities and/or at different addresses and/or having different names;

   $i = 3$ (C(5, 3) = 10)

1. *Name • Location • Address* is already known to be a key;

2. *Name • Location • SubsidiaryOf* is not a key, as there may be partners having same names, located in same cities, and being subsidiaries of a same partner or not being subsidiaries, but being located at different addresses and/or having different types;

3. *Name • Location • Bank?* is not a key, as there may be partners having same names and types, and being located in same cities, but being located at different addresses and/or being subsidiaries of other partners;

4. *Name • Address • SubsidiaryOf* is not a key, as there may be partners having same names, located at same addresses within their cities, and being subsidiaries of a same partner or not being subsid-

iaries, but being located in different cities and/or having different types;

5. *Name • Address • Bank?* is not a key, as there may be partners having same names and types, and located at same addresses within their cities, but being located in different cities and/or being subsidiaries of other partners;

6. *Name • SubsidiaryOf • Bank?* is not a key, as there may be partners having same names and types, and being subsidiaries of a same partner or not being subsidiaries, but being located in different cities and/or at different addresses;

7. *Location • Address • SubsidiaryOf* is not a key, as there may be partners located in a same city at a same address and being subsidiaries of a same partner or not being subsidiaries, but having different types and/or names;

8. *Location • Address • Bank?* is not a key, as there may be partners located in a same city at a same addresses and having same types, but being subsidiaries of different partners or not being subsidiaries and/or having different names;

9. *Location • SubsidiaryOf • Bank?* is not a key, as there may be partners having same types, located in same cities, and being subsidiaries of a same partner or not being subsidiaries, but having different names and/or being located at different addresses;

10. *Address • SubsidiaryOf • Bank?* is not a key, as there may be partners having same types, located at same addresses within their cities, and being subsidiaries of a same partner or not being subsidiaries, but having different names and/or being located in different cities;

$i = 4$ $(C(5, 4) = 5)$

1-2. *Name • Location • Address • SubsidiaryOf* and *Name • Location • Address • Bank?* are superkeys;

3. *Name • Location • SubsidiaryOf • Bank?* is not a key, as there may be several partners having same names and types, located in a same city, and being subsidiaries of a same partner or not being subsidiaries, but being located at different addresses within the city;

4. *Name • Address • SubsidiaryOf • Bank?* is not a key, as there may be several partners having same names and types, located at same addresses within their cities, and being subsidiaries of a same partner or not being subsidiaries, but being located in different cities;

5. *Location • Address • SubsidiaryOf • Bank?* is a key, as there may not be several partners located in a same city, at a same address, having same type, and being subsidiaries of a same partner or not being subsidiaries, but having different names;

K' = {x, Name • Location • Address, Location • Address • SubsidiaryOf • Bank?}; l = 3;

$i = 5$ (C(5, 5) = 1)

Name • Location • Address • SubsidiaryOf • Bank? is a superkey;

To conclude with, *K' = {x, Name • Location • Address, Location • Address • SubsidiaryOf • Bank?}; l = 3.*

(*v*) Table *ACCOUNTS*

$m = 5$; *K' = K = {x, BankSubsidiary • No.}; T' = {BankSubsidiary, No., Holder, Amount}*; $n = 4$; $l = k = 2$;

Amount is not a key (as there may be several accounts having same amount, but managed by different bank subsidiaries or even by a same one, but having different numbers) and, moreover, it is not prime, as it cannot even help uniquely distinguishing accounts together with other columns; it follows that *T' = {BankSubsidiary, No., Holder}*; $n = 3$; $k_{max} = C(3, 1) = 3$;

$i = 1$ (C(3, 1) = 3)

1. *BankSubsidiary* is not a key, as any bank subsidiary manages several accounts;
2. *No.* is not a key, as there may be several accounts having same numbers, but managed by different bank subsidiaries;
3. *Holder* is not a key, as anybody may have several accounts managed by different bank subsidiaries, and even by a same one, but having different numbers;

$i = 2$ (C(3, 2) = 3)

1. *BankSubsidiary • No.* is already known to be a key;
2. *BankSubsidiary • Holder* is not a key, as anybody may hold several accounts at a same bank subsidiary, but with different numbers;
3. *No. • Holder* is not a key, as anybody may hold several accounts having same numbers, but at different bank subsidiaries;

$i = 3$ (C(3, 3) = 1)

BankSubsidiary • No. • Holder is a superkey;

To conclude with, *K' = K = {x, BankSubsidiary • No.}; l = 2.*

(*vi*) Table *CARDS*

$m = 8$; $K' = K = \{x, Type \bullet No. \bullet Year\}$; $T' = \{Type, No., Month, Year,$ *SecurityCode, D/C, Account*$\}$; $n = 7$; $l = k = 2$;

Month is not a key, as there are several cards having same expiration month, but of different types and/or having different numbers and/or expiration years; moreover, *Month* is not prime, as cards have at least one year validity, and hence it cannot even help uniquely distinguishing cards together with other columns.

SecurityCode is not a key (as there may be several cards having same security code, but managed by different bank subsidiaries or even by a same one, but having different numbers and/or types, etc.) and, moreover, it is not prime, as it cannot even help uniquely distinguishing cards together with other columns.

D/C is not a key (as there may be several debit or credit cards, but managed by different bank subsidiaries or even by a same one, but having different numbers and/or types, etc.) and, moreover, it is not prime, as it cannot even help uniquely distinguishing cards together with other columns.

It follows that $T' = \{Type, No., Year, Account\}$; $n = 4$; $k_{max} = C(4, 2) = 6$; $i = 1$ (C(4, 1) = 4)

1. *Type* is not a key, as there are several cards of any type;
2. *No.* is not a key, as there are several cards having same numbers, but of different types;
3. *Year* is not a key, as there are several cards having same expiration year, but of different types and/or having different numbers;
4. *Account* is not a key, as there may be several cards having same bank account, but having different expiration years;
    $i = 2$ (C(4, 2) = 6)

1-3. *Type* • *No.*, *Type* • *Year*, and *No.* • *Year*, are subproducts of the key *Type* • *No.* • *Year*;

4. *Type* • *Account* is not a key, as there are several cards having same type and associated to a same bank account, but having different expiration years;
5. *No.* • *Account* is not a key, as there are several cards having same number and associated to a same bank account, but having different expiration years;

6. *Year • Account* is a key, as, according to the simplifying assumption that there may be no several cards associated to a same account and simultaneously valid, there may not be several cards having same expiration year and associated to a same bank account;

$K' = \{x, Type • No. • Year, Year • Account\}; l = 3;$

$i = 3\ (C(4, 3) = 4)$

1. *Type • No. • Year* is already known as a key;

2-3. *Type • Year • Account* and *No. • Year • Account* are superkeys;

4. *Type • No. • Account* is not a key, as there may be cards of a same type, having a same number, and being associated to a same bank account, but having different expiration years;

$i = 4\ (C(4, 4) = 1)$

Type • No. • Year • Account is a superkey;

To conclude with, $K' = \{x, Type • No. • Year, Year • Account\}; l = 3.$

(*vii*) Table *INVOICES*

$m = 6; K' = K = \{x, No. • Date • Supplier\}; T' = \{No., Date, Supplier, Customer, *TotAmount\}; n = 5; l = k = 2;$

*TotAmount is not a key (as there may be invoices having same total amount) and non prime either, as it cannot even help uniquely distinguishing cards together with other columns.

It follows that $T' = \{No., Date, Supplier, Customer\}; n = 4; k_{max} = C(4, 2) = 6;$

$i = 1\ (C(4, 1) = 4)$

1-3. *No.*, *Date*, and *Supplier*, are not keys as they are members of the key *No. • Date • Supplier*;

4. *Customer* is not a key, as there may be several invoices for a same customer (established even by a same supplier, and even at a same date, but with different numbers);

$i = 2\ (C(4, 2) = 6)$

1-3. *No. • Date*, *No. • Supplier*, and *Date • Supplier* are subproducts of the key *No. • Date • Supplier*;

4. *No. • Customer* is not a key, as there may be several invoices for a same customer having a same number (established even by a same supplier, but at different dates);

5. *Date • Customer* is not a key, as there may be several invoices for a same customer having a same date (established even by a same supplier, but with different numbers);

6. *Supplier • Customer* is not a key, as there may be several invoices for a same customer established by a same supplier, but with different numbers and/or dates;

$i = 3$ (C(4, 3) = 4)

1. *No. • Date • Supplier* is already known to be a key;

2. *No. • Date • Customer* is not a key, as there may be several invoices for a same customer having a same number and date, but established by different suppliers;

3. *No. • Supplier • Customer* is not a key, as there may be several invoices for a same customer established by a same supplier, with a same number, but at different dates;

4. *Date • Supplier • Customer* is not a key, as there may be several invoices for a same customer established by a same supplier, at a same date, but with different numbers;

$i = 4$ (C(4, 4) = 1)

No. • Date • Supplier • Customer is a superkey;

To conclude with, $K' = K = \{x, No. • Date • Supplier\}; l = 2$.

(*viii*) Table *PAYM_DOCS*

$m = 9$; $K' = K = \{x, No. • Date • Debtor • Type\}$; $T' = \{No., Date, Debtor, Type, From, To, Card, *TotAmount\}$; $n = 8$; $l = k = 2$;

TotAmount is not a key (as there may even be payment documents of a same debtor, having same number and amount, but different dates) and non prime either, as it cannot even help uniquely distinguishing cards together with other columns.

It follows that $T' = \{No., Date, Debtor, Type, From, To, Card\}$; $n = 7$; $k_{max} = C(7, 3) = 35$;

$i = 1$ (C(7, 1) = 7)

1-4. *No., Date, Debtor,* and *Type* are not keys, as they are members of the key *No. • Date • Debtor • Type*;

5. *From* is not a key, as there may be several payments from a same bank account;

6. *To* is not a key, as there may be several payments to a same bank account;

7. *Card* is not a key, as there may be several payments done with a same card;

$i = 2$ (C(7, 2) = 21)

1-6. *No.* • *Date, No.* • *Debtor, No.* • *Type, Date* • *Debtor, Date* • *Type,* and *Debtor* • *Type* are not keys, as they are subproducts of the key *No.* • *Date* • *Debtor* • *Type*;

7. *No.* • *From* is not a key, as there may be several payments from a same bank account with a same number, but having different dates and/or debtors, etc.;

8. *No.* • *To* is not a key, as there may be several payments to a same bank account with a same number, but having different dates and/or debtors, etc.;

9. *No.* • *Card* is not a key, as there may be several payments done with a same card and having a same number, but different dates and/or debtors, etc.;

10. *Date* • *From* is not a key, as there may be several payments from a same bank account at a same date, but having different numbers and/or types, etc.;

11. *Date* • *To* is not a key, as there may be several payments to a same bank account at a same date, but having different numbers and/or types, etc.;

12. *Date* • *Card* is not a key, as there may be several payments done with a same card at a same date, but having different numbers and/or destination bank accounts, etc.;

13. *Debtor* • *From* is not a key, as there may be several payments from a same bank account done by a same debtor, but having different numbers and/or types, etc.;

14. *Debtor* • *To* is not a key, as there may be several payments to a same bank account done by a same debtor, but having different numbers and/or types, etc.;

15. *Debtor* • *Card* is not a key, as there may be several payments done with a same card by a same debtor, but having different numbers and/or destination bank accounts, etc.;

16. *Type* • *From* is not a key, as there may be several payments of a same type and from a same bank account, but having different numbers and/or dates, etc.;

17. *Type* • *To* is not a key, as there may be several payments of a same type and to a same bank account, but having different numbers and/or dates, etc.;

18. *Type* • *Card* is not a key, as there may be several payments done with a same card, but having different numbers and/or destination bank accounts, etc.;

19. *From* • *To* is not a key, as there may be several payments from a same bank account to a same bank account, but having different numbers and/or dates, etc.;

20. *From* • *Card* is not a key, as there may be several payments done with a same card, but having different numbers and/or destination bank accounts, etc.;

21. *To* • *Card* is not a key, as there may be several payments done with a same card to a same bank account, but having different numbers and/or dates, etc.;

$i = 3$ ($C(7, 3) = 35$)

1-4. *No.* • *Date* • *Debtor*, *No.* • *Date* • *Type*, *No.* • *Debtor* • *Type*, and *Date* • *Debtor* • *Type* are not keys, as they are subproducts of the key *No.* • *Date* • *Debtor* • *Type*;

5. *No.* • *Date* • *From* is a key, as there may not be several payments from a same bank account with a same number and date;

$K' = \{x, No.$ • *Date* • *Debtor* • *Type*, *No.* • *Date* • *From*$\}; l = 3$;

6. *No.* • *Date* • *To* is a key, as there may not be several payments to a same bank account with a same number and date;

$K' = \{x, No.$ • *Date* • *Debtor* • *Type*, *No.* • *Date* • *From*, *No.* • *Date* • *To*$\}; l = 4$;

7. *No.* • *Date* • *Card* is a key, as there may not be several payments done with a same card at a same date and with a same number;

$K' = \{x, No.$ • *Date* • *Debtor* • *Type*, *No.* • *Date* • *From*, *No.* • *Date* • *To*, *No.* • *Date* • *Card*$\}; l = 5$;

8. *No.* • *Debtor* • *From* is not a key, as there may be several payments done by a same debtor from a same bank account with a same number, but having different dates and/or destination bank accounts, etc.;

9. *No.* • *Debtor* • *To* is not a key, as there may be several payments done by a same debtor to a same bank account with a same number, but having different dates and/or source bank accounts, etc.;

10. *No.* • *Debtor* • *Card* is not a key, as there may be several payments done by a same debtor with a same card with a same number, but at different dates;

11. *No.* • *Type* • *From* is not a key, as there may be several payments having a same type, number, and from a same bank account, but different dates and/or destination bank accounts, etc.;

12. *No.* • *Type* • *To* is not a key, as there may be several payments having a same type, number, and to a same bank account, but different dates and/or source bank accounts, etc.;

13. *No.* • *Type* • *Card* is not a key, as there may be several payments having a same type and number, done with a same card, but at different dates;

14. *No.* • *From* • *To* is not a key, as there may be several payments having a same number, from a same bank account to a same bank account, but at different dates, etc.;

15. *No.* • *From* • *Card* is not a key, as there may be several payments having a same number, from a same bank account, done with a same card, but at different dates;

16. *No.* • *To* • *Card* is not a key, as there may be several payments having a same number, to a same bank account, done with a same card, but at different dates;

17. *Date* • *Debtor* • *From* is not a key, as there may be several payments done by a same debtor from a same bank account at a same date, but having different numbers and/or destination bank accounts, etc.;

18. *Date* • *Debtor* • *To* is not a key, as there may be several payments done by a same debtor to a same bank account at a same date, but having different numbers and/or source or destination bank accounts, etc.;

19. *Date* • *Debtor* • *Card* is not a key, as there may be several payments having a same date, done with a same card by a same debtor, but having different numbers and/or destination bank accounts;

20. *Date • Type • From* is not a key, as there may be several payments of a same type, from a same bank account, and at a same date, but having different numbers and/or destination bank accounts, etc.;

21. *Date • Type • To* is not a key, as there may be several payments of a same type, to a same bank account, and at a same date, but having different numbers and/or source bank accounts, etc.;

22. *Date • Type • Card* is not a key, as there may be several payments having a same date, done with a same card, but having different numbers and/or destination bank accounts;

23. *Date • From • To* is not a key, as there may be several payments done at a same date, from a same bank account to a same bank account, but with different numbers, etc.;

24. *Date • From • Card* is not a key, as there may be several payments done at a same date, from a same bank account, with a same card, but with different numbers and/or to different destination bank accounts, etc.;

25. *Date • To • Card* is not a key, as there may be several payments done at a same date, to a same bank account, with a same card, but with different numbers, etc.;

26. *Debtor • Type • From* is not a key, as there may be several payments of a same type, from a same bank account, done by a same debtor, but having different numbers and/or dates and/or destination bank accounts;

27. *Debtor • Type • To* is not a key, as there may be several payments of a same type, to a same bank account, done by a same debtor, but having different numbers and/or dates and/or source bank accounts;

28. *Debtor • Type • Card* is not a key, as there may be several payments done by a same debtor with a same card, but having different numbers and/or dates and/or destination bank accounts;

29. *Debtor • From • To* is not a key, as there may be several payments from a same bank account to a same bank account, done by a same debtor, but having different numbers and/or dates;

30. *Debtor • From • Card* is not a key, as there may be several payments done by a same debtor with a same card, but having different numbers and/or dates and/or destination bank accounts;

31. *Debtor • To • Card* is not a key, as there may be several payments done by a same debtor with a same card to a same bank account, but having different numbers and/or dates;

32. *Type • From • To* is not a key, as there may be several payments of a same type, from a same bank account to a same bank account, but with different numbers and/or dates, etc.;

33. *Type • From • Card* is not a key, as there may be several payments done from a same bank account, with a same card, but with different numbers and/or dates and/or to different destination bank accounts;

34. *Type • To • Card* is not a key, as there may be several payments done to a same bank account, with a same card, but with different numbers and/or dates;

35. *From • To • Card* is not a key, as there may be several payments done to a same bank account, with a same card, but with different numbers and/or dates;

$i = 4$ (C(7, 4) = 35)

1. *No. • Date • Debtor • Type* is already known to be a key;

2-10. *No. • Date • Debtor • From, No. • Date • Debtor • To, No. • Date • Debtor • Card, No. • Date • Type • From, No. • Date • Type • To, No. • Date • Type • Card, No. • Date • From • To, No. • Date • From • Card*, and *No. • Date • To • Card* are superkeys;

11. *No. • Debtor • Type • From*, is not a key, as there may be several payments of a same type done by a same debtor from a same bank account with a same number, but having different dates and/or destination bank accounts;

12. *No. • Debtor • Type • To*, is not a key, as there may be several payments of a same type done by a same debtor to a same bank account with a same number, but having different dates and/or source bank accounts;

13. *No. • Debtor • Type • Card*, is not a key, as there may be several payments done by a same debtor with a same card and having a same number, but different dates and/or destination bank accounts;

14. *No. • Debtor • From • To*, is not a key, as there may be several payments done by a same debtor from a same bank account to a same bank account with a same number, but having different dates;

15. *No.* • *Debtor* • *From* • *Card*, is not a key, as there may be several payments done by a same debtor with a same card and having a same number, but different dates and/or destination bank accounts;

16. *No.* • *Debtor* • *To* • *Card*, is not a key, as there may be several payments done by a same debtor with a same card to a same bank account and having a same number, but different dates;

17. *No.* • *Type* • *From* • *To*, is not a key, as there may be several payments of a same type from a same bank account to a same bank account with a same number, but having different dates;

18. *No.* • *Type* • *From* • *Card*, is not a key, as there may be several payments done with a same card and having a same number, but different dates and/or destination bank accounts;

19. *No.* • *Type* • *To* • *Card*, is not a key, as there may be several payments done with a same card to a same bank account and having a same number, but different dates;

20. *No.* • *From* • *To* • *Card*, is not a key, as there may be several payments done with a same card to a same bank account and having a same number, but different dates;

21. *Date* • *Debtor* • *Type* • *From*, is not a key, as there may be several payments of a same type done by a same debtor from a same bank account at a same date, but having different numbers and/or destination bank accounts;

22. *Date* • *Debtor* • *Type* • *To*, is not a key, as there may be several payments of a same type done by a same debtor to a same bank account at a same date, but having different numbers and/or source bank accounts;

23. *Date* • *Debtor* • *Type* • *Card*, is not a key, as there may be several payments done by a same debtor with a same card at a same date, but having different numbers and/or destination bank accounts;

24. *Date* • *Debtor* • *From* • *To*, is not a key, as there may be several payments done by a same debtor from a same bank account to a same bank account at a same date, but having different numbers;

25. *Date* • *Debtor* • *From* • *Card*, is not a key, as there may be several payments done by a same debtor with a same card at a same date, but having different numbers and/or destination bank accounts;

26. *Date • Debtor • To • Card*, is not a key, as there may be several payments done by a same debtor with a same card to a same bank account at a same date, but having different numbers;

27. *Date • Type • From • To*, is not a key, as there may be several payments of a same type from a same bank account to a same bank account at a same date, but having different numbers;

28. *Date • Type • From • Card*, is not a key, as there may be several payments done with a same card at a same date, but having different numbers and/or destination bank accounts;

29. *Date • Type • To • Card*, is not a key, as there may be several payments done with a same card at a same date to a same bank account, but having different numbers;

30. *Date • From • To • Card*, is not a key, as there may be several payments done with a same card at a same date to a same bank account, but having different numbers

31. *Debtor • Type • From • To*, is not a key, as there may be several payments of a same type from a same bank account to a same bank account done by a same debtor, but having different numbers and/or dates;

32. *Debtor • Type • From • Card*, is not a key, as there may be several payments done by a same debtor with a same card, but having different numbers and/or dates and/or destination bank accounts;

33. *Debtor • Type • To • Card*, is not a key, as there may be several payments done by a same debtor with a same card to a same bank account, but having different numbers and/or dates;

34. *Debtor • From • To • Card*, is not a key, as there may be several payments done by a same debtor with a same card to a same bank account, but having different numbers and/or dates;

35. *Type • From • To • Card*, is not a key, as there may be several payments done with a same card to a same bank account, but having different numbers and/or dates;

$i = 5$ ($C(7, 5) = 21$)

1-10. *No. • Date • Debtor • Type • From, No. • Date • Debtor • Type • To, No. • Date • Debtor • Type • Card, No. • Date • Debtor • From • To, No. • Date • Debtor • From • Card, No. • Date • Debtor • To • Card, No. • Date • Type • From • To, No. • Date • Type • From •*

Card, No. • *Date* • *Type* • *To* • *Card*, and *No.* • *Date* • *From* • *To* • *Card* are superkeys;

11. *No.* • *Debtor* • *Type* • *From* • *To* is not a key, as there may be several payments of a same type done by a same debtor from a same bank account to a same bank account and having a same number, but different dates;

12. *No.* • *Debtor* • *Type* • *From* • *Card* is not a key, as there may be several payments by a same debtor with a same card and having a same number, but different dates and/or destination bank accounts;

13. *No.* • *Debtor* • *Type* • *To* • *Card* is not a key, as there may be several payments by a same debtor with a same card to a same bank account and having a same number, but different dates;

14. *No.* • *Debtor* • *From* • *To* • *Card* is not a key, as there may be several payments by a same debtor with a same card to a same bank account and having a same number, but different dates;

15. *No.* • *Type* • *From* • *To* • *Card* is not a key, as there may be several payments with a same card to a same bank account and having a same number, but different dates;

16. *Date* • *Debtor* • *Type* • *From* • *To* is not a key, as there may be several payments of a same type done by a same debtor from a same bank account to a same bank account and having a same date, but different numbers;

17. *Date* • *Debtor* • *Type* • *From* • *Card* is not a key, as there may be several payments by a same debtor with a same card and having a same date, but different numbers and/or destination bank accounts;

18. *Date* • *Debtor* • *Type* • *To* • *Card* is not a key, as there may be several payments by a same debtor with a same card to a same bank account and having a same date, but different numbers;

19. *Date* • *Debtor* • *From* • *To* • *Card* is not a key, as there may be several payments by a same debtor with a same card to a same bank account and having a same date, but different numbers;

20. *Date* • *Type* • *From* • *To* • *Card* is not a key, as there may be several payments with a same card to a same bank account and having a same date, but different numbers;

21. *Debtor • Type • From • To • Card* is not a key, as there may be several payments done by a same debtor with a same card to a same bank account, but having different numbers and/or dates;

$i = 6$ ($C(7, 6) = 7$)

1-5. *No. • Date • Debtor • Type • From • To, No. • Date • Debtor • Type • From • Card, No. • Date • Debtor • Type • To • Card, No. • Date • Debtor • From • To • Card*, and *No. • Date • Type • From • To • Card* are superkeys;

6. *No. • Debtor • Type • From • To • Card* is not a key, as there may be several payments done by a same debtor with a same card to a same bank account and having a same number, but different dates;

7. *Date • Debtor • Type • From • To • Card* is not a key, as there may be several payments done by a same debtor with a same card to a same bank account and having a same date, but different numbers;

$i = 7$ ($C(7, 7) = 1$)

No. • Date • Debtor • Type • From • To • Card is a superkey;

To conclude with, $K' = \{x, \text{No. • Date • Debtor • Type, No. • Date • From, No. • Date • To, No. • Date • Card}\}$; $l = 5$.

(*ix*) Table *PAYMENTS*

$m = 4$; $K' = K = \{x, \text{Invoice • PaymDoc}\}$; $T' = \{\text{Invoice, PaymDoc, Amount}\}$; $n = 3$; $l = k = 2$;

Amount is not a key (as there may even be payment documents including partial or total payments for invoices having same amount) and non prime either, as it cannot even help uniquely distinguishing payments together with other columns.

It follows that $T' = \{\text{Invoice, PaymDoc}\}$; $n = 2$; $k_{max} = C(2, 1) = 2$;

$i = 1$ ($C(2, 1) = 2$)

1-2. *Invoice* and *PaymDoc* are not keys as they are members of the key *Invoice • PaymDoc*;

$i = 2$ ($C(2, 2) = 1$)

Invoice • PaymDoc is already known to be a key;

To conclude with, $K' = K = \{x, \text{Invoice • PaymDoc}\}$; $l = 2$.

In conclusion, the following 5 newly discovered keys should be added to the tables from Figure 3.26 above:

COUNTRIES (x, Country)

x	Country
autonumber(3)	ASCII(128)
NOT NULL	NOT NULL
1	Romania
2	Moldavia
3	Serbia
4	Bulgaria
5	Hungary
6	Ukraine
7	Greece
8	Malta

NEIGHBORS (x, Country • Neighbor)

x	Country	Neighbor
autonumber(4)	Im(COUNTRIES.x)	Im(COUNTRIES.x)
NOT NULL	NOT NULL	NOT NULL
1	1	2
2	1	3
3	1	4
4	1	5
5	1	6
6	2	6
7	3	4
8	3	5
9	4	7

FIGURE 3.27   Rdb for Exercise 3.1.

-    *Location • Address • SubsidiaryOf • Bank?* to *PARTNERS*;

-    *Year • Account* to *CARDS*;

-    *No. • Date • From, No. • Date • To,* and *No. • Date • Card* to *PAYM_DOCS*.

3.1. (*An actual example of basic test for hiring SQL developers*) Consider a RDBMS that does not provide autonumbering and an rdb having the following two tables and associated instances (see Fig. 3.27).

(a) Design, develop, and test ANSI-92 standard *SQL* statements for creating and populating these two tables.

(b) Consider that triggers exist for enforcing irreflexivity and asymmetry of the *NEIGHBORS* binary relation (i.e., rejecting any attempts to store in columns *Country* and *Neighbors* pairs of type <x, x>, as well as <y, x>, whenever <x, y> already is stored) and design, develop, and test *SQL* statements for computing:

1. The number of neighbors of Romania. What is the corresponding result?

2. The number of neighbors of Moldova. What is the corresponding result?

3. The set of countries without neighbors, in ascending order of their names. What is the corresponding result?

4. The set of countries having at least k neighbors (k natural), in descending order of the number of their neighbors and then ascending on their names. What is the corresponding result for $k = 3$?

5. The set of countries that do not appear in column *Country* of table *NEIGHBORS*, in ascending order of their names, without using subqueries. What is the corresponding result?

6. The set of pairs <country name, neighbors number>, in descending order of neighbors number and then ascending on country name, also including countries that do not have neighbors. What is the corresponding result?

7. Adding the fact that Ukraine is also neighbor to Russia, without knowing anything on the two table instances, except for the fact that Ukraine exists in *COUNTRIES*, while Russia does not. What lines will be added according to the current test instance?

8. Changing to uppercase the names of the countries having the property that they are neighbors to at least one neighbor of one of their neighbors (i.e., country x will have its name modified if it is neighbor both to countries y and z, where z is a neighbor of y). What are the countries whose names will be modified?

9. Discarding all neighborhood data for all countries whose names start with the letter 'G'. What lines will disappear?

10. Adding to table *COUNTRIES* a column labeled *Code* for storing the international country codes (those also used on vehicle plates; e.g. USA, RO, MD, UK, etc.), having at most 8 characters, and whose values are mandatory and distinct between them (i.e., storing two countries with the same code must not be allowed); add then corresponding code values, in uppercase, by extracting only the first 2 letters of each country name. How will the extended table look like, sorted in ascending order by country names, taking into account all modifications that were done to it by 1 to 9 above?

Solution:

(a)

```
CREATE TABLE COUNTRIES (
    X INT(3) PRIMARY,
    COUNTRY VARCHAR(128 BYTE) NOT NULL UNIQUE
);
Insert into COUNTRIES (X, COUNTRY) values
(1, 'Romania');
Insert into COUNTRIES (X, COUNTRY) values
(2, 'Moldova');
Insert into COUNTRIES (X, COUNTRY) values
(3, 'Serbia');
Insert into COUNTRIES (X, COUNTRY) values
(4, 'Bulgaria');
Insert into COUNTRIES (X, COUNTRY) values
(5, 'Hungary');
Insert into COUNTRIES (X, COUNTRY) values
(6, 'Ukraine');
Insert into COUNTRIES (X, COUNTRY) values
(7, 'Greece');
Insert into COUNTRIES (X, COUNTRY) values
(8, 'Malta');
CREATE TABLE NEIGHBORS (
    X INT(4) PRIMARY,
    COUNTRY INT(3) NOT NULL,
    NEIGHBOR INT(3) NOT NULL,
```

```
 CONSTRAINT NEIGHBORS_Key UNIQUE (COUNTRY, NEIGH-
BOR),
 CONSTRAINT  NEIGHBORS_COUNTRY_FK  FOREIGN  KEY
(COUNTRY) REFERENCES COUNTRIES,
 CONSTRAINT  NEIGHBORS_NEIGHBOR_FK  FOREIGN  KEY
(NEIGHBOR) REFERENCES COUNTRIES
);
Insert into NEIGHBORS (X, COUNTRY, NEIGHBOR)
values (1,1,2);
Insert into NEIGHBORS (X, COUNTRY, NEIGHBOR)
values (2,1,3);
Insert into NEIGHBORS (X, COUNTRY, NEIGHBOR)
values (3,1,4);
Insert into NEIGHBORS (X, COUNTRY, NEIGHBOR)
values (4,1,5);
Insert into NEIGHBORS (X, COUNTRY, NEIGHBOR)
values (5,1,6);
Insert into NEIGHBORS (X, COUNTRY, NEIGHBOR)
values (6,2,6);
Insert into NEIGHBORS (X, COUNTRY, NEIGHBOR)
values (7,3,4);
Insert into NEIGHBORS (X, COUNTRY, NEIGHBOR)
values (8,3,5);
Insert into NEIGHBORS (X, COUNTRY, NEIGHBOR)
values (9,4,7);
```

(b)

```
1. SELECT COUNT(*) AS NEIGHBORS# FROM NEIGHBORS
   WHERE COUNTRY = 1;
   Result: 5
2. SELECT COUNT(*) AS N# FROM NEIGHBORS
   WHERE COUNTRY = 2 OR NEIGHBOR = 2;
   Result: 2
3. SELECT COUNTRY FROM COUNTRIES WHERE X NOT IN
   (SELECT DISTINCT COUNTRY FROM NEIGHBORS
   UNION
   SELECT DISTINCT NEIGHBOR FROM NEIGHBORS)
```

```
ORDER BY COUNTRY;
Result: Malta
```
4. ```
SELECT COUNTRIES.COUNTRY, NEIGHBORS#
FROM COUNTRIES INNER JOIN
(SELECT COUNTRY, SUM (N#) AS NEIGHBORS# FROM
    (SELECT COUNTRY, COUNT(*) AS N#
    FROM NEIGHBORS GROUP BY COUNTRY
    UNION ALL
    SELECT NEIGHBOR, COUNT(*) AS N#
    FROM NEIGHBORS GROUP BY NEIGHBOR)
GROUP BY COUNTRY
HAVING SUM (N#) > :K — 1) S
ON COUNTRIES.X = S.COUNTRY
ORDER BY NEIGHBORS# DESC, COUNTRIES.COUNTRY;
```
Result:

| COUNTRY | NEIGHBORS# |
|---------|-----------|
| Romania | 5 |
| Bulgaria | 3 |
| Serbia | 3 |

5. ```
SELECT COUNTRIES.COUNTRY
FROM COUNTRIES LEFT JOIN NEIGHBORS
ON COUNTRIES.X = NEIGHBORS.COUNTRY
WHERE NEIGHBORS.COUNTRY IS NULL
ORDER BY COUNTRIES.COUNTRY;
```

Result: Greece
Hungary
Malta
Ukraine

6. ```
SELECT COUNTRIES.COUNTRY, SUM (N#) AS NEIGHBORS#
FROM COUNTRIES INNER JOIN
(SELECT COUNTRY, COUNT(*) AS N# FROM NEIGHBORS
GROUP BY COUNTRY
UNION ALL
SELECT NEIGHBOR, COUNT(*) AS N# FROM NEIGHBORS
GROUP BY NEIGHBOR) S
```
```

```
  ON COUNTRIES.X = S.COUNTRY
GROUP BY COUNTRIES.COUNTRY
UNION
SELECT COUNTRY, 0 FROM COUNTRIES WHERE X NOT IN
  (SELECT DISTINCT COUNTRY FROM NEIGHBORS
  UNION
  SELECT DISTINCT NEIGHBOR FROM NEIGHBORS)
ORDER BY NEIGHBORS# DESC, COUNTRY;
```
   Result:

COUNTRY	NEIGHBORS#
Romania	5
Bulgaria	3
Serbia	3
Hungary	2
Moldova	2
Ukraine	2
Greece	1
Malta	0

```
7. INSERT INTO COUNTRIES
   SELECT max(X) + 1, 'Russia' FROM COUNTRIES;
```
      Result: line < 9, Russia> is appended to table *COUNTRIES*

```
CREATE TABLE T AS
SELECT X FROM COUNTRIES
WHERE COUNTRY IN ('Ukraine', 'Russia');
INSERT INTO NEIGHBORS
SELECT max(NEIGHBORS.X) + 1, min(T.X), max(T.X)
FROM NEIGHBORS, T;
```
   Result: line < 10, 6, 9 > is appended to table *NEIGHBORS*

```
DROP TABLE T;
```

   Note that, for example, the equivalent, much more elegant statement:

```
INSERT INTO NEIGHBORS(X, Country, Neighbor)
SELECT max(X) + 1, (SELECT X FROM COUNTRIES WHERE
```

```
Country = 'Ukraine'), (SELECT X
   FROM COUNTRIES WHERE Country = 'Russia')
FROM NEIGHBORS;
```

is accepted by MS *Access*, but not by *Oracle* (which is not able to notice that X is a primary key, so at most only one value is returned by each of the subqueries)!

```
8. UPDATE COUNTRIES SET COUNTRY = UPPER(COUNTRY)
   WHERE X IN
     (SELECT DISTINCT A.COUNTRY FROM
       (SELECT COUNTRY, NEIGHBOR FROM NEIGHBORS
       UNION
       SELECT NEIGHBOR, COUNTRY FROM NEIGHBORS)
     A INNER JOIN (SELECT DISTINCT A.COUNTRY,
                           B.NEIGHBOR FROM
     (SELECT COUNTRY, NEIGHBOR FROM NEIGHBORS
     UNION
     SELECT NEIGHBOR, COUNTRY FROM NEIGHBORS)
     A INNER JOIN
     (SELECT COUNTRY, NEIGHBOR FROM NEIGHBORS
     UNION
     SELECT NEIGHBOR, COUNTRY FROM NEIGHBORS)
     B ON A.NEIGHBOR = B.COUNTRY AND
       A.COUNTRY <> B.NEIGHBOR) B
     ON A.NEIGHBOR = B.COUNTRY AND
       A.COUNTRY = B.NEIGHBOR);
```

Result: Romania, Moldova, Serbia, Bulgaria, Hungary, and Ukraine will be changed to ROMANIA, MOLDOVA, SERBIA, BULGARIA, HUNGARY, and UKRAINE, respectively

```
9. DELETE FROM NEIGHBORS WHERE COUNTRY IN
(SELECT X FROM COUNTRIES WHERE COUNTRY LIKE 'G%')
OR NEIGHBOR IN
(SELECT X FROM COUNTRIES WHERE COUNTRY LIKE 'G%');
```

Result: line < 9, 4, 7 > will be deleted from *NEIGHBORS*

```
10. CREATE TABLE C AS SELECT * FROM COUNTRIES;
    CREATE TABLE N AS SELECT * FROM NEIGHBORS;
```

```
DELETE FROM NEIGHBORS;
DELETE FROM COUNTRIES;
ALTER TABLE COUNTRIES
ADD (CODE VARCHAR(8) NOT NULL UNIQUE);
INSERT INTO COUNTRIES
SELECT X, COUNTRY, UPPER(SUBSTR(COUNTRY,1,2))
FROM C;
INSERT INTO NEIGHBORS SELECT * FROM N;
DROP TABLE T;
DROP TABLE N;
```
Result:

X	COUNTRY	CODE
4	BULGARIA	BU
7	Greece	GR
5	HUNGARY	HU
8	Malta	MA
2	MOLDOVA	MO
1	ROMANIA	RO
9	Russia	RU
3	SERBIA	SE
6	UKRAINE	UK

3.2. Prove that the FD math relation is a partial preorder.

Proof:

Let R, A, B, C be any sets, $f : R \to A$, $g : R \to B$, and $h : R \to C$ be any attributes of R with g functionally dependent on f and h functionally dependent on g, that is $f \to g$ and $g \to h$ hold in any R's instance, so, by definition, $ker(f) \subseteq ker(g)$ and $ker(g) \subseteq ker(h)$;

(*partiality*) trivially, by the FD definition, not even two functions defined on a same set need to be functionally dependent;

(*reflexivity*) trivially, from the set inclusion reflexivity, $\forall f$, $ker(f) \subseteq ker(f) \Rightarrow f \to f$;

(*transitivity*) obviously, from the set inclusion transitivity, $(\forall f, g, h)$ $(ker(f) \subseteq ker(g) \wedge ker(g) \subseteq ker(h) \Rightarrow ker(f) \subseteq ker(h)) \Rightarrow f \to h$.

Q.E.D.

3.3. A relation satisfies $X \to Y \Leftrightarrow$ it satisfies $X \to A$, $\forall A \in Y$.

Proof:

Let R, $B = C_1 \times \ldots \times C_n$ ($n > 1$, natural), D be any sets, with R and C_i atomic, while D might be a Cartesian product too, $Y : R \to B$, $X : R \to D$, $A_i : R \to C_i$, for any $1 \leq i \leq n$, and $Y = A_1 \bullet \ldots \bullet A_n$ be any attributes and attribute products of R;

($\Rightarrow$) By definition, $X \to Y \Rightarrow ker(X) \subseteq ker(Y)$; trivially, as $ker(f \bullet g) = ker(f) \cap ker(g)$, for any function product $f \bullet g$, and as set intersection is associative, $ker(Y) = ker(A_1) \cap \ldots \cap ker(A_n)$; obviously, as $ker(A_1) \cap \ldots \cap ker(A_n) \subseteq ker(A_i)$, for any $1 \leq i \leq n$, it follows that $ker(X) \subseteq ker(A_i)$, so $X \to A$, $\forall A \in Y$.

($\Leftarrow$) By definition, $X \to A$, $\forall A \in Y \Rightarrow ker(X) \subseteq ker(A_i)$, for any $1 \leq i \leq n$, which obviously means that $ker(X) \subseteq ker(A_1) \cap \ldots \cap ker(A_n)$; as $ker(Y) = ker(A_1) \cap \ldots \cap ker(A_n)$, it follows that $ker(X) \subseteq ker(Y)$, so $X \to Y$.

Q.E.D.

3.4. Consider a table $r(U)$ and X_1, X_2, $X \subset U$ such that $X_1 \cup X_2 = U$ and $X = X_1 \cap X_2$; $r(U)$ decomposition over X_1, X_2 is lossless $\Leftrightarrow r(U)$ satisfies MVD $X \to\to X_1$ (or, equivalently, $X \to\to X_2$).

Proof:

($\Rightarrow$) Let $r = \pi_{X1}(r) |><| \pi_{X2}(r)$ and t_1, $t_2 \in r$ such that $t_1[X] = t_2[X]$; it has to be shown that $\exists t \in r$ such that $t[X_1] = t_1[X1]$ and $t[X_2] = t_2[X_2]$, which would confirm the fact that r satisfies $X \to\to X_1$. Let $t_1' = t[X_1]$ and $t_2' = t[X_2]$; it follows that $t_1' \in \pi_{X1}(r)$, and $t_2' \in \pi_{X2}(r)$. By join definition, t is a tuple of $\pi_{X1}(r) |><| \pi_{X2}(r)$ that comes out of t_1' and t_2'. Finally, as $r = \pi_{X1}(r) |><| \pi_{X2}(r)$, it follows that $t \in r$, and that r satisfies $X \to\to X_1$.

($\Leftarrow$) Let r be an instance satisfying $X \to\to X_1$ and $t \in \pi_{X1}(r) |><| \pi_{X2}(r)$; for proving this implication, it has to be shown that $t \in r$ (because $r \subseteq \pi_{X1}(r) |><| \pi_{X2}(r)$ always). From the join and projection operators definitions, $\exists t_1, t_2 \in r$ such that $t[X_1] = t_1[X_1]$ and $t[X_2] = t_2[X_2]$; from the MVD definition, as $X = X_1 \cap X_2$, so $t_1[X] = t_2[X]$, $\exists t' \in r$ such that $t'[X_1] = t_1[X_1]$ and $t'[X_2] = t_2[X_2]$. This obviously means that $t = t'$, which proves that $t \in r$.

Q.E.D.

3.5.* (*DKNF characterization theorem*)

A table is in DKNF if and only if it does not have insertion or deletion anomalies.

Proof:

($\Rightarrow$) Let R be a DKNF table, r a valid instance of R, and t a tuple compatible with r; then, $r \cup \{t\}$ satisfies all constraints of R as, from the compatibility definition, primitive constraints are satisfied, and, from the DKNF definition, primitive constraints imply all other constraints.

Similarly, for any $t \in r$, $r - \{t\}$ satisfies all constraints of R as, obviously, deleting a tuple cannot violate any primitive constraint, and, from the DKNF definition, primitive constraints imply all other constraints.

Consequently, R does not have update anomalies.

($\Leftarrow$) Let R be a table free of update anomalies; let us assume that R is not in DKNF; as R is not in DKNF, there is at least one constraint c that is not implied by the primitive constraints; this means that there is an instance r of R in which c does not hold, although all primitive constraints are holding; as the constraint set associated to R is coherent, it follows that there is at least an instance r^* of R that is valid (and, consequently, c holds in it too); let us build r from r^* in the following two steps: first, deleting from r^* all rows, one by one, and secondly inserting in r^*, again one by one, all rows of r (which is obviously possible, as both instances are finite, and all constraints on R are static); as r^* is valid and r is not, then, during this building process, there has to be a tuple t and two instances r' and r'' such that:

1. r' valid
2. r'' invalid and
3. $r'' = r' \cup \{t\}$ or $r'' = r' - \{t\}$.

Let us consider both above possible cases:

- if $r'' = r' \cup \{t\}$, then $r' \subset r$; as r satisfies the primitive constraints, t is compatible with r'; as r'' is invalid, it follows that R has an insertion anomaly;
- if $r'' = r' - \{t\}$, then $r' \subset r^*$, so r' is valid; as r'' is not valid, it follows that R has a deletion anomaly.

Trivially, as in any above possible cases the assumption that R does not have update anomalies is violated, it follows that R is in DKNF.

Q.E.D.

3.6. Prove that C_5 from 3.7 (*ConstrRelation* $\circ$ *Key* = *Relation* $\circ$ *Attribute*) implies ($\forall x, y \in KEYS\_COLUMNS$, $x \neq y$) (*Key*(x) = *Key*(y)) $\Rightarrow$ *Relation* $\circ$

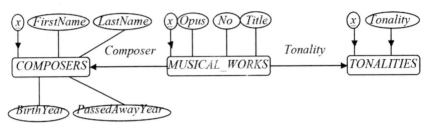

FIGURE 3.28   E-RD for the rdb in Figure 3.2.

There may be at most 10^8 composers characterized by their compulsory last names (at most 32 characters) and birth years (between 1200 and three years ago), as well as optional first names (at most 32 characters) and passed away years (between 1200 and current year). There may not be two composers having same first and last names. Nobody can compose before he/she is three or after his/her death. Composers may not live more than 120 years.

There may be at most 100 tonalities characterized by their compulsory names (at most 13 characters). There may not be two tonalities having same names.

There may be at most 10^{12} musical works characterized by their compulsory composer and opus number (at most 16 characters), as well as optional number within opus (a natural between 1 and 255), title (at most 255 characters), and tonality. There may not be two musical works by a same composer having same opuses and numbers.

FIGURE 3.29   Description of the rdb scheme in Figure 3.2.

$Attribute(x) = Relation \circ Attribute(y))$ (all attributes of a key should belong to a same relation) and that the converse is not true.

Proof:

$(\Rightarrow) Relation \circ Key = Relation \circ Attribute \Rightarrow \forall x,y \in KEYS\_COLUMNS$, $x \neq y$, $Relation(Key(x)) = Relation(Attribute(x)) \wedge Relation(Key(y)) = Relation(Attribute(y))$; let's consider that $Key(y) = Key(x)$; as *Relation* is a function, it follows that $Relation(Key(y)) = Relation(Key(x))$ and,

1. **COMPOSERS**

   a. Range restrictions:

   - *x* (cardinality): 10^8 (*RC*1)

   - *FirstName, LastName*: ASCII(32)   (*RC*2)

   - *BirthYear*: [1200, *currentYear*() — 3]   (*RC*3)

   - *PassedAwayYear*: [1200, *currentYear*()]   (*RC*4)

   b. Compulsory data: x, LastName, BirthYear   (*RC*5)

   c. Uniquenesses:

   - *x* (surrogate key)   (*RC*6)

   - *FirstName • LastName* (there may not be two
     composers having same first and last names)   (*RC*7)

   d. Other restrictions:

   - Nobody can compose before he/she is three or
     after his/her death.   (*RC*8)

   - Composers may not live more than 120 years.   (*RC*9)

2. **MUSICAL_WORKS**

   a. Range restrictions:

   - *x* (cardinality): 10^{12} (*RM*1)

   - *Opus*: ASCII(16)   (*RM*2)

   - *Title*: ASCII(255)   (*RM*3)

   - *No*: [1, 255]   (*RM*4)

   b. Compulsory data: x, Composer, Opus   (*RM*5)

   c. Uniquenesses:

   - *x* (surrogate key)   (*RM*6)

   - *Composer • Opus • No* (there may not be two
     musical works of a same composer having same
     opus and number)   (*RM*7)

3. **TONALITIES**

   a. Range restrictions:

   - *x* (cardinality): 100   (*RT*1)

   - *Tonality*: ASCII(13)   (*RT*2)

   b. Compulsory data: x, Tonality   (*RT*3)

FIGURE 3.30   Restriction set for the rdb in Figure 3.2.

---

c. *Uniquenesses:*
- *x* (surrogate key)   (*RT4*)
- *Tonality* (there may not be two tonalities having
   same names)   (*RT5*)

FIGURE 3.30   (Continued)

by equality transitivity, that *Relation*(*Attribute*(*x*)) = *Relation*(*Key*(*x*)) = *Relation*(*Key*(*y*)) = *Relation*(*Attribute*(*y*)) $\Rightarrow$ *Relation* $\circ$ *Attribute*(*x*) = *Relation* $\circ$ *Attribute*(*y*)

($\Longleftarrow$) Let us consider $\forall x,y \in KEYS\_COLUMNS$, $x \neq y$, such that *Key*(*x*) = *Key*(*y*); it follows that *Relation* $\circ$ *Attribute*(*x*) = *Relation* $\circ$ *Attribute*(*y*); obviously, as *Relation* is a function, it follows that *Relation*(*Key*(*y*)) = *Relation*(*Key*(*x*)); let us denote u = *Relation* $\circ$ *Attribute*(*x*) = *Relation* $\circ$ *Attribute*(*y*) and v = *Relation*(*Key*(*y*)) = *Relation*(*Key*(*x*)), $u,v \in RELATIONS$; trivially, it does not follow that $u = v$ for any *x* and *y*, so the converse is not true.   *Q.E.D.*

3.7. Apply algorithm *REA*1–2 to the rdb from Fig. 3.2.
Solution:
Figure 3.28 presents the corresponding E-RD, Fig. 3.29 its associated restriction set, and Fig. 3.30 its full description.

3.8. Apply algorithm *A*7/8–3 to table *CITIES* from Fig. 3.5.
   Solution:
$T = CITIES$, $m = 6$, $C_1 = x$, $C_2 = City$, $C_3 = Commune$, $C_4 = *Country$, $C_5 = Zip$, $C_6 = Population$,
   $K = \{x, City \bullet Commune, *Country \bullet Zip\}$, $k = 3$;
   $K' = \{x, City \bullet Commune, *Country \bullet Zip\}$;
   - *x* is a surrogate autonumber key;
   - *City* $\bullet$ *Commune* is a key, as there may not be two cities of a same commune having same names;

- *Country • Zip* is a key, as there may not be two identical zip codes in a same country;
- *City* and *Commune* are prime, but are not keys as they are part of the key *City • Commune*;
- **Country* and *Zip* are prime, but are not keys as they are part of the key **Country • Zip*;

$SK = \{ x, City, Commune, *Country, Zip \}$;

$T' = \{ Population \}$;

- *Population* is not a key, as there may be two cities having same population; moreover, it is nonprime, as population cannot help with uniquely identifying cities.

$T' = \emptyset$;

$T' = \{ City, Commune, *Country, Zip \}$;

$n = 4$;

- as $4 > 1$:

$l = 3$;

$kmax = C(4, 2) = 4!/(2! * 2!) = 4 * 3 * 2/(2 * 2) = 3 * 2 = 6$;

$i = 2$;

$allSuperkeys = false$;

- as $2 \leq 4$ *and* $3 < 6$ *and true* = *true*:

$allSuperkeys = true$;

- *City • *Country* is not a key, as there may be two cities of a same country (but of different communes) having same names;
- *City • Zip* is not a key, as there may be two cities (but of different countries) having same zip codes;
- *Commune • *Country* is not a key, as, generally, communes have several cities;
- *Commune • Zip* is not a key, as there may be two cities of a same commune having same zip codes;

$allSuperkeys = false$;

$i = 3$;

- as $3 \leq 4$ *and* $3 < 6$ *and true* = *true*:

$allSuperkeys = true$;

- as all products of arity 3 are superkeys, the algorithm halts without discovering any new keys, so that $K' = \{x, City • Commune, *Country • Zip\}$.

3.14.2 PROPOSED EXERCISES

*Some problems are so complex that you have to be highly intelligent
and well-informed just to be undecided about them.*
—Laurence J. Peter

3.9. Apply to the following E-R data models the algorithm $A1$–7 (for
obtaining corresponding relational schemas and nonrelational constraint
sets) and provide a plausible valid instance of at least two rows for each
table:

    a. the one from Fig.2.10 and restriction set specified in Subsections
       2.7.1 to 2.7.4;

    b. the ones obtained in Exercises 2.1 to 2.5 above.

3.10. Apply the algorithm $REA1$–2 to the following rdbs:

    a. the ones from Figs. 3.4 and 3.5;

    b. the one presented in section 3.7.

3.11. Apply algorithm $A7/8$–3 to the Tables 3.8 to 3.13 and 3.18,
3.19, 3.20, 3.21, and 3.22, as well as to all tables from Figs. 3.13 and 3.26.

3.12. Design and develop needed standard ANSI-92 *SQL* statements
for creating and populating the tables shown in Figs. 3.4 and 3.5, as well as
those from subsection 3.7 (Tables from 3.7 to 3.13). You can do it directly
in the *SQL* of a RDBMS version of your choice, so as to be able to also
test them.

3.13. Design, develop, and test *SQL* queries for the ones from section
1.6 (on the corresponding rdb obtained from Exercise 3.9.b).

3.14. Design, develop, and test *SQL* statements for computing the following sets:

    a. from Fig. 3.2:

    – <*First* and *LastName, Title, TonalityName, Opus, No*> for all
      musical works, in the ascending order of the first and last names,
      then tonality, opus, and number.

    – <*First* and *LastName, LifeYears*> for all composers that lived
      at least k years, k natural parameter, in descending order of the
      numbers of years they lived and then ascending on first and last
      names.

    – <*First* and *LastName, WorksNo, TonalitiesNo*> for all composers
      that wrote in at least k tonalities and for each such tonality at least

n works, *k* and *n* natural parameters, in descending order of the numbers of tonalities, of works, and then ascending on first and last names; give (and compare their execution times after adding at least 100 other works) two solutions: one with and one without subqueries.

b. from Fig. 3.4:

– *<CountryName, CapitalName, Population, StatesNo, CitiesNo>* for all countries, including those for which the capital is not known, in the descending order of population, cities and states numbers, and then ascending on country names.

– *< CityName, StateName, CountryName, Population>* for all cities that are both country and state capitals, in the descending order of population and then ascending on country name.

c. from Fig. 3.5:

– *<CountryName, StateName, CommuneName, CityName, Population>* for all cities and communes, in the descending order of their population and then ascending on country, state, commune, and city names.

– *<CountryName, StateName, CommunesNo, CitiesNo, TotCitiesPopul.>* for all states of all countries, including those that have no communes and/or cities, in the descending order of the total city populations, number of communes, then cities, and then ascending on country and state name.

d. from Table 3.3: *<FirstandLastNames, SSN, CoordEmployeesNo>* for all managers, in descending order of the number of corresponding coordinated employees and then ascending on their first and last names.

e. from Fig. 3.6: *<CountryName, CapitalName, SubordCountriesNo, MinSubordYear>*, including countries with not known capitals, in descending order of the number of subordinated countries and then ascending on the minimum subordination year and, finally, country name.

3.15. Design, develop, and test *SQL* statements for:

a. adding all remaining U.S. states data in the *STATES* table of Fig. 3.4.

      b. replacing country names from the *COUNTRIES* table of Fig. 3.4 with their corresponding uppercase names.

      c. deleting from table *TONALITIES* of Fig. 3.2 all lines for which there is not at least one musical work composed in those tonalities.

    3.16. Design, develop, and test *SQL* statements for replacing the instance of a temporary table *CAPITALS(CountryName, StateName, CapitalName)* with the corresponding data from the tables in Fig. 3.4, in ascending order of countries, states, and capital names, only for cities that are both country and state capitals.

    3.17. Design and develop needed standard ANSI-92 *SQL* statements for the following queries on the metacatalog rdb presented in section 3.7 and compute the corresponding results (alternatively, you can do it directly in the *SQL* of a RDBMS version of your choice, so that to may also test them):

      a. Compute the set of quadruples <db name, no. of relations, total no. of attributes, total no. of constraints>, in the descending order of the numbers of relations, then on the one of constraints, then of attributes, and then ascending on the db names.

      b. Compute the set of quadruples <db name, relation name, no. of attributes per relation, no. of constraints per relation> only for those dbs that have at least 3 relations, in the descending order of the numbers of attributes, then on the one of constraints, and then ascending on the db and relation names.

      c. Compute the set of quintuples <db name, relation name, attribute name, no. of keys containing current attribute, no. of foreign keys containing current attribute> only for those relations that have at least n keys and m foreign keys (n and m natural parameters), in the descending order of the numbers of keys, then on the one of foreign keys, and then ascending on the db, relation, and attribute names.

      d. Compute the set of quintuples <db name, relation name, attribute name, no. of keys containing current attribute, no. of foreign keys containing current attribute> only for the prime attributes, in the descending order of the numbers of keys, then on the one of foreign keys, and then ascending on the db, relation, and attribute names.

e. Compute the set of quintuples <db name, relation name, constraint name, constraint type, no. of attributes of the current constraint> only for those relations that have at least n constraints (n natural parameter), in the descending order of the numbers of attributes, and then ascending on the db, relation, and constraint types and names.

3.18. Prove that the join operator is commutative and associative.

3.19. Compute minimum and maximum cardinalities for the joins of n tables, $n > 1$, natural, based on its operands' cardinalities.

3.20. Prove that the natural join of two tables having identical schemas is equal to their intersection.

3.21. Is the full outer join equal to the Cartesian product? Why? Please exemplify.

3.22. Design, develop, and test equivalent LEFT, RIGHT, and FULL OUTER JOIN SELECT statements by using only INNER JOIN and UNION operators.

3.23. Provide necessary and sufficient conditions for SELECT statements to always compute same results regardless of the rdb instance when they only differ by DISTINCTROW instead of DISTINCT. Give an example and a counterexample.

3.24. a. Design, develop, and test a parameterized *SQL* statements family template for computing the difference between instances of two tables having identical schemas, by using the *IsNull* predicate.

b. Do same thing as above for the symmetrical difference.

3.25. Prove that division is a sort of Cartesian product inverse, as, for any tables t_1 and t_2, $(t_1 \times t_2) \div t_2 = t_1$.

3.26. (a) Try to design, develop, and test a SELECT having in its WHERE clause a parameterized IN predicate. Why do you think it does not execute? (*hint*: revise RDBMSs data types).

(b) Design, develop, and test, however, a pure *SQL* (*ANSI-92*) solution to this problem (*hint*: use a temporary table for storing actual parameter values).

3.27. a. Add to Table 3.3 at least 10 new rows such as to store in *ReportsTo* a hierarchy of at least 5 levels.

b. Design, develop, and test a *SQL* query for computing the set of all "leaf" employees (i.e., those to whom nobody is reporting to).

c. Design, develop, and test a *SQL* query for computing the set of all pairs <"leaf" employee, employee to whom the "leaf" employee reports to>.

d. Design, develop, and test a *SQL* query for computing the set of all similar pairs for the first and the second hierarchical levels (*hint*: if 1 reports to 2 and 2 reports to 3, then the answer should include the set {<1,2>, <2,3>, <1,3>}).

e.* Design, develop, and test a high level programming language program (e.g., *C#*, *Java*, *C++*, *VBA*, etc.) that generates the *SQL* query for computing the set of all such pairs up to the k-th hierarchical level, k being a natural parameter; test the resulting *SQL* output for $k = 4$.

f.* Design, develop, and test for $k = 4$ and k greater than the *ReportsTo* tree height a high level programming language program (e.g., *C#*, *Java*, *C++*, *VBA*, etc.) with embedded *SQL* to compute same thing, by using a temporary table for storing the result; if $k \leq 0$, then compute the subset of all employees reporting to nobody (*hint*: pad second column with nulls); if k greater than the highest hierarchical level h, stop immediately after adding level h pairs, that is, when the transitive closure of *ReportsTo* was computed (*hint*: add to the temporary table a *Level* numeric column; when $k > 0$, initialize it with the result of query (c) above, all of them having 1 for *Level*; loop until k or until no more lines are added to the temporary result table; in each step i, increase *Level* values by 1 and add to the temporary tables the pairs of the current level, which can be obtained by a join between the rows of the temporary table having the *Level* values equal to $i - 1$ and Table 3.3).

g.** Compute the transitive closure of *ReportsTo* in extended *SQL* (e.g., in *T-SQL*, *PL/SQL*, etc.); establish a top of execution speeds for computing it with solutions (e), (f), and (g), when Table 3.3 has at least 100 rows and 16 levels.

3.28. Consider the subuniverse of a simple OS network file management system that is storing only file names, extensions, sizes, folders (which are files too, having size 0!), virtual, logic, and physical drives, as well as computers.

a. Using algorithms $A0$, $A1$–7 and $A7/8$–3, as well as the best practice rules from sections 2.10 and 3.11 design and implement its
DKNF rdb.

b. Populate it with at least 2 computers, each of which having at
least 2 physical, 4 logic, and 2 virtual drives; each logic drive
should have at least 16 non folder files and 8 folders structured
in hierarchies of at least 4 levels.

*c.** Design, develop, and test a program for computing the total
size of all files of a subtree of folders, the parameter being the
path of its root folder (*hint*: see (*f*) and (*g*) of exercise 3.27
above, as this is also a transitive closure type problem).

*d.** Design, develop, and test a program for computing the set
<computer name, physical drive name, logic drive letter, full
folder path, file name and extension, size> for all computers
having at least k logic drives, each of which having at least
m folders with at least n files each, each such file having size
at least s, where k, m, n, and s are natural parameters, in the
ascending order of computer names, logic drive letters, full
folder paths, then descending on size and ascending on file
names and extensions (*hint*: also see (*f*) and (*g*) of exercise
3.27 above, as computing full paths is also a transitive closure
type problem).

3.29.** Prove that transitive closures are not the only meaningful
queries that complete relational languages cannot express, although their
results contain only rdb stored data[87] (*hint*: consider Table 3.3 and try to
design *SQL* statements for computing pairs of employees reporting to a
same person such that the result does not contain any pair (*f, g*) if it already
contains the pair (*g, f*). Design, develop, and test a program in a high level
programming language at your choice, which embeds *SQL*, for computing
this desired result. Generalize!).

3.30. Translate all remaining *SQL* queries from subsection 3.9.2,
as well as those from exercises 3.12 to 3.14 into corresponding RA
expressions.

---

[87] And, moreover, they are invariant with respect to any automorphism of its instance (where *automor-
phisms* are autofunctions renaming values such that the corresponding set remains the same).

3.31. Prove that the RA selection and projection operators are monotonic. Extend monotonicity definition to n-ary operators, $n > 1$, natural, and prove that the RA join operator is monotonic too.

3.32. Consider any m tables $t_1(X_1), \ldots, t_m(X_m)$, $m \geq 1$, natural (where X_j-s are their schemas); prove that the projection operator is idempotent, as:
$\pi_{Xj}(\bowtie|_{k=1}^{m}(\pi_{Xk}(\bowtie|_{i=1}^{m} t_i))) = \pi_{Xj}(\bowtie|_{i=1}^{m} t_i)$, $\forall j$, $1 \leq j \leq m$.

3.33. Consider same tables as above; prove that there is a duality between projection and join, as:

(a) $\pi_{Xj}(\bowtie|_{i=1}^{m} t_i) \subseteq tj$, $\forall j$, $1 \leq j \leq m$.

(b) $\pi_{Xj}(\bowtie|_{i=1}^{m} t_i) = tj$, $\forall j$, $1 \leq j \leq m \Leftrightarrow t_1, \ldots, t_m$ have a full join.

3.34. Consider table $t(X)$ and $X_1, \ldots, X_m$, $m \geq 1$, natural, attribute sets such that $\cup_{j=1}^{m} Xj=X$; prove the following somewhat dual results of the above ones:

(a) $\bowtie|_{j=1}^{m}(\pi_{Xj}(t)) \supseteq t$.

(b) $\bowtie|_{k=1}^{m}(\pi_{Xk}(\bowtie|_{j=1}^{m}(\pi_{Xj}(t)))) = \bowtie|_{j=1}^{m}(\pi_{Xj}(t))$.

3.35. Prove that any RA expression may be transformed into an equivalent one in which selections have only atomic conditions, projections eliminate only one attribute, and the rest of the needed operators are unions, renaming, differences, and Cartesian products.

3.36. Prove that any RA expression may be transformed into an equivalent one whose constant tables are all defined over only one attribute and contain only one row.

3.37. Prove that the proposition corresponding to no-information null values (i.e., $R_{-k}(t[A_1], \ldots, t[A_{k-1}], t[A_{k+1}], \ldots, t[A_n]))$) is equivalent to the disjunction of the formulas associated to the other two types of null values: $x(R(t[A_1], \ldots, t[A_{k-1}], x, t[A_{k+1}], \ldots, t[A_n])) \vee (x(R(t[A_1], \ldots, t[A_{k-1}], x, t[A_{k+1}], \ldots, t[A_n])) = R_{-k}(t[A_1], \ldots, t[A_{k-1}], t[A_{k+1}], \ldots, t[A_n]))$.

3.38. Prove that $\forall Y \subset X$, $card(X) = card(\pi_Y(X)) \Leftrightarrow \exists K \subseteq Y$, K key.

3.39. Apply algorithm $A7/8-3$ to the following table schemas:

a. RULERS (x, Title, Name, Sex, Dynasty, Father, Mother, BirthDay, BirthPlace, PassedAwayDay, PassedAwayPlace, BurrialPlace, KilledBy, Nationality, Notes)

b. REIGNS (x, Ruler, Country, FromDate, ToDate)

c. BATTLES (x, BattleYear, BattleSite, Ruler, Opponent, RuledArmy, EnnemyArmy, Victory?, Notes)

d. MARRIAGES (x, Husband, Wife, WeddingDate, DivorceDate)

e. *MONUMENTS_PICTURES* (*x*, *Monument*, *JPEGPicture*, *Picture-Size*, *PictureDescription*)

3.40. Let R be any rdb table and c_1, ..., c_n its columns, $n > 1$, natural, $1 \leq i, j \leq n$; if $\exists j$ such that, $\forall i$, $c_i \to c_j$, c_j is *minimal*; dually, if $\exists i$ such that, $\forall j$, $c_i \to c_j$, c_i is *maximal*; prove that:

(*i*) c is minimal $\Leftrightarrow$ card($Im(c)$) = 1

(*ii*) c is maximal $\Leftrightarrow$ c is one-to-one.

3.41. Let S be any set and f, g, h, f_1, ..., f_n any functions defined on it, $n > 1$, natural; let K and K' be any function (Cartesian) products defined on S; prove that:

(*i*) $ker(f \bullet g) = ker(f) \cap ker(g)$

(*ii*) $f_1 \bullet ... \bullet f_n \to f_i$, $\forall i \in \{1, ..., n\}$

(*iii*) $g \to f_i$, $\forall i \in \{1, ..., n\} \Rightarrow g \to f_1 \bullet ... \bullet f_n$.

(*iv*) K is a key $\Rightarrow K \to f$, for any f

(*v*) K is a key and $K' \to K \Rightarrow K'$ is one-to-one.

3.42. Let R be any rdb relation and f, g, h some of its columns; if $f \to g$ and $g \to f$ then f is *functionally equivalent* to g[88], denoted $f \leftrightarrow g$; prove that:

(*i*) the *functional equivalence* (*FED*) "$\leftrightarrow$" is an equivalence relation

(*ii*) all minimal columns of a relation are functional equivalent

(*iii*) all maximal columns of a relation are functional equivalent

(*iv*) all keys of a relation are functional equivalent

(*v*) $(f \bullet g) \bullet h \leftrightarrow f \bullet (g \bullet h)$

(*vi*) $f \bullet g \leftrightarrow g \bullet f$

(*vii*) $f \bullet f \leftrightarrow f$.

3.43. Prove that when discarding a's one-to-oneness from Proposition 3.2, 3.2(*i*) and 3.2(*ii*) still hold, while 3.2(*iv*) does not hold anymore, and 3.2(*iii*) becomes card($Im(fk)$) $\leq$ card($Im(f)$).

3.44. Prove that FD $X \to Y$ is trivial $\Leftrightarrow X \supseteq Y$.

3.45. Let $R(U)$ be a table and $n = $ card(U), $n > 1$, natural; prove that the number of trivial FDs over $R(U)$ is $3^n - 2^n$.

3.46. Prove that in any instance of a table $T(U)$, if $X \bullet Y = U$, then $X \to Y \Leftrightarrow X$ is a key.

3.47. Prove that any table satisfies FD $X \to Y$ if and only if it satisfies $X \to A$, $\leftrightarrow A \in Y$.

---

[88] Note that, trivially, the kernels of two equivalent functions are equal and any function is equivalent to itself.

3.48. Let $t(U)$ be a table and X_1, X_2, $X \subset U$ such that $X_1 \bullet X_2 = U$ and $X = X_1 \cap X_2$; $t(U)$ has a lossless decomposition over X_1, X_2 if it satisfies at least one of the following FDs: $X \to X_1$ or $X \to X_2$.

3.49. Prove that if instance i of a table satisfies $X \to Y$, $Z \subseteq X$ and $W \supseteq Y$, then i satisfies $Z \to W$ as well.

3.50. Prove that if instance i of a table satisfies $X \to Y$ and $Z \to W$, $Z \subseteq Y$, then i satisfies $X \to YW$ as well.

3.51. Prove that (*unikey FD schemas characterization theorem*): a scheme $[R, F]$, such that $R(U)$ and $F = \{X_1 \to Y_1, \ldots, X_k \to Y_k\}$, $k > 0$, natural, has only one key if and only if $U - Z_1 \bullet \ldots \bullet Z_k$ is a superkey, where $Z_i = Y_i - X_i$, $1 \le i \le k$.

3.52. Prove that BCNF $\Rightarrow$ 3NF $\Rightarrow$ 2NF $\Rightarrow$ 1NF (any scheme in BCNF is in 3NF too, any one in 3NF is in 2NF too, and any one in 2NF is in 1NF too).

3.53. By definition, a table scheme S with FDs is in *(3,3)-NF* if for any logic proposition F, a nontrivial FD $X \to A$ holds in $\sigma_F(s)$ for any valid instance s of S if and only if X is a superkey of S; prove that (3,3)-NF $\Rightarrow$ BCNF (a scheme in (3,3)-NF is in BCNF too).

3.54.* A scheme $[R, F]$ includes an *elementary key* K if there is at least an attribute $A \in R$ such that $\neg \exists K' \subset K$ with $K' \to A \in F^+$; $[R, F]$ is in the *elementary keys NF (EKNF)* if for any nontrivial FD $X \to A \in F$, either X is an elementary key or A is part of such a key. Prove that EKNF is strictly stronger than 3NF, but strictly weaker than BCNF.

3.55. Prove that (*MVD complementary property*) if $R(U)$ satisfies MVD $X \to\!\!\to Y$, then it satisfies MVD $X \to\!\!\to U - XY$ too.

3.56. Prove that (*trivial MVDs characterization theorem*) $X \to\!\!\to U$ of T is trivial if and only if $U \subseteq X$ or $T = X \bullet U$.

3.57. Consider table $t(U)$ and $XY \subseteq U$; prove that:
   (a) t satisfies FD $X \to Y \Rightarrow t$ satisfies MVD $X \to\!\!\to Y$
   (b) the reverse is not true.

3.58. Prove that a relation may satisfy MVD $X \to\!\!\to YZ$, although it violates FD $X \to Y$.

3.59. Prove that $X \to\!\!\to Y \wedge X \to\!\!\to Z \Rightarrow X \to\!\!\to Y - Z$ (and consequently, due to complementarity, it also implies $X \to\!\!\to Y \cap Z$).

3.60. Prove that:
   (a) 4NF $\Rightarrow$ BCNF (a scheme in 4NF is in BCNF too)

(b) above implication would not be true were 4NF only referring MVDs from Γ instead of Γ^+.

3.61. Prove that a table in 4NF cannot have more than one pair of complementary nontrivial MVDs.

3.62. (*i*) What is the truth value of the following sentence: "Any table with two columns is in PJNF"?

   (*ii*) Propose and prove a proposition that provides necessary and sufficient conditions for a two columns table to be in PJNF.

3.63. Prove that 5NF $\Rightarrow$ 4NF (a scheme in 5NF is in 4NF too).

3.64. Prove that:

   (*i*) a scheme [R, F] is in BCNF (4NF, respectively) if for any FD (MVD, respectively) $d \in F$ there is only one key dependency implying d.

   (*ii*) a similar result for PJNF (i.e., d is a JD) is not true.

3.65.* Prove that, generally, given a set F of FDs, it is not possible to find a set M of MVDs such that an instance satisfies F if and only if it satisfies M.

3.66.* Let $R(U)$ be a relation scheme, D its associated domain constraint set, K its key dependency set, and M an associated set of FDs and MVDs;

   (*i*) if, for any attribute $A \in U$, $card(D_A) \geq 2$, then, for any FD or MVD d of M, d is implied by $D \cup M$ if and only if it is implied by M;

   (*ii*) Prove that in the absence of $card(D_A) \geq 2$, (a) above does not hold anymore;

   (*iii*) if, for any attribute $A \in U$, $card(D_A) \geq n$, $n > 1$, natural, then, for any JD j over U having n components (i.e., of form $|><| [X_1, X_2, ..., X_n]$, where $X_i \subseteq U$, $\leftrightarrow i$, $1 \leq i \leq n$), j is implied by $D \cup K$ if and only if it is implied by K.

3.67. (*DKNF is the highest RDM normal form theorem*) Consider a scheme [R, Γ] such that all domain constraints allow at least k distinct values, where natural $k > 1$ is the components number of the JD having the greatest number of components from all JDs of a cover J of the JDs subset of Γ^+, and, when this subset is empty, there are at least two possible distinct values for any attribute of any table from R; prove that:

   (*i*) DKNF $\Rightarrow$ PJNF (any scheme in DKNF is in PJNF too)

   (*ii*) DKNF $\Rightarrow$ (3,3)-NF (any scheme in DKNF is in (3,3)-NF too).

3.68. Give an example of an incoherent constraint set made out only from coherent constraints.

3.69. Prove that a constraint set Γ does not imply a constraint Γ if and only if there exists a db instance that satisfies Γ, but not Γ.

3.70.* Consider any two constraint sets Γ and Δ; prove that:

(i) $\Gamma \subseteq \Gamma^+$ (any constraint set is included in its closure)

(ii) $(\Gamma^+)^+ = \Gamma^+$ (the closure operator is idempotent)

(iii) $\Gamma \subseteq \Delta \Rightarrow \Gamma^+ \subseteq \Delta^+$ (the closure operator is monotonic)

(iv) $\Gamma^+ \cup \Delta^+ \neq (\Gamma \cup \Delta)^+$ (the closure operator is not additive)

3.71. Prove that C_9 from Fig. 3.21 (*ConstrRelation ° ForeignKey = Relation ° FkAttribute*) implies the constraint ($\forall x,y \in FOREIGN\_KEYS\_COLS$, $x \neq y$) (*ForeignKey*(x) = *ForeignKey*(y) $\Rightarrow$ *ConstrRelation ° FkAttribute*(x) = *ConstrRelation ° FkAttribute*(y)) (all attributes of a foreign key should belong to a same relation) and that the converse is not true.

3.72. Prove that the referential integrity constraint (C_{10} from Fig. 3.21: ($\forall x \in FOREIGN\_KEYS\_COLUMNS$)(*Im*(*FKAttribute*($x$)) $\subseteq$ *Im*(*RefAttribute*(x)))) implies the constraint ($\forall x \in FOREIGN\_KEYS\_COL$-*UMNS*)(*Im*(*Domain ° FKAttribute*(x)) $\subseteq$ *Im*(*Domain ° RefAttribute*(x))) (foreign keys' domains should always be included in the corresponding referenced attributes' domains) and that the converse is not true.

3.73*. Considering the relations from Fig. 3.21, prove the following proposition: ($\forall x \in KEYS\_COLUMNS$)($\forall y \in FOREIGN\_KEYS\_COLS$)(*PrimaryKey*(*Relation*(*Attribute*(x))) = *Key*(x) $\wedge$ *Attribute*(x) = *FKAttribute*(y) $\Leftrightarrow$ *Relation*(*Attribute*(x)) $\subseteq$ *Relation*(*RefAttribute*(y))) (a primary key of a relation is a foreign key if and only if that relation corresponds to a set that is included in the set to which corresponds the relation referenced by the primary key).

3.74. Prove that the metacatalog scheme presented in Section 3.7 is in DKNF.

3.75. Design a DKNF extension to the metacatalog scheme presented in Section 3.7 for also storing tuple constraints. Provide a nonvoid, valid, and plausible instance for it.

3.76. Consider the scheme *STUDENTS*(x, *Course, Credits, Student, Sex, Age*). Identify its *FD*s and keys. Design an equivalent BCNF scheme. Is a DKNF equivalent one obtainable too? Justify your answer.

3.77. Consider the scheme *STUDENTS*(*#Student*, *#Lab*, *#PC*, *Student*, *Lab*, *PC*) having following *FD*s: *#Student* → *Student*, *#Student* → *Lab*, *#Student* → *#Lab*, *#Student* → *PC*, *#Lab*→ *PC*, *#Lab* → *Lab*, *#PC* → *PC*, *#PC* → *#Lab*, and *Lab* → *#Lab*. Analyze if it has anomalies and design an equivalent DKNF one.

3.78. Consider a "technological" db for storing data on items (spare parts, submodules, modules, products, etc.) that are made out of other items. The technological item structures are of tree type.

a. Design and implement in any DBMS of your choice a corresponding DKNF db, only containing item code, description, price, and structure. Provide a valid instance with at least 4 tree levels.

b. Design and test for this instance SPJR expressions and corresponding *SQL* statements for computing the subsets of:

(*i*). structures (<item code and description, immediately superior item code and description>) for all items having at least three embedded levels;

(*ii*). items (<code, description, price>) for which price is at most 10% of the price of its immediately superior item;

3.79. Consider a HR subuniverse interested in employees' first and last name, sex, birth date and place, home address, hire and end working dates, and working place (subsidiary), as well as subsidiaries' name, address, and subordination (e.g., *Microsoft Romania srl*, from Bucharest, Romania, is subordinated to *Microsoft Central and Eastern Europe GmbH*, from Munich, Germany, which is subordinated to *Microsoft Corporation*, from Redmond, U.S.A.).

a. Design and implement in any DBMS of your choice a corresponding DKNF db and provide a valid instance with at least four rows per table.

b. Design and test for this instance SPJR expressions and corresponding *SQL* statements for computing the subsets of:

(*i*) stored subordinations (working place name, direct subsidiary name), for all working places coordinating at least three levels of subsidiaries;

(*ii*) employees that worked in their birth city at a working place directly subordinated to an entity having its head-quarters in the same city;

3.80. Consider the set of courses offered by a University; each of them has a (possibly empty) set of prerequisite ones (i.e., courses that have to be completed prior to enrollment to the current one); only titles are of interest for courses. For each course several books may be recommended and some books may be even recommended for several courses; only the title and the (co)authors are of interest for books. For students, only first, middle, and last name, birth date and place, completion degrees, and current enrollments are of interest.

a.   Design a DKRINF db scheme for this subuniverse and provide a plausible instance of it with at least four rows per table.

b.   Design and test for this instance SPJR expressions and corresponding *SQL* statements for computing the subsets of:

(*i*)   courses (title, precondition title) having at least two preconditions;

(*ii*) courses (title) for which at least four students are enrolled;

(*iii*) students (first, middle, and last name, birth date and place) that enrolled to courses for which books containing "Mathematical Logic" in their titles are recommended;

(*iv*) completions (student first, middle, and last name, birth date and place, course title) for which course title is identical or included in the title of at least one recommended book for that course.

3.81.

a.   Refine the rdb scheme from Fig. 3.5, by adding a *STREETS* table to store street names (which, generally, are repeatedly found in dozens, if not hundreds of cities) and all needed foreign keys, such as to obtain a DKNF scheme. Provide a plausible instance of it with at least two rows per city from *CITIES*.

b.   Apply the algorithm *A*7/8–3 for finding all corresponding *STREETS* keys.

c.   Design and test for this instance SPJR expressions and corresponding *SQL* statements for computing the subsets of:

(*i*)   countries having at least two states each of which having at least two cities each having at least 1,000,000 inhabitants;

(*ii*) states (name, country name, capital city name) to which capital cities having at least 1,000,000 inhabitants belong to;

(*iii*) Romania's states (name, vehicle plate code, telephone prefix, state capital city name) whose names are different from their capital city name;

   (*iv*) countries (name, vehicle plates code, telephone prefix, capital city name) having same name as their capital city name;

   (*v*) cities (country name, capital city name, state name, state capital city name) that are country capitals, but not state capitals;

   (*vi*) streets (country name, state name, city name, street name) that have names existing in at least two cities, in the ascending order of street, country, state, and city names.

  3.82. Build a directed graph having as nodes all RDM concepts from this chapter, including exercises, and with arrows (implicitly) labeled "directly based on" (*hint*: some definitions are leaves, whereas propositions, theorems, corollaries, and the rest of the definitions are nonterminal nodes).

  a. Which are its sets of minimal and maximal elements?

  *b.** Design the corresponding DKNF db scheme and provide its full instance (*hint*: extend the RDM metamodel from Subsection 3.7.2).

  3.83. Consider the following table scheme: *PEOPLE* (*x*, *FirstName*, *LastName*, *Sex*, *BirthDate*, *Mother*, *Father*), where *Mother* and *Father* are acyclic autofunctions (i.e., foreign keys referencing *PEOPLE.x* whose column values are not allowed storing cycles, i.e. the fact that somebody is his/her own ancestor/descendant).

  a. Provide at least three different valid instances, each having at least 4 rows per table.

  b. Design and test for every instance a SPJR expression and its corresponding *SQL* statement for computing the set <*x*, *First&LastName*, *Sex*, *BirthDate*, *xSibling*, *SiblingFirst&LastName*, *SiblingBirthDate*, *xCommonGrandFather*, *xCommonGrandMother*>.

  3.84. Consider the table *FILES* from Fig. 3.31.

  a. Apply algorithm *A*7/8–3 for discovering all of its semantic keys.

  *b.** Discover all of its other (both relational and non-relational) constraints.

  *c.*** Design and develop in any high-level event-driven programming language of your choice (e.g., *Java*, *C#*, *VBA*, etc.) needed methods for enforcing all found non-relational constraints.

  d. Design the *SQL* statements (both with and without subqueries) for computing the set of all files from the root folder, its subfolders, and their subsubfolders; what is the corresponding result?

  e. Same as (*c*) above, but also adding the fourth folder hierarchical level of files.

x	FileName	FileExt	Folder	FileSize	Folder?
0				0	Y
1	Bootmgr		0	399	N
2	Swapfile	sys	0	266,144	N
3	Windows		0	0	Y
4	Explorer	exe	3	2,336	N
5	System32		3	0	Y
6	Shutdown	exe	5	35	N
7	WinMetadata		5	0	Y
8	Windows.Security	winmd	7	45	N

FIGURE 3.31   Some files and folders from a MS *Win8* computer.

 f. Design and develop a parameterized method in any high-level programming language of your choice (e.g., *Java, C#, VBA*, etc.) that
accepts as parameters a table name *T*, one of its columns name *C*,
and a natural value *k*, and generates and stores in a (e.g., .txt or
.sql) file the corresponding ANSI-92 *SQL* standard statements for
computing the *k*-th level *C*'s partial closure. Run this method for
T = "FILES", *C* = "Folder", and *k* = 7. Run the corresponding generated *SQL* statement over table *FILES'* instance.

 3.85. Consider the RDM model of E-RDM presented in Subsection
3.3.2.

 a. Apply the algorithm *A7/8–3* to each of its tables, thus proving that
all keys are correctly declared and that there are no other keys.

 *b.** Automatically (re)compute **Arity* values and enforce the non-relational constraints (*RU5*) and (*RU6*) (*hint*: use either extended or
embedded *SQL*).

 3.86.* Prove that constraint C_{11} from Fig. 3.21 implies the following constraint: C_{14}: $(\forall x \in KEYS\_COLUMNS)(\forall y \in FOREIGN\_KEYS\_$
$COLS)(PrimaryKey(Relation(Attribute(x)))$ $=$ $Key(x) \wedge Attribute(x)$ $=$
$FKAttribute(y))$ $\Rightarrow$ $Relation(Attribute(x)) \neq Relation(RefAttribute(y)))$ (a
primary key of a relation must not reference that relation, that is trivial
inclusions $R \subseteq R$ should be banned).

 3.87. Consider the constraint set from Fig. 3.21 (Subsection 3.7.2).

 a. Prove that it is coherent.

*b.** Prove that it does not imply constraint C: $(\forall x \in CONSTRAINTS)$ $(\forall y_1, \ldots, y_n \in KEYS\_COLUMNS, n > 0, natural)(ConstrType(x) =$ 'K' $\wedge Key(y_1) = \ldots = Key(y_n) = x \Rightarrow Attribute(y_1) \bullet \ldots \bullet Attribute(y_n)$ is onto) (keys are onto)

*c.*** Prove that it is minimal.

d. Add a constraint to it such that it becomes incoherent.

3.88. Consider the http://www.databaseanswers.org/data_models/index. htm data models and:

a. apply algorithm $A7/8–3$ to all of their tables (in order to discover all of their keys);

b. apply algorithm $REA1–2$ to all models (in order to infer their corresponding E-R data models);

c. analyze each such E-R data model and correct it if necessary;

d. for each corrected E-R data model, apply algorithm $A1–7$ and compare its results with the corresponding initial data model.

3.89. Consider the extension from problem 2.11 and modify accordingly the solution of problem 3.0.

3.15   PAST AND PRESENT

Einstein's space is no closer to reality than Van Gogh's sky. The glory of science is not in a truth more absolute than the truth of Bach or Tolstoy, but in the act of creation itself. The scientist's discoveries impose his own order on chaos, as the composer or painter imposes his; an order that always refers to limited aspects of reality, and is based on the observer's frame of reference, which differs from period to period as a Rembrandt nude differs from a nude by Manet.

—Arthur Koestler

There are very many books on RDM. Not only from my point of view, the current "Bible" in this domain still remains (Abiteboul et al., 1995).

Chris Date's books (the latest being Date (2013, 2012, 2011, 2003)), are a must too. Other latest not to miss ones are Garcia-Molina et al. (2014), Hernandez (2013), Churcher (2012), Lightstone et al. (2011), and Halpin and Morgan (2008). Jeffrey D. Ullman's ones (Ullman (1988, 1989), Ullman and Widom (2007)) are still of very great value.

3.15.1   DEPENDENCY (CONSTRAINT) THEORY

Although the big word on the left is 'compassion', the big agenda on
the left is dependency.
—Thomas Sowell

3.15.1.1   Dependencies and Normal Forms

Form follows function.
—Louis Henri Sullivan

RDM and RA were introduced by Edgar F. Codd (Codd, 1970), to whom we all are especially indebted, among others, for introducing and studying RC (Codd, 1971a), keys, FDs, and the first three NFs (Codd, 1971b, 1972, 1974)), as well as for extending RDM with semantic concepts too (Codd, 1979) and RA, by including a three valued logic and a so-called *nulls substitution principle*[89] (Codd, 1975). Although BCNF was also introduced in Codd (1974), it was actually first presented in Heath (1971).

The much more elegant definition of FDs used in this book is due to Val (Breazu-)Tannen (Breazu and Mancas, 1980 and 1981).

MVDs were independently introduced by Zaniolo (1976), Fagin (1977), and Delobel (1978), and then generalized by Rissanen (1977), Nicolas (1978), and Aho et al. (1979).

JDs were introduced by Rissanen (1979).

We are all especially indebted to Ronald Fagin too, who, besides MVDs, introduced 4NF (Fagin, 1977), 5NF (Fagin, 1979), DKNF and the primitive (domain and key) constraints (Fagin, 1981).

Anomalies formal study is due to Bernstein and Goodman (1980), LeDoux and Parker (1982), and, especially, to Chan (1989).

(3,3)-NF and *EKNF* (see Exercises 3.53 and 3.54, respectively) were proposed by Smith (1978) and Zaniolo (1982), respectively.

---

[89] which simply states that unknown nulls may take on values only in a finitely restricted attribute domain

Orlowska and Zhang (1992) proved that the PJNF (generally thought of as being equivalent to 5NF) is stronger than 5NF. Date and Fagin (1992) proved that any 3NF table only having single keys is also in PJNF.

Unknown nulls were introduced by Grant (1977); Vassiliou (1979) and Vassiliou (1980a) consider both unknown and nonexistent ones, in the framework of denotational semantics; no-information ones were introduced by Zaniolo (1977) and then studied by lot of researchers for more than a decade.

Lipski (1979) and Lipski (1981) proposed a first null values generalization, by using *partially specified values* (nonvoid subsets of attribute domains from which nulls may take values). Wong (1982) advanced another generalization, based on assigning probabilities to each value of any domain: 1 to specified ones, 0 to all others, 0 too for all values corresponding to a nonexistent null, $1/n$ to all unknown nulls for a domain having n values[90], $1/k$ to partially specified nulls for a subset having k values, and again 0 for all of the rest of the values.

Korth and Ullman (1980), Maier (1980), Sagiv (1981), Imielinski and Lipski (1981) proposed and studied *marked nulls*: numbered null values that allow for storing the fact that certain ones should coincide.

Interaction between FDs and nulls was studied by (Vassiliou, 1980b), (Honeyman, 1982) (who considered FDs as being interrelational constraints), Lien (1982), Atzeni and Morfuni (1984) and Atzeni and Morfuni (1986).

Existence constraints are due to Maier (1980a). Other constraints on nulls were proposed by Maier (1980a) (*disjunctive existence constraints*), then studied in depth by Goldstein (1981) and Atzeni and Morfuni (1986) and Sciore (1981) (*objects*).

Although referential integrity was already well-known in the 70s and despite their crucial role in any RDBMS, inclusion dependencies were first considered much later, starting with Casanova (1981) and Casanova and Fagin (1984). IND interaction with other dependency types is studied, for example, in Casanova and Vidal (1983).

Tuple (check) constraints were very rarely formally studied. Mancas et al. (2003) introduced a generalization of tuple (check) constraints called *object constraints*, which also presents and studies a first order *object calculus*,

---

[90] When domains are infinite, distributions should be considered instead.

strictly more powerful than the relational ones and less powerful than the predicates one, designed to express both domain and object constraints.

More general constraint classes were introduced and studied by Fagin (1982) (where also studied is domain independence—see Subsection 3.15.2), Yannakakis and Papadimitriou (1982), Beeri and Vardi (1984a), and Beeri and Vardi (1984b).

General dependency theory was addressed by very many papers: for example, Fagin and Vardi (1986), Vardi (1987), Kanellakis (1991), and Thalheim (1991); Nicolas (1978), Gallaire and Minker (1978), and Nicolas (1982) approaches it from the first order logic perspective; Imielinski and Lipski (1983) had an interesting and singular point of view in this field too.

The set of all known dependencies may be partitioned in the following two classes (Beeri and Vardi (1984a) and Beeri and Vardi (1984b)):

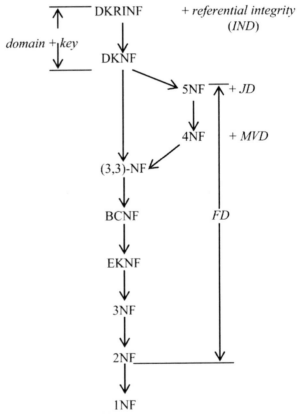

FIGURE 3.32 RDM normal forms and corresponding dependency types' hierarchy.

1. *equalities generating dependencies* (which includes FDs) that are always equalizing column values on certain rows;
2. *tuples generating dependencies* (which includes MVDs and JDs) that are always adding new rows to tables' instances.

Figure 3.32 synoptically shows NFs hierarchy and corresponding constraint types.

Lot of effort in RDM's theory was deployed for solving the *normalization problem*: given a class of constraints (generally FD, or FD and MVD), design an as fast as possible algorithm that accepts any set of constraints of that class and outputs a db scheme in a desired NF and, if possible, also displays other desired properties.

Besides LD, another desirable property, called *constraint preservation* (*CP*) or *representation*, was first introduced in RDM for FDs (Beeri and Rissanen, 1980): given a relation scheme R with an associated set F of constraints and a decomposition $R_1, \dots, R_n$ of it with corresponding associated constraint sets $F_1, \dots, F_n$, the decomposition is said to preserve dependencies if $(F_1 \cup \dots \cup F_n)^+ = F^+$.

Two equivalent strategy types were considered towards this goal: *synthesis* (*bottom-up*), which assumes no relations, but only a set of attributes (which is suited only for FDs and not for MVDs) and *decomposition* (*top-down*), which assumes either several relations or even only one (said "universal").

Beeri and Honeyman (1981) includes a polynomial algorithm for FD preservation testing, which is fundamental for the top-down approach.

Table 3.25 contains the main normalization algorithms (where *S/D* indicates the strategy type –S(ynthesis) or D(ecomposition), *NF* –the output

TABLE 3.25 Main Normalization Algorithms

No.	Authors	S/D	NF	Class	O(?)	Prop.
1	(Bernstein, 1976)	S	3	FD	P	CP
2	(Biskup et al., 1979)	S	3	FD	P	LD+CP
3	(Fagin, 1977)	D	4	FD+MVD	E	LD
4	(Tsou and Fischer, 1982)	D	4	FD+MVD	P	LD
5	(Ullman, 1982)	D	4	FD+MVD	E	LD
6	(Zaniolo and Melkanoff, 1981)	D	3	FD+MVD	E	LD+CP

NF, *Class* —the considered constraints class, $O(?)$ stores the algorithm complexity –P(olynomial) or E(xponential)–, and *Prop.* the normalization properties —C(onstraint)P(reservation) and L(ossless)D(ecomposition)):

Initially, normalization aimed to provide automatic design of rdb schemas. At least starting with the 90s (see, for example, Atzeni and de Antonellis (1993)), more and more researchers are convinced that such an approach is not adequate and that it is preferable to design db schemas in a higher, semantic data model that provides an algorithm for translating then its schemas into the desired relational NFs. In this context, RDM normalization theory only provides milestones and characterizes the targets of these translating algorithms.

3.15.1.2   The RDM Constraints Implication Problem

> *Whatever may happen to you was prepared for you from all eternity;*
> *and the implication of causes was from eternity spinning*
> *the thread of your being.*
> —Marcus Aurelius

3.15.1.2.1   *Derivation Rules*

> *Omnis cellula e cellula. (Every cell is derived from another cell.)*
> —François-Vincent Raspall

Derivation rules was the first used technique for studying RDM constraints implication problem: Armstrong (1974) introduced a sound and complete set of inference rules for FDs. Of course that, generally, besides being a "nice" theoretical result *per se*, the existence of such a set is very useful for solving the corresponding implication problem.

Note, however, that, surprisingly, for decades, the existence of such a sound and complete set of inference rules for any constraint class (the so-called *complete axiomatization* of the corresponding class) was considered more important than the implication problem decidability, although the existence or nonexistence of complete axiomatizations and the decidability (generally, the complexity) of the corresponding implication problem are orthogonal.

For example, for some constraint classes, there is no *limited complete axiomatization* (i.e., a complete one in which each rule has to have at most a given maximum number of antecedents), even if the corresponding implication problem is decidable: see (Kanellakis et al., 1983) for the unary inclusion dependencies class and (Casanova et al., 1984) for FD ∪ IND.

Dually, there are, for example, two classes for which limited complete axiomatization exist, but the corresponding implication problem is undecidable: the *encapsulated implicational dependencies* (Chandra et al. (1981) and Beeri and Vardi (1984a)) and the *typed template dependencies* (Sadri and Ullman (1982) and Vardi (1984)).

FDs' implication problem was "classically" studied in Beeri and Bernstein (1979) and Maier (1980b), and, alternatively, in the propositional logic framework, in Sagiv et al. (1981), Cosmadakis and Kanellakis (1985), and Cosmadakis et al. (1986).

A sound and complete set of inference rules for FD ∪ IND finite implication was proposed in Mitchell (1983b). A seminal paper on INDs theory is Casanova et al. (1984), which not only introduces their inference rules, but it is also proved that FD ∪ IND finite and unrestricted implications axiomatization is impossible. Related results are presented in Laver et al. (1983) and Johnson and Klug (1984) (where finite and unrestricted implications are generalized, also used in FD ∪ IND chasing, see below, and for characterizing inclusion between conjunctive queries).

The undecidability of FD ∪ IND finite implication was also independently proved in Mitchell (1983a) and Chandra and Vardi (1985) (which is also providing an elegant proof of the fact that decidability of FD ∪ IND unrestricted implication cannot be generalized, because the finite corresponding implication is undecidable; the proof uses reduction to the undecidable problems of monoid words (Post, 1947) and finite monoid words (Gurevich, 1966)).

This negative result, corroborated with the fact that IND implication problem complexity is PSPACE-complete, lead to restricting to only unary INDs, for which, unioned with FDs, the decision problem for both implication types is in PTIME (see, Cosmadakis et al. (1990), where additional results on unary INDs unioned with other, more general dependency classes are provided too).

Similar to FD ∪ IND, MVDs do interact with FDs too: the axiomatization of FD ∪ MVD is due to Beeri et al. (1977), which also solved the MVD implication problem, by using inference rules.

After one decade of unfruitful research, Petrov (1989) proved that there is no complete axiomatization for JDs. However, such axiomatizations exist for larger constraint sets that include JDs (e.g., Sciore, 1982), but their implication problem is not decidable.

3.15.1.2.2   The Chase

> *Gratitude is the most exquisite form of courtesy.*
> —Jacques Maritain

The second technique used for studying the RDM constraints implication problem was the *chasing tableaux algorithms* (or, simply, *chase*): algorithms that start with a *tableaux* (i.e., a table whose instance contains not only constants, but variables too, thus abstracting several possible (sub) instances) and a constraint set, try to build a counterexample (of the fact that every set is included in its closure) tableaux and force it to satisfy all given constraints, in order to obtain a tableaux which satisfies the given set of constraints and is as similar as possible with the original one.

Note that a given constraint is implied by the given set of constraints *iff* it holds in the obtained tableaux too. Also note that chase's appeal comes not only from its very intuitive nature, but also from the fact that it is somewhat parameterizable with respect to the considered constraint class.

A chase foremother was already prefigured in Aho et al. (1979a); chase was introduced in Maier et al. (1979) (that developed ideas from Aho et al. (1979a) and Aho et al. (1979b)), which also uses it both for solving logic implication problems and tableaux queries, as well as for studying the FD and MVD implication problem.

Chang and Lee (1973) and Beeri and Vardi (1980) compare the chase with the *paramodulation resolution* theorem proofing technique. Beeri and Vardi (1984b) extends chase use to the more general *data dependencies*.

Typed INDs are studied using the chase and equational theories in Cosmadakis and Kanellakis (1986).

Abiteboul et al. (1995) used chase both for JDs and JD ∪ FD, proving that they are NP-complete and that the implication problem JDs ⊢— MVD is NP-hard.

3.15.2 RELATIONAL LANGUAGES THEORY. *SQL, QUEL, QBE*

Blessings flow toward gratitude.
—Jonathan Lockwood Huie

As we have already seen, E. F. Codd introduced both RA (Codd, 70) and RC (Codd, 71a). Final results of RA's expressive power are due to (Paredaens, 1978); similar ones were obtained for RC by Bancilhon (1978).

Codd (71a) proposed the TRC language *ALPHA*, which has an embedded values domain declarations form that laid the foundation for the RDBMSes one's. Pirotte (1978) made the distinction between TRC and DRC, while providing a detailed analysis of various RC forms, including issues occurring when transforming a generic applied calculus into a query language.

One year before RDM's introduction, Di Paola (1969) proposed the notion of *domain independence*, in a broader context. Unfortunately, in db theory, this notion[91] was assimilated very slowly, although undesirable expressions whose result might have been the whole domain were detected as soon as TRC and DRC introduction, which immediately lead to the notions of *type* (Codd, 71a) and *value domain* (Bancilhon, 1978).

Fagin (1982) finally recognized the importance of domain independence as a semantic notion (in the context of formulae describing integrity constraints); its undecidability was inferred from previous results (Di Paola, 1969; Vardi, 1981), whereas domains' *restraint, safety, evaluability, permission*, and *interpretation limitations* were considered as sufficient conditions for it.

The equivalence between RA and CR was proved first by Codd (1972); alternative proofs were then proposed by Ullman (1982) and Maier (1983). This equivalence determined E. F. Codd to also introduce the notion of *completeness* (Codd, 1972); note that, before, for *ALPHA* already, he felt like needing more than completeness and, consequently, introduced tuple and aggregation functions. Completeness was subsequently characterized by Paredaens (1978) and Bancilhon (1978). Klug (1982) proved the equivalence between RA and CR, when both are extended with aggregation functions.

A more general completeness notion (*CH-completeness*) was introduced by Chandra and Harel (1980), called *computable queries*, which

---

[91] An *expression* is *domain independent* iff it represents a same query, regardless of its subjacent domains; a *language* is *domain independent* iff all of its expressions are domain independent.

moves the focus from query results for a given db instance to queries viewed as functions: a language L is *CH-complete iff*, for any db scheme and generic query Q, there exists a L-expression E such that, for any corresponding db instance i, $Q(i) = E(i)$.

Moreover, this paper also proposed QL, a canonical *CH*-complete language: informally, QL includes RA extended with variables over relations, an assignment statement, and a looping one equipped with a test for reaching the empty set, the kernel of today's extended *SQL*s.

Related notions over generic queries were studied by Hull (1986).

Note that, as we have already discussed, actual commercial *SQL* and *Quel* are generally extended in the *CH*-completeness sense[92] and/or embedded in high-level host programming languages[93] that provide full computing power.

Aho and Ullman (1979) proved that transitive closures cannot be expressed in RA.

Chandra (1988) is an excellent overview of query languages, presenting hierarchies of such languages based on expressive power, complexity, and programming primitives; moreover, it provides an impressive bibliography, including mathematical logic previously obtained results (in larger contexts) that apply to rdb querying as well.

There are three approaches for dealing with nulls in rdb querying:

1. based on derived RA operators: for example, *augmentation* (unary, computing a relation over the operand's attributes, plus other ones, and whose tuples are obtained from those of the operand's, but extending them with nulls) and *external (natural) join* (or full outer join) from Gallaire and Minker (1978), and *total projection* (the usual projection followed by elimination of all tuples containing nulls) from Kandzia (1979);

2. based on constraints on nulls: see Imielinski and Lipski (1981, 1983, 1984), (Abiteboul et al., 1987), and (Grahne, 1989);

3. based on the *IsNULL* predicate, which does not originate from the db research, but from the RDBMS industry.

SQL stemmed from IBM's *SEQUEL* ("*Structured English QUEry Language*"), introduced by Chamberlin and Boyce (1974), while work-

[92] See, for example, IBM *DB2 SQL PL*, *Oracle PL/SQL*, MS *T-SQL*, etc.
[93] See, for example, IBM embedded *SQL* for *Java, C, C++, Cobol*, etc., MS *VBA* and *.NET* languages, etc.

ing for the world's first RDBMS prototype, *System R* (Astrahan et al., 1976; Chamberlin et al., 1981). *SQL*'s formal semantics is provided by Negri et al. (1991). *SQL* is the relational both *de facto* and *de jure* (ISO/IEC, 1991; ISO/IEC, 1992) standard language; its O-O extension (called O_2SQL) was proposed in Cattell (1997).

Starting with the *SQL*-99 standard (ISO/IEC, 2003), which extended *SQL*, among others, with recursion, both linear, nonlinear, and mutual recursion are allowed, but recursive subqueries and aggregation are disallowed. The current standard is (ISO/IEC, 2011).

Lot of good books on *SQL* programming are available, from those offered by the RDBMS providers (e.g., Zamil et al. (2004), Lorentz et al. (2013), Moore et al. (2013)), part of their online documentation, and up to the commercial ones (e.g., Rockoff (2011), Ben-Gan (2012), Berkowitz (2008), Murach (2012)).

A similar TRC-type language, *Quel*, was introduced for the RDBMS *INGRES* (Stonebraker et al., 1976).

QBE is due to another IBM researcher, Moshe Zloof (Zloof, 1977). Ozsoyoglu and Wang (1993) presents all of its important extensions, as well as all of its flavors. *QBE*'s exceptional graphic power is exploited by several RDBMSs: for example, IBM embeds *QBE* into its *DB2 QMF* (*Query Management Facility*), Microsoft provides it as its queries' *Design View*, in both *Access* and *SQL Server*, and also in *Excel*, as *MS Query*, *PostgreSQL* offers its *OBELisQ*, etc.

Selinger et al. (1979) describes *SystemR*'s query evaluation optimizations. Ullman (1982) presents a widely used today heuristic optimization algorithm in six steps. An exhaustive discussion of query evaluation optimizations is presented in Graefe (1993). Sun et al. (1993) contains both a proposal of an estimation technique based on series approximating instances values distribution and the regressions analysis for estimating queries with joins results cardinality, as well as an impressive list of references dealing with query evaluation plans' cost estimations.

For a couple of decades, db instances modifications were almost neglected, as it was believed that they may be trivially formalized with three simple operators (insert, update, and delete of tuples satisfying desired criteria) that might be reducible to basic set operators. Abiteboul and Vianu (1988a, 1988b, 1989, 1990) proved that there are, however, depth problems in this direction too, which opened a new study field still active. An overview of the fundamentals in this area is provided by (Abiteboul, 1988).

3.15.3 MISCELLANEA

> *Philosophy of science without history of science is empty;*
> *history of science without philosophy of science is blind.*
> —Imre Lakatos

This chapter is mainly an abstract of Mancas (2007) and Mancas (2002a).

There is lot of algorithms for translating E-RDs into relational schemas (e.g., see the one presented in Levene and Loizou (1999), which is also citing other 7 ones). *A*1–7 (presented in Subsection 3.3.1), which is also taking into considerations associated restriction sets and informal descriptions, as well as the RDM model of the E-RDM come from Mancas (2001).

There are also other reverse engineering algorithms for translating relational schemas into E-RDs (e.g., Chiang et al. (1994) presents a methodology for extracting an extended E-R model from a rdb, through a combination of data schema and instance analysis, which derives a semantically richer and more comprehensible for maintenance and design purposes model than the original rdb). The simple *REA*1–2 (presented in Section 3.6), which is, however, taking into considerations associated restriction sets and informal descriptions too, comes also from Mancas (2001).

The keys discovery assistance algorithm *A*7/8–3 (see Section 3.7), together with proposition 3.1, were initially proposed in the framework of the (E)MDM, first for relationship structural keys (see second chapter of the second volume) (Mancas and Dragomir, 2003) and then for all possible keys (see *A*3 and *A*3' in the second chapter of the second volume) (Mancas and Crasovschi, 2003). Its current English version for the RDM comes from (Mancas, 2008).

Proposition 3.0, on which *A*3, *A*3', and *A*7/8–3 heavily rely, was published earlier (Mancas, 1990). Its simplified proof from (Mancas and Crasovschi, 2003) was inspired by the one from Lubell (1966).

The RDBMS metacatalog kernel scheme and the corresponding relational metamodel of the RDM (see Section 3.7) were published first in Mancas (2002b); the E-R data metamodel of the RDM comes from Mancas (2001).

Note that the relational one from (Mancas, 2002b) is slightly simpler than this one (coming from Mancas (2008)), as it was in fact designed for *MatBase*, which only uses single surrogate primary keys and single foreign keys always referencing primary keys.

DKRINF and the best practice rules (see Section 3.11) come from Mancas (2001).

Definition 3.3 and proposition 3.3, as well as Exercises 3.2, 3.40, 3.41 and 3.42 are due to Val Tannen (Breazu and Mancas, 1980, 1981, and 1982).

The key propagation principle formalization (proposition 3.2), comes from Mancas (2001) and was first published outside Romania in Mancas and Crasovschi (2003), in the (E)MDM framework.

For the theory of NP-completeness, see, for example, Garey and Johnson (1979). For the complexity theory, see, for example, the excellent (Calude, 1988).

Exercises 3.0, 3.1, 3.6 to 3.17, 3.26 to 3.30, 3.39, 3.43, 3.68, and 3.71 to 3.87 are original (most of them are from Mancas (2001), Mancas (2002), Mancas (2007), and Mancas (2008)). All other exercises are from other books (mainly from Abiteboul et al. (1995) and Atzeni and de Antonellis (1993)).

Cons on relational db design (see also, for example, Atzeni and de Antonellis (1993)) were first presented in Kent (1979). The ones from Subsection 3.13.2 come also from Mancas (1985, 1997, 2007).

For ISO 3166 country (and their subdivisions) codes international standard, see http://www.iso.org/iso/home/standards/country_codes.htm.

All URLs mentioned in this section were last accessed on August 15th, 2014.

KEYWORDS

- **2nd Normal Form (2NF)**
- **3rd Normal Form (3NF)**
- **4th Normal Form (4NF)**
- **5th Normal Form (5NF)**
- **ACID**

- **acyclicity constraint**
- **aggregation function**
- **ALL**
- **ALTER TABLE**
- **anti join**
- **ANY**
- **arity**
- **AS**
- **attribute**
- **attribute concatenation**
- **attribute domain**
- **attribute name**
- **attribute product**
- **augmentation**
- **autonumber**
- **AVG**
- **BEGIN TRANS**
- **Boyce-Codd Normal Form (BCNF, 3.5NF)**
- **candidate key**
- **cardinal**
- **Cartesian product**
- **CH-completeness**
- **chase**
- **CHECK**
- **check constraint**
- **column**
- **COMMIT**
- **complementary property**
- **complementation**
- **complete axiomatization**
- **complete join**
- **completeness**
- **composed key**

- **compulsory attribute**
- **computable queries**
- **computed attribute**
- **computed relation**
- **concatenated key**
- **constraint**
- **constraint name**
- **constraint preservation**
- **constraint representation**
- **COUNT**
- **CREATE TABLE**
- **cross tabulation**
- **cursor**
- **dangling pointer**
- **data definition language (DDL)**
- **data dependency**
- **data manipulation anomalies**
- **data manipulation language (DML)**
- **data set**
- **decidability**
- **decomposition**
- **DELETE**
- **delete anomaly**
- **dependency**
- **derivation rules**
- **dirty null**
- **DISTINCT**
- **DISTINCTROW**
- **Division**
- **Domain-Key Normal Form (DKNF)**
- **Domain-Key Referential Integrity Normal Form (DKRINF)**
- **domain constraint**
- **domain independence**

- **domain relational calculus (DRC)**
- **DROP CONSTRAINT**
- **DROP SCHEMA**
- **DROP TABLE**
- **dynamic SQL**
- **dynamic subquery**
- **embedded SQL**
- **equalities constraint**
- **equijoin**
- **existence constraint**
- **EXISTS**
- **expressive power**
- **extended SQL**
- **field**
- **finite implication**
- **First Normal Form (1NF)**
- **FOREIGN KEY**
- **foreign key**
- **FROM**
- **full join**
- **functional dependency**
- **function kernel (ker)**
- **fundamental relation**
- **GRANT**
- **GROUP BY**
- **HAVING**
- **header**
- **IN**
- **inner join**
- **INSERT**
- **insert anomaly**
- **instance**
- **intersection**

- **IsNULL predicate**
- **join**
- **join dependency**
- **kernel equivalence relation**
- **key**
- **key constraint**
- **Key Propagation Principle (KPP)**
- **left join**
- **LIMIT**
- **limited complete axiomatization**
- **lossless decomposition**
- **mandatory attribute**
- **marked null**
- **MAX**
- **metacatalog**
- **metadata catalog**
- **MIN**
- **minimally one-to-one**
- **minimally uniqueness**
- **multivalued dependency**
- **natural join**
- **nested relation**
- **no information null**
- **non-relational constraint**
- **non prime attribute**
- **normalization**
- **normalization problem**
- **not applicable null**
- **NOT EXISTS**
- **NOT IN**
- **NOT NULL**
- **not null**
- **not null constraint**

- null value
- object constraint
- **ORDER BY**
- outer join
- partial functional dependency
- primary key
- prime attribute
- primitive constraint
- projection
- Projection-Join Normal Form (PJNF)
- **QBE**
- **QL**
- Quel
- query
- quotient set
- range
- rdb instance
- rdb name
- rdb scheme
- record set
- referential integrity
- relation
- relational algebra (RA)
- relational calculus (RC)
- relational completeness
- relational constraint
- relational db (rdb)
- relation name
- renaming
- required attribute
- **REVOKE**
- right join
- **ROLLBACK**

- **row**
- **scheme**
- **SELECT**
- **Selection**
- **semantic key**
- **semi join**
- **simple key**
- **SOME**
- **soundness**
- **SQL**
- **static SQL**
- **static subquery**
- **stored function**
- **stored procedure**
- **subquery**
- **SUM**
- **superkey**
- **surrogate key**
- **syntactic key**
- **synthesis**
- **table**
- **tableaux**
- **table name**
- **temporarily unknown null**
- **temporary relation**
- **theta join**
- **TOP**
- **total attribute**
- **totality constraint**
- **total projection**
- **transaction**
- **transitive closure**
- **transitive functional dependency**

- **trigger**
- **tuple**
- **tuple constraint**
- **tuple relational calculus (TRC)**
- **tuples generating constraint**
- **typed inclusion constraint**
- **type domain**
- **undecidability**
- **union**
- **UNION**
- **UNION ALL**
- **UNIQUE**
- **unique attribute**
- **uniqueness constraint**
- **universal relation**
- **UPDATE**
- **update anomaly**
- **value domain**
- **view**
- **WHERE**

REFERENCES

I have a lot of cultural references that have amassed in my brain like shrapnel over the years that are meaningful to me.
—John Hodgman

Abiteboul, S. (1988). Updates, a new frontier. ICDT'88 2nd Intl. Conf. on DB Theory, 1–18, Springer-Verlag, Berlin, Germany.

Abiteboul, S., Hull, R., Vianu, V. (1995). *Foundations of Databases;* Addison-Wesley: Reading, MA.

Abiteboul, S., Kanellakis, P. C., Grahne, G. (1987). On the representation and querying of possible worlds. ACM SIGMOD Intl. Conf. on Manag. of Data, 34–48.

Abiteboul, S., Vianu, V. (1988a). Equivalence and optimization of relational transactions. JACM, 35(1), 70–120.

Abiteboul, S., Vianu, V. (1988b). Procedural and declarative database update languages. 7th ACM SIGACT SIGMOD SIGART Symp. Princ. DB Syst., 240–250.

Abiteboul, S., Vianu, V. (1989). A transaction-based approach to relational database specification. JACM, 36(4), 758–789.

Abiteboul, S., Vianu, V. (1990). Procedural languages for database queries and updates. J. Comp. Syst. Sci., 41(2), 181–229.

Aho, A. V., Beeri C., Ullman J. D. (1979). The theory of joins in relational databases. ACM TODS 4(3), 297–314.

Aho, A. V., Sagiv, Y., Ullman, J. D. (1979a). Efficient optimization of a class of relational expressions. ACM TODS 4(3), 435–454.

Aho, A. V., Sagiv, Y., Ullman, J. D. (1979b). Equivalence of relational expressions. J. Comp. Syst. Sci., 8(2), 218–246.

Aho, A. V., Ullman, J. D. (1979). Universality of data retrieval languages. 6th ACM Symp. on Princ. of Progr. Lang., 110–117.

Allison, C., Berkowitz, N. (2008). *SQL for Microsoft Access,* 2nd edition: Wordware Publishing, Inc. Plano, TX.

ANSI/X3/SPARC Study Group on Database Management Systems. (1975). Interim Report 75–02–08. FDT-Bulletin ACM SIGMOD, 7(2).

Armstrong, W. W. (1974). Dependency structure of database relationships. IFIP Congress, 580–583.

Astrahan, M. M., et al. (1976). System R: A Relational Approach to Database Management. ACM TODS, 1(2), 97–137.

Atzeni, P., de Antonellis, V. (1993). *Relational Database Theory*. Benjamin/ Cummings: Redwood, CA.

Atzeni, P., Morfuni, N. M. (1984). Functional dependencies in relations with null values. Inf. Processing Letters, 18(4), 233–238.

Atzeni, P., Morfuni, N. M. (1986). Functional dependencies and constraints on null values in database relations. Information and Control, 70(1), 1–31.

Bancilhon, F. (1978). On the completeness of query languages for relational databases. *LNCS Math. Found. of Comp. Sci.*, 64, 112–124, Springer-Verlag: Berlin, Germany.

Beeri, C., Bernstein, P. A. (1979). Computational problems related to the design of normal form relational schemas. ACM TODS 4(1), 30–59.

Beeri, C., Fagin, R., Howard, J. H. (1977). A complete axiomatization for functional and multivalued dependencies. ACM SIGMOD Intl. Symp. Manag. Data, 47–61.

Beeri, C., Honeyman, P. (1981). Preserving functional dependencies. *SIAM Journal on Computing,* 10(3), 647–656.

Beeri, C., Rissanen, J. (1980). Faithful representation of relational database schemata. Report RJ 2722. IBM Research. San Jose.

Beeri, C., Vardi, M. Y. (1980). A proof procedure for data dependencies (preliminary report). Tech. Rep., Hebrew Univ.: Jerusalem, Israel.

Beeri, C., Vardi, M. Y. (1984a). Formal systems for tuple and equality-generating dependencies. SIAM J. Computing, 13(1), 76–98.

Beeri, C., Vardi, M. Y. (1984b). A proof procedure for data dependencies. *JACM* 31(4), 718–741.

Ben-Gan, I. (2012). *Microsoft SQL Server 2012. T-SQL Fundamentals* (*Developer Reference*). O'Reilly Media, Inc.: Sebastopol, CA.

Bernstein, P. A. (1976). Synthesizing third normal form relations from functional dependencies. *ACM TODS*, 1(4), 277–298.

Bernstein, P. A., Goodman N. (1980). What does Boyce-Codd form do?. 6th Intl. Conf. on VLDB, 245–259, Montreal, Canada.

Biskup, J., Dayal, U., Bernstein, P. A. (1979). Synthesizing independent database schemes. ACM SIGMOD Intl. Conf. on Manag. of Data, 143–151.

Breazu, V., Mancas, C. (1980). On a Functional Data Model in the database mathematical theory. Comparison to the Relational Data Model (in Romanian). Unpublished communication at the 6th Symposium on Informatics for Management CONDINF'80, Cluj-Napoca, Romania.

Breazu, V., Mancas, C. (1981). On the description power of a Functional Database Model. Proc. 4th Intl. Conf. on Control Systems and Computer Science, 4:223–226, *Politehnica* University, Bucharest, Romania.

Breazu, V., Mancas, C. (1982). Normal forms for schemas with functional dependencies in the Relational Database Model (in Romanian). Proc. 12^{ve} Conf. on Electronics, Telecommunications and Computers, 152–1566, *Politehnica* University, Bucharest, Romania.

Calude, C. (1988). *Theories of Computational Complexity*. North Holland: Amsterdam, Holland.

Casanova, M. A. (1981). The theory of functional and subset dependencies over relational expressions. Tech. Rep., 3/81, Rio de Janeiro, Brazilia.

Casanova, M. A., Fagin R., Papadimitriou C. H. (1984). Inclusion dependencies and their interaction with functional dependencies. J. Comp. Syst. Sci., 28(1), 29–59.

Casanova, M. A., Vidal, V. M. P. (1983). Towards a sound view integration technology. ACM SIGACT SIGMOD Symp. Princ. DB Syst., 36–47.

Cattell, R. G. G., ed. (1994). *The Object Database Standard: ODMB-93*. Morgan Kaufman: Los Altos, CA.

Chamberlin, D. D., Boyce, R. F. (1974). SEQUEL—A Structured English QUEry Language. SIGMOD Workshop, 1, 249–264.

Chamberlin, D.D., et al. (1981). A History and Evaluation of System R. *CACM*, 24(10), 632–646.

Chan, E. P. F. (1989). A design theory for solving the anomalies problem. SIAM J. Computing, 18(3), 429–448.

Chandra, A. K. (1988). Theory of *database queries*. *7th ACM SIGACT SIGMOD SIGART* Symp. Princ. DB Syst., 1–9.

Chandra, A. K., Harel, D. (1980). Computable queries for relational databases. J. Comp. Syst. Sci., 21, 333–347.

Chandra, A. K., Lewis, H. R., Makowsky, J. A. (1981). Embedded implicational dependencies and their inference problem. 13th ACM SIGACT Symp. on Theory of Computing, 342–354.

Chandra, A. K., Vardi, M. Y. (1985). The implication problem for functional and inclusion dependencies is undecidable. SIAM J. Computing, 14(3), 671–677.

Chang, C. L., Lee, R. M. (1973). Symbolic Logic and Mechanical Theorem Proving. Academic Press: New York, NY.

Chiang, R. H. L., Barron, T. M., Storey, V. C. (1994). Reverse engineering of relational databases: Extraction of an EER model from a relational database. *Data & Knowledge Engineering* 12(2): 107–142, Elsevier B.V.

Churcher, C. (2012). *Beginning Database Design: From Novice to Professional*, 2nd ed., Apress Media LLC: New York, NY.

Codd, E. F. (1970). A relational model for large shared data banks. *CACM*, 13(6), 377–387.

Codd, E. F. (1971a). A database sublanguage founded on the relational calculus. *SIGFI-DET'71*, Proceedings of the ACM SIGFIDET (now SIGMOD) Workshop on Data Description, Access, and Control, San Diego, CA, November 11–12, (1971); ACM: New York, NY, 35–68.

Codd, E. F. (1971b). *Further normalization of the data base relational model*; Research Report RJ909; IBM: San Jose, CA.

Codd, E. F. (1972). *Relational completeness of data base sublanguages*; Research Report RJ987; IBM: San Jose, CA.

Codd, E. F. (1974). *Recent investigations into relational database systems*; Research Report RJ1385; IBM: San Jose, CA.

Codd, E. F. (1975). Understanding Relations. FDT (ACM SIGMOD Records), Vol. 7, No. 3–4, 23–28.

Codd, E. F. (1979). Extending the database relational model to capture more meaning. *ACM TODS*, 4(4), 397–434.

Cosmadakis, S., Kanellakis, P. C. (1985). Equational theories and database constraints. ACM SIGACT Symp. on the Theory of Computing, 73–284.

Cosmadakis, S., Kanellakis, P. C. (1986). Functional and inclusion dependencies: A graph theoretic approach. Adv. in Computing Res., 3, 164–185, JAI Press.

Cosmadakis, S., Kanellakis, P. C., Spyratos, N. (1986). Partition semantics for relations. J. Comp. Syst. Sci., 32(2), 203–233.

Cosmadakis, S., Kanellakis, P. C., Vardi, M. Y. (1990). Polynomial-time implication problems for unary inclusion dependencies. JACM, 37(1), 15–46.

Date, C. J. (2003). *An Introduction to Database Systems*, 8th ed., Addison-Wesley: Reading, MA.

Date, C. J. (2011). *SQL and Relational Theory: How to Write Accurate SQL Code*, 2nd ed., Theories in Practice; O'Reilly Media, Inc.: Sebastopol, CA.

Date, C. J. (2012). *Database Design & Relational Theory: Normal Forms and All That Jazz;* Theories in Practice; O'Reilly Media, Inc.: Sebastopol, CA.

Date, C. J. (2013). *Relational Theory for Computer Professionals: What Relational Databases are Really All About;* Theories in Practice; O'Reilly Media, Inc.: Sebastopol, CA.

Date, C. J., Fagin, R. (1992). Simple conditions for guaranteeing higher normal forms in relational databases. ACM TODS 17(3), 465–476.

Delobel, C. (1978). Normalization and hierarchical dependencies in the relational data model. ACM TODS 3(3), 201–222.

Di Paola, R. (1969). The recursive unsolvability of the decision problem for the class of definite formulas. JACM, 16(2), 324–327.

Fagin, R. (1977). Multivalued dependencies and a new normal form for relational databases. ACM TODS 2(3), 226–278.

Fagin, R. (1979). Normal forms and relational database operators. ACM SIGMOD Intl. Conf. on Manag. of Data, 123–134.

Fagin, R. (1981). A normal form for relational databases that is based on domains and keys. ACM TODS 6(3), 387–415.

Fagin, R. (1982). Horn clauses and database dependencies. JACM, 29(4), 952–983.

Fagin, R., Vardi, M. Y. (1986). The theory of data dependencies: A survey. Math. of Information Processing: Proc. of Symp. In Applied Math., 34, 19–71, American Math. Soc., Providence, RI.

Gallaire, H., Minker, J. (1978). *Logic and Databases*. Plenum Press: New York, NY.

Garcia-Molina, H., Ullman, J.D., Widom, J. (2014). *Database Systems: The Complete Book*, 2nd ed., Pearson Education Ltd.: Harlow, U.K., (Pearson New International Edition)

Garey, M. R., Johnson, D. S. Victor Klee. ed. (1979). *Computers and Intractability: A Guide to the Theory of NP-Completeness*. A Series of Books in the Mathematical Sciences. W. H. Freeman and Co.

Goldstein, B. S. (1981). *Constraints on null values in relational databases*. 7th Intl. Conf. on VLDB, 101–111.

Graefe, G. (1993). Query evaluation techniques for large databases. ACM Computing Surveys, 25(2), 73–170.

Grahne, G. (1989). Horn tables—an efficient tool for handling incomplete information in databases. 8th ACM SIGACT SIGMOD SIGART Symp. Princ. DB Syst., 75–82.

Grant, J. (1977). Null values in a relational database. Inf. Processing Letters, 6(5), 156–159.

Groff, J. R. (1990). *Using SQL*. Osborne McGraw-Hill, New York, NY.

Gurevich, Y. (1966). The word problem for certain classes of semigroups. Algebra and Logic, 5, 25–35.

Halpin, T., Morgan, T. (2008). *Information Modeling and Relational Databases, 2nd Edition*. Morgan Kauffman: Burlington, MA.

Heath, I. (1971). Unacceptable File Operations in a Relational Database. *SIGFIDET'71*, Proceedings of the ACM SIGFIDET (now SIGMOD) Workshop on Data Description, Access, and Control, San Diego, CA, November 11–12, ACM: New York, NY, 19–33.

Hernandez, M. J. (2013). *Database Design for Mere Mortals: A Hands-on Guide to Relational Database Design*, 3rd ed., Addison-Wesley: Reading, MA.

Honeyman, P. (1982). Testing satisfaction of functional dependencies. JACM, 29(3), 668–677.

Hull, R. B. (1986). Relative information capacity of simple relational schemata. SIAM J. Computing, 15(3), 856–886.

Imielinski, T., Lipski, W. (1981). On representing incomplete information in relational databases. 7th Intl. Conf. on VLDB, 388–397, Cannes, France.

Imielinski, T., Lipski, W. (1983). Incomplete information and dependencies in relational databases. ACM SIGMOD Intl. Conf. on Manag. of Data, 178–184.

Imielinski, T., Lipski, W. (1984). Incomplete information in relational databases. JACM, 761–791.

ISO/IEC. Database language SQL (SQL-2011). 9075: 2011, 2011.

ISO/IEC. Database language SQL (SQL-99). 9075-2: 1999, 2003.

ISO/IEC. Database language SQL (SQL3). JTC1/SC21 N 6931, 1992.

ISO/IEC. Database language SQL. JTC1/SC21 N 5739, 1991.

Johnson, D. S., Klug, A. (1984). Testing containment of conjunctive queries under functional and inclusion dependencies. J. Comp. Syst. Sci., 28, 167–189.

Kandzia, P., Klein, H. (1979). On equivalence of relational databases in connection with normalization. Workshop on Formal Bases for DB, ONERA-CERT: Toulouse, France.

Kanellakis, P. C. (1991). Elements of relational database theory. *Handbook of Theoretical Comp. Sci.*, 1074–1156, Elsevier: Amsterdam, Holland.

Kanellakis, P. C., Cosmadakis, S., Vardi, M. Y. (1983). Unary inclusion dependencies have polynomial-time inference problems. 15th ACM SIGACT Symp. on Theory of Computing, 264–277.

Kent, W. (1979). Limitations of record-based information models. ACM TODS 4(1).

Klug, A. (1982). Equivalence of relational algebra and relational calculus query languages having aggregate functions. JACM, 29(3), 699–717.

Korth, H. F., Ullman, J. D. (1980). *System/U: A system based on the universal relation assumption.* Stonybrook, NY.

Laver, K., Mendelzon, A. O., Graham, M. H. (1983). Functional dependencies on cyclic database schemes. ACM SIGMOD Intl. Symp. Manag. Data, 79–91.

LeDoux, C. H., Parker, D. S. (1982). Reflections on Boyce-Codd normal form. 8th Intl. Conf. on VLDB, 131–141, Mexico City, Mexico.

Levene, M., Loizou, G. (1999). *A Guided Tour of Relational Databases and Beyond*: Springer Verlag, London, U.K.

Lien, Y. E. (1982). On the equivalence of database models. JACM, 29(2), *333–362.*

Lightstone, S. S., Teorey, T. J., Nadeau, T., Jagadish, H. V. (2011). *Database Modeling and Design: Logical Design, 5th ed.*, Data Management Systems; Morgan Kaufmann: Burlington, MA.

Lipski, W. (1979). On semantic issues connected with incomplete information databases. ACM TODS 4(3), 262–296.

Lipski, W. (1981). On databases with incomplete information. JACM, 28(1), 41–70.

Lorentz, D., Roeser, M. B., et al. (2013). *SQL Language Reference 11 g Release 2*: Oracle Corp.

Lubell, D. (1966). A short proof of Sperner's lemma. J. Combinatorial Theory, 1, 299.

Maier, D. (1980a). *Discarding the universal relation assumption: Preliminary report.* Stonybrook, NY.

Maier, D. (1983). *The Theory of Relational Databases.* Computer Science Press: Rockville, MD.

Maier, D., Mendelzon, A. O., Sagiv Y. (1979). Testing implications of data dependencies. ACM TODS 4(4), 455–468.

Maier D. (1980b). Minimum covers in the relational database model. JACM, 27(4), 664–674.

Mancas, C. (1985). Some db design counterexamples proving that data modeling should not be approached relationally. Proc. INFO IAŞI'85, 303–313, A.I. Cuza University, Iaşi, Romania (in Romanian).

Mancas, C. (1990). A Deeper Insight into the Mathematical Data Model. 13th Intl. Seminar on DBMS, ISDBMS, 122–134, Mamaia, Romania.

Mancas, C. (1997). *Conceptual Data Modeling.* PhD Thesis: *Politehnica* University, Bucharest, Romania (in Romanian).

Mancas, C. (2001). *Databases for Undergraduates Lecture Notes*: Computer Science Dept., *Ovidius* State University, Constanta, Romania.

Mancas, C. (2002a). *Programming in SQL ANSI-92 with applications in MS JetSQL 4*: Ovidius University Press, Constanta, Romania (in Romanian).

Mancas, C. (2002b). On Modeling the Relational Domain-Key Normal Form Using an Elementary Mathematical Data Model. Proc. IASTED SEA 2002 MIT Conf. on Software Eng. and App., 767–772, Acta Press, Cambridge, MA.

Mancas, C. (2007). Theoretical foundations of the Relational Data Model: Ovidius University Press, Constanta, Romania (in Romanian).

Mancas, C., Crasovschi L. (2003). An Optimal Algorithm for Computer-Aided Design of Key Type Constraints. Proc. 1st Balkan Inf. Techn. BIT 2003 Conf., 574–584, Aristotle University Press, Thessaloniki, Greece.

Mancas, C., Dragomir, S., Crasovschi, L. (2003). On modeling First Order Predicate Calculus using the Elementary Mathematical Data Model in *MatBase* DBMS. Proc. IASTED AI 2003 MIT Conf. on Applied Informatics, 1197–1202, Acta Press, Innsbruck, Austria.

Mancas, C., Dragomir S. (2003). An Optimal Algorithm for Structural Keys Design. *Proc.* IASTED SEA 2003 Conf. on Software Eng. and App., 328–334, Acta Press, Marina del Rey, CA.

Mancas C. (2008). *Database Management Systems for Undergraduates Lecture Notes*: Engineering in Foreign Languages Dept., *Politehnica* University, Bucharest, Romania.

Mitchell, J. C. (1983a). The implication problem for functional and inclusion dependencies. Information and Control, 56(1), 154–173.

Mitchell, J. C. (1983b). Inference rules for functional and inclusion dependencies. ACM SIGACT SIGMOD Symp. Princ. DB Syst., 58–69.

Moore, S., et al. (2013). *PL/SQL Language Reference 11 g Release 2*: Oracle Corp.

Murach, J. (2012). *Murach's MySQL*. Mike Murach & Associates, Inc.: Fresno, CA.

Negri, M., Pelagatti, S., Sbattella, L. (1991). Formal semantics of SQL queries. ACM TODS 16(3), 513–535.

Nicolas, J.-M. (1978). First order logic formalization for functional, multivalued, and mutual dependencies. ACM SIGMOD Intl. Symp. Manag. Data, 40–46.

Nicolas, J.-M. (1982). Logic for improving integrity checking in relational databases. Acta Informatica, 18(3), 227–253.

Orlowska, M., Zhan, Y. (1992). Understanding the 5th normal form (5NF). Australian Computer Science Communications 14(1), 631–639.

Ozsoyoglu, G., Wang, H. (1993). A survey of QBE languages. Computer, 26.

Paredaens, J. (1978). On the expressive power of the relational algebra. Inf. Processing Letters, 7(2).

Petrov, S. V. (1989). Finite axiomatization of languages for representation of system properties. Inf. Sci., 47, 339–372.

Pirotte, A. (1978). High-level database languages. Logic and Databases, 409–435, Plenum Press: New York, NY.

Post, E. L. (1947). Recursive unsolvability of a problem of Thue. J. of Symb. Logic, 12, 1–11.

Rissanen, J. (1977). Independent components of relations. ACM TODS, 2(4), 317–325.

Rissanen, J. (1979). Theory of joins for relational databases—a tutorial survey. LNCS Math. Found. of Comp. Sci., 64, 537–551, Springer-Verlag, Berlin, Germany.

Rockoff, L. (2011). *The Language of SQL*: Course Technology, Boston, MA.

Sadri, F., Ullman, J. D. (1982). Template dependencies: A large class of dependencies in the relational model and its complete axiomatization. JACM, 29(2), 363–372.

Sagiv, Y. (1981). Can we use the universal instance assumption without using nulls? ACM SIGMOD Intl. Conf. Manag. Data, 108–120.

Sagiv, Y., Delobel, C., Parker, D. S., Fagin, R. (1981). An equivalence between relational database dependencies and a fragment of propositional logic. JACM, 28, 435–453.

Sciore, E. A complete axiomatization of full join dependencies. JACM, 29(2), 373–393, (1982).

Sciore, E. *The Universal Instance and Database Design*. Ph.D. dissertation, Princeton Univ. (Dept. Of EECS), (1980).

Selinger, P., Astrahan, M. M., Chamberlin, D. D., Lorie, R. A., Price, T. G. (1979). Access path selection in a relational database management system. ACM SIGMOD Intl. Symp. on Manag. Data, 23–34.

Smith, J. M. (1978). A normal form for abstract syntax. 4th Intl. Conf. on VLDB, 156–162, Berlin, Germany.

Stonebraker, M., Wong, E. A., Kreps, P., Held, G. (1976). The design and implementation of Ingres. ACM TODS, 1(3), 189–222.

Sun, W., Ling, Y., Rishe, N., Deng, Y. (1993). An instant and accurate size estimation method for joins and selection in a retrieval-intensive environment. ACM SIGMOD Intl. Symp. Manag. Data, 79–88.

Thalheim, B. (1991). *Dependencies in Relational Databases*. Teubner Verlagsgesellschaft: Sttutgart, Germany.

Tsou, D.-M., Fischer, P. C. (1982). Decomposition of a relation scheme into Boyce-Codd normal form. SIGACT News, 14(3), 23–29.

Ullman, J. D. (1982). *Principles of Database Systems*, *2nd ed.* Computer Science Press: Rockville, MD.

Ullman, J. D. (1988). *Principles of Database and Knowledge-Based Systems—Volume 1: Classical Database Systems.* Computer Science Press: Rockville, MD.

Ullman, J. D. (1989). *Principles of Database and Knowledge-Based Systems—Volume 2: The New Technologies.* Computer Science Press: Rockville, MD.

Ullman, J. D., Widom, J. (2007). *A First Course in Database Systems*, 3rd ed. Prentice Hall: Upper Saddle River, NJ.

Vardi, M. Y. (1981). The decision problem for database dependencies. Inf. Processing Letters, 12(5), 251–254.

Vardi, M. Y. (1984). The implication and finite implication problems for typed template dependencies. J. Comp. Syst. Sci., 28(1), 3–28.

Vardi, M. Y. (1987). Fundamentals of dependence theory. *Trends in Theoretical Comp. Sci.*, 171–224, Computer Science Press: Rockville, MD.

Vassiliou, Y. (1979). Null values in database management: A denotational semantics approach. ACM SIGMOD Intl. Conf. Manag. Data, 162–169.

Vassiliou, Y. (1980a). Functional dependencies and incomplete information. 6th Intl. Conf. on VLDB, 260–269, Montreal, Canada.

Vassiliou, Y. (1980b). A *Formal Treatment of Imperfect Information in Database Management*. Ph.D. thesis (2nd Ed.): University of Toronto.

Wong, E. (1982). A statistical approach to incomplete information in database systems. *ACM TODS*, 7(3), 470–488.

Yannakakis, M., Papadimitriou, C. H. (1982). Algebraic dependencies. J. Comp. Syst. Sci., 21(1), 2–41.

Zamil, J., et al. (2004). *DB2 SQL PL: Essential Guide for DB2 UDB on Linux, Unix, Windows, i5/OS, and z/OS*, 2nd edition: IBM Press, Indianapolis, IN.

Zaniolo, C. (1976). *Analysis and Design of Relational Schemata for Database Systems*. Ph.D. thesis, Eng-7669, UCLA, Los Angeles, U.S.A.

Zaniolo, C. (1977). Relational views in a database system; support for queries. IEEE Intl. Conf. on Comp. Software App., 267–275.

Zaniolo, C. (1982). A new normal form for the design of relational database schemas. ACM TODS 7(3), 489–499.

Zaniolo, C., Melkanoff, M. A. (1981). On the design of relational database schemata. ACM TODS, 6(1), 1–47.

Zloof, M. M. (1977). Query-by-example: A Database Language. IBM Syst. Journal, 16(4), 324–343.

CHAPTER 4

RELATIONAL SCHEMAS IMPLEMENTATION AND REVERSE ENGINEERING

If God were romantic, He would have created other gods.
As He is classic, He created the world.
—Lucian Blaga

CONTENTS

4.1 The Algorithm for Translating Relational Schemas into
SQL DDL ANSI-92 Scripts (*A8*)... 347

4.2 Relevant Differences Between IBM *DB2*, Oracle *Database*
and *MySQL*, Microsoft *SQL Server* and *Access*.......................... 348

4.3 Case Study: Implementing the Public Library rdb into
DB2, *Oracle*, *MySQL*, *SQL Server*, and *Access*.......................... 425

4.4 The Reverse Engineering Algorithm for Translating
Access 2013 rdb Schemas into *SQL* ANSI *DDL* Scripts
(*REA2013A0*), A Member of *REAF0'*....................................... 473

4.5 The Algorithms for Translating E-R Data Models into rdbs
and Associated Non-Relational Constraint Sets (*AF1-8*)............. 488

4.6 The Reverse Engineering Family of Algorithms for
Translating rdb Schemas into E-R Data Models (*REAF0-2*)....... 490

4.7 Case Study: Reverse Engineering of An *Access Stocks*
db Scheme into both An ANSI Standard *SQL DDL* Script
and An E-R Data Model.. 497

4.8 Best Practice Rules 529

4.9 The Math Behind the Algorithms Presented In This Chapter...... 534

4.10   Conclusion 534

4.11   Exercises 535

4.12   Past and Present 577

Keywords 578

References... 579

After studying this chapter you should be able to:
- Concisely and precisely define, and exemplify the following key terms: *database* (creation, reverse engineering), *relational scheme*, *tablespace*, *table* (fundamental, temporary –both local and global), *column* (prime, non-prime, computed, virtual), *key* (uniqueness, primary, foreign, superkey), *constraint* ((co-)domain, not null, referential integrity, minimal uniqueness, check, tuple, relational, non-relational), *index file* (*unique*, not unique), *E-R data model*, *trigger*, *data* (event, compulsory, range, instance), *data type* (*SQL* ANSI standard, *DB2 10.5*, *Oracle 12c*, *MySQL 5.7*, *SQL Server 2014*, *Access 2013*, built-in, extended, user defined, strongly/weakly typed, geospatial), *identifier* (ordinary, delimited), *type hierarchies* (root, super, sub, target, and recursive –direct and indirect), *number* (infinity, NaN –both quiet and signaling–, subnormal, *sequence*), *array* (ordinary, associative), *view* (materialized, *query*), *identifier* (ordinary, delimited, *global transaction –GTID*).
- Correctly choose the data types you need for any RDBMS version considered in this book (IBM DB2, Oracle Database and MySQL, Microsoft SQL Server and Access) among those that they offer (ANCHOR, ANYDATA, ANYDATASET, ANYTYPE, ATTACHMENT, AUTO_INCREMENT, BFILE, BIGINT, BIT, BINARY, BINARY_DOUBLE, BINARY_FLOAT, BLOB, BOOLEAN, BYTE, CHAR, CLOB, COUNTER, CURRENCY, DATE, DATETIME2, DATETIMEOFFSET, DBCLOB, DBURIType, DECFLOAT, DECIMAL, DISTINCT, DOUBLE, DOUBLEPRECISION, ENUM, FLOAT, GEOGRAPHY, GEOMETRY, GRAPHIC, GUID, HIERARCHYID, HTTPURIType, IDENTITY, INT, INTEGER, INTERVAL DAY TO SECOND, INTERVAL YEAR TO MONTH, LOGICAL, LONG, LONGBLOB, LONG

RAW, LONGTEXT, MEDIUMBLOB, MEDIUMINT, MEDI-
UMTEXT, MONEY, MULTILINESTRING, MULTIPLE BYTE,
MULTIPOINT, MULTIPOLYGON, NCHAR, NCLOB, NESTED
TABLE, NEWID, NUMBER, NVARCHAR, NVARCHAR2,
OBJECT, POINT, POLYGON, RAW, REAL, REF, REFERENCE,
ROW BEGIN, ROW CHANGE TIMESTAMP, ROW END,
ROWGUIDCOL, ROWID, ROWVERSION, SDO_GEOMETRY,
SDO_GEORASTER, SDO_TOPO_GEOMETRY, SERIAL, SET,
SHORT, SHORT TEXT, SI_AverageColor, SI_Color, SI_Colo-
rHistogram, SI_FeatureList, SI_PositionalColor, SI_StillIm-
age, SI_Texture, SINGLE, SMALLDATETIME, SMALLINT,
SMALLMONEY, SQL_VARIANT, STRUCTURED, TIME-
STAMP, TIMESTAMP WITH LOCAL TIME ZONE, TIME-
STAMP WITH TIME ZONE, TINYBLOB, TINYINT, TINYTEXT,
TEXT, UNIQUEIDENTIFIER, URIFactory, URIType, UROWID,
VARBINARY, VARCHAR, VARCHAR2, VARGRAPHIC, VAR-
RAYS, XMLType, YESNO).

➢ Understand and apply algorithm A8 (translating relational schemes into standard ANSI *SQL DDL* scripts for creating rdbs), which is equivalent to understanding exactly how RDBMSes that provide translation of the rdbs they manage into *SQL DDL* scripts are generating these scripts.

➢ Understand, explain, and use the knowledge on the differences and similarities between the *SQL* ANSI standards and IBM DB2 10.5, Oracle Database 12c and MySQL 5.7, MS SQL Server 2014 and Access 2013 SQL idioms.

➢ Understand the five *SQL* DDL scripts for creating the public library case study rdb in all above idioms.

➢ Design and use algorithms of the family AF8', for translating standard ANSI *SQL DDL* scripts into any RDBMS version SQL DDL idiom.

➢ Design and use algorithms of the family REAF0, to reverse engineer any rdbs in order to obtain corresponding standard ANSI *SQL DDL* scripts.

➢ Understand and apply algorithm REA0 (reverse engineering of *SQL DDL* scripts, in order to obtain the corresponding relational schemes), which is equivalent to understanding exactly how RDBMSes create rdbs when executing *SQL DDL* scripts.

➢ Design and use algorithms of the family REAF0–2, to reverse engineer any rdbs in order to obtain corresponding E-R data models.

➢ Understand and apply the best practice rules for correctly creating and populating rdbs, as well as for reverse engineering them.

➢ Better understand the most usual forms of the DDL and DML SQL statements ALTER SESSION, ALTER TABLE, COMMENT, CREATE DATABASE, CREATE FUNCTION, CREATE SEQUENCE, CREATE TABLE, CREATE TRIGGER, CREATE TYPE, CREATE USER, DROP, INSERT INTO, SELECT, UNION ALL, and of their clauses and subclauses CONSTRAINT, CHECK, DELETE RESTRICT, DETERMINISTIC, FOREIGN KEY, GROUP BY, NOT NULL, ON DELETE CASCADE (CASCADE DELETE), ON UPDATE CASCADE (CASCADE UPDATE), ORDER BY, REFERENCES, UNIQUE (UNIQUE INDEX), UPDATE OF, UPDATE RESTRICT, and VALUES, in all of the five RDBMS versions considered in this book, as well some of their non-standard extensions (DESCRIBE, DO, EXPLAIN, FLUSH, LOAD DATA INFILE, PIVOT, REPLACE, RESET, SET, SHOW, TRANSFORM, UNPIVOT).

➢ Get acquainted with some data events (AFTER UPDATE, AS ROW BEGIN, AS ROW CHANGE, AS ROW END, BEFORE INSERT, BEFORE UPDATE, VALIDATED, VALIDATING) and associated programming techniques.

➢ Understand the solved exercises and solve all of the proposed ones.

➢ Better understand the relationships between the algorithms and family of algorithms A1-7, A2014SQLS1-8, A7/8-3, A8, AF1-8, AF8', REA0, REA1, REA2, REA2013A0-2, REAF0', REAF0-1, and REAF0-2.

➢ Understand the meaning of a plethora of DBMS jargon terms and abbreviations (@@DBTS, @@IDENTITY, ALL_OBJECTS, Allow Zero Length, BasicFile LOB, BITEMPORAL, BOOL, BUSINESS_TIME, CAST, CCSID, CODEUNITS16, CODEUNITS32, CONCAT, COUNT, CurDate, CurTime, CURRENT DATE,

CURRENT_DATE, CURRENT LOCALE, CURRENT TIME, CURRENT_TIME, CURRENT TIMESTAMP, CURRENT_ TIMESTAMP, CURRENT TIMEZONE, CURRENT_TIME-ZONE, DANGLING, DATALINK, DATE_ADD, DATE_SUB, DBCS, DBMS_XMLSCHEMA, DBTimeZone, DB2SECURITY-LABEL, DB2SQLSTATE, DBA_OBJECTS, DEREF, Design View, DETERMINED BY, dummy date, EBCDIC, Enforce Referential Integrity, Extract, FOR BIT DATA, FOR EACH ROW, GENERATE_UNIQUE, GetDate, GetUTCDate, grbit, HEX, Indexed, LAST_INSERT_ID, LC_TIME, LIMIT, Limit to List, Local_Timestamp, Lookup, Lookup Wizard, MAX_STRING_ SIZE, MAXDB, MBCS, MIXED DATA, MIXED DECP, Mixed Based Replication (MBR), MSysRelationships, Now, OID, ON COMMIT PRESERVE ROWS, PARAMETERS, Query Design, QUOTED_IDENTIFIER, REF, REGEXP, Required, Row Based Replication (RBR), SecureFile LOB, SBCS, Statement Based Replication (SBR), STDDT, SessionTimeZone, STRAIGHT_JOIN, SUM, SysDate, SysDateTime, SysDateTimeOffset, SysTime-stamp, SysUTCDateTime, SYSCAT, SYSDUMMY, SYSFUN, SYSIBM, SYSPROC, SYSTEM_TIME, szColumn, szObject, szReferencedColumn, szReferencedObject, szRelationship, Time, TOP, TRANSACTION START ID, TRANSLATE, TRIM, UDT, Universal Naming Convention (UNC), USER_OBJECTS, UTF-16, UTF-8, Validation Rule, Validation Text, WTDDT, Yes (No Duplicates)).

Any relational scheme may be at least partially implemented as an rdb, by using any RDBMS version. Generally, their objects are grouped in *tablespaces* (typically, there are several system ones, for the metacalogs, the temporary tables, etc. and at least a user one).

Uniqueness is implemented through *unique index files*, also used for speeding-up data search. Index files are using several data structures (e.g., trees, bitmaps, hash tables, linked lists, etc.) to essentially store pairs of the type <data value, pointer to the location where that data value is stored>. Non-unique index files are used to speed-up data search (that should be balanced with the cost of maintaining them when data is modified—see Chapter 4 of the second volume).

Sometimes, due to limitations in RDBMS versions implemented *SQL*, in order to implement domain and tuple constraints (but also non-relational ones), it is necessary to also use their extended *SQL triggers*: special stored procedures associated to a table and a (or several) data manipulation associated event(s) that are automatically called by the system whenever the corresponding data event(s) occur(s). These *data events* (*events* for short), which may have different names for different platforms[94], are generally associated to moments in time corresponding to immediately before and after inserting, updating, or deleting a row, to user's clicks, double-clicks, etc.

Although *SQL* is standardized by both ANSI and ISO, due to historical and commercial competition reasons, its actual implementations differ among RDBMSes, especially at the *DDL* level (while at the *DML* one differences are minor). This is why, on one hand, section 4.1 presents only the algorithm *A*8, for translating RDM schemes into *SQL* ANSI-92 scripts (ignoring any RDBMS implementation differences), Section 4.2 is focusing on the differences among the five RDBMSs chosen to illustrate the approach proposed by this book, section 4.3 shows actual *SQL DDL* scripts (as well as additional MS *Access* GUI techniques needed for enforcing domain and tuple/check constraints) for implementing the public library case study RDM scheme obtained in Section 3.4, and, on the other, in order to actually implement any RDM scheme in any given RDBMS version, you need the additional family of algorithms *AF*8', which translates ANSI standard *SQL DDL* scripts into RDBMSs corresponding ones.

In fact, algorithm *A*8 also embeds the ANSI-2003 standard of auto-generated (autonumber) columns; moreover, as this is the only way to implement autonumbering in *Oracle* versions prior to *12c* and tuple/check constraints in *MySQL* (but also in *Oracle* for the ones that need nondeterministic functions), the scripts that implement the public library rdb from Subsections 4.3.2 and 4.3.3 also use triggers, which were standardized only in the *SQL* ANSI-99.

As shown in Section 4.2, most of the considered RDBMS versions also provide either other higher *SQL* ANSI standards or/and own nonstandard facilities; exercise 4.4 proposes to readers designing corresponding extensions to algorithm *A*8.

---

[94] For example, even for products of a same manufacturer: in .NET, MS refers as *Validating* and *Validated* the events that in *Access* are known (just like in *Oracle*) as *BeforeUpdate* and *AfterUpdate*, respectively.

As they are very easy to design and even implement, especially after studying Sections 4.2 and 4.3, *AF8'* algorithms are left to the reader (see Exercise 4.3). Similarly, the dual *REAF0'* algorithms are very simple, except for *Access*, which is dealt with in section 4.4; consequently, they are left to the reader too (see Exercise 4.5). Same thing goes for the dual of *A*8, the reverse engineering algorithm *REA*0 (see Exercise 4.6).

Please note that *A*8 only considers single foreign keys referencing (single) primary surrogate keys, as you should always use. If, however, you also use the totally not recommended concatenated foreign keys provided too by the RDM, then you need a slight variation of this algorithm (see Exercise 4.7).

Section 4.5 discusses the family of algorithms *AF*1–8, probably the most important set of forward "shortcuts" for actually implementing E-R data models into rdbs as fast as advisable, and presents, as an example, one of its members (for MS *SQL Server* dbs).

Section 4.6 discusses the *REAF*0–2 family, dual to the above one, and at least as important, especially for decrypting the architecture and the semantics of legacy (but not only), undocumented (or poorly documented) rdbs, and presents too, as an example, one of its members (for MS *Access* dbs).

Section 4.7 presents (as a rdb reverse engineering case study) the results of applying for an *Access* db (*Access* again, as it is the only one out of the five RDBMSes considered in this book for which reverse engineering is not that obvious) the corresponding algorithms of the families *REAF0'* (composed with *REA*0) and *REAF*0–2.

Finally, this chapter ends like all other ones of this book with sections dedicated to the corresponding best practice rules, the math behind, conclusion, exercises, and past, present and references.

4.1 THE ALGORITHM FOR TRANSLATING RELATIONAL SCHEMAS INTO *SQL DDL* ANSI-92 SCRIPTS (*A*8)

Thinking things through is hard work and it sometimes seems safer to follow the crowd. That blind adherence to such group thinking is, in the long run, far more dangerous than independently thinking things through.
                                                        —Thomas Watson, Jr.

Figure 4.1 presents algorithm *A8*, for translating relational schemas into *SQL DDL* ANSI standard corresponding scripts. For the column domains *D'*, see Table 4.1, which is presenting the *SQL* ANSI-92 data types (with the addition of the autonumbers, collections, and *XML* that were introduced in later ANSI standards).

Variants of the algorithm *A8* (which are targeting as output language their own *SQL* idioms instead of the ANSI standard ones) are implemented by all RDBMS versions that provide automatic generation of *SQL DDL* scripts of the rdbs they manage (which includes *DB2 10.5, Oracle 12c, MySQL 5.7*, and *SQL Server 2014*).

4.2   RELEVANT DIFFERENCES BETWEEN IBM *DB2*, ORACLE *DATABASE* AND *MYSQL*, MICROSOFT *SQL SERVER* AND *ACCESS*

> *Technology, which is generally believed to kill the imagination freedom, the beauty, and the soul, has, among its latest consequences, the chance to bring to the Universe the fairytale, the poetry, and the immortality.*
> —Lucian Blaga

Although most of both commercial and open source RDBMSes adhere at least to the pure ANSI-92 *SQL* standard, there are, however, significant differences between them especially in the data types they provide, but also in naming conventions, and sometimes, even in *SQL* syntax.

This is why this section is presenting the most relevant of these differences, in order to help with implementing ANSI standard *SQL* scripts in each of these 5 flagship RDBMSs (and fully understand the differences between their implementations of the library db case study presented in Section 4.3).

For those in a hurry, at least the tables at the end of each *Data Types* subsections below should be of interest, as they summarize what data type is best to choose, depending on your actual needs and target RDBMS.

In all data type definitions that follow, square brackets are embracing optional items; for example, DECIMAL[(*n*[,*m*])] means that both *n*

Algorithm A8 (Translation of RDM schemas into SQL ANSI-92 scripts)
Input: an rdb scheme *S*;
Output: a *SQL* ANSI-92 script for creating a corresponding rdb *S*;
Strategy: open new script (text file) *S*.sql;
add to *S*.sql line "CREATE DATABASE S;";
loop for all tables T (having maximum cardinal *n*) *from S*
  if T's primary key *x* (of type NAT(*n*)) is not a foreign key *then*
    add to *S*.sql line "CREATE TABLE S.T (x NUMERIC(n,0)
            GENERATED ALWAYS AS IDENTITY(1,1) PRIMARY);";
  else -- T' is the table referenced by x
    add to *S*.sql line "CREATE TABLE S.T (x NUMERIC(n,0) PRIMARY,
      CONSTRAINT FKT-x FOREIGN KEY (x) REFERENCES T');";
  end if;
  loop for all columns C of T, except for its primary key *x*, having domain *D*
    compute *D'*, the ANSI equivalent of *D*, based on Table 4.1;
    s = "ALTER TABLE S.T ADD C D' ";
    if C should not be null *then s = s &* " NOT NULL ";
    if C should be unique *then s = s &* " UNIQUE ";
    add to *S*.sql line *s &* "; ";
  end loop for all columns C of T;
  loop for all concatenated (unique) keys CK of T
    s = "ALTER TABLE S.T ADD CONSTRAINT T_CK UNIQUE (";
    first = true;
    loop for all CK component attributes A
      if first then first = false; else s = s & ",   "; *end if; s = s & A;*
    end loop;
    add to *S*.sql line *s &* ") ; ";
  end loop (for all concatenated (unique) keys CK of T);
  loop for all foreign keys FK of T referencing *R*
    add to *S*.sql line "ALTER TABLE S.T ADD CONSTRAINT T_FK
                    FOREIGN KEY (FK) REFERENCES R;";
  end loop;
  loop for all tuple (check) constraints CC of T
    add to *S*.sql line "ALTER TABLE S.T ADD CONSTRAINT T_CC
                    CHECK (CC);";
  end loop;
end loop (for all tables T from S);
save and close script *S*.sql;
End Algorithm A8 (Translation of RDM schemas into SQL ANSI-92 scripts);

FIGURE 4.1   Algorithm A8 (Translation of RDM Schemas into *SQL* ANSI-92 Scripts).

TABLE 4.1   The *SQL* ANSI Standard Data Types

Data Type	Description
CHARACTER(n)	Character string, fixed length n. A string of text in an implementer-defined format. The size argument is a single nonnegative integer that refers to the maximum length of the string. Values for this type must be enclosed in single quotes.
VARCHAR(n)	Variable length character string, maximum length n.
BINARY(n)	Fixed length binary string, maximum length n.
BOOLEAN	Stores truth values—either *TRUE* or *FALSE*.
VARBINARY(n)	Variable length binary string, maximum length n.
INTEGER(p)	Integer numerical, precision p.
SMALLINT	Integer numerical precision 5.
INTEGER	Integer numerical, precision 10. It is a number without decimal point with no digits to the right of the decimal point, that is, with a scale of 0.
BIGINT	Integer numerical, precision 19.
IDENTITY(s, i)	Integer numerical auto-generated, starting value s, increment value i (added in *SQL2003*).
DECIMAL(p, s)	Exact numerical, precision p, scale s. A decimal number is a number that can have a decimal point in it. The size argument has two parts: precision and scale. The scale cannot exceed the precision. Precision comes first, and a comma must separate from the scale argument. Mantissa's length is $p - s$.
NUMERIC(p, s)	Exact numerical, precision p, scale s. (Same as DECIMAL).
FLOAT(p)	Approximate numerical, precision p. A floating number in base 10 exponential notation. The size argument for this type consists of a single number specifying the minimum precision.
REAL	Approximate numerical, mantissa precision 7.
FLOAT	Approximate numerical, mantissa precision 16.
DOUBLE PRECISION	Approximate numerical, mantissa precision 16.
DATE, TIME, TIMESTAMP	Composed of a number of integer fields, representing an absolute point in time, depending on subtype.

TABLE 4.1   (Continued)

Data Type	Description
INTERVAL	Composed of a number of integer fields, representing a period of time, depending on the type of interval.
COLLECTION (ARRAY, MULTISET)	*ARRAY* (added in *SQL99*) is a set-length and ordered collection of elements, *MULTISET* (added in *SQL2003*) is a variable-length and unordered collection of elements. Both the elements must be of a predefined data type.
XML	Stores *XML* data (added in *SQL99*). It can be used wherever a *SQL* data type is allowed, such as a column of a table.

and *m* parameters are optional, so all of (and only) the following expressions are valid: DECIMAL, DECIMAL(n), and DECIMAL($n,$ m) .

4.2.1   IBM *DB2 10.5*

> *Good design is good business.*
> —Thomas J. Watson

DB2 10.5 is a very robust, well-performing and fully featured RDBMS, also available in the cloud and supporting both *SQL* and *XQuery* (*DB2* has native implementation of *XML* data storage, where *XML* data is stored as XML (not as relational or CLOB data) for faster access using *XQuery*).

DB2 10.5 can also be accessed through *ODBC*, *ADO*, and *OLEDB* and has APIs for *Java*, *.NET CLI*, *Python*, *Perl*, *PHP*, *Ruby*, *C++*, *C*, *Delphi*, *REXX*, *PL/I*, *COBOL*, *RPG*, *FORTRAN*, and many other programming languages. *DB2* also supports integration into the *Eclipse* and *Visual Studio* integrated development environments.

In the clustered DBMS arena, where databases can grow to many terabytes, IBM offers two approaches (that compete with *Oracle Real Application Clusters* (*RAC*)): *DB2 pureScale* and *DB2 Database Partitioning Feature* (*DPF*). *DB2 pureScale* is a shared-disk db cluster solution that is ideal for high-capacity OLTP workloads. *DB2 DPF* lets you partition your db across multiple servers or within a large SMP server, which is ideal for OLAP workloads.

DB2 DPF is sold as part of the IBM *InfoSphere Warehouse*, which is the *DB2*'s name when sold in data warehouse environments. It includes several BI features, such as ETL, data mining, OLAP acceleration, and in-line analytics.

As market share, according to Gartner (probably the most objective source in the field), until 2013, *DB2* was slightly surpassed only by *Oracle*, which lead it by some 33% at 30%.[95] In 2013, *Oracle* rose at more than 40%, whereas *DB2* was for the first time surpassed by the MS *SQL Server* (mainly due to its native bridge to MS *Excel*, but also to the one to *Hadoop*, as well as to the MS *Azure* cloud). Corresponding data for 2014 will certainly be very interesting too.

Besides several commercial versions, there is also a free to download *DB2 Express-C* version (for 15 TB dbs maximum that should kill any current competition, which only allows for some 10 GB!) that can also be redistributed with your products; it allows for 16 GB RAM (again remarkable, as competition currently allows for only 1 GB) and 2 CPU cores.

DB2 10.5 was the base for all *SQL* ANSI standards, also including support for temporal dbs.

DB2 10.5 runs on *Linux*, *Unix*, MS *Windows*, *zOS* (mainframes), *iSeries* (formerly *OS/400*) and *VM/VSE* OSs. Its *Express-C* free edition runs on MS *Windows* (*x86/x64*), *Linux* (*x86, x64, Power Systems*), *Solaris*, and Mac *OS X*.

4.2.1.1   Identifiers and Literals

To be successful, you have to have your heart in your business and
your business in your heart.
—Thomas J. Watson

DB2 object names may generally have at most 128 bytes (i.e., between 32 and 128 characters, depending on the chosen alphabet and encoding) and are not case sensitive: you can freely use both upper and lower case (and *DB2* stores them exactly as you defined them, if they were enclosed in

---

[95] Third place, with some 17%, was occupied by the MS *SQL Server*; all other competitors, including the very popular open source *MySQL* that was leading the pack, are fighting to survive in or emerge from the rest of some 20%; surprisingly, at that time, the only *NoSQL* DBMS ranked among the top ten as popularity was *MongoDB*; again as popularity, *Access* ranked 7th, only two seats behind *DB2*.

double quotes, case in which you should always refer to them alike, or in uppercase, if they were not delimited with double quotes).

The so-called *ordinary identifiers* should start with a letter and can contain only letters (with all lowercase ones automatically translated to uppercase), digits, and '_'. The *delimited identifiers*, which should be declared and used enclosed in double quotes, can use almost anything, including lowercase letters and even leading spaces.

Date and time constants are enclosed in single quotation marks. Text strings are enclosed in double quotation marks. You can use double quotation marks within text strings by doubling them; for example, the text string """"quote"""""'s value is "quote".

4.2.1.2 Databases and Tables

> *If you want to achieve excellence, you can get there today.*
> *As of this second, quit doing less-than-excellent work.*
> —Thomas J. Watson

DB2 provides the *SQL* CREATE DATABASE statement too, but, mainly due to security and physical disk space management, this is typically done only by DBAs, who are generally using for this task too the GUI of the *IBM Data Studio* tool. However, you can do it too in its *Express* editions, as there you are the DBA too.

Computed tables are called *views* and can be created with the CRE-ATE VIEW ... AS SELECT ... *SQL DDL* statement; they can be created graphically through the *IBM DB2 Query Management Facility*'s (*QMF*) *QBE*-type GUI too. Unfortunately, *QMF* (which also provides analytics) currently has only a commercial *Enterprise* edition.

DB2 was the first RDBMS to introduce in-memory tables, very useful especially for OLAP processing (but not only) as it speeds up data analysis even more than 25 times, as part of the *BLU* technologies (a development code name standing for "*big data, lightning fast, ultra-easy*"), which is a bundle of novel techniques for columnar processing, data deduplication, parallel vector processing, and data compression.

Temporary tables, either local or global are created in user temporary tablespaces with the CREATE [GLOBAL] TEMPORARY TABLE ... *SQL*

DDL statement. The local ones exist during the current user session or the lifetime of the procedure in which they are declared. The global ones have persistent definitions, but data is not persistent (as it is automatically deleted by the system at the end of the db sessions) and they generate no redo or rollback information. If, however, you would like their instances be persistent too, you can add the ON COMMIT PRESERVE ROWS clause at the end of the above statement.

Moreover, *DB2* also provides temporary tables which only exist in a FROM or WHERE clause during the execution of a single *SQL* statement, as well as temporary tables that are part (and exist only during the execution) of a WITH *SQL* statement.

You can associate indexes, triggers, statistics, and views to temporary tables too. Dually, for temporary tables you cannot define referential integrity, or specify default values for their columns, or have columns declared as a ROWID or LOB (V6), or create such tables LIKE a declare global temporary table, or issue LOCK TABLE statements on them, or use *DB2* parallelism, etc.

DB2 also provides table hierarchies, in which subtables may inherit the privileges of their supertables.

4.2.1.3   Data Types

All the problems of the world could be settled easily if men were only willing to think.
The trouble is that men very often resort to all sorts of devices in order not to think, because thinking is such hard work.
—Thomas J. Watson

4.2.1.3.1   Numeric Data Types

The great accomplishments of man have resulted from the transmission of ideas and enthusiasm.
—Thomas J. Watson

DB2 has a BOOLEAN data type.

There are 7 (signed) numeric data types:

> ➢ SMALLINT = [−32,768; 32,767];
> ➢ INTEGER (or INT) = [−2,147,483,648; 2,147,483,647];
> ➢ BIGINT   =   [−9,223,372,036,854,775,808; 9,223,372,036,854,775,807];
> ➢ DECIMAL (or DEC)[(n[, m])]   = [−10³¹+1; 10³¹−1]; $0 \leq n \leq 31$, $0 \leq m \leq 31$, $1 \leq n + m \leq 31$, default values: $n = 5$, $m = 0$; almost equivalent is the NUMERIC (or NUM): the only difference between them is in the internal storage mode on AS/400 systems;
> ➢ REAL = [−3.4028234663852886 × 10³⁸; 3.4028234663852886 × 10³⁸]; precision is 7 digits after the decimal point;
> ➢ DOUBLE = [−1.7976931348623158 × 10³⁰⁸; 1.7976931348623158 × 10³⁰⁸]; precision is 15 digits after the decimal point;

A virtual type FLOAT(n) can also be used: $0 < n < 25$ means REAL, $24 < n < 54$ means DOUBLE.

> ➢ DECFLOAT(n) = [−9.999999999999999999999999999999999 × 10⁶¹⁴⁴; 9.999999999999999999999999999999999 × 10⁶¹⁴⁴]; $n \in \{16, 34\}$.

Decimal floating-point numbers (DECFLOAT) can have a precision of 16 or 34 and an exponent range of 10⁻³⁸³ to 10³⁸⁴ or 10⁻⁶¹⁴³ to 10⁶¹⁴⁴, respectively: the minimum exponent *Emin* for DECFLOAT values are −383 for DECFLOAT(16) and −6,143 for DECFLOAT(34); the maximum exponent *Emax* values are 384 for DECFLOAT(16) and 6,144 for DECFLOAT(34).

Decimal floating-point supports both exact representations and approximations of real numbers and so is not considered either an exact or an approximate numeric type.

In addition to finite numbers, DECFLOAT numbers are able to represent the following named decimal floating-point special values:

> ✓ *Infinity* (a value that represents a number whose magnitude is infinitely large);
> ✓ *Quiet NaN* (a value that represents undefined results and that does not cause an invalid number warning);
> ✓ *Signaling NaN* (a value that represents undefined results and that causes an invalid number warning if used in any numeric operation).

When a number has one of these special values, its coefficient and exponent are undefined.

It is possible to have positive or negative infinity. The sign of *NaN* values has no meaning for arithmetic operations.

Nonzero numbers whose adjusted exponents are less than *Emin* are called *subnormal numbers*. These subnormal numbers are accepted as operands for all operations and can result from any operation. For a subnormal result, the minimum values of the exponent become *Emin*—(*precision*—1), called *Etiny*, where *precision* is the working precision. If necessary, the result is rounded to ensure that the exponent is no smaller than *Etiny*. If the result becomes inexact during rounding, an underflow warning is returned. A subnormal result does not always return the underflow warning.

When a number underflows to zero during a calculation, its exponent will be *Etiny*. The maximum value of the exponent is unaffected. The maximum value of the exponent for subnormal numbers is the same as the minimum value of the exponent that can arise during operations that do not result in subnormal numbers. This occurs when the length of the coefficient in decimal digits is equal to the precision.

Autonumbering is obtained by declaring the corresponding column as being `GENERATED ALWAYS AS IDENTITY`. A table can have only one identity column. The `IDENTITY` keyword can only be specified for columns of exact numeric or user-defined `DISTINCT` types (see Subsection 4.2.1.2.3) for which the source type is an exact numeric type, with a scale of zero. An identity column is implicitly `NOT NULL`. Obviously, an identity column cannot have a `DEFAULT` clause.

Only `SMALLINT`, `INTEGER`, `BIGINT`, `DECIMAL` with a scale of zero, and a `DISTINCT` type based on one of these types are exact numeric types. `REFERENCE` types (see Subsection 4.2.1.2.3 below), even if represented by an exact built-in numeric type, cannot be defined as identity columns.

Explicit start, stop, (no) minimum, (no) maximum, increment, and (no) caching values, as well as (no) order direction and (no) cycling can also be specified for `IDENTITY` columns. Implicitly, they are 1 for start and increment, corresponding minimum and maximum data type values, 20 for cache, ascending for order, and *no* for cycling.

There are 3 DATE/TIME data types:
- ➢ `DATE` = [1/1/1; 12/31/9999];
- ➢ `TIME` = [00:00:00; 24:00:00];
- ➢ `TIMESTAMP`[(*fsp*)] = [1/1/1 00:00:00.000000000000; 12/31/9999 00:00:00.000000000000]; $0 \leq fsp \leq 12$, default 6.

You can get the current system date, time, and timestamp directly from the registers CURRENT DATE (or CURRENT_DATE), CURRENT TIME (or CURRENT_TIME), and CURRENT TIMESTAMP (or CURRENT_TIMESTAMP), respectively. For getting the current system locale (by default, the "en_US" one, for English U.S.), you can use the CURRENT LOCALE LC_TIME register. Similarly, you can get the local offset with respect to the current UTC time from the CURRENT TIMEZONE (or CURRENT_TIMEZONE) register. You can get the year part of a date or timestamp d with the function *Year*(d).

There is no CURRENCY data type: during data integrations, *DB2* translates such data types into DECIMAL.

4.2.1.3.2 String Data Types

> *Whenever an individual or a business decides that success has been*
> *attained, progress stops.*
> —Thomas J. Watson

There are seven (signed) numeric data types in *DB2*:
➢ CHAR[(n)] = ASCII/UNICODE(n), fixed length, $1 \leq n \leq 254$, for ASCII (OCTET), and $1 \leq n \leq 63$, for UNICODE (CODEUNITS32, that is UTF-32); default n value: 1; synonym: CHARACTER;
➢ VARCHAR(n) = ASCII/UNICODE(n), variable length, $1 \leq n \leq 32{,}672$, for ASCII (OCTET) and $1 \leq n \leq 8{,}168$, for UNICODE (CODEUNITS32, that is UTF-32); synonym: CHAR (or CHARACTER) VARYING;
➢ CLOB[(n)] = ASCII/UNICODE(n), variable length, $1 \leq n \leq 2{,}147{,}483{,}647$, for ASCII (OCTET), and $1 \leq n \leq 536{,}870{,}911$, for UNICODE (CODEUNITS32, that is UTF-32); default n value: 1 M (one megabyte); synonym: CHAR (or CHARACTER) LARGE OBJECT;
➢ GRAPHIC[(n)]: equivalent of CHAR, but with encodings of 2 bytes per character (UTF-16); corresponding maximum values for n are 127 and 63;
➢ VARGRAPHIC(n): equivalent of VARCHAR, but with encodings of 2 bytes per character (UTF-16); corresponding maximum values for n are 16,336 and 8,168, respectively;

➢ DBCLOB[(n)]: equivalent of CLOB, but with encodings of 2 bytes per character (UTF-16); corresponding maximum values for n are 1,073,741,823 and 536,870,911;

➢ BLOB[(n)]: variable length binary strings of up to 2 GB ($1 \leq n \leq 2,147,483,647$ bytes), not associated with a code page; default n value: 1 M (one megabyte).

BLOB, CLOB, and DBCLOB sizes (i.e., their parameter n) can be expressed either in bytes, kilobytes (K), megabytes (M), or gigabytes (G), specified by n's suffix (nothing, 'K', 'M', or 'G', respectively).

NULL-terminated character strings that are found in C are handled differently, depending on the standards level of the precompile option. Each character string is further defined as one of:

✓ *BIT* data (binary string data not associated with a code page);

✓ Single-byte character set (*SBCS*) data (data in which every character is represented by a single byte);

✓ · Mixed data (data that might contain a mixture of characters from a single-byte character set and a multibyte character set (*MBCS*)).

If you need to store UNICODE data, please note the following eight facts:

✓ In the corresponding CREATE DATABASE, CREATE TABLESPACE, or/and CREATE TABLE statements, specify the CCSID UNICODE clause.

✓ By default, the encoding scheme of a table is the same as the encoding scheme of its tablespace. Also by default, the encoding scheme of a tablespace is the same as the encoding scheme of its db. You can override the encoding scheme with the CCSID clause in the CREATE TABLESPACE or CREATE TABLE statement. However, all tables within a tablespace must have the same CCSID.

✓ In the CREATE TABLE statements, for each column definition, specify the appropriate data type, subtype, and length value.

✓ Use one of the following data types:
   • For UTF-8 data: CHAR, VARCHAR, or CLOB.
   • For UTF-16 data: GRAPHIC, VARGRAPHIC, or DBCLOB.
   • For binary data: BINARY, VARBINARY, or BLOB.

✓ For character columns, optionally specify one the following subtypes, by adding the FOR *subtype* DATA clause to the column definition:

- SBCS: columns contain only those UTF-8 characters that are stored as 1 byte (the first 128 ones in the UNICODE code page); data stored in SBCS columns of UNICODE tables has CCSID 367.
- MIXED (default value: if you do not specify a subtype, *DB2* assumes by default the FOR MIXED DATA one): columns contain any UTF-8 data that is more than 1 byte; character data in UNICODE tables is stored as mixed data by default, even if your subsystem is defined with a MIXED DECP value of NO; data stored in MIXED columns of UNICODE tables has CCSID 1208.
- BIT: columns contain BIT (i.e., binary, not string) data; CCSID 66534 is associated with FOR BIT DATA columns.

✓ Do not use FOR BIT DATA columns for the sole purpose of handling international data: only use FOR BIT DATA columns if you have a specific reason, such as encryption; otherwise, this data type can cause problems. For example, if you have a string of length 10 and put it in a FOR BIT DATA column of length 12, *DB2* pads the string with two blanks. The hexadecimal value that is used for those blanks is system specific. For example, X'40' is used for ASCII/EBCDIC and X'20' is used for UNICODE. These different hexadecimal values can potentially cause problems when you convert this data.

✓ To determine the appropriate length value, follow these instructions:
  - For UTF-8 data, allocate three times the column size that you would allocate for a non UNICODE table. For example, if you use CHAR(10) for a name column in an ASCII/EBCDIC table, use VARCHAR(30) for the same column in a UNICODE table. Such columns can contain 30 bytes or ten 3-byte characters. Use VARCHAR instead of CHAR, because the length 30 is greater than 18 (where 18 is traditionally the length from which VARCHAR should be used instead of CHAR.) This estimate allows for the worst-case expansion of UTF-8 data. The worst case for SBCS data is that 1 byte in ASCII/EBCDIC expands to 3 bytes in UTF-8. For mixed data, such as Chinese, Japanese, or Korean characters, the same worst-case scenario

applies: you might have 2, 3, and 4-byte characters, depending on the encoding, that expand to a four-byte UTF-8 character in the worst case; however, because these characters use more than one byte in ASCII or EBCDIC, the worst-case expansion in UTF-8 is still three times the original size.

◆ For UTF-16 data, allocate two times the column size that you would allocate for a non UNICODE table, and use the GRAPHIC or VARGRAPHIC data types. For example, if you use CHAR(10) for a name column in an ASCII/EBCDIC table, use VARGRAPHIC(10) for the same column in a UNICODE table: CHAR(10) is 10 bytes long and VARGRAPHIC(10) is 20 bytes long, or the equivalent of 10 two-byte characters.

✓ DB2 associates certain CCSIDs with columns, depending on the data type that you specify. Table 4.2 summarizes the possible column data types in UNICODE tables and their associated CCSIDs:

TABLE 4.2   CCSIDs Associated by DB2 with Columns in UNICODE Tables

Column data type	Associated CCSID	Format in which data is stored
CHAR	1208	UTF-8
CHAR FOR SBCS DATA	367	7-bit ASCII
CHAR FOR MIXED DATA	1208	UTF-8
CHAR FOR BIT DATA	66534	NA
VARCHAR	1208	UTF-8
VARCHAR FOR SBCS DATA	367	7-bit ASCII
VARCHAR FOR MIXED DATA	1208	UTF-8
VARCHAR FOR BIT DATA	66534	NA
CLOB	1208	UTF-8
CLOB FOR SBCS DATA	367	7-bit ASCII
CLOB FOR MIXED DATA	1208	UTF-8
GRAPHIC	1200	UTF-16
VARGRAPHIC	1200	UTF-16
DBCLOB	1200	UTF-16

BLOBs are generally used to store non-traditional data such as pictures, voice, or mixed media, but they can also hold structured data for processing by user-defined data types and functions. Character strings of the FOR BIT DATA subtype may be used for similar purposes.

In UNICODE dbs, when using CODEUNITS16 or CODEUNITS32, CHAR and GRAPHIC may also be referred as NCHAR, VARCHAR and VARGRAPHIC as NVARCHAR, whereas CLOB and DBCLOB as NCLOB (all of them having the prefix 'N' from National character sets).

CLOBs are used to store large SBCS or mixed (SBCS and MBCS) character-based data (such as documents written with a single character set) and, therefore, has an SBCS or mixed code page that is associated with it.

The double-byte character large object (DBCLOB) data type is used to store large double-byte character set (*DBCS*) based data (such as documents written with a single character set) and, therefore, has a DBCS code page that is associated with it.

4.2.1.3.3 Other Data Types

> *Every time we've moved ahead in IBM, it was because someone was*
> *willing to take a chance,*
> *put his head on the block, and try something new.*
> —Thomas J. Watson

There are also six other *DB2* data types: one built-in (XML), one extended (ANCHOR), and four user-defined (DISTINCT, STRUCTURED, REFERENCE, and ARRAY).

> ➤ XML stores well-formed XML documents; such values share the first 5 restrictions of the CLOB, DBCLOB, and BLOB data types above.
>
> ➤ ANCHOR is a data type based on another *SQL* object such as a column, global variable, *SQL* variable or parameter, or a table or view row. A data type defined using an anchored type definition maintains a dependency on the object to which it is anchored. Any change in the data type of the anchor object will impact the

anchored data type. If anchored to the row of a table or view, the anchored data type is ROW[96], with the fields defined by the columns of the anchor table or view.

User-defined data types (*UDT*) are derived from existing ones, for extending the built-in ones and create your own customized data types:

> DISTINCT shares its internal representation with an existing built-in data type (its "source" type) and includes qualified identifiers.

If the schema name is not used to qualify the distinct type name when used in other than the CREATE TYPE (DISTINCT), DROP, or COMMENT statements, the *SQL* path is searched in sequence for the first schema with a distinct type that matches.

Distinct types that are sourced on LOB types are subject to the same restrictions as their source type.

Not all built-in data types can be used to define DISTINCT types: the source types cannot be BOOLEAN, XML, ARRAY, CURSOR, or ROW.

The specified distinct type cannot have any data type constraints and the source type cannot be an ANCHORed data type.

A distinct type is defined to use either strong or weak typing rules, with the strong ones being the default.

A *strongly typed* DISTINCT *data type* (*STDDT*) is considered to be a separate and incompatible type for most operations. For example, you would use it to define a picture, text, or audio type, as each of these types has its own different semantics, but all of them use the built-in data type BLOB for their internal representation.

The *AUDIO* following example is a STDDT:

CREATE TYPE AUDIO AS BLOB(1 M);

This definition allows the creation of functions that are written specifically for *AUDIO*, and assures that these functions are not applied to values of any other data type (e.g., pictures or text ones): STDDTs support strong typing by ensuring that only those functions and operators that are explicitly defined on the distinct type can be applied to its instances. For this reason, a STDDT does not automatically acquire the functions and

---

[96] ROW is a user-defined data type that can be used only within the IBM's extended *SQL Procedural Language* (*SQL PL*) applications; it is a structure composed of multiple fields, each with its own name and data type, which can be used to store the column values of a row in a result set or other similarly formatted data.

operators of its source type, because these functions and operators might not be meaningful.

For example, a *LENGTH* function could be defined to support a parameter with the data type *AUDIO* that returns the length of the object in seconds instead of bytes.

A *weakly typed DISTINCT data type* (*WTDDT*) is considered to be the same as its source type for all operations, except when it applies constraints on values during assignments, casts, and function resolutions.

The *NATURALS* following example is a WTDDT:

```
CREATE TYPE NATURALS AS INTEGER
WITH WEAK TYPE RULES CHECK (VALUE >= 0);
```

Weak typing means, in this case, that except for accepting only positive integer values, *NATURALS* operates in the same way as its underlying INTEGER data type.

WTDDTs can be used as alternative methods of referring to built-in data types within application code: the ability to define constraints on the values that are associated with them provides a method for checking values during assignments and casts.

There are two distinguished built-in system DISTINCT types that can be used as table column ones:

- ✓ SYSPROC.DB2SECURITYLABEL must be used to define the row security label column of a protected table; its underlying data type is VARCHAR(128) FOR BIT DATA; a table can have at most one column of type DB2SECURITYLABEL; for such a column, NOT NULL WITH DEFAULT is implicit and cannot be explicitly specified; the default value is the session authorization ID's security label for write access.
- ✓ SYSPROC.DB2SQLSTATE is used to store *DB2* errors, warnings, and information codes; its underlying data type is INTEGER.

Using DISTINCT types provides benefits in the following areas:

- ✓ *Extensibility*: increasing the set of data types available to support your applications.
- ✓ *Flexibility*: any semantics and behavior can be specified for these data types by using user-defined functions to augment the diversity of the data types available in the system.

✓ *Consistent and inherited behavior*: strong typing guarantees that only functions defined on your distinct type can be applied to instances of the distinct type; weak typing ensures that the distinct type behaves the same way as its underlying data type and so can use all the same functions and methods available to that underlying type.

✓ *Encapsulation*: using a WTDDT makes it possible to define data type constraints in one location for all usages within application code for that distinct type.

✓ *Performance*: distinct types are highly integrated into the db manager; because they are internally represented the same way as the built-in ones, they share the same efficient code that is used to implement components such as built-in functions, comparison operators, and indexes for built-in data types.

➢ STRUCTURED contains a sequence of named attributes, each of which has a data type and also includes a set of method specifications. A structured type can be used as the type of a table, view, or column. When used as a type for a table or view, that table or view is called a *typed* one.

For typed tables and views, the names and data types of the attributes of the structured type become the names and data types of the columns of that typed table or view. Rows of the typed table or view can be thought of as a representation of instances of the structured type. When used as a data type for a column, the column contains values of that structured type (or values of any of the subtypes for that type, as defined immediately below). Methods are used to retrieve or manipulate attributes of a structured column object.

A *supertype* is a structured type for which other structured types, called *subtypes*, are defined. A subtype inherits all the attributes and methods of its supertype and can have additional attributes and methods defined. The set of structured types that is related to a common supertype is called a *type hierarchy* and the type that does not have any supertype is called the *root type* of the type hierarchy.

The term *subtype* applies to a user-defined structured type and all user-defined structured types that are below it in the type hierarchy. Therefore, a subtype of a structured type T is T and all structured types below it in the hierarchy (i.e., when the hierarchy is a tree, the subtree rooted in *T*). A proper subtype of a structured type *T* is a structured type below *T* in the type hierarchy.

Recursive type definitions in type hierarchies are subject to some restrictions, for which it is necessary to develop a shorthand way of referring to the specific type of recursive definitions that are allowed; the following definitions are used:

✓ *Directly usage*: a type *A* is said to *directly use* another type *B*, if and only if one of the following statements is true:

1. Type *A* has an attribute of type *B*.
2. Type *B* is a subtype of *A* or a supertype of *A*.

✓ *Indirectly usage*: A type *A* is said to *indirectly use* a type *B*, if one of the following statements is true:

1. Type *A* directly uses type *B*.
2. Type *A* directly uses some type *C* and type *C* indirectly uses type *B*.

A type cannot be defined so that one of its attribute types directly or indirectly uses itself (i.e., STRUCTURED type definitions should be acyclic). If it is, however, necessary to have such a configuration, consider using a REFERENCE (see below) as the attribute. For example, with structured type attributes, there cannot be an instance of "employee" with an attribute of "manager" when "manager" is of type "employee". There can, however, be an attribute of "manager" with a type of REF(employee).

A type cannot be dropped if certain other objects use the type, either directly or indirectly. For example, a type cannot be dropped if a table or view column makes direct or indirect use of the type.

➢ REFERENCE (or REF) is a companion type to a structured type. Similar to a DISTINCT type, a reference one is a scalar type that shares a common representation with one of the built-in data types. This same representation is shared for all types in the type hierarchy. The reference type representation is defined when the root type of a type hierarchy is created. When using a reference type, a structured type is specified as a parameter of the type. This parameter is called the *target type* of the reference.

➢ ARRAY is a data type that is defined as an array with elements of another data type. An array is a structure that contains an ordered collection of data elements in which each element can be referenced by its index value in the collection. All elements in an array have the same data type. The *cardinality* of an array is the number of its elements. There are two types of arrays: ordinary and associative.

An *ordinary array* has a defined upper bound on the number of elements, known as the *maximum cardinality*. Each element in the array is referenced by its ordinal position as the index value. If n is the number of elements in an ordinary array, the ordinal position associated with each element is an integer value greater than or equal to 1 and less than or equal to n. Every ordinary array type has an INTEGER index.

The maximum cardinality of an ordinary array is not related to its physical representation, unlike the ones of some PLs such as *C*. Instead, the maximum cardinality is used by the system at run time to ensure that subscripts are within bounds. The amount of memory required to represent an ordinary array value is not proportional to the maximum cardinality of its type either.

An *associative array* has no specific upper bound on the number of elements (i.e., no defined maximum cardinality). Each element is referenced by its associated index value. The data type of the index value can be either INTEGER or VARCHAR, but is the same one for the entire array. The amount of memory required to represent an associative array value is usually proportional to its cardinality.

When an array is being referenced, all of the values in the array are stored in main memory. Therefore, arrays that contain a large amount of data will consume large amounts of main memory.

The ARRAY type is not supported for multirow insert, update, or delete.

Optionally, you can buy a license for IBM *DB2 Data Links Manager*, with which, by using the data type DATALINK, you can connect *DB2* db table columns to *Windows*, *AIX*, and/or *Solaris* OSs managed files.

Optionally too, you can buy the IBM *DB2 Spatial Extender*, which provides geospatial data, as well as a complimentary geobrowser.

4.2.1.3.4 *Computed Columns*

> *Machines might give us more time to think but will*
> *never do our thinking for us.*
> —Thomas Watson, Jr.

Computed columns can be defined with the clause GENERATED ALWAYS AS (*generation-expression*).

If the expression for a GENERATED ALWAYS column includes a user-defined external function, changing the executable for the function (such that the results change for given arguments) can result in inconsistent data. This can be avoided by using the SET INTEGRITY statement to force the generation of new values.

The generation-expression cannot contain any of the following:
- ✓ Subqueries
- ✓ XMLQUERY or XMLEXISTS expressions
- ✓ Column functions
- ✓ Dereference operations or *DEREF* functions
- ✓ User-defined or built-in functions that are nondeterministic
- ✓ User-defined functions using the EXTERNAL ACTION option
- ✓ User-defined functions that are not defined with NO SQL
- ✓ Host variables or parameter markers
- ✓ Special registers and built-in functions that depend on the value of a special register
- ✓ Global variables
- ✓ References to columns defined later in the column list
- ✓ References to other generated columns
- ✓ References to columns of type XML

The data type for the column is based on the result data type of the *generation-expression*. A CAST specification can be used to force a particular data type and to provide a scope (for a reference type only). If the data type is specified, values are assigned to the column according to the appropriate assignment rules.

A generated column is implicitly considered nullable, unless the NOT NULL constraint is used too. The data type of a generated column and the result data type of the *generation-expression* must have equality defined between them: this excludes columns and generation expressions of types LOB, XML, STRUCTURED, and DISTINCT based on any of these types.

For defining temporal tables (that may be SYSTEM_TIME, BUSINESS_TIME, or BI-TEMPORAL, which is the combination of the former two ones), *DB2* provides the following four system generated column types:

> ➢ FOR EACH ROW ON UPDATE AS ROW CHANGE TIME-STAMP: generated values are assigned by the db manager in each row that is inserted and for any row in which any column is updated. The value that is generated for such a column is a time-stamp that corresponds to the insert or update time for that row. If multiple rows are inserted or updated with a single statement, the value of the ROW CHANGE TIMESTAMP column might be different for each row.

If the data type is specified, it must be TIMESTAMP or TIMESTAMP(6).

> ➢ AS ROW BEGIN: generated values are assigned by the db manager whenever a row is inserted into the table or any column in the row is updated.

Values are generated using a reading from the time-of-day clock during execution of the first data change statement in the transaction that requires a value to be assigned to the ROW BEGIN or TRANSACTION START ID (see the fourth type below) column in the table, or a row in a system-period temporal table is deleted.

For a system-period temporal table, the db manager ensures uniqueness of the generated values for a ROW BEGIN column across transactions. The timestamp value might be adjusted to ensure that rows inserted into an associated history table have the end timestamp value greater than the begin timestamp one. This can happen when a conflicting transaction is updating the same row in the system-period temporal table. The db configuration parameter *systime_period_adj* must be set to *Yes* for this adjustment to the timestamp value to occur.

If multiple rows are inserted or updated within a single *SQL* transaction and an adjustment is not needed, the values for the ROW BEGIN column are the same for all the rows and are unique from the values generated for the column for other transactions. A ROW BEGIN column is required as the begin column of a SYSTEM_TIME period, which is the intended use for this type of generated columns.

> ➢ AS ROW END: generated values are assigned by the db manager whenever a row is inserted or any column in the row is updated. The assigned value is TIMESTAMP '9999–12–30–00.00.00.000000000000'.

A ROW END column is required as the second column of a SYS-TEM_TIME period, which is the intended use for this type of generated columns.

> ➤ AS TRANSACTION START ID: generated values are assigned by the db manager whenever a row is inserted into the table or any column in the row is updated. Such values are unique timestamp ones per transaction or the null value.

The null value is assigned to the TRANSACTION START ID columns if they are nullable and if there is a ROW BEGIN column in the table for which the value did not need to be adjusted; otherwise, the value is generated using a reading of the time-of-day clock during execution of the first data change statement in the transaction that requires a value to be assigned to a ROW BEGIN or TRANSACTION START ID column in the table, or a row in a system-period temporal table is deleted.

If multiple rows are inserted or updated within a single *SQL* transaction, the values for the TRANSACTION START ID column are the same for all the rows and are unique from the values generated for the column for other transactions.

A TRANSACTION START ID column is required for a system-period temporal table, which is the intended use for this type of generated columns.

These four system generated column types share the following common restrictions and properties:

> ✓ tables can only have one column of each such type (ROW CHANGE TIMESTAMP, ROW BEGIN, ROW END, and TRANSACTION START ID);
> ✓ they cannot have a DEFAULT clause;
> ✓ except for the TRANSACTION START ID, columns of all other three types must be defined as NOT NULL;
> ✓ except for the ROW CHANGE TIMESTAMP, columns of all other three types are not updatable;
> ✓ except for the ROW CHANGE TIMESTAMP, if the data type of the columns of all other three types is not specified, columns are defined as TIMESTAMP(12); otherwise, it must be TIME-STAMP(12).

4.2.1.3.5   Data Types Usage Overview

Wisdom is the power to put our time and our knowledge to the proper use.
—Thomas J. Watson

Table 4.3 shows how to implement in *DB2* your most frequently needed data types.

To conclude this subsection, Table 4.4 presents the rest of the data types provided by *DB2*.

4.2.1.4   Other DB2 10.5 Particularities

IBM's future is in the hands of its people. Our future is unlimited.
—Thomas Watson, Jr.

The closest to GUIDs (or UUIDs) unique values are those obtainable by calls to the *GENERATE_UNIQUE*() function; as they are numeric, not string ones, in order to obtain their string GUID-like equivalent also use the *HEX*(), *CHAR*(), and *TRIM*() functions, as follows:

```
SELECT TRIM(CHAR(HEX(GENERATE_UNIQUE())))
FROM SYSIBM.SYSDUMMY1;
```

However, you do not need GUIDs in *DB2*: data replication is managed by the dedicated powerful tool IBM *InfoSphere Data Replication*.

DB2 also provides an improved syntax of the INSERT ... VALUES *SQL DML* statement that allows for inserting as many rows as desired with only one such statement.

The functions in the SYSFUN schema taking a VARCHAR as an argument will not accept VARCHARs greater than 4,000 bytes long as an argument. However, many of these functions also have an alternative signature accepting a CLOB(1 M). For these functions, the user can explicitly cast the greater than 4,000 VARCHAR strings into CLOBs and then recast the result back into VARCHARs of the required length.

Special restrictions apply to expressions that result in a CLOB, DBCLOB, and BLOB data types, as well as to structured type columns; such expressions and columns are not allowed in:

TABLE 4.3 Most Frequently Needed *DB2* Data Types

Data type	*DB2 10.5 implementation*
AUTONUMBER(n), $n < 5$	SMALLINT GENERATED ALWAYS AS IDENTITY
AUTONUMBER(n), $n \in [5; 9]$	INT GENERATED ALWAYS AS IDENTITY
AUTONUMBER(n), $n \in [10; 18]$	BIGINT GENERATED ALWAYS AS IDENTITY
AUTONUMBER(n), $n \in [19; 31]$	DECIMAL (n) GENERATED ALWAYS AS IDENTITY
BOOLEAN	BOOLEAN
NAT(n), INT(n), $n < 5$	SMALLINT
NAT(n), INT(n), $n \in [5; 9]$	INTEGER (INT)
NAT(n), INT(n), $n \in [10; 18]$	BIGINT
RAT(n, m), NAT(p), INT(p), CURRENCY(n), $n < 32$, $p \in [19; 31]$	DECIMAL (DEC, NUMERIC)
RAT(n, m), $n < 39$ (fastest, but only 7 digits scale precision)	REAL
RAT(n, m), $n \in [39; 308]$ (faster, but only 15 digits scale precision)	DOUBLE
RAT(n, m), NAT(n), INT(n), CURRENCY(n), $n \in [32; 385]$	DECFLOAT(16)
RAT(n, m), NAT(n), INT(n), CURRENCY(n), $n \in [386; 6,145]$	DECFLOAT(34)
DATE/TIME [1/1/1; 12/31/9999]	DATE (no time), TIMESTAMP (date and time, user defined scale of maximum 12 digits, default 6)
TIME [00:00:00; 24:00:00]	TIME
ASCII(n), $n \leq 254$, fixed length	CHAR(n) FOR SBCS DATA
UNICODE(n), $n \leq 63$, fixed length	CHAR(n) [FOR MIXED DATA]
ASCII(n), $n \leq 32,767$, variable length	VARCHAR(n) FOR SBCS DATA
UNICODE(n), $n \leq 8,168$, variable length	VARCHAR(n) [FOR MIXED DATA]
ASCII(n), $n \in [32,768; 2,147,483,647]$, variable length (up to 2 GB)	CLOB(n) FOR SBCS DATA
UNICODE(n), $n \in [8,169; 536,870,911]$, variable length (up to 0.5 GB)	CLOB(n) [FOR MIXED DATA]
User defined data types	DISTINCT
Computed column	GENERATED ALWAYS AS

TABLE 4.4   Other *DB2* Data Types

Data type	DB2 10.5 implementation
BLOBs (pictures, voice, mixed media, structured user-defined) up to 2 GB	BLOB(*n*)
equivalent of CHAR, but with encodings of at least 2 bytes per character; corresponding maximum values for *n* are 127 and 63	GRAPHIC(*n*)
equivalent of VARCHAR, but with encodings of at least 2 bytes per character; corresponding maximum values for *n* are 16,336 and 8,168, respectively	VARGRAPHIC(*n*)
equivalent of CLOB, but with encodings of at least 2 bytes per character; corresponding maximum values for *n* are 1,073,741,823 and 536,870,911	DBCLOB(*n*)
File handlers	DATALINK (only with IBM *DB2 Data Links Manager*)
Homogeneous vectors	ARRAYS
XML (well-formatted) files	XML
Dependencies on *SQL* objects	ANCHOR
Temporal tables (dbs)	ROW CHANGE TIMESTAMP, ROW BEGIN, ROW END, and TRANS-ACTION START ID
GPS coordinates, imagery data, geometric descriptions of 3-D objects.	Only with the data types provided by the IBM *DB2 Spatial Extender*
Nested (not 1NF) tables (object-relational)	STRUCTURED, REFERENCE

- ✓ SELECT lists preceded by the DISTINCT clause
- ✓ GROUP BY and ORDER BY clauses
- ✓ A subselect of a set operator other than UNION ALL
- ✓ Basic, quantified, BETWEEN, and IN predicates
- ✓ Aggregate functions
- ✓ VARGRAPHIC, TRANSLATE, and date/time scalar functions
- ✓ The pattern operand in a LIKE predicate, or the search string operand in a POSSTR function
- ✓ The string representation of a date time value.

A ROW CHANGE TIMESTAMP column cannot be used as part of a primary key.

Although the LONG VARCHAR and LONG VARGRAPHIC data types continue to be supported, do not use them anymore, as they are deprecated and might be removed in a future release.

DB2 also includes as check (tuple) constraints the functional dependencies: *column-name* DETERMINED BY *column-name* or (*column-name*, ...) DETERMINED BY (*column-name*, ...) defines a functional dependency between columns.

The parent set of columns contains the identified columns that immediately precede the DETERMINED BY clause. The child set of columns contains the identified columns that immediately follow the DETERMINED BY clause.

All of the restrictions on the search conditions apply to both parent and child set columns, and only simple column references are allowed in the set of columns.

The same column must not be identified more than once in the functional dependency. The data type of the column must not be a LOB data type, a DISTINCT type based on a LOB data type, an XML data type, or a STRUCTURED type.

No column in the child set of columns can be a nullable column.

Functional dependencies are not enforced by the db manager during normal operations such as insert, update, delete, or set integrity. They might be used during query rewrite to optimize queries. Incorrect results might be returned if the integrity of a functional dependency is not maintained.

4.2.2   ORACLE *DATABASE 12C*

> *When you innovate, you've got to be prepared for people*
> *telling you that you are nuts.*
> —Larry Ellison

Oracle 12c is a robust, well-performing and fully featured RDBMS, designed for the cloud (the first one in this respect, as Oracle Corp. claims). Standard components include db language and storage server, developer and ETL tools, schedulers, network encryption, and basic replication.

There are dozens of other components too, including OLAP, BI, spatial and graph data, and the new multitenant architecture, which allows user to centrally manage any number of dbs. *Oracle 12c* can also be accessed through *ODBC*, *ADO*, and *OLEDB*.

Besides several commercial versions, there is also a free to download *Oracle Express* (*XE*) one (for 11 GB dbs maximum), which can also be redistributed with your products. It allows for 1 GB RAM and 1 CPU core.

Oracle 12c partially adheres to the *SQL* 2011 ANSI standard (which includes the 1999 object-relational extensions, recursive SELECT queries, triggers, and support for procedural and control-of-flow statements in its extended *SQL PL/SQL* language, as well as the 2003 and 2006 *SQL/XML*), plus its own extensions (that can be detected with its *FIPS Flagger*, provided you execute the *SQL* statement ALTER SESSION SET FLAGGER = ENTRY).

Oracle 12c runs on (*z*)*Linux*, MS *Windows*, HP-*UX*, *AIX*, *Solaris*, and *OpenVMS* OSs.

4.2.2.1   Identifiers and Literals

> *I am so disturbed by kids who spend all day playing videogames.*
> —Larry Ellison

Oracle object names may have at most 30 characters, are not case sensitive (and *Oracle* stores them in uppercase), except for diagrams, and, among other reserved characters, do not accept '-' (that you have to replace with '_'). Apart letters and numbers, you may only use '$', '#', '@', and '_'. Only for parameters you can use any of them as the first one: for the rest of the identifiers, the first one should be a letter. However, you may use any number of ASCII not allowed characters in names, including for the first one, provided you embed them in double quotation marks.

Text strings have to be embedded in apostrophes (single quotation marks). Calendar date values have to be entered as the result of the system conversion function *To_Date* applied to their text string values and system calendar date template (e.g., in the U.S., 'MM/DD/YYYY').

You can use apostrophes within text strings by doubling them; for example, the text string '''quote'''s value is 'quote'.

Comments in *Oracle SQL* start immediately after "--" and end at the end of the line; alternatively, you can embed any number of lines in a single comment by embracing it in "/*" (for start) and "*/" (for end).

4.2.2.2 Databases and Tables

*They don't call it the Internet anymore, they call it cloud computing.
I'm no longer resisting the name. Call it what you want.*
—Larry Ellison

In *Oracle*, dbs are curiously named *users*.[97] *Oracle* provides a *SQL* CREATE USER statement too, but, similarly to *DB2*, mainly due to security and physical disk space management, this is typically done only by DBAs, who are generally using for this task too the GUI of the *Oracle Enterprise Console* tool. However, in the free *Express* edition, you can do it too, as you are also the DBA.

Computed tables are called *views* and can be created with the CREATE VIEW ... AS SELECT ... *SQL DDL* statement; they cannot be created through the *Oracle SQL Developer*'s GUI too, as it doesn't have a *QBE*-type facility.

You can also define in-memory tables, very useful for OLAP processing, but only for some additional $23,000 per CPU.

Temporary tables, either local or global are created in user temporary tablespaces with the CREATE [GLOBAL] TEMPORARY TABLE ... *SQL DDL* statement. The local ones exist during the current user session or the lifetime of the procedure in which they are declared. The global ones have persistent definitions, but data is not persistent (as it is automatically deleted by the system at the end of the db sessions) and they generate no redo or rollback information. If, however, you would like their instances be persistent too, you can add the ON COMMIT PRESERVE ROWS clause at the end of the above statement.

You can associate indexes, triggers, statistics, and views to temporary tables too.

---

[97] Reminiscence of its beginnings, when they did not anticipate that a user might need to manage several dbs: this is another great counterexample of how not to name objects, especially conceptual ones.

4.2.2.3   Data Types

> *I have had all of the disadvantages required for success.*
> —Larry Ellison

4.2.2.3.1   Numeric Data Types

> *Number rules the universe.*
> —Pythagoras

Unfortunately, first of all, *Oracle* did not provide autonumbering up to its *12c* version. The standard workaround for each table in which you need an autonumber primary key is to define it as a NUMBER one, to declare a *sequence* and a trigger associated to the event BEFORE INSERT of that table, which uses the sequence for automatically assigning its current value to the primary key for the currently to be inserted row (see examples in Subsection 4.3.2 below).

An *Oracle sequence* is a special type (namely, arithmetic progression) of stored parameterized numeric function (parameters: minimum, maximum, and ratio); any call to it returns the current value in the progression, if any (otherwise it returns a system error), starting with the minimum value (for the first call) and up to the maximum one.

There is no data type for replication either: replication (as well as heterogeneous data integration with various other platforms) is dealt with the powerful (and expensive) *Oracle GoldenGate* add-on.

There is no Boolean data type in *Oracle* either: instead, you have to use numeric or string data types (of at least 1 byte) and your own codifications for *True* and *False* (preferably defined as a user data type).

There is no CURRENCY data type either: you have to use the numeric types instead.

There are four numeric data types:

- ➢ NUMBER[(p[, s])] = [-10^{38}; 10^{38}] ⊂ RAT(38). Precision $p \in$ [1; 38], default 38 and scale $s \in$ [-84; 127], default 0;
- ➢ BINARY_FLOAT = [-3.40282×10^{38}; 3.40282×10^{38}];

> BINARY_DOUBLE = [$-1.79769313486231 \times 10^{308}$;
> $1.79769313486231 \times 10^{38}$];

Both BINARY_FLOAT and BINARY_DOUBLE support the distinguished values *infinity* and *NaN* (Not a Number).

> FLOAT[(p)] = [-8.5×10^{37}; 8.5×10^{37}]. Precision $p \in [1; 126]$, default
> 126 (binary); scale is interpreted from the data; used internally,
> when converting ANSI FLOAT values; not recommended to be
> used explicitly too, due to truncations; use instead the more robust
> NUMBER, BINARY_FLOAT, and BINARY_DOUBLE data types.

There are 4 DATE/TIME data types (DATE, TIMESTAMP, TIMESTAMP
WITH TIME ZONE, and TIMESTAMP WITH LOCAL TIME ZONE), plus
2 additional *interval* (INTERVAL YEAR TO MONTH and INTERVAL
DAY TO SECOND) data types:

> DATE = [1/1/4712 BC; 12/31/9999];
> TIMESTAMP [(*fractional_seconds_precision*)]; extension of DATE
> with fractional second precision between 0 and 9 digits (default: 6);
> TIMESTAMP [(*fractional_seconds_precision*)] WITH TIME ZONE;
> variant of TIMESTAMP that includes in its values a time zone
> region name or a time zone offset (i.e., the difference (in hours
> and minutes) between local time and UTC — formerly Greenwich
> Mean Time).
> TIMESTAMP [(*fractional_seconds_precision*)] WITH LOCAL
> TIME ZONE; variant of TIMESTAMP WITH TIME ZONE: it differs
> in that data stored in the db is normalized to the db time zone, and
> the time zone information is not stored as part of the corresponding
> columns data. When a user retrieves such data, *Oracle* returns it in
> the user's local session time zone. This data type is useful for date
> information that is always to be displayed in the time zone of the
> client system, in two-tier applications;
> INTERVAL YEAR [(*year_precision*)] TO MONTH = [−4712/1;
> 9999/12]; *year_precision* is the number of digits in the YEAR field
> and has the default value 2;
> INTERVAL DAY [(*day_precision*)] TO SECOND [(*fractional_
> seconds_precision*)] = [0 0:0:0; 9 23:59:59.999999999]; *day_pre-
> cision* is the number of digits in the DAY field, and accepts values
> between 0 and 9, the default being 2; *fractional_seconds_precision*

is the number of digits in the fractional part of the SECOND field, and accepts values between 0 and 9, the default being 6.

All of the above DATE/TIME data types use the following 10 fields:

- ✓ YEAR = [–4712; 9999] — {0};
- ✓ MONTH = [1; 12];
- ✓ DAY = [1; 31];
- ✓ HOUR = [0; 23];
- ✓ MINUTE = [0; 59];
- ✓ SECOND = [0; 59.9(n)]; 9(n) is the precision of fractional seconds;
- ✓ TIMEZONE_HOUR = [–12; 14];
- ✓ TIMEZONE_MINUTE = [0; 59];
- ✓ TIMEZONE_REGION: The result of SELECT TZNAME FROM V$TIMEZONE_NAMES;
- ✓ TIMEZONE_ABBR: The result of SELECT TZABBREV FROM V$TIMEZONE_NAMES;.

You can obtain the current system date and/or time, in various formats, with the functions SysDate, SysTimestamp, Current_Date, Current_Timestamp, and Local_Timestamp. Function EXTRACT extracts any desired field from a DATE/TIME value; for example, in order to extract years, you should use EXTRACT(year FROM date_value); functions DBTimeZone and SessionTimeZone return the current corresponding time zone offsets.

4.2.2.3.2 String Data Types

There is geometry in the humming of the strings.
There is music in the spacing of the spheres.
—Pythagoras

Oracle provides the following 11 string data types:

- ➢ VARCHAR2(n [BYTE | CHAR]) = ASCII(n), $0 \le n \le 4,000$[98]; n should be at least 1; implicitly, n is in bytes, but you can specify it in characters (with CHAR);

---

[98] 4,000 is the standard default maximum value (for MAX_STRING_SIZE = STANDARD); you can set MAX_STRING_SIZE = EXTENDED, in which case VARCHAR2, as well as NVARCHAR2, and ROW have maximum length 32,767. For such extended data types, however, restrictions apply. Moreover, such columns are stored out-of-line, leveraging *Oracle*'s LOB technology. In tablespaces managed with *ASSM*, extended data type columns are stored as SecureFiles LOBs. Otherwise, they are internally stored as BasicFiles LOBs.

> ➢ NVARCHAR2(n) = UNICODE(n), $0 \leq n \leq 4,000$[99];
> ➢ CHAR[(n [BYTE | CHAR])] = ASCII(n), $1 \leq n \leq 2,000$; fixed size variant of VARCHAR2 (right-padded with blanks if shorter than n); default n is 1;
> ➢ NCHAR[(n)] = UNICODE($n/2$), $1 \leq n \leq 2,000$; fixed size (in bytes) variant of NVARCHAR2 (right-padded with blanks if shorter than n); default size is 1; when the AL16UTF16 encoding is used (2 bytes per character), maximum 1000 characters can be stored; for the UTF8 one (3 bytes per character), only 666 may be stored;
> ➢ RAW(n) = binary data of at most n bytes, $0 \leq n \leq 2,000$[100];
> ➢ LONG RAW: binary data of at most n bytes, $0 \leq n \leq 2^{32}-1$; *Oracle* strongly recommends that you convert LONG RAW columns to binary LOB (BLOB) columns (see below), as LOB columns are subject to far fewer restrictions than LONG ones;
> ➢ ROWID: Base 64 string representing the unique address of a row in its table; this data type is primarily for values returned by the ROWID pseudocolumn: you can define columns having this data type, but *Oracle* does not guarantee that they contain valid row ids;
> ➢ UROWID[(n)]: Base 64 string representing the logical address of a row of an index-organized table; the optional n is the size of a column of type UROWID; its maximum size and default are 4,000 bytes;
> ➢ BLOB: binary large object; maximum size is (4 GB–1) * (db block size);
> ➢ CLOB: ASCII large object; maximum size (bytes) is (4 GB–1) * (db block size);
> ➢ NCLOB: UNICODE large object; maximum size (bytes) is (4 GB–1) * (db block size);

Do not use the LONG data type (provided only for backward compatibility) anymore: use the LOB ones instead.

The rows in heap-organized tables that are native to *Oracle* have row addresses called *rowids*. You can examine a rowid row address by querying the pseudocolumn ROWID. Values of this pseudocolumn are strings

---

[99] 4,000 is the standard default maximum value, but you can set it to 32,767 (see above footnote).
[100] 2,000 is the standard default maximum value, but you can set it to 32,767 (see above two footnotes). However, extended RAW values are stored as out-of-line LOBs only if their size is greater than 4,000 bytes; otherwise, they are stored inline.

representing the address of each row. These strings have the data type `ROWID`.

Rowids contain the following information:

✓ The *data block* of the data file containing the row; the length of this string depends on your operating system.

✓ The *row* in the data block.

✓ The *database file* containing the row; the first data file has the number 1; the length of this string depends on your OS.

✓ The *data object number*, which is an identification number assigned to every db segment. You can retrieve the data object number from the data dictionary views `USER_OBJECTS`, `DBA_OBJECTS`, and `ALL_OBJECTS`. Objects that share the same segment (clustered tables in the same cluster, for example) have the same object number.

Rowids are stored as base 64 values that can contain the characters A-Z, a-z, 0–9, and the plus sign (+) and forward slash (/). Rowids are not available directly: you can use the supplied package `DBMS_ROWID` to interpret rowid contents. The package functions extract and provide information on the four rowid elements listed above.

The rows of some tables have addresses that are not physical or permanent or were not generated by *Oracle*. For example, the row addresses of index-organized tables are stored in index leaves, which can move. Rowids of foreign tables (such as *DB2* tables accessed through a gateway) are not standard *Oracle* rowids.

Oracle uses universal rowids (*urowids*) to store the addresses of index-organized and foreign tables. Index-organized tables have logical urowids and foreign tables have foreign urowids. Both types of urowid are stored in the `ROWID` pseudocolumn (as are the physical rowids of heap-organized tables).

Oracle creates logical rowids based on the primary key of the table. The logical rowids do not change as long as the primary key does not change. The `ROWID` pseudocolumn of an index-organized table has a data type of `UROWID`. You can access this pseudocolumn as you would the `ROWID` pseudocolumn of a heap-organized table (i.e., by using a `SELECT...ROWID` statement). If you want to store the rowids of an index-organized table, then you can define a column of type `UROWID` for the table and retrieve the value of the `ROWID` pseudocolumn into that column.

4.2.2.3.3 Other Data Types

Most men and women, by birth or nature, lack the means to advance in wealth or power, but all have the ability to advance in knowledge.
                                                                    —Pythagoras

Oracle provides the following additional 23 data types too:
- ➢ BFILE: locators to large binary files stored outside the db; enables byte stream I/O access to external LOBs residing on the db server; maximum size is 4 GB; when deleting such a value from the db, the corresponding OS file is not deleted;
- ➢ VARRAYS: homogeneous vectors (i.e., all of their elements are of a same data type) of variable size that should be declared before using them (as table columns, but also as an object type attribute, or a *PL/SQL* variable, parameter, or function return type); stored either in-line (as part of the row data) or out of line (in a LOB), depending on its size; however, if you specify separate storage characteristics for a varray, then *Oracle* stores it out of line, regardless of its size;
- ➢ OBJECT: user defined object-relational abstract data types, modeling real-world entities (e.g., purchase orders, addresses, apartments, etc.) that application programs deal with; any object type is a *schema object* with the following three components:
  - ✓ *name*, which identifies the object type uniquely within that schema;
  - ✓ *attributes*, which are *Oracle* built-in or other user-defined types modeling the structure of the real-world entity;
  - ✓ *methods*, which are db stored functions or procedures written in *PL/SQL*, or written in some PL (e.g., *C* or *Java*) and stored externally, implementing the operations that the application can perform on that real-world entity;
- ➢ REF: container for an object identifier[101] (pointer to an object[102]);

---

[101] An *object identifier* (*OID*) uniquely identifies an *Oracle* object and enables you to reference it from other objects or from db table columns.

[102] When a REF value points to a nonexistent object, it is said to be "dangling"; a dangling REF is different from a null REF; to determine whether a REF is dangling or not, use the condition IS [NOT] DANGLING.

> ➢ NESTED TABLE: whenever a nested table appears as the type of a column in a relational table or as an attribute of the underlying object type of an object table, *Oracle* stores all of the nested table data in a single table, which it associates with the enclosing relational or object table; you can use both built-in and user-defined data types for defining nested table columns;

> ➢ XMLType: stores *XML* data[103]; you can choose to store it in a CLOB column, as binary *XML* (stored internally as a CLOB), or object relationally; you can also register the schema (using the DBMS_XMLSCHEMA package) and create a table or column conforming to it: in this case, *Oracle* stores the *XML* data in underlying object-relational columns by default, but you can specify storage in a CLOB or binary *XML* column even for schema-based data; queries and *DML* on XMLType columns operate the same regardless of the storage mechanism;

> ➢ URIType: object supertype with three subtypes (DBURIType, HTTPURIType, and XDBURIType, which are related by an inheritance hierarchy); you can create columns of this type and store in them any of its subtypes instances; *Oracle* provides a URIFactory package, which can create and return instances of the various subtypes of the URITypes; the package analyzes the URL string, identifies the type of URL (HTTP, DBURI, etc.), and creates an instance of the corresponding subtype;

>> ✓ DBURIType: stores DBURIRef values, which reference data inside or outside the db and access it consistently; DBURIRef values use an *XPath*-like representation to reference data[104]: if you imagine the db as an *XML* tree, then you would see the tables, rows, and columns as elements in the *XML* document and the DBURIRef is an *XPath* expression over this virtual *XML* document; using this model, you can reference, for example, data stored in CLOB columns and expose them as URLs to

---

[103] You can also create tables and views of XMLType.

[104] For example, in order to reference the *Salary* value of an employee having $x = 205$, stored in an *HR XML* file under the tags *EMPLOYEES/ROW*, the corresponding value is: /HR/EMPLOYEES/ ROW[x=205]/Salary

the external world; to create a `DBURI` instance, the URL must begin with the prefix `/oradb`;[105]

✓ `HTTPURIType`: stores URLs to external web pages or files; *Oracle* accesses these files using HTTP;

✓ `XDBURIType`: expose documents in the *XML* database hierarchy as URIs that can be embedded in any `URIType` column in a table; `XDBURIType` consists of a URL, which comprises the hierarchical name of the *XML* document to which it refers and an optional fragment representing the *XPath* syntax; the fragment is separated from the URL part by a pound sign (#)[106].

The following 3 spatial data types are available only if the *Oracle Spatial and Graph* is installed[107]:

➢ `SDO_GEOMETRY`: the geometric description of a spatial object is stored in a single row, in a single column of object type `SDO_GEOMETRY`, in a user-defined table; any table that has a column of type `SDO_GEOMETRY` must have a primary key; such tables are called *geometry tables*;

➢ `SDO_TOPO_GEOMETRY`: describes a topology geometry, which is stored in a single row, in a single column of object type `SDO_TOPO_GEOMETRY`, in a user-defined geometry table;

➢ `SDO_GEORASTER`: in the `GeoRaster` object-relational model, a raster grid or image object is stored in a single row, in a single column of object type `SDO_GEORASTER`, in a user-defined table; such tables are called *GeoRaster tables*.

The following 7 media data types provide compliance with the ISO-IEC 13249–5 Still Image standard, commonly referred to as *SQL/MM StillImage*:

➢ `SI_StillImage`: represents digital images with inherent image characteristics such as height, width, and format;

➢ `SI_Color`: encapsulates color values;

---

[105] For example, `URIFactory.getURI('/oradb/HR/EMPLOYEES')` would create a `DBURIType` instance, `URIFactory.getUri('/sys/ schema')` would create an `XDBURIType` instance, etc.

[106] For example, `/home/oe/doc.xml#/orders/order_item`

[107] *Oracle Spatial and Graph* is designed to make spatial data management easier and more natural to users of location-enabled, geographic information system (GIS), and geoimaging applications. After spatial data is stored in an *Oracle* db, you can easily manipulate, retrieve, and relate it to all other data stored in that db.

> ➢ SI_AverageColor: characterizes an image by its average color;
> ➢ SI_ColorHistogram: characterizes an image by the relative frequencies of the colors exhibited by samples of the raw image;
> ➢ SI_PositionalColor: given an image divided into *n* by *m* rectangles, the *SI_PositionalColor* object type represents the feature that characterizes an image by the *n* by *m* most significant colors of the rectangles;
> ➢ SI_Texture: characterizes an image by the size of repeating items (*coarseness*), brightness variations (*contrast*), and predominant direction (*directionality*);
> ➢ SI_FeatureList: list containing up to four of the image features represented by the preceding object types (SI_AverageColor, SI_Color-Histogram, SI_PositionalColor, and SI_Texture), where each feature is associated with a feature weight.

The following three ANY data types provide highly flexible modeling of procedure parameters and table columns where the actual type is not known. These data types let you dynamically encapsulate and access type descriptions, data instances, and sets of data instances of any other *SQL* type. These types have both *OCI* and *PL/SQL* interfaces for construction and access.

> ➢ ANYTYPE: type descriptions of any named *SQL* type or unnamed transient type;
> ➢ ANYDATA: instances of a given data type, containing both data and a description of its actual type; allows storing heterogeneous values in a single column; values can be both built-in and user-defined types;
> ➢ ANYDATASET: descriptions of a given data type plus a set of data instances of that type; can be used as a procedure parameter data type where such flexibility is needed; values can be both built-in and user-defined types.

4.2.2.3.4 *Computed Columns*

> *Above the cloud with its shadow is the star with its light.*
> —Pythagoras

Oracle computed columns are referred to as *virtual* and are defined using the predicate AS. Their data type may be explicitly stated or implicitly derived by *Oracle* from the corresponding expression. Here are the main characteristics of and restrictions on virtual columns:

- Indexes defined against virtual columns are equivalent to function-based indexes.
- Virtual columns can be referenced in the WHERE clause of updates and deletes, but they cannot be manipulated by *SQL DML* statements.
- Tables containing virtual columns can still be eligible for result caching.
- Functions in expressions must be deterministic at the time of table creation, but can subsequently be recompiled and made nondeterministic without invalidating the corresponding virtual columns. In such cases the following steps must be taken after the function is recompiled:
  - ✓ Constraints on the virtual column must be disabled and reenabled.
  - ✓ Indexes on the virtual column must be rebuilt.
  - ✓ Materialized views that access the virtual column must be fully refreshed.
  - ✓ The result cache must be flushed if cached queries have accessed the virtual column.
  - ✓ Table statistics must be regathered.
- Virtual columns are not supported for index-organized, external, object, cluster, or temporary tables.
- Expressions used in virtual column definitions have the following restrictions:
  - ✓ cannot refer to another virtual column by name;
  - ✓ can only refer to columns defined in the same table;
  - ✓ if they refer to a deterministic user-defined function, they cannot be used as a partitioning key column;
  - ✓ their output must be a scalar value: it cannot return an *Oracle* supplied data type, a user-defined type, or LOB or LONG RAW.

4.2.2.3.5 Data Types Usage Overview

> *Do not say a little in many words, but a great deal in few!*
> —Pythagoras

Table 4.5 shows how to implement in *Oracle* your most frequently needed data types.

To conclude this subsection, Table 4.6 presents the rest of the data types provided by *Oracle*.

4.2.2.4   Other Oracle 12c Particularities

> *Be careful about virtual relationships with artificially intelligent pieces of software.*
> —Larry Ellison

Beware that *Oracle* triggers are very slow. For example, as *Oracle* does not accept using the *SysDate* system function (which returns the current system calendar date and time) in constraints, although such constraints might be enforced through triggers, we prefer to use instead additional virtual columns and associated functions, as they are four times faster than equivalent triggers.

For example, in order to detect whether or not the values that users would like to store for years is greater than the current year, we first declare a *Year_in_future* function (based on *SysDate*, see Subsection 4.3.2); then, this function is used to define virtual columns in all tables that need such a domain or tuple constraint, for rejecting storing implausible data in the associated actual columns (see examples in Subsection 4.3.2).

Note that, fortunately, this type of solution is possible only due to the existence of a bug in *Oracle*: virtual columns can be based only on deterministic functions, whereas, in fact, *Year_in_future* is not deterministic, because is based on *SysDate*; the helping bug is that, however, *Oracle* accepts the false DETERMINISTIC definition keyword for it. Obviously, same thing is true for the other two such needed functions, *Day_in_future* and *Day_in_far_future*.

TABLE 4.5 Most Frequently Needed *Oracle* Data Types

Data type	Oracle 12c implementation
AUTONUMBER(n), $n < 39$ (no replication)	NUMBER(n) GENERATED [ALWAYS | BY DEFAULT [ON NULL]] AS IDENTITY [(identity_options)] (For all previous versions: NUMBER(n) and associated sequence and BEFORE INSERT trigger)
AUTONUMBER(n), $n \in$ [39; 308] (no replication)	same as above, but using BINARY_DOUBLE instead of NUMBER
AUTONUMBER(n) for replication	NUMBER(n) or BINARY_DOUBLE and the *Oracle GoldenGate* add-on
BOOLEAN	NUMBER(1) or CHAR and your own corresponding codifications
NAT(n), INT(n), $n < 39$	NUMBER(n)
RAT(n, m), CURRENCY(n), $n < 39$	NUMBER(n, m)
RAT(n, m), INT(n), CURRENCY(n), $n \in$ [39; 308]	BINARY_DOUBLE
DATE/TIME [1/1/4712 BC; 31/12/9999]	DATE (no scale, no time zone), TIMESTAMP (user defined scale of maximum 9 digits, no time zone), TIMESTAMP WITH TIME ZONE (similar to TIMESTAMP, but including time zone with respect to UTC too), TIMESTAMP WITH LOCAL TIME ZONE (similar to TIMESTAMP, but of local time zone), depending on desired scale and time zone (not) awareness
ASCII(n), $n \leq 2{,}000$, fixed length	CHAR(n)
UNICODE(n), $n \leq 2{,}000$, fixed length	NCHAR(n)
ASCII(n), $n \leq 32{,}767$, variable length	VARCHAR2(n)
UNICODE(n), $n \leq 32{,}767$, variable length	NVARCHAR2(n)
ASCII(n), $n \in$ [32,768; (4 GB—1) * (db block size)]	CLOB
UNICODE(n), $n \in$ [32,768; (4 GB—1) * (db block size)]	NCLOB
User defined data type	REF (TYPE)
Computed column	AS

TABLE 4.6   Other *Oracle* Data Types

Data type	Oracle 12c implementation
DATE/TIME intervals (only years and month) [–4712/1; 9999/12]	INTERVAL YEAR [(*year_precision*)] TO MONTH
DATE/TIME intervals (only days, hours, minutes, and seconds, including fractions of seconds) [0 0:0:0; 9 23:59:59.999999999]	INTERVAL DAY [(*day_precision*)] TO SECOND [(*fractional_seconds_precision*)]
Binary objects of up to 32767bytes	RAW(*n*)
BLOBs up to (4 GB—1) * (db block size)	BLOB
File handlers	BFILE
Homogeneous vectors	VARRAYS
XML (well-formatted) files	XMLType
URLs of documents of a XML db	XDBURIType
URLs of (sub)objects of a rdb (e.g., CLOB fragments)	DBURIType
URLs to external Web pages or files	HTTPURIType
Union of XDBURIType, HTTPURIType, and DBURIType	URIType
GPS coordinates, imagery data, etc.	SDO_GEORASTER
Geometric descriptions of 3-D objects	SDO_GEOMETRY
Topologic descriptions of 3-D objects	SDO_TOPO_GEOMETRY
Images data	SI_StillImage
Image colors data	SI_Color
Image average colors data	SI_AverageColor
Raw image relative frequencies of the colors exhibited by samples	SI_ColorHistogram
Image *n* by *m* rectangles most significant colors	SI_PositionalColor
Image size of repeating items (*coarseness*), brightness variations (*contrast*), and predominant direction (*directionality*)	SI_Texture
Lists of weighted values of types SI_AverageColor, SI_ColorHistogram, SI_PositionalColor, and SI_Texture	SI_FeatureList
Abstract data types	OBJECT
Nested (not 1NF) tables (object-relational)	NESTED TABLE
Type descriptions of any named *SQL* type or unnamed transient type	ANYTYPE
Instances of pairs <data, description of corresponding actual type>	ANYDATA
Descriptions of a given data type plus a set of data instances of that type	ANYDATASET

LOB columns cannot be used either in keys (be them primary or not), or in SELECT DISTINCT, joins, GROUP BY or ORDER BY clauses, or in the UPDATE OF clauses of AFTER UPDATE triggers. You cannot define VARRAYs or ANYDATAs using LOB columns either.

However, you can specify a LOB attribute of an object type column in a SELECT...DISTINCT statement, a query that uses the UNION or a MINUS set operator, if the object type of the column has a MAP or ORDER function defined on it.

LOB columns have full transactional support; however, you cannot save a LOB locator in a *PL/SQL* or *OCI* variable in one transaction and then use it in another transaction or session.

Binary file LOBs (pointed by the BFILE data type) do not participate in transactions and are not recoverable: the underlying OS should provide their integrity and durability. DBAs must ensure that the corresponding external file exists and that *Oracle* processes have OS read permissions on them: BFILE enables only read-only support of such large binary files (you cannot modify or replicate such a file: *Oracle* provides only APIs to access file data). The primary interfaces that can be used to access file data are the DBMS_LOB package and the *OCI*.

4.2.3 ORACLE *MYSQL* 5.7

> *Everyone thought the acquisition strategy was extremely risky*
> *because no one had ever done it successfully.*
> *In other words, it was innovative.*
> —Larry Ellison

MySQL 5.7 is robust, well-performing, and well-featured, although it is a free product. Since it has been acquired by Oracle, there are also extensions to it, as well as a much more powerful *Enterprise Edition* that are commercial products.

MySQL 5.7 handles terabyte-sized dbs (but the code can also be compiled in a reduced version suitable for hand-held and embedded devices), which are also accessible through *ODBC* and *ADO*.

Generally, *MySQL 5.7* adheres to the pure *SQL'99* ANSI standard, with both higher standards (like the geometry spatial data types, the high-availability db clustering using the *NDBCLUSTER* storage engine, the *XML* functions that support most of the W3C *XPath* standard) and nonstandard extensions (like the STRAIGHT_JOIN operator, which is an INNER one, but forcing the read of its left operand first, the DO statement, which is a slightly faster SELECT as it does not return any results, the FLUSH, RESET, SET, SHOW, LOAD DATA INFILE, and EXPLAIN SELECT/DELETE/INSERT/REPLACE/UPDATE statements, the RENAME, CHANGE, and IGNORE clauses of the ALTER TABLE statement, the REPLACE statement that can be used instead of a DELETE followed by an INSERT, the capability of inserting several lines with one INSERT ... VALUES and of dropping several tables with one DROP, the REGEXP and NOT REGEXP extended regular expression operators, the CONCAT() and CHAR() functions with only one or more than two arguments, the ORDER BY and LIMIT clauses of the UPDATE and DELETE statements, the DELAYED clause of the INSERT and REPLACE statements, the LOW_PRIORITY clause of the INSERT, REPLACE, DELETE, and UPDATE statements, etc.).

MySQL 5.7 runs on the *Unix/Linux*, MS *Windows*, Apple Mac *OS X*, *Solaris*, *OpenSolaris*, and *FreeBSD* OSs.

4.2.3.1   Identifiers and Literals

> *Nothing is particularly hard if you divide it into small jobs.*
> —Henry Ford

MySQL object names may have at most 64 UNICODE BMP characters, generally are not case sensitive for tables (depending on the host OS: in *Linux*, they are) and all other object types, except for triggers; you can freely use both upper and lower case (and *MySQL* stores them in UTF-8 UNICODE, exactly as you defined them), but always consistently in a same statement.

The only other reserved characters for names are the ASCII NULL (U+0000) and the supplementary characters (U+10000 and higher). If they are quoted, you may use the full UNICODE BMP, except for NULL, as well as naturals; otherwise, you can only use the ASCII basic Latin letters, digits, '$', '_', and the extended characters between U+0080 and U+FFFF.

Date and time constants are generally enclosed in single quotation marks, but they can also be used simply as numbers wherever the context is clear to *MySQL* (e.g., 20150714 is interpreted as July 14, 2015).

Text strings are enclosed in either apostrophes (single) or double quotation marks. You can use quotation marks within text strings by prefixing them with the escape character '\' or by doubling them; for example, the text string """"quote"""'s value is "quote".

4.2.3.2 Databases and Tables

> *The only real security that a man can have in this world is*
> *a reserve of knowledge, experience and ability.*
> —Henry Ford

You can define temporary tables with CREATE TEMPORARY TABLE…

Computed tables are called *views* and can be created with the CRE-ATE VIEW … AS SELECT … *SQL DDL* statement; they cannot be created through the *MySQL Workbench*'s GUI too, as it doesn't have a *QBE*-type facility. You cannot use temporary tables within views. Views do not accept parameters either.

4.2.3.3 Data Types

> *Thinking is the hardest work there is, which is probably the reason*
> *why so few engage in it.*
> —Henry Ford

4.2.3.3.1 Numeric Data Types

> *All is Number.*
> —Pythagoras

BOOL and BOOLEAN are aliases for TINYINT.

MySQL provides the following 9 numeric data types (if not otherwise specified, $n = 1$ by default):

➢ TINYINT[(*n*)] [UNSIGNED] = [–128; 127] (signed) or [0; 255] (unsigned);

➢ SMALLINT [(*n*)] [UNSIGNED] = [–32,768; 32,767] (signed) or [0; 65535] (unsigned);

➢ MEDIUMINT [(*n*)] [UNSIGNED] = [–8,388,608; 8,388,607] (signed) or [0; 16,777,215] (unsigned);

➢ INT [(*n*)] [UNSIGNED] = [–2,147,483,648; 2,147,483,647] (signed) or [0; 4,294,967,295] (unsigned); synonym: INTEGER;

➢ BIGINT [(*n*)] [UNSIGNED] = [–9,223,372,036,854,775,808; 9,223,372,036,854,775,807](signed)or[0;18,446,744,073,709,551,615] (unsigned); synonym: INT8;

➢ DECIMAL [(*n*[, *m*])] [UNSIGNED] = [-10^{65}; 10^{65-1}] (signed) or [0; 10^{65-1}] (unsigned), $0 \le n \le 65$, $0 \le m \le 30$; by default, $n = 10$, $m = 0$; precision for arithmetic calculations is 65 digits; synonyms: NUMERIC, FIXED;

➢ FLOAT [(*n*[, *m*])] [UNSIGNED] = [$-3.402823466 \times 10^{38}$; $-3.402823466 \times 10^{38}$] (signed) or [0; $-3.402823466 \times 10^{38}$] (unsigned), $0 \le n$, $m \le 38$ ($3.402823466 \times 10^{38}$ is only theoretical: actually, depending on the host hardware and software platform, it can be less than that); no default: if *n* and/or *m* are not specified, their actual values depend on the host hardware platform; precision for arithmetic calculations is some 7 scale digits; unfortunately, sometimes, you might get unexpected results, because calculations are never actually done with FLOAT;

➢ DOUBLE [(*n*[, *m*])] [UNSIGNED] = [$-1.7976931348623157 \times 10^{308}$; $1.7976931348623157 \times 10^{308}$] (signed) or [0; $1.7976931348623157 \times 10^{308}$] (unsigned), $0 \le n$, $m \le 308$ ($1.7976931348623157 \times 10^{308}$ is only theoretical: actually, depending on the host hardware and software platform, it can be less than that); no default: if *n* and/or *m* are not specified, their actual values depend on the host hardware platform; precision for arithmetic calculations is some 15 scale digits; synonyms: DOUBLE PRECISION, REAL;

➢ FLOAT (*n*) [UNSIGNED] = (*iif n* > 53, *unexpected results*, (*iif* 0 ≤ *n* ≤ 24 FLOAT [UNSIGNED], DOUBLE [UNSIGNED])); provided only for *ODBC* compatibility.

Actually, all numeric calculations are done with either BIGINT or DOUBLE; consequently, you should not use unsigned big integers larger than 9,223,372,036,854,775,807 (63 bits), except with bit functions. If you do that, some of the last digits in the result may be wrong because of rounding errors when converting a BIGINT value to a DOUBLE.

However, *MySQL* correctly handles BIGINT in the following 4 cases:

✓ When using integers to store large unsigned values in a BIGINT column.
✓ In MIN(*col_name*) or MAX(*col_name*), where *col_name* refers to a BIGINT column.
✓ When both operands are integers.
✓ When storing it using a string: in such cases, *MySQL* performs a string-to-number conversion that involves no intermediate double-precision representation.

Note that when the result of arithmetic operations on two integers would be outside the BIGINT range, unexpected values within this range are obtained instead.

Autonumber surrogate keys can be obtained from integer data types by adding the AUTO_INCREMENT reserved word in their definition.

SERIAL is an alias for BIGINT UNSIGNED NOT NULL AUTO_INCREMENT UNIQUE.

SERIAL DEFAULT VALUE in the definition of an integer column is an alias for NOT NULL AUTO_INCREMENT UNIQUE.

You can get the latest auto-increment value that was generated by the system with SELECT LAST_INSERT_ID();

There are 5 DATE/TIME data types (date format is 'YYYY-MM-DD'; the optional, default 0 *fsp* value –that can range from 0 to 6 digits– specifies the fractional seconds precision):

➢ DATE = [1000–01–01; 9999–12–31];
➢ DATETIME[(*fsp*)] = [1000–01–01 00:00:00; 9999–12–31 23:59:59.999999];
➢ TIMESTAMP[(*fsp*)] = [1970–01–01 00:00:01.000000 UTC; 2038–01–19 03:14:07.999999 UTC];

> ➤ TIME[(*fsp*)] = [−838:59:59; 838:59:59];
> ➤ YEAR[(4)] = [1000; 9999].

Automatic initialization and updating to the current date and time for DATETIME columns can be specified using the DEFAULT and ON UPDATE column definition clauses.

The way the server handles TIMESTAMP definitions depends on the value of the *explicit_defaults_for_timestamp* system variable; by default, this variable is disabled and the server handles TIME-STAMP as follows:

   ✓ Unless specified otherwise, the first TIMESTAMP column in a table is defined to be automatically set to the date and time of the most recent modification, if not explicitly assigned a value. This makes TIMESTAMP useful for recording timestamps of INSERT and UPDATE operations.

   ✓ You can also set any TIMESTAMP column to the current date and time by assigning it a NULL value, unless it has been defined with the NULL attribute to permit NULL values.

   ✓ Automatic initialization and updating to the current date and time can be specified using DEFAULT CURRENT_TIMESTAMP and ON UPDATE CURRENT_TIMESTAMP column definition clauses. By default, the first TIMESTAMP column has these properties, as previously noted. However, any TIMESTAMP column in a table can be defined to have these properties.

If *explicit_defaults_for_timestamp* is enabled, there is no automatic assignment of the DEFAULT CURRENT_TIMESTAMP or ON UPDATE CURRENT_TIMESTAMP attributes to any TIMESTAMP column. They must be included explicitly in the column definition. Also, any TIMESTAMP not explicitly declared as NOT NULL permits NULL values.

If the *MySQL* server is run with the MAXDB *SQL* mode enabled, TIME-STAMP is identical with DATETIME; as a result, such columns use the DATETIME display format, have the same range of values, and there is no automatic initialization or updating to the current date and time.

Do not use two digits years (or the legacy data type YEAR(2) either), as they are ambiguous; if, however, you do it, note that *MySQL* interprets two-digit year values using these rules:

✓ Year values in the range 70–99 are converted to 1970–1999.

✓ Year values in the range 00–69 are converted to 2000–2069.

MySQL allows storing dates where the day or month and day are zero in DATE or DATETIME columns, which is very useful for applications that need to store dates for which you may not know the exact month and/ or date. Obviously, if you store such date values, you should not expect to get correct results from functions such as *DATE_SUB*() or *DATE_ADD*() (that require complete dates). Moreover, *MySQL* also allows storing "zero" values ('0000–00–00') as "dummy dates", which is in some cases more convenient than using NULL values and uses less data and index space. To disallow zero year, month, and day parts in dates, enable strict *SQL* mode.

You can get the current system date and time by calling the functions *Now*(), *CurDate*(), and *CurTime*(). You can extract the year from a date *d* with the function EXTRACT(YEAR FROM *d*).

There are three types of replication formats: *Statement Based Replication* (*SBR*), which replicates entire *SQL* statements, *Row Based Replication* (*RBR*), which replicates only the changed rows, and *Mixed Based Replication* (*MBR*), which combines the above two ones.

SBR of AUTO_INCREMENT and TIMESTAMP values is done correctly, subject to a couple of exceptions (see bugs #45670, #45677, and #11754117).

MySQL recommends using the new RBR with transaction-based replication using *global transaction identifiers* (*GTIDs*). When using GTIDs, each transaction can be identified and tracked as it is committed on the originating server and applied by any slaves. RBR with GTID replication has only the following three restrictions:

✓ updates to tables using nontransactional storage engines such as MyISAM cannot be made in the same statement or transaction as updates to tables using transactional storage engines such as InnoDB;

✓ CREATE TABLE ... SELECT statements are not supported;

✓ CREATE/DROP TEMPORARY TABLE statements are supported only outside of any transaction and with autocommit = 1.

4.2.3.3.2 String Data Types

> *Wisdom thoroughly learned will never be forgotten.*
> —Pythagoras

There are 14 types of strings in *MySQL* (using, for CHAR, VARCHAR, and TEXT ones either ASCII or UNICODE encodings, at your choice, depending on their CHARSET attribute and/or SET NAME and/or SET CHARACTER SET statements execution or the default, which is ASCII; except for CHAR and VARCHAR, for which n is in characters, n is in bytes):

- CHAR(n) = UNICODE(n), $0 \leq n \leq 30$, fixed length (right padding);
- VARCHAR(n) = UNICODE(n), $0 \leq n \leq 65,535$;
- TINYTEXT = UNICODE(n), $0 \leq n \leq 255$;
- TEXT = UNICODE(n), $0 \leq n \leq 65,535$;
- MEDIUMTEXT = UNICODE(n), $0 \leq n \leq 16,777,215$;
- LONGTEXT = UNICODE(n), $0 \leq n \leq 4,294,967,295$;
- BINARY(n): equivalent to CHAR(n), except that there is no character set, n is in bytes (not characters) and padding is done with binary 0 (not space);
- VARBINARY(n): equivalent to VARCHAR(n), except that there is no character set and n is in bytes (not characters);
- TINYBLOB: binary equivalent of TINYTEXT;
- BLOB: binary equivalent of TEXT;
- MEDIUMBLOB: binary equivalent of MEDIUMTEXT;
- LONGBLOB: binary equivalent of LONGTEXT;
- ENUM: static sets of (maximum under 3,000) string literals (that are stored in numeric coding, but showed as defined); only one of them is stored per table cell; (example: ENUM('red', 'orange', 'yellow', 'green', 'blue', 'indigo', 'violet') is stored as TINYINT values 1, 2, 3, 4, 5, 6, and 7, respectively);
- SET: static sets of (maximum 64) string literals (that are stored in numeric coding, but showed as defined); none or any combination of them may be stored per table cell.

BLOB and TEXT data types differ from VARBINARY and VARCHAR, respectively, in the following respects:

✓ For indexes on BLOB and TEXT columns, you must specify an index prefix length. For CHAR and VARCHAR, a prefix length is optional.

✓ BLOB and TEXT columns cannot have DEFAULT values.

✓ Because BLOB and TEXT values can be extremely long, there are some constraints in using them:

- Only the first *max_sort_length* bytes of such columns are used when grouping and sorting; the default value of *max_sort_length* is 1,024; you can make more bytes significant in sorting or grouping by increasing the value of *max_sort_length* at server startup or runtime; any client can change the value of its session *max_sort_length* variable with the *SQL* statement SET *max_sort_length = n*;

- Instances of BLOB or TEXT columns in the result of a query that is processed using a temporary table causes *MySQL* to use a table on disk rather than the internal memory (because the MEMORY storage engine does not support these data types); obviously, use of disks incurs performance penalties, so include BLOB or TEXT columns in the query result only if they are really needed (e.g., avoid using SELECT * when you don't need all columns).

- The maximum size of a BLOB or TEXT object is determined by its type, but the largest value you can actually transmit between clients and servers is determined by the amount of available memory and the size of the communications buffers; you can change the message buffer size by changing the value of the *max_allowed_packet* variable, but you must do so for both the server and your client program (e.g., both mysql and mysqldump enable you to change the client-side *max_allowed_packet* value).

A table can have no more than 255 unique element list definitions among its ENUM and SET columns (that are considered as a group).

4.2.3.3.3 Other Data Types

> *Quality means doing it right when no one is looking.*
> —Henry Ford

Similarly to *Oracle*, *MySQL* 5.7 also provides 8 extended (planar) geometry (geospatial) spatial data types, for both single values (GEOMETRY) and value sets (GEOMETRYCOLLECTION):

➢ GEOMETRY: geometry values of any type;
➢ POINT: geometry values of type point (0-dimensional);
➢ LINESTRING: geometry values of type line (1-dimensional);
➢ POLYGON: geometry values of type polygon (2-dimensional);
➢ GEOMETRYCOLLECTION: collections of geometry values of any type;
➢ MULTIPOINT: collections of geometry values of type point;
➢ MULTILINESTRING: collections of geometry values of type line;
➢ MULTIPOLYGON: collections of geometry values of type polygon.

4.2.3.3.4 Computed Columns

Failure is simply the opportunity to begin again, this time more intelligently.
—Henry Ford

There are no computed columns in *MySQL*. If you need materialized ones, the only solution is to define columns of the corresponding type, to design and implement all needed triggers for computing the corresponding values whenever needed, and to make sure that db users do not write in such columns.

You can obviously simply compute with *SQL* SELECT statements, whenever needed, not stored desired calculable data.

4.2.3.3.5. Data Types Usage Overview

The oldest, shortest words—"yes" and "no"—are those which require the most thought.
—Pythagoras

Table 4.7 shows how to implement in *MySQL* your most frequently needed data types. To conclude this subsection, Table 4.8 presents the rest of the data types provided by *MySQL*.

TABLE 4.7 Most Frequently Needed *MySQL* Data Types

Data type	MySQL 5.7 implementation
AUTONUMBER(n), $n \in \{1, 2\}$	TINYINT(n) AUTO_INCREMENT
AUTONUMBER(n), $n \in \{3, 4\}$	SMALLINT(n) AUTO_INCREMENT
AUTONUMBER(n), $n \in \{5, 6\}$	MEDIUMINT(n) AUTO_INCREMENT
AUTONUMBER(n), $n \in [7, 9]$	INT(n) AUTO_INCREMENT
AUTONUMBER(n), $n \in [10, 19]$	SERIAL or BIGINT(n) UNSIGNED AUTO_INCREMENT
AUTONUMBER(n), $n \in [20, 65]$	DECIMAL(n) AUTO_INCREMENT
AUTONUMBER(n), $n \in [66, 308]$	DOUBLE(n) AUTO_INCREMENT
BOOLEAN	BOOL, BOOLEAN
NAT(n), $n \in \{1, 2\}$	TINYINT(n) UNSIGNED
NAT(n), $n \in \{3, 4\}$	SMALLINT(n) UNSIGNED
NAT(n), $n \in \{5, 6\}$	MEDIUMINT(n) UNSIGNED
NAT(n), $n \in \{7, 8, 9\}$	INT(n) UNSIGNED
NAT(n), $n \in [10; 19]$	BIGINT(n) UNSIGNED
NAT(n), $n \in [20; 65]$	DECIMAL(n) UNSIGNED
NAT(n), $n \in [66; 308]$	DOUBLE(n) UNSIGNED
INT(n), $n \in \{1, 2\}$	TINYINT(n)
INT(n), $n \in \{3, 4\}$	SMALLINT(n)
INT(n), $n \in \{5, 6\}$	MEDIUMINT(n)
INT(n), $n \in [7, 9]$	INT(n)
INT(n), $n \in [10; 19]$	BIGINT(n)
INT(n), $n \in [20; 65]$	DECIMAL(n)
INT(n), $n \in [66; 308]$	DOUBLE(n)
RAT(n, m), $n < 67$	DECIMAL(n, m)
RAT(n, m), CURRENCY(n), $n \in [67, 308]$	DOUBLE(n, m)
CURRENCY(n), $n < 66$	DECIMAL(n, 2)
CURRENCY(n), $n \in [66, 308]$	DOUBLE(n, 2)
DATE/TIME (1000–01–01; 9999–12–31)	DATE (no scale) or DATETIME[(fsp)] (user defined scale of maximum 6 digits, default 0)

TABLE 4.7   (Continued).

Data type	MySQL 5.7 implementation
DATE/TIME (1970–01–01 00:00:01.000000 UTC; 2038–01–19 03:14:07.999999 UTC)	TIMESTAMP[(*fsp*)] (user defined scale of maximum 6 digits, default 0)
TIME (–838:59:59; 838:59:59)	TIME[(*fsp*)] (user defined scale of maximum 6 digits, default 0)
YEAR (1000; 9999)	YEAR(4)
ASCII(n), $n \leq 30$, fixed length	CHAR(n) CHARSET CP850 (or HP8, LATIN2, etc.)
UNICODE(n), $n \leq 30$, fixed length	CHAR(n) CHARSET UCS2 (or UTF8, UTF16, UTF32, UTF8 MB4, UTF16LE)
ASCII/UNICODE(n), $n \leq 65,535$; variable length	VARCHAR(n)[CHARSET *see above*]
ASCII/UNICODE(n), $n \in [65,536; 16,777,215]$	MEDIUMTEXT [CHARSET *see above*]
ASCII/UNICODE(n), $n \in [16,77,216; 4,294,967,295]$	LONGTEXT [CHARSET *see above*]
Static sets of strings (only one of them is stored per table cell)	ENUM
Static sets of strings (none or any combination of them may be stored per table cell; not 1NF tables anymore)	SET

4.2.3.4   Other MySQL 5.7 Particularities

There are no big problems, there are just a lot of little problems.
—Henry Ford

In some cases, *MySQL* silently changes column specifications (data types, and/or data type attributes, and/or index specifications) from those given in CREATE TABLE or ALTER TABLE *SQL DDL* statements. All changes are subject to the internal row-size limit of 65,535 bytes, which may cause some attempts at data type changes to fail; for example:

TABLE 4.8   Other *MySQL* Data Types

Data type	MySQL 5.7 implementation
Binary objects of up to 30 bytes, fixed length	BINARY(n)
Binary objects between 30 bytes and 64 KB	VARBINARY(n)
BLOBs between 64 KB and 16 MB	MEDIUMBLOB
BLOBs between 16 MB and 4 GB	LONGBLOB
Geometry values of type point	POINT
Geometry values of type line	LINESTRING
Geometry values of type polygon	POLYGON
Geometry values of any type (union of POINT, LINESTRING, and POLYGON above)	GEOMETRY
Geometry value sets of type point	MULTIPOINT
Geometry value sets of type line	MULTILINESTRING
Geometry value sets of type polygon	MULTIPOLYGON
Geometry value sets of any type (union of MULTIPOINT, MULTILINE-STRING, and MULTIPOLYGON)	GEOMETRYCOLLECTION

✓ Columns that are part of a PRIMARY KEY are made NOT NULL, even if not declared that way.

✓ Trailing spaces are automatically deleted from ENUM and SET member values when the table is created.

✓ Certain data types used by other *SQL* db vendors are mapped to corresponding *MySQL* types.

✓ When you include a USING clause to specify an index type that is not permitted for a given storage engine, but there is another index type available that the engine can use without affecting query results, the engine uses the available type.

✓ If strict *SQL* mode is not enabled, a VARCHAR column with a length specification greater than 65,535 is converted to TEXT, and a VARBINARY column with a length specification greater than 65,535 is converted to BLOB. Otherwise, an error occurs in either of these cases.

✓ Specifying the CHARACTER SET binary attribute for a character data type causes the column to be created as the corresponding binary data type: CHAR becomes BINARY, VARCHAR becomes VARBINARY, and TEXT becomes BLOB. For the ENUM and SET data types, this does not occur: they are created as declared.

✓ Certain other data type changes can occur if you compress a table using *myisampack*.

To see whether *MySQL* used a data type other than the one you specified, issue a DESCRIBE or SHOW CREATE TABLE statement after creating or altering tables.

MySQL, just like *DB2*, also provides an improved syntax of the INSERT ... VALUES *SQL DML* statement that allows for inserting as many rows as desired with only one such statement; for example, the following statement inserts (the first) 2 lines in table TEST_UNSIGNED_ DECIMAL (and rejects the third as only unsigned decimal numbers are valid for this table):

```
INSERT INTO TEST_UNSIGNED_DECIMAL
VALUES (9.37), (+1.23), (-4.56);
```

Unfortunately, *MySQL* does not enforce check constraints: consequently, you have to enforce them through BEFORE INSERT and UPDATE triggers.

4.2.4 MS *SQL SERVER* 2014

The first rule of any technology used in a business is that automation applied to an efficient operation will magnify the efficiency. The second is that automation applied to an inefficient operation will magnify the inefficiency.

—Bill Gates

SQL Server 2014 is a robust, well-performing and fully featured RDBMS too. Standard components include db language and storage server, developer and ETL tools, schedulers, and replication. Other components include OLAP, reporting, and parallel computation. Components run as *Windows NT Services*. It can also be accessed through *ODBC*, *ADO*, and *OLEDB*. Both MS *Visual Studio.NET* and *Access* may easily be used to develop db applications based on *SQL Server* dbs.

Besides several commercial versions, there is also a free to download *SQL Server Express* version (for 10 GB dbs maximum), which can also be redistributed with your products; it allows for 1 GB RAM and 4 CPU cores.

SQL Server 2014 partially adheres to the *SQL* 2008 ANSI standard (which includes the 1999 recursive SELECT queries, triggers, and support for procedural and control-of-flow statements in its extended *SQL T-SQL* language, as well as the 2003 and 2006 *SQL/XML* and identity columns), plus its own extensions (such as its *PIVOT* and *UNPIVOT* clauses of the SELECT statement, for (un)crosstabulations).

SQL Server main limitation is that it only runs on MS *Windows* and Mac *OS X* OSs.

4.2.4.1   Identifiers and Literals

> *I really had a lot of dreams when I was a kid, and I think a great deal*
> *of that grew out of the fact that*
> *I had a chance to read a lot.*
> —Bill Gates

SQL Server object names may have at most 128 characters (116 for temporary tables), are not case sensitive, but you can freely use both upper and lower case (and *SQL Server* stores them exactly as you defined them). They should start with a letter or one of the special characters '@', '#', and '_'.

Those starting with '@' are considered as local variables; moreover, '@' cannot be used in any other name position.

Those starting with '#' are considered as temporary objects, only existing during the current work session; all those starting with "##" are global, that is accessible to all concurrent users, while the others are local, accessible only to the current user.

Money constants are prefixed (after sign, if any) with '$' or some other over 30 currency symbols available (including '€', '£', '¥', etc.).

The only other reserved character for names is the space. However, you may use any number of spaces in names provided you embed them in either square brackets or double quotation marks (if portability is not a concern, always prefer square brackets, in order to avoid confusions with text strings).

Text strings, as well as date and time constants are enclosed in apostrophes (or in double quotation marks, if the QUOTED_IDENTIFIER option is set to OFF). You can use apostrophes within text strings by doubling them; for example, the text string '"quote"''s value is 'quote'.

4.2.4.2   Databases and Tables

> *If GM had kept up with technology like the computer industry has,*
> *we would all be driving $25 cars that got 1,000 MPG.*
> —Bill Gates

Computed tables are called *views* and can be created with the CREATE VIEW ... AS SELECT ... *SQL DDL* statement; they can also be created and stored through the *SQL Server Management Studio*'s *QBE*-type *Query Design* GUI too.

You can define temporary tables as variables, but they only exist during the execution of the corresponding script (batch, transaction) and you cannot use the SELECT ... INTO *SQL* statement for populating them.

You can also define in-memory tables, very useful for OLTP processing.

Temporary tables, either local (names prefixed by '#') or global (names prefixed by "##") exist during the current user session or the lifetime of the procedure in which they are declared.

4.2.4.3   Data Types

> *Success is a lousy teacher. It seduces smart people*
> *into thinking they can't lose.*
> —Bill Gates

4.2.4.3.1. Numeric Data Types

> *Governments will always play a huge part in solving big problems.*
> *They set public policy and are uniquely able to provide the resources*
> *to make sure solutions reach everyone who needs them. They also fund*
> *basic research, which is a crucial component of the innovation*
> *that improves life for everyone.*
> —Bill Gates

The Boolean type is called BIT.

There are only four subsets of the integers:
➢ TINYINT = [0; 255];
➢ SMALLINT = [−32,768; 32,767];
➢ INT = [−2,147,483,648; 2,147,483,647];
➢ BIGINT = [−9,223,372,036,854,775,808; 9,223,372,036,854,775,807].

In order to define an autonumber column, you have to choose for it an integer data type and add to its declaration the keyword IDENTITY.

If you need replication, you have to use instead the UNIQUEIDENTI-FIER data type, which stores GUIDs (16-byte binary); such values are unique worldwide, as they are obtained from the corresponding (world-wide unique) network card number and the current system timestamp. You can obtain new values for such a column through the *T-SQL* function NEWID, used either as the default value (recommended choice) of the column or the corresponding inserted value; for example:

```
CREATE TABLE T
x UNIQUEIDENTIFIER PRIMARY KEY DEFAULT NEWID(),
TextCol VARCHAR(10));
INSERT INTO T(TextCol) VALUES ('abc');
```
or:
```
CREATE TABLE T (x UNIQUEIDENTIFIER PRIMARY KEY,
TextCol VARCHAR(10));
INSERT INTO T VALUES (NEWID(),'abc');
```

Note that the IDENTITY property is not available for such columns, so that you should declare them as keys (be them primary or not), in order for the *SQL Server* to prevent storing duplicates for them, as there may be several UNIQUEIDENTIFIER columns in any table. Hopefully, only one of them may also have the ROWGUIDCOL property, used for replication.[108]

The UNIQUEIDENTIFIER data type has a major disadvantage: it is relatively large compared to numeric data types, such as 4-byte integers, which means that indexes built on them are slower. Consequently, always use integers and the IDENTITY property when global uniqueness is not necessary.

There are only three subsets of the rationals (both of them also stored in binary formats):

> DECIMAL (DEC, NUMERIC) = $[-10^{38} + 1; 10^{38} - 1]$;
> REAL (FLOAT24) = $[-3.402823 \times 10^{38}; 3.402823 \times 10^{38}]$;
> FLOAT (DOUBLEPRECISION, FLOAT(53)) =
  $[-1.79769313486232 \times 10^{308}; 1.79769313486232 \times 10^{308}]$.

The default (fixed) precision for DECIMAL is 18. In *T-SQL*, any numeric constant including a decimal point is considered to be decimal; for example, 123.45 is of type DECIMAL(5, 2), where 5 is the precision (total number of digits) and 2 is the scale (total number of digits after the decimal point).

REAL and FLOAT are approximate (not fixed) precision numeric data types. By default, FLOAT's precision is 15 (with a mantissa of 53 digits); you can specify any mantissa m, $1 \le m \le 53$, as FLOAT(m), but for any $m \ge 24$, it is treated like the default 53, while for any $m < 25$ it is treated like 24, which has a precision of only 7 digits. *SQL Server* uses roundup to approximate such values. Avoid using float or real columns in WHERE clause search conditions, especially with the = and <> operators; it is best to limit them to only > and < comparisons.

There are two CURRENCY data types:

> SMALLMONEY = $[-214{,}748.3648; 214{,}748.3647]$;
> MONEY = $[-922{,}337{,}203{,}685{,}477.5808;$
  $922{,}337{,}203{,}685{,}477.5807]$.

Both of them have precision of a ten thousands (of the monetary unit). Some argue that there are significant rounding errors (especially when dividing money to money, which makes sense only when establishing currency rates, but which is avoidable with proper conversions) when

---

[108] Do not get confused, however: the ROWGUIDCOL property indicates that the values of that column uniquely identify the corresponding table rows, but does not do anything to enforce uniqueness.

using these two data types; moreover, as MONEY is one byte less than a large DECIMAL, with up to 19 digits of precision and as most real-world monetary calculations (up to $9.99 M) can fit in a DECIMAL(9, 2), which requires just five bytes, you can save size, worry less about rounding errors, and make your code more portable by using DECIMAL instead of MONEY.

There are several DATE/TIME data types (except for DATETIMEOFF-SET, all others date/time types ignore time zones):

➢ DATE = [01/01/1; 31/12/9999], with 1 day accuracy, 10 digits precision, no scale;

➢ DATETIME2 = [01/01/1; 31/12/9999], with 100 nanoseconds accuracy, precision and (user defined) scale of maximum 7 digits;

➢ DATETIME = [01/01/1753; 31/12/9999], with accuracy rounded to increments of .000, .003, or .007 seconds and fixed scale of 3 digits;

➢ SMALLDATETIME = [01/01/1900; 06/06/2079], with 1 min accuracy and seconds always :00; 23:59:59 is rounded to 00:00:00 of the next day;

➢ TIME = [00:00:00; 23:59:59.9999999], with 100 nanoseconds accuracy, precision and (user defined) scale of maximum 16 and 7 digits, respectively;

➢ DATETIMEOFFSET = [01/01/1; 31/12/9999], with 100 nanoseconds accuracy, precision and (user defined) scale of maximum 7 digits, and from −14:00 to +14:59 user defined offset (depending on desired time zone);

➢ ROWVERSION (TIMESTAMP) is an automatically generated relative date and time stamp used for row version-stamping (avoid using TIMESTAMP, which will soon be not supported anymore).

A nonnullable ROWVERSION column is semantically equivalent to a BINARY(8) one (see Section 4.2.4.3.2). A nullable ROWVERSION column is semantically equivalent to a VARBINARY(8) one (see Section 4.2.4.3.2).

A table can have only one ROWVERSION column. Every time that a row of such a table is modified or inserted, the incremented db rowversion value is inserted into its ROWVERSION column. To get the current rowversion value for the current db (as a VARBINARY!), use SELECT @@DBTS.

You can add a ROWVERSION column to a table to help maintain the integrity of the db when multiple users are updating rows simultaneously. You may also want to know how many rows and exactly which ones were updated, without requerying the table.

You can use the ROWVERSION column of a table to easily determine whether any value in its rows has changed since the last time rows were read: if any change is made to a row, its rowversion value is updated; if no change was made to a row, its rowversion value is the same as when it was previously read.

ROWVERSION columns are very poor candidate for prime attributes and especially for primary key ones: any update made to a row changes the rowversion value; if a ROWVERSION column is (part of) a primary key, the old key value is no longer valid, and foreign keys referencing the old value are no longer valid; moreover, if a table is referenced in a dynamic cursor, all updates change the position of the rows in the cursor and if the ROWVERSION column is in an index key, all updates to the data row also generate updates of the index.

Duplicate rowversion values can be generated by using the SELECT INTO statement in which a ROWVERSION column is in the SELECT list. MS does not recommend using ROWVERSION in this manner.

You can obtain the system date and time from the function *SysDateTime*(), a DATETIME2(7) one (or *SysDateTimeOffset*(), a DATETIMEOFFSET(7) one, if you need time zones to be also included or *SysUTCDateTime*(), also a DATETIME2(7), if you need the corresponding UTC date and time).

If only a DATETIME is needed, you can get it from the functions *Current_Timestamp*() (or *GetDate*()) and *GetUTCDate*().

The year from a date can be obtained with the *Year*() function.

4.2.4.3.2 String Data Types

> *Good engineering and good business are one in the same.*
> —Bill Gates

There are nine types of strings:
➢ CHAR[(*n*)]—fixed length ASCII strings of maximum 8000 characters (default 1);
➢ NCHAR[(*n*)]—fixed length UNICODE strings of maximum 4000 characters (default 1);
➢ VARCHAR[(*n*)]—variable length ASCII strings of maximum 8000 characters (default 1);
➢ VARCHAR(max)—variable length ASCII strings of more than 8000 and less than $2^{31}-1$ characters (2 GB storage maximum);
➢ NVARCHAR[(*n*)]—variable length UNICODE strings of maximum 4000 characters (default 1);
➢ NVARCHAR(max)—variable length UNICODE strings of more than 4000 and less than $2^{30}-1$ characters (2 GB storage maximum);
➢ BINARY[(*n*)]—fixed length binary strings of maximum 8000 bytes (default 1);
➢ VARBINARY[(*n*)]—variable length binary strings of maximum 8000 characters (default 1);
➢ VARBINARY(max)—variable length binary strings of more than 8000 and less than $2^{31}-1$ characters (2 GB storage maximum).

Avoid using the TEXT, NTEXT, and IMAGE data types, as they will not be supported anymore in the near future: use VARCHAR, NVARCHAR, and VARBINARY instead, respectively.

4.2.4.3.3 *Other Data Types*

> *I believe in innovation and that the way you get innovation is*
> *you fund research and you learn the basic facts.*
> —Bill Gates

The following other six data types are provided too:
➢ HIERARCHYID
➢ XML
➢ GEOGRAPHY
➢ GEOMETRY
➢ SQL_VARIANT
➢ User-defined data types

The HIERARCHYID data type is a variable length, system data type, for representing positions in trees. However, a column of type HIERARCHYID does not automatically represent a tree: it is up to the application to generate and assign HIERARCHYID values, in such a way that the desired relationship between rows is reflected in the corresponding values.

Values for HIERARCHYID have the following properties:

✓ are *extremely compact*: the average number of bits that are required to represent a node in a tree with n nodes depends on the average fan-out (i.e., the average number of children per node); for small fan outs (0–7), the size is about $6*\log_{An}$ bits, where A is the average fan-out; for example, a node in an organizational hierarchy of 100,000 people with an average fan-out of 6 levels takes about 38 bits; this is rounded up to 40 bits, or 5 bytes, for storage;

✓ *comparison is in depth-first order*: given two HIERARCHYID values v and w, $v < w$ means that v comes before w in a depth-first traversal of the tree; indexes on HIERARCHYID data type columns are in depth-first order, and nodes close to each other in a depth-first traversal are stored near each other; for example, all children of a record are stored adjacent to that record;

✓ *support for arbitrary insertions and deletions*: by using the *GetDescendant*() method, it is always possible to generate a sibling to the right of any given node, to the left of any given node, or between any two siblings; the comparison property is maintained when an arbitrary number of nodes is inserted or deleted from the tree; most insertions and deletions preserve the compactness property; however, insertions between two nodes will produce HIERARCHYID values with a slightly less compact representation;

✓ the *encoding used in the HIERARCHYID type is limited to 892 bytes*; consequently, nodes which have too many levels in their representation to fit into 892 bytes cannot be represented by the HIERARCHYID type.

The HIERARCHYID type logically encodes information about a single node in a tree, by encoding the path from the root of the tree to that node. Such a path is logically represented as a sequence of node labels of all children orderly visited after the root.

Just like for tree-like file management systems, a slash starts the representation and a path that only visits the root is represented by a single slash; for levels underneath the root, each label is encoded as a sequence of integers separated by dots; comparison between children is performed by comparing the integer sequences separated by dots, in dictionary order; each level is followed by a slash (i.e., slash separates parents from their children).

For example, the following are valid HIERARCHYID paths of lengths 1, 2, 2, 3, and 3 levels respectively:

- /
- /1/
- /0.3/7/
- /1/3/
- /0.1/0.2/

Nodes can be inserted in any location. For example, nodes inserted after /1/2/ but before /1/3/ can be represented as /1/2.5/. Nodes inserted before 0 have logical representations as negative numbers. For example, a node that comes before /1/1/ can be represented as /1/-1/. Nodes cannot have leading zeros. For example, /1/1.1/ is valid, but /1/1.01/ it is not. To prevent errors, insert nodes by using the *GetDescendant*() method.

Columns of type HIERARCHYID can be used on any replicated table. The requirements for your application depend on whether replication is a directional or bidirectional one, and on the versions of *SQL Server* that are used.

XML has the following syntax:

XML([content | document] xml_schema_collection),
where:

- ✓ content restricts the XML instance to be a well-formed *XML* fragment; *XML* data can contain multiple zero or more elements at the top level; text nodes are also allowed at the top level (and this is the default behavior);
- ✓ document restricts the XML instance to be a well-formed *XML* document; *XML* data must have one and only one root element; text nodes are not allowed at the top level;
- ✓ xml_schema_collection is the name of an *XML* schema collection; to create a typed XML column or variable, you can optionally specify the *XML* schema collection name.

The stored representation of XML data type instances size cannot exceed 2 GB. The *content* and *document* facets apply only to typed *XML*.

GEOGRAPHY represents data in a round-earth coordinate system and stores it ellipsoidally (round-earth), using GPS latitude and longitude coordinates.

GEOMETRY is a planar spatial data type, representing data in a Euclidean (flat) coordinate system.

SQL_VARIANT is a data type that may store values of any *SQL Server*-supported data type, except for the following 14 ones: DATE-TIMEOFFSET, GEOGRAPHY, GEOMETRY, HIERARCHYID, IMAGE, NTEXT, NVARCHAR(MAX), ROWVERSION, SQL_VARIANT, TEXT, USER_DEFINED, VARBINARY(MAX), VARCHAR(MAX), and XML.

SQL_VARIANT can have a maximum length of 8,016 bytes. This includes both the base type information and the base type value. The maximum length of the actual base type value is 8,000 bytes.

A SQL_VARIANT data type value must first be cast to its base data type one, before participating in operations such as addition and subtraction. SQL_VARIANT cannot have another SQL_VARIANT as its base type.

SQL_VARIANT can be assigned a default value. This data type can also have NULL as its underlying value, but the NULL values will not have an associated base type.

A unique, primary, or foreign key may include columns of type SQL_VARIANT, but the total length of the data values that make up the key of a specific row should not be more than the maximum length of an index, which is 900 bytes.

A table can have any number of SQL_VARIANT columns.[109]

Finally, *SQL Server*'s users can define their own data types. Such types should be created with the *sp_addtype* table definitions. They are based on the system data types and are recommended to be used when several tables must store the same type of data in a column and you must ensure that these columns have exactly the same data type, length, and nullability.

When a user-defined data type is created, you must supply the following three parameters: new data type name (unique across the db), system data type upon which the new data type is based, and nullability (i.e.,

---

[109] However, SQL_VARIANT cannot be used in either CONTAINSTABLE or FREETEXTTABLE.

whether the data type allows null values or not). When nullability is not explicitly defined, it will be assigned based on the ANSI null default setting for the db or connection.

User-defined data types created in the *model* db exist in all new user-defined dbs; if a data type is created in a user-defined db, then it exists only in that db.

For example, a user-defined data type called `Address` could be created based on the varchar data type in the *model* db (the "supreme" model) as follows:

```
USE model
EXEC  sp_addtype  Address,  'VARCHAR(128)',  'NOT
NULL'
```

User-defined data types are not supported in table variables.

4.2.4.3.4 Computed Columns

> *If your culture doesn't like geeks, you are in real trouble.*
> —Bill Gates

You can also add computed columns to tables: columns that are not physically stored in the table, unless the column is marked `PERSISTED`. A computed column expression can use data from other columns of its table to calculate a value for that column.

A computed column cannot be used as a `DEFAULT` or `FOREIGN KEY` constraint definition or with a `NOT NULL` constraint definition. However, if the computed column value is defined by a deterministic expression and the data type of the result is allowed in index columns, a computed column can be used as a key column in an index or as part of any `PRIMARY KEY` or `UNIQUE` constraint.

For example, if the table has integer columns A and B, the computed column $C = A + B$ may be indexed, but computed column $D = A$ `+ DATEPART(dd, GETDATE())` cannot be indexed, because the value might change in subsequent invocations.

Computed columns can be created either through the GUI of the *SQL Server Management Studio* or with the `AS` predicate of *SQL*. For example, the following *SQL DDL* statement creates a table *PRODUCTS* having the primary key autonumber column *x*, a column *Product* for storing product names, a *QtyAvailable* column for storing corresponding quantities in stock, a *UnitPrice* one, and a virtual computed column *InventoryValue* that calculates, for each row, the corresponding value of the product between *QtyAvailable* and *UnitPrice*:

```
CREATE TABLE PRODUCTS (
    x INT IDENTITY (1,1) PRIMARY KEY,
    Product VARCHAR(64) NOT NULL,
    QtyAvailable SMALLINT NOT NULL
        CHECK (QtyAvailable BETWEEN 0 AND 100),
    UnitPrice SMALLMONEY NOT NULL
        CHECK (UnitPrice BETWEEN 0 AND 10000),
    InventoryValue AS QtyAvailable * UnitPrice);
```

4.2.4.3.5 Data Types Usage Overview

> *Treatment without prevention is simply unsustainable.*
> —Bill Gates

Table 4.9 shows how to implement in *SQL Server* your most frequently needed data types. To conclude this subsection, Table 4.10 presents the rest of the data types provided by the *SQL Server*.

4.2.4.4   Other SQL Server 2014 Particularities

> *Information technology and business are becoming inextricably interwoven. I don't think anybody can talk meaningfully about one without talking about the other.*
> —Bill Gates

SQL Server stores nulls in indexes and considers that there is only one null value; consequently, it does not accept in keys columns that may store nulls.

TABLE 4.9 Most Frequently Needed *SQL Server* Data Types

Data type	*SQL Server 2014 implementation*
AUTONUMBER(*n*), *n* < 3 (no replication)	`TINYINT IDENTITY`
AUTONUMBER(*n*), *n* ∈ {3, 4} (no replication)	`SMALLINT IDENTITY`
AUTONUMBER(*n*), *n* ∈ [5; 9] (no replication)	`INT IDENTITY`
AUTONUMBER(*n*), *n* ∈ [10; 18] (no replication)	`BIGINT IDENTITY`
AUTONUMBER(*n*), *n* ∈ [19; 38] (no replication)	`DECIMAL (n, 0) IDENTITY`
AUTONUMBER(*n*), *n* ∈ [39; 308] (no replication)	`FLOAT (n)` and an associated `BEFORE INSERT` trigger
AUTONUMBER(*n*), *n* < 38 (for replication)	`UNIQUEIDENTIFIER` and `NEWID()`
BOOLEAN	`BIT`
NAT(*n*), *n* < 3	`TINYINT`
NAT(*n*), *n* ∈ {3, 4}	`SMALLINT`
NAT(*n*), INT(*n*), *n* ∈ [5; 9]	`INT`
NAT(*n*), INT(*n*), *n* ∈ [10; 18]	`BIGINT`
NAT(*n*), INT(*n*), *n* ∈ [19; 38]	`DECIMAL (n, 0)`
NAT(*n*), INT(*n*), *n* ∈ [39; 308]	`FLOAT(53)`
INT(*n*), *n* < 5	`SMALLINT`
RAT(*n*, *m*), *n* < 39	`DECIMAL (n, m)`
RAT(*n*, *m*), INT(*n*), CURRENCY(*n*), *n* ∈ [39; 308]	`FLOAT(53)`
CURRENCY(*n*), *n* < 8	`DECIMAL (n, 2)`, `SMALLMONEY`
CURRENCY(*n*), *n* ∈ [9; 38]	`DECIMAL (n, 2)`, `MONEY`
DATE/TIME (01/01/1; 31/12/9999)	`DATE` (1 day accuracy, no scale, no time zone), `DATETIME2` (100 ns accuracy, user defined scale of maximum 7 digits, no time zone), `DATETIMEOFFSET` (similar to `DATETIME2`, but user defined time zone), depending on desired accuracy, scale, and time zone (not) awareness
DATE/TIME (01/01/1753; 31/12/9999)	`DATETIME2`, if accuracy rounded to increments of .000, .003, or .007 seconds and fixed scale of 3 digits are enough

TABLE 4.9 (Continued).

Data type	SQL Server 2014 implementation
DATE/TIME (01/01/1900; 06/06/2079)	SMALLDATETIME, if seconds do not matter and 1 min accuracy is enough
ASCII(n), $n \leq 8,000$, fixed length	CHAR(n)
UNICODE(n), $n \leq 4,000$, fixed length	NCHAR(n)
ASCII(n), $n \leq 8,000$, variable length	VARCHAR(n)
UNICODE(n), $n \leq 4,000$, variable length	NVARCHAR(n)
ASCII(n), $n \in [8,001; 2^{31} - 1]$	VARCHAR(MAX)
UNICODE(n), $n \in [4,001; 2^{30} - 1]$	NVARCHAR(MAX)
User defined data type	EXEC sp_addtype
Computed column	AS (PERSISTED)

The simplest workaround for it is to augment the ranges of these string-type columns with as many explicit "*unknownX*" values as needed, and add the corresponding NOT NULL constraints.

For example, in the *CITIES* table from Fig. 3.5, if you would like to store data for small cities without actually knowing their zip code, while enforcing the key **Country • Zip*, you may use in its *Zip* column values as "unknown1", "unknown2", etc.

Sometimes, fortunately, you need only one "unknown" value per range. For example, in order to allow storing in the *COMMUNES* table of the same Fig. 3.5 data for communes without actually knowing their state, while enforcing the key *State • Country*, you may add in the *STATES* table from Fig. 3.4 one row per country having "unknown" in its *State* column. Similarly, in order to be able to enforce its *Capital* key while being able to store data even on states without actually knowing their capitals, you may add to the *CITIES* table of Fig. 3.4 one row per state having "unknown" in its *City* column.

Unfortunately, for numeric columns similar solutions are interfering with the corresponding domain constraints, which is unacceptable. For example, in table *CO-AUTHORS* from Fig. 3.13, in order to store data for coauthors without actually knowing their positions in the corresponding book coauthor lists, while enforcing the key *Book • PosInList* (that you have to discover when solving Exercise 3.11, as, for any book, there may

TABLE 4.10   Other *SQL Server* Data Types

Data type	SQL Server 2014 implementation
Binary objects of up to 8,000 bytes, fixed length	BINARY(n)
Binary objects of up to 8,000 bytes, variable length	VARBINARY(n)
BLOBs up to 2 GB	VARBINARY(MAX)
Insert and update transactions' unique ids	ROWVERSION
Tree nodes	HIERARCHYID
XML (well-formatted) files	XML
GPS coordinates	GEOGRAPHY
Planar Euclidean coordinates	GEOMETRY
Union of all numeric (except for DATETIMEOFFSET) and string (except for the MAX ones) data types	SQL_VARIANT

not be more than one coauthor in any of its positions), a similar to the above solution would imply relaxing the domain constraint *PosInList* ∈ [2; 16], which is unacceptable (as then you were not able to distinguish between plausible and not plausible *PosInList* values).

Fortunately, in such cases no workaround is needed: for example, it doesn't actually matter whether values entered in *PosInList* are correct or not, as long as they are plausible. For example, even if, to begin with, a user would not exactly know whether Prof. Dan Suciu is the second or the third coauthor of the book *Data on the Web. From Relations to Semistructured Data and XML*, he/she might enter him as the second one and then enter Prof. Peter Buneman as its third one, which can be corrected later, by interchanging their list positions.

4.2.5   MS ACCESS 2013

Technology is just a tool. In terms of getting the kids working together and motivating them, the teacher is the most important.
—Bill Gates

Access 2013 is a robust, well-performing, and well-featured commercial product, well suited for small businesses.

Generally, *Access 2013* adheres to the pure *SQL* ANSI 1992 standard, with some extensions (both standard, as the identity autonumber columns, and nonstandard, as the very powerful *Lookup* feature for foreign keys or the TRANSFORM *SQL* statement).

Access dbs are limited to 2 GB (but can be linked between them in any number, so that from one db you can access a huge number of other dbs, and not only *Access* ones) and are also accessible through *ODBC*, *ADO*, and *OLEDB*. Besides its rdb engine, *Access* comes equipped with the high level PL *VBA*, which embeds *SQL*, *ADO*, and *DAO*, with which you can develop db applications. By also using the MS *SharePoint 2013* server or the *Office 365* website as hosts, you can also develop web-based *Access 2013* applications.

There are no free *Access* versions, but there is a freely download-able *Access 2013 Runtime* rdb engine that you can distribute together with your *Access 2013* desktop applications to users that do not own full *Access* versions.

Access main limitation is that it only runs on MS *Windows* and Mac *OS X* OSs.

4.2.5.1   Identifiers and Literals

I believe that if you show people the problems and you show them the solutions they will be moved to act.
—Bill Gates

Access object names may have at most 255 characters, are not case sensi-tive, but you can freely use both upper and lower case (and *Access* stores them exactly as you defined them).

Among other reserved characters, they do not accept '?' and '#'. However, you may use any reserved characters in their names (as well as reserved words), provided you embed them in square brackets.

Text strings have to be embedded in double quotation marks, whereas date and time constants within sharp (numeric: '#') signs. You can use double quotation marks within text strings by doubling them; for example, the text string """"quote"""""'s value is "quote".

4.2.5.1 Databases and Tables

> *Your most unhappy customers are your greatest source of learning.*
> —Bill Gates

As *Access* manages only one rdb at a time[110], its *SQL* does not include a `CREATE DATABASE` statement. The simplest way to create an *Access* rdb is to use its GUI for opening a new blank db, selecting the desired path, giving it the desired name, and saving it.

Computed tables (views) are called *queries*, which cannot be materialized (only their definition is stored): they are evaluated each time when invoked. They can be created in *SQL* only through *VBA* and *ADOX* with the `CREATE VIEW ... AS SELECT ...` statement; they can also be created and stored through the *Access*' GUI *QBE*-type *Design View* too.

You can create temporary tables too (which exists only during the work session in which they were created; at the end of these sessions, they are automatically deleted by the system) with `CREATE TEMPORARY TABLE ...`

4.2.5.2 Data Types

> *Innovation is moving at a scarily fast pace.*
> —Bill Gates

4.2.5.2.1 Numeric Data Types

> *Software is a great combination between artistry and engineering.*
> —Bill Gates

The Boolean type is called `BIT` (`YESNO`, `LOGICAL`, `LOGICAL1`).

There are only three subsets of the integers (all of them binary-stored):

---

[110] In fact, through table linking, from any one such rdb you can work with data from as many other rdbs (not necessarily MS *Access* or *SQL Server* ones, for which there is native support: any OLEDB or ODBC data source for which there is a compatible driver is accepted); you can even have read-only access to the linked table schemas.

- ➢ BYTE (INTEGER1) = [0; 255];
- ➢ SHORT (SMALLINT, INTEGER2) = [–32,768; 32,767];
- ➢ LONG (INTEGER, INT, INTEGER4) = [–2,147,483,648; 2,147,483,647].

Autonumbers may be of either the incremental, random, or replication type. When you don't need unicity across several platforms, choose for your surrogate key columns the COUNTER (AUTOINCREMENT) integer data type (which is an autonumbering LONG, either incremental –the default– or random, at your choice; for the incremental one, by default, both the initial value and the increment ones are 1; you can set them to any other values either through GUI or with a *DDL SQL* statement of the form ALTER TABLE *table name* ALTER COLUMN *column name* COUNTER([*seed*], [*increment*]). You can switch to the random type only through the GUI or through *VBA* and *DAO/ADO*). For replication purposes, you might use the GUID (16 bytes binary value) data type. However, you have no support at all for automatic generation of unique GUID values: you should use custom *VBA* methods to do it.

Moreover, *Access* is the only well-known RDBMS that provides (through his GUI) a remarkable *Lookup Wizard* for attaching to LONG-type foreign keys combo-boxes that may hide actual pointers (preferably to surrogate primary key values) and display instead corresponding semantically relevant data, computed with attached SELECT *SQL* statements or statically enumerated matrices. Lookups can also be attached/modified/deleted to/from columns manually (through the *Lookup* tab in its GUI table *Design View*) or programmatically (through *VBA* and *ADOX* or/and *DAO*). Unfortunately, corresponding semantically relevant data may be anything desired, so it is the db designer task to choose a (super)key for it, in order for the users to be able to uniquely identify each referenced row.

There are only two subsets of the rationals (both of them also stored in binary formats):

- ➢ SINGLE (IEEESINGLE, REAL, FLOAT4) = [– 3.402823×10^{38}; 3.402823×10^{38}];
- ➢ DOUBLE (IEEEDOUBLE, NUMBER, NUMERIC, FLOAT, FLOAT8) = [– $1.79769313486232 \times 10^{308}$; $1.79769313486232 \times 10^{308}$].

There is a CURRENCY (MONEY) data type, stored on 8 bytes (just like DOUBLE and DATE/TIME), which can store rationals with up to 15 digits before and 4 after the decimal point. By default, USD is the used currency, but it can be changed to any other world one (either through *Access* option settings or/and through the host OS' ones).

The DATETIME (DATE, TIME, TIMESTAMP) data type is a subset of DOUBLE and can store any second since January 1st, 100 and up to December 31st, 9999. You can obtain the system date and time from the function *Now*(), only the date from the function *Date*() (only the time from *Time*()), and the year from a date with the *Year*() function.

4.2.5.2.2 String Data Types

> *The advance of technology is based on making it fit in so that you don't really even notice it, so it's part of everyday life.*
> —Bill Gates

There are two types of strings (both of them stored using corresponding actual (variable) lengths and UTF-8, for both UNICODE and ASCII):

> ➤ TEXT (VARCHAR, CHAR, STRING, ALPHANUMERIC) = UNI-CODE(255);
> ➤ LONGTEXT (LONGCHAR, MEMO, NOTE) = UNICODE(4096).

You can search within LONGTEXT values, but you cannot group or order by them.

4.2.5.2.3 Other Data Types

The Internet is becoming the town square for the global village of tomorrow.
—Bill Gates

There are other two data types stored in binary format as well, namely:

> ➤ BINARY (VARBINARY); *Length*(BINARY) ∈ [0; 510 B];
> ➤ LONGBINARY(OLEOBJECT, GENERAL); *Length*(LONGBINARY) ∈ [0; 1 GB].

LONGBINARY is used only for storing OLE objects (embedded pictures, audio, video, or other Binary Large OBjects (*BLOBs*) not over 1 GB each).

You can also define HYPERLINK columns (stored as LONGTEXTs) for storing (absolute) hyperlinks to local or remote files (of the *Universal Naming Convention* (UNC) type *\\server\share\path\filename.extension*), URLs (including file transfer type ones, be it through browser downloading or the File Transfer Protocol (*FTP*)), e-mail addresses, and/or (relative) *Access*' data access page addresses.

Hyperlinks are stored in the following format: *displaytext#address# subaddress#screentip*, where *displaytext* is the text that you want to be displayed underlined, *address* (the only mandatory part) is the address of the file, URL, etc., *subaddress* is a location within the document found at *address*, and *screentip* is the text that is displayed when the mouse hovers over the hyperlink.

In fact, you can use *Access*' GUI to add, edit, and delete hyperlinks, which you invoke by right-clicking the corresponding table cell and then choosing the desired action.

You can define ATTACHMENT-type columns too, for storing lists of files that users desire to be attached to table rows. A maximum of 2 GB of data (which is the maximum size for *Access* dbs) can be attached per db. Size of any individual attached file cannot exceed 256 MB. Only files of the following types can be attached (and the ones which are not natively compressed are compressed by *Access*):

> ➢ MS *Office* (.docx, .xlsx, .accdb, .pptx, etc.);
> ➢ text and zipped (.txt, .log, .zip, etc.);
> ➢ image (.bmp, .jpg, .jpeg, .jpe, .gif, .tiff, .tif, .ico, .png, .dib, .rle, .exif, .wmf, .emf).

Some 50 file types are rejected (among which executable and temporarily ones, some interpretable ones, etc.).

Names of attached files cannot exceed 255 UNICODE characters (including extensions) and cannot contain the ones reserved by the NTFS file system.

Finally, besides ATTACHMENT, there are also 8 other *Access* data types that can store lists (namely, MULTIPLE BYTE, DECIMAL, DOUBLE, GUID, INT16, INT32, SINGLE, and TEXT), all of them obviously violating the 1NF.

4.2.5.2.4 Computed Columns

> *Until we're educating every kid in a fantastic way,*
> *until every inner city is cleaned up, there is no shortage of things to do.*
> —Bill Gates

You can also add computed columns to tables, but only through *Access'* GUI. Moreover, you can also add a *Total* row for any table and specify for each of its columns with what aggregation function would you like *Access* to compute the corresponding total (the only available function for text-type ones being COUNT).

4.2.5.2.5 Data Types Usage Overview

> *Research shows that there is only half as much variation*
> *in student achievement between schools*
> *as there is among classrooms in the same school.*
> *If you want your child to get the best education possible, it is actually*
> *more important to get him assigned to a great teacher than to a great school.*
> —Bill Gates

Table 4.11 shows how to implement in *Access* your most frequently needed data types. To conclude this subsection, Table 4.12 presents the rest of the data types provided by *Access*.

4.2.5.3 Other Access 2013 Particularities

> *It's fine to celebrate success, but it is more important to*
> *heed the lessons of failure.*
> —Bill Gates

The ANSI *SQL* standard wildcards '_' (any character) and '%' (any character string, including the empty one) are different in *Access*: '?' and '*', respectively.

TABLE 4.11   Most Frequently Needed *Access* Data Types

Data type	Access 2013 implementation
AUTONUMBER(*n*), *n* < 10 (no replication)	COUNTER (AUTOINCREMENT)
AUTONUMBER(*n*), *n* ∈ [10 ;38] (no replication)	SINGLE (IEEESINGLE, REAL, FLOAT4) and an associated *VBA* BEFORE INSERT trigger
AUTONUMBER(*n*), *n* ∈ [39; 308] (no replication)	DOUBLE (IEEEDOUBLE, NUMBER, NUMERIC, FLOAT, FLOAT8) and an associated *VBA* BEFORE INSERT trigger
AUTONUMBER(*n*), *n* < 38 (for replication)	GUID and an associated *VBA* BEFORE INSERT trigger
BOOLEAN	BIT (YESNO, LOGICAL, LOGICAL1)
NAT(*n*), *n* < 3	BYTE (INTEGER1)
NAT(*n*), *n* ∈ {3, 4}	SHORT (SMALLINT, INTEGER2)
NAT(*n*), INT(*n*), *n* ∈ [5; 9]	LONG (INTEGER, INT, INTEGER4)
NAT(*n*), *n* ∈ [10; 38]	SINGLE (IEEESINGLE, REAL, FLOAT4)
NAT(*n*), *n* ∈ [39; 308]	DOUBLE (IEEEDOUBLE, NUMBER, NUMERIC, FLOAT, FLOAT8)
INT(*n*), *n* < 5	SHORT (SMALLINT, INTEGER2)
INT(*n*), *n* ∈ [10; 38]	SINGLE (IEEESINGLE, REAL, FLOAT4)
RAT(*n*, *m*), *n* < 39	SINGLE (IEEESINGLE, REAL, FLOAT4)
RAT(*n*, *m*), INT(*n*), CURRENCY(*n*), *n* ∈ [39; 308]	DOUBLE (IEEEDOUBLE, NUMBER, NUMERIC, FLOAT, FLOAT8)
CURRENCY(*n*), *n* < 15	CURRENCY (MONEY)
CURRENCY(*n*), *n* ∈ [15; 38]	SINGLE (IEEESINGLE, REAL, FLOAT4)
DATE/TIME (01/01/100; 31/12/9999)	DATETIME (DATE, TIME, TIMESTAMP)
ASCII/UNICODE(*n*), *n* < 256	TEXT (VARCHAR, CHAR, STRING, ALPHANUMERIC)
ASCII/UNICODE(*n*), *n* ∈ [256; 4,096]	LONGTEXT (LONGCHAR, MEMO, NOTE)
Static sets of strings (only one of them is stored per table cell)	Combo-box value list *Lookup*, with *Allow Multiple Values = No* (only through GUI or *VBA*)
Static sets of strings (none or any combination of them may be stored per table cell; not 1NF tables anymore)	Combo-box value list *Lookup*, with *Allow Multiple Values = Yes* (only through GUI or *VBA*)
Computed column	CALCULATED (only through GUI or *VBA*)

TABLE 4.12   Other *Access* Data Types

Data type	Access 2013 implementation
Binary objects of up to 510 bytes	`BINARY (VARBINARY)`
BLOBs (including OLE) up to 1 GB	`LONGBINARY (OLEOBJECT, GENERAL)`
File/URL/e-mail address	`HYPERLINK`
Embedded file lists	`ATTACHMENT`
List of various data type values	`MULTIPLE BYTE, DECIMAL, DOUBLE, GUID, INT16, INT32, SINGLE,` and `TEXT`

In the logical atoms of the form `expr [NOT] BETWEEN value1 AND value2`, *Access* also accepts `value1` values greater than `value2`.

Access accepts grouping and ordering by expressions too.

You can declare and use in *Access* `PARAMETERS` for queries.

Besides `SELECT`, you can also query in *Access* by using a `TRANSFORM` statement, which computes crosstabulations (just like with the MS *Excel* pivoting).

The predicate `DISTINCT` cannot be used with the *SQL* aggregate functions (e.g., instead of directly computing `SELECT SUM(DISTINCT c) FROM T;` you have first to define a query *Q*: `SELECT DISTINCT c AS e FROM T;` and then to compute the desired result on it, with `SELECT SUM(e) FROM Q;`).

There is no `LIMIT TO n ROWS` clause; you can use instead the `TOP n` predicate.

Check (tuple) constraints can be enforced only through GUI or/and *VBA* and *ADO*. Same thing applies for triggers and (explicit) transactions.

You cannot attach comments to *Access SQL* code.

4.3   CASE STUDY: IMPLEMENTING THE PUBLIC LIBRARY RDB INTO *DB2, ORACLE, MYSQL, SQL SERVER,* AND *ACCESS*

> *The world plays between a chaos full of symmetries and*
> *a cosmos full of asymmetries.*
> —Lucian Blaga

In fact, generally, almost nobody is using *SQL* for creating or populating tables, except for the temporary ones, as this is done much easier through the RDBMS GUIs. However, this section is needed not only for completeness, but also because, seldom enough, we are forced in this field to also dynamically create table schemes and manipulate their instances through *SQL*.

Please note that, in fact, all powerful RDBMSs like, for example, *DB2*, *Oracle*, *SQL Server*, are adding to the below requests much more other default settings for the actual, physical implementation of tables and indexes (also see the chapter on optimizations from the second volume of this book). For example, here is the complete *SQL* statement for the creation of the *BOOKS* table in *Oracle* (see subsection 4.3.1 for the concise statements needed in fact):

```
CREATE TABLE "LIBRARY_DB"."BOOKS"(
      "X" NUMBER(16,0) NOT NULL ENABLE,
   "FIRST_AUTHOR" NUMBER(6,0) NOT NULL ENABLE,
   "BTITLE" VARCHAR2(255 BYTE) NOT NULL ENABLE,
   "BYEAR" NUMBER(4,0),
   "BYEAR_IN_PAST_IND" VARCHAR2(1 BYTE) GENERATED
ALWAYS AS
      (CAST("LIBRARY_DB"."YEAR_IN_FUTURE" ("BYEAR")
AS VARCHAR2(1)))
      VIRTUAL VISIBLE,
   CONSTRAINT "BOOKS_PK" PRIMARY KEY ("X")
      USING INDEX PCTFREE 10 INITRANS 2 MAXTRANS 255
COMPUTE
      STATISTICS STORAGE(INITIAL 65536 NEXT 1048576
MINEXTENTS 1
      MAXEXTENTS 2147483645 PCTINCREASE 0 FREELISTS
1 FREELIST GROUPS
      1 BUFFER_POOL DEFAULT FLASH_CACHE DEFAULT
CELL_FLASH_CACHE
      DEFAULT) TABLESPACE "USERS" ENABLE,
   CONSTRAINT "BOOKS_UK1" UNIQUE ("FIRST_AUTHOR",
"BTITLE", "BYEAR")
```

```
        USING INDEX PCTFREE 10 INITRANS 2 MAXTRANS 255
COMPUTE
        STATISTICS STORAGE(INITIAL 65536 NEXT 1048576
MINEXTENTS 1
        MAXEXTENTS 2147483645 PCTINCREASE 0 FREELISTS
1 FREELIST GROUPS
        1  BUFFER_POOL  DEFAULT  FLASH_CACHE  DEFAULT
CELL_FLASH_CACHE
        DEFAULT) TABLESPACE "USERS" ENABLE,
  CONSTRAINT "BOOKS_CHK1" CHECK (BYEAR >= -2500)
ENABLE,
  CONSTRAINT "BOOKS_CHK2" CHECK (BYEAR_IN_PAST_IND
= 'Y') ENABLE,
  CONSTRAINT   "BOOKS_PERSONS_FK1"   FOREIGN   KEY
("FIRST_AUTHOR")
        REFERENCES "LIBRARY_DB"."PERSONS" ("X") ENABLE)
  SEGMENT CREATION IMMEDIATE PCTFREE 10 PCTUSED 40
INITRANS 1
        MAXTRANS 255 NOCOMPRESS LOGGING STORAGE(INITIAL
65536 NEXT
        1048576  MINEXTENTS  1  MAXEXTENTS  2147483645
PCTINCREASE 0
        FREELISTS  1  FREELIST  GROUPS  1  BUFFER_POOL
DEFAULT FLASH_CACHE
        DEFAULT CELL_FLASH_CACHE DEFAULT)
  TABLESPACE "USERS";
CREATE OR REPLACE TRIGGER "LIBRARY_DB"."BOOKS_AUTON"
before insert on BOOKS for each row
begin
  if (:new.X is null) then
      select  BOOKS_SEQ.nextval  into  :new.X  from
dual;
  end if;
end;
ALTER TRIGGER "LIBRARY_DB"."BOOKS_AUTON" ENABLE;
```

The subsections of this section are presenting how to create and populate rdbs corresponding to the public library relations from Fig. 3.13 (Subsection 3.4.2) in IBM *DB2 10*, Oracle *Database 12c* and *MySQL 5.7*, MS *SQL Server 2014* and *Access 2013*, by applying algorithm *A*8 and taking into consideration the peculiarities of these RDBMS versions described in Section 4.2.

Due to differences between these RDBMS versions, table and column names slightly differ: for example, because *Oracle* does not accept '-' in object names, table *CO-AUTHORS* has been renamed as *CO_AUTHORS* and because *Oracle* stores all such names only using uppercase, column *DueReturnDate* has been renamed as *DUE_RETURN_DATE*.

4.3.1   *IBM* DB2 10.5 *PUBLIC LIBRARY RDB*

> *If you only read the books that everyone else is reading,*
> *you can only think what everyone else is thinking.*
> —Haruki Murakami

```
CREATE DATABASE Library_DB;

-- DDL statements for creating the Library_DB
tables

CREATE TABLE PERSONS (
   x INT GENERATED ALWAYS AS IDENTITY,
   FName VARCHAR(128),
   LName VARCHAR(64) NOT NULL,
   "e-mail" VARCHAR(255),
   "Author?" BOOLEAN,
   PRIMARY KEY (x),
   CONSTRAINT PKey UNIQUE (FName, LName,
"e-mail"));

CREATE TABLE BOOKS (
```

```
  x INT GENERATED ALWAYS AS IDENTITY,
  First_Author INT NOT NULL,
  BTitle VARCHAR(255) NOT NULL,
  BYear INT NULL CHECK(BYear BETWEEN -5000 AND
  Year(CURRENT DATE)),
  PRIMARY KEY (x),
  CONSTRAINT BKey UNIQUE (First_Author, BTitle,
  BYear),
  CONSTRAINT fkFA FOREIGN KEY (First_Author)
  REFERENCES PERSONS (x));

CREATE TABLE CO_AUTHORS (
  x INT GENERATED ALWAYS AS IDENTITY,
  Co_Author INT NOT NULL,
  Book INT NOT NULL,
Pos_In_List SMALLINT NOT NULL CHECK
  (Pos_In_List > 1 AND Pos_In_List < 17),
  PRIMARY KEY (x),
  CONSTRAINT CAAKey UNIQUE (Co_Author, Book),
  CONSTRAINT CAPKey UNIQUE (Book, Pos_In_List),
CONSTRAINT fkP FOREIGN KEY (Co_Author)
REFERENCES PERSONS (x),
  CONSTRAINT fkB FOREIGN KEY (Book) REFERENCES
  BOOKS (x));

CREATE TABLE PUBLISHERS (
  x SMALLINT GENERATED ALWAYS AS IDENTITY,
  Pub_Name VARCHAR(128) NOT NULL UNIQUE,
  PRIMARY KEY (x));

CREATE TABLE EDITIONS (
  x INT GENERATED ALWAYS AS IDENTITY,
  Publisher INT NOT NULL,
  First_Book INT NOT NULL,
  ETitle VARCHAR(255) NOT NULL,
  EYear INT CHECK (EYear BETWEEN -2500 AND
```

```
    Year(CURRENT DATE)),
    PRIMARY KEY (x),
    CONSTRAINT fkPub FOREIGN KEY (Publisher)
    REFERENCES PUBLISHERS,
CONSTRAINT fkBk FOREIGN KEY (First_Book)
    REFERENCES BOOKS,
    CONSTRAINT EKey UNIQUE (First_Book, Publisher,
    EYear));

CREATE TABLE VOLUMES (
    x INT GENERATED ALWAYS AS IDENTITY,
    Edition INT NOT NULL,
    VNo SMALLINT NOT NULL CHECK
    (VNo BETWEEN 1 AND 255),
    VTitle VARCHAR(255),
    ISBN VARCHAR(16) UNIQUE,
    Price DECIMAL(8,2) NOT NULL CHECK
    (Price BETWEEN 0 AND 100000),
    PRIMARY KEY (x),
CONSTRAINT fkE FOREIGN KEY (Edition)
    REFERENCES EDITIONS,
    CONSTRAINT VNKey UNIQUE (Edition, VNo),
    CONSTRAINT VTKey UNIQUE (Edition, VTitle));

CREATE TABLE VOLUMES_CONTENTS (
    x INT GENERATED ALWAYS AS IDENTITY,
    Volume INT NOT NULL,
    Book INT NOT NULL,
    Book_Pos SMALLINT NOT NULL CHECK
    (Book_Pos BETWEEN 1 AND 16),
    PRIMARY KEY (x),
    CONSTRAINT fkV FOREIGN KEY (Volume)
    REFERENCES VOLUMES,
    CONSTRAINT fkVB FOREIGN KEY (Book)
    REFERENCES BOOKS,
    CONSTRAINT VCBKey UNIQUE (Volume, Book),
```

```
   CONSTRAINT VCPKey UNIQUE (Volume, Book_Pos));
CREATE TABLE COPIES (
   x INT GENERATED ALWAYS AS IDENTITY,
   Volume INT NOT NULL,
   Inv_No VARCHAR(32) NOT NULL UNIQUE,
   PRIMARY KEY (x),
   CONSTRAINT fkS FOREIGN KEY (Volume)
   REFERENCES VOLUMES);

CREATE TABLE BORROWS (
   x BIGINT GENERATED ALWAYS AS IDENTITY,
   Subscriber INT NOT NULL,
   Borrow_Date DATE NOT NULL CHECK (Borrow_Date
      BETWEEN '6/1/2000' AND CURRENT DATE),
   PRIMARY KEY (x),
   CONSTRAINT fkSubscriber FOREIGN KEY(Subscriber)
   REFERENCES PERSONS,
   CONSTRAINT BRKey UNIQUE (Borrow_Date,
      Subscriber));

CREATE TABLE BORROWS_LISTS (
   x BIGINT GENERATED ALWAYS AS IDENTITY,
   Borrow BIGINT NOT NULL,
   Copy INT NOT NULL,
   Due_Return_Date DATE NOT NULL CHECK (
   Due_Return_Date BETWEEN '6/1/2000' AND
   CURRENT DATE + 300 DAYS),
   Actual_Return_Date DATE CHECK (
   Actual_Return_Date BETWEEN '6/1/2000' AND
   CURRENT DATE),
   PRIMARY KEY (x),
   CONSTRAINT fkBw FOREIGN KEY (Borrow)
      REFERENCES BORROWS,
   CONSTRAINT fkCy FOREIGN KEY (Copy)
   REFERENCES COPIES,
   CONSTRAINT BLKey UNIQUE (Copy, Borrow));
```

```
-- DML statements for populating the Library_DB db

INSERT INTO PERSONS (FName, LName, "e-mail",
"Author?")
VALUES (NULL, 'Homer', NULL, true),
  ('William', 'Shakespeare', NULL, true),
  ('Peter', 'Buneman', 'opb@inf.ed.ac.uk',
true),
  ('Serge', 'Abiteboul', NULL, true),
  ('Dan', 'Suciu', 'suciu@washington.edu',
true);

INSERT INTO BOOKS(First_Author, BTitle, BYear)
VALUES (1, 'Odyssey', -700),
  (2, 'As You Like It', 1600),
  (4, 'Data on the Web: From Relations to
  Semi-structured Data and XML', 1999);

INSERT INTO CO_AUTHORS(Book, Co_Author,
                          Pos_In_List)
VALUES (3, 3, 2), (3, 5, 3);

INSERT INTO PUBLISHERS(Pub_Name)
VALUES ('Apple Academic Press'),
  ('Springer Verlag'),
  ('Morgan Kaufmann'), ('Penguin Books'),
  ('Washington Square Press');

INSERT INTO EDITIONS(Publisher, First_Book,
                       ETitle, EYear) VALUES
  (4, 1, 'The Odyssey translated by Robert
  Fagles', 2012),
  (5, 2, 'As You Like It', 2011),
  (3, 3, 'Data on the Web: From Relations to
  Semi-structured Data and XML', 1999);
```

```
INSERT INTO VOLUMES(Edition, VNo, VTitle, ISBN,
Price)
VALUES (1, 1, NULL, '0-670-82162-4', 12.95),
   (2, 1, NULL, '978-1613821114', 9.99),
   (3, 1, NULL, '978-1558606227', 74.95);

INSERT INTO VOLUMES_CONTENTS(Volume, Book,
                                   Book_Pos)
VALUES (1, 1, 1), (2, 2, 1), (3, 3, 1);

INSERT INTO COPIES(Inv_No, Volume) VALUES ('H-O-
1', 1),
   ('H-O-2', 1), ('S-AYL1-1', 2), ('ABS-W-1', 3),
   ('ABS-W-2', 3);

INSERT INTO BORROWS(Subscriber, Borrow_Date)
VALUES (5, '10/29/2012');

INSERT INTO BORROWS_LISTS(Borrow, Copy,
   Due_Return_Date, Actual_Return_Date)
VALUES (1, 1, '11/29/2012', '11/23/2012'),
   (1, 3, '12/29/2012', NULL);
```

4.3.2 ORACLE DATABASE 12C PUBLIC LIBRARY RDB

> *I find television very educating.*
> *Every time somebody turns on the set, I go into the other room and*
> *read a book.*
> —Groucho Marx

Note that, as no desired tablespace is asked, the default user one (generally called USERS) is used by *Oracle* to create tables and needed indexes into. As 12c is, for the time being, not so widely spread, the following code does not use its newest autonumber feature, but the "classic" trigger method needed for all of its previous versions.

```
-- DDL statements for creating the Library_DB
tables
CREATE SEQUENCE PERSONS_SEQ MINVALUE 1 MAXVALUE
999999
   INCREMENT BY 1 NOCYCLE;
CREATE SEQUENCE BOOKS_SEQ MINVALUE 1 MAXVALUE
999999
   INCREMENT BY 1 NOCYCLE;
CREATE SEQUENCE CO_AUTHORS_SEQ MINVALUE 1
MAXVALUE 9999999 INCREMENT BY 1 NOCYCLE;
CREATE SEQUENCE PUBLISHERS_SEQ MINVALUE 1
MAXVALUE 9999 INCREMENT BY 1 NOCYCLE;
CREATE SEQUENCE EDITIONS_SEQ MINVALUE 1
MAXVALUE 999999 INCREMENT BY 1 NOCYCLE;
CREATE SEQUENCE VOLUMES_SEQ MINVALUE 1
MAXVALUE 9999999 INCREMENT BY 1 NOCYCLE;
CREATE SEQUENCE VOLUMES_CONTENTS_SEQ MINVALUE 1
MAXVALUE 9999999 INCREMENT BY 1 NOCYCLE;
CREATE SEQUENCE COPIES_SEQ MINVALUE 1
MAXVALUE 99999999 INCREMENT BY 1 NOCYCLE;
CREATE SEQUENCE BORROWS_SEQ MINVALUE 1
MAXVALUE 99999999999 INCREMENT BY 1 NOCYCLE;
CREATE SEQUENCE BORROWS_LISTS_SEQ MINVALUE 1
MAXVALUE 999999999999 INCREMENT BY 1 NOCYCLE;

CREATE FUNCTION Year_in_future (year NUMBER)
RETURN VARCHAR2 DETERMINISTIC IS
  begin
    return case when year > extract(year from
           sysdate) then 'N' else 'Y' end;
  end Year_in_future;

CREATE FUNCTION Day_in_future (day varchar2)
RETURN VARCHAR2 DETERMINISTIC IS
```

```
  begin
return case when to_date(day, 'MM/DD/YYYY') >
     sysdate then 'N' else 'Y' end;
  end Day_in_future;
CREATE FUNCTION Day_in_far_future (day varchar2)
RETURN VARCHAR2
DETERMINISTIC IS
  begin
    return case when to_date(day, 'MM/DD/YYYY') >
sysdate + 300
                then 'N' else 'Y'
           end;
  end Day_in_far_future;

CREATE TABLE PERSONS (
  X NUMBER(6,0) PRIMARY KEY,
  FNAME VARCHAR2(128 BYTE),
  LNAME VARCHAR2(64 BYTE) NOT NULL,
  E_MAIL VARCHAR2(255 BYTE),
  CONSTRAINT PKEY UNIQUE (E_MAIL, FNAME,
LNAME));
CREATE TRIGGER PERSONS_AUTON before insert on
PERSONS for each row
  begin
    if (:new.X is null) then
      select PERSONS_SEQ.nextval into :new.X from
dual;
    end if;
  end;

CREATE TABLE BOOKS (
  X NUMBER(16,0) PRIMARY KEY,
  FIRST_AUTHOR NUMBER(6,0) NOT NULL,
  BTITLE VARCHAR2(255 BYTE) NOT NULL,
  BYEAR NUMBER(4,0) CHECK (BYEAR >= -5000),
```

```
    CONSTRAINT BKEY UNIQUE (FIRST_AUTHOR, BTITLE,
BYEAR),
    CONSTRAINT FK_FA FOREIGN KEY (FIRST_AUTHOR)
REFERENCES PERSONS);

ALTER TABLE BOOKS ADD (BYEAR_IN_PAST_IND AS
    (cast(Year_in_future(BYEAR) AS VARCHAR2(1))));
ALTER TABLE BOOKS ADD CONSTRAINT BOOKS_CHK2 CHECK
    (BYEAR_IN_PAST_IND = 'Y');

CREATE TRIGGER BOOKS_AUTON before insert on BOOKS
for each row
  begin
    if (:new.X is null) then
      select BOOKS_SEQ.nextval into :new.X from
dual;
    end if;
  end;

CREATE TABLE CO_AUTHORS (
  X NUMBER(7,0) PRIMARY KEY,
  POS_IN_LIST NUMBER(2,0) NOT NULL CHECK (
    POS_IN_LIST BETWEEN 2 AND 16),
  CO_AUTHOR NUMBER(6,0) NOT NULL,
  BOOK NUMBER(6,0) NOT NULL,
  CONSTRAINT CO_A_KEY UNIQUE (BOOK, CO_AUTHOR),
  CONSTRAINT FK_BOOK FOREIGN KEY (BOOK)
      REFERENCES BOOKS,
 CONSTRAINT FK_CO_AUTHOR FOREIGN KEY(CO_AUTHOR)
REFERENCES PERSONS);

CREATE TRIGGER CO_AUTHORS_AUTON before insert on
CO_AUTHORS
    for each row
```

```
  begin
    if (:new.X is null) then
        select CO_AUTHORS_SEQ.nextval into :new.X
from dual;
    end if;
  end;

CREATE TABLE PUBLISHERS (
  X NUMBER(4,0) PRIMARY KEY,
  PUB_NAME VARCHAR2(128 BYTE) NOT NULL UNIQUE);
CREATE TRIGGER PUBLISHERS_AUTON before insert on
PUBLISHERS
  for each row
  begin
    if (:new.X is null) then
      select PUBLISHERS_SEQ.nextval into :new.X from
dual;
    end if;
  end;

CREATE TABLE EDITIONS (
  X NUMBER(7,0) PRIMARY KEY,
  PUBLISHER NUMBER(4,0) NOT NULL,
  FIRST_BOOK NUMBER(6,0) NOT NULL,
  ETITLE VARCHAR2(255 BYTE) NOT NULL,
  EYEAR NUMBER(4,0) CHECK (EYEAR >= -2500),
 CONSTRAINT E_KEY UNIQUE (PUBLISHER, FIRST_BOOK,
EYEAR),
 CONSTRAINT FK_FIRST_BOOK FOREIGN KEY
  (FIRST_BOOK) REFERENCES BOOKS,
 CONSTRAINT FK_PUBLISHER FOREIGN KEY (PUBLISHER)
REFERENCES PUBLISHERS);

ALTER TABLE EDITIONS ADD (EYEAR_IN_PAST_IND AS
  (cast(Year_in_future(EYEAR) AS VARCHAR2(1))));
```

```
ALTER TABLE EDITIONS ADD CONSTRAINT EDITIONS_CHK2
CHECK
  (EYEAR_IN_PAST_IND = 'Y');

CREATE TRIGGER EDITIONS_AUTON before insert on
EDITIONS for each row
  begin
    if (:new.X is null) then
      select EDITIONS_SEQ.nextval into :new.X from
dual;
    end if;
  end;

CREATE TABLE VOLUMES (
  X NUMBER(7,0) PRIMARY KEY,
  EDITION NUMBER(6,0) NOT NULL,
  VNO NUMBER(3,0) NOT NULL CHECK (VNO BETWEEN 1
AND 255),
  VTITLE VARCHAR2(255 BYTE),
  ISBN VARCHAR2(16 BYTE) UNIQUE,
  PRICE NUMBER(8,2) NOT NULL CHECK (Price
BETWEEN 0 AND 100000),
  CONSTRAINT FK_EDITION FOREIGN KEY (EDITION)
    REFERENCES EDITIONS);

CREATE TRIGGER VOLUMES_AUTON before insert on
VOLUMES for each row
  begin
    if (:new.X is null) then
      select VOLUMES_SEQ.nextval into :new.X from
dual;
    end if;
  end;

CREATE TABLE VOLUMES_CONTENTS (
  X NUMBER(7,0) PRIMARY KEY,
```

```
   VOLUME NUMBER(7,0) NOT NULL,
   BOOK NUMBER(6,0) NOT NULL,
   BOOK_POS NUMBER(3,0) NOT NULL CHECK (BOOK_POS
BETWEEN 1 AND 16),
   CONSTRAINT VC_KEY UNIQUE (VOLUME, BOOK),
   CONSTRAINT VOLUMES_FK_BOOKS FOREIGN KEY (BOOK)
REFERENCES BOOKS,
   CONSTRAINT FK_VOLUME FOREIGN KEY (VOLUME)
       REFERENCES VOLUMES);

CREATE TRIGGER VOLUMES_CONTENTS_AUTON before
insert on
   VOLUMES_CONTENTS for each row
   begin
     if (:new.X is null) then
       select VOLUMES_CONTENTS_SEQ.nextval into
:new.X from dual;
     end if;
   end;

CREATE TABLE COPIES (
   X NUMBER(8,0) PRIMARY KEY,
   INV_NO VARCHAR2(32 BYTE) NOT NULL UNIQUE,
   VOLUME NUMBER(7,0) NOT NULL,
  CONSTRAINT FK_COPIES_VOLUME FOREIGN KEY (VOLUME)
REFERENCES VOLUMES);

CREATE TRIGGER COPIES_AUTON before insert on
       COPIES for each row
   begin
     if (:new.X is null) then
       select COPIES_SEQ.nextval into :new.X from
dual;
     end if;
   end;
```

```
CREATE TABLE BORROWS (
  X NUMBER(11,0) PRIMARY KEY,
  SUBSCRIBER NUMBER(6,0) NOT NULL,
  BORROW_DATE DATE NOT NULL CHECK (BORROW_DATE >=
          TO_DATE('6/1/2000', 'MM/DD/YYYY')),
  CONSTRAINT BORROWS_KEY UNIQUE (BORROW_DATE,
SUBSCRIBER),
  CONSTRAINT FK_SUBSCRIBER FOREIGN KEY
      (SUBSCRIBER) REFERENCES PERSONS);

CREATE TRIGGER BORROWS_AUTON before insert on
BORROWS for each row
  begin
    if (:new.X is null) then
      select BORROWS_SEQ.nextval into :new.X from
dual;
    end if;
  end;

ALTER TABLE BORROWS ADD (BDATE_IN_PAST_IND AS
  (cast(Day_in_future(TO_CHAR(BORROW_DATE, 'MM/
DD/YYY')) AS VARCHAR2(1))));

ALTER TABLE BORROWS ADD CONSTRAINT BORROWS_CHK2
      CHECK (BDATE_IN_PAST_IND = 'Y');

CREATE TABLE BORROWS_LISTS (
  X NUMBER(12,0) PRIMARY KEY,
  BORROW NUMBER(11,0) NOT NULL,
  COPY NUMBER(8,0) NOT NULL,
  DUE_RETURN_DATE DATE NOT NULL CHECK (
  DUE_RETURN_DATE >= TO_DATE('6/1/2000',
      'MM/DD/YYYY')),
  ACTUAL_RETURN_DATE DATE CHECK (
  ACTUAL_RETURN_DATE >= TO_DATE('6/1/2000',
      'MM/DD/YYYY')),
```

```
   CONSTRAINT BLKEY UNIQUE (BORROW, COPY),
   CONSTRAINT FK_BORROW FOREIGN KEY (BORROW)
                    REFERENCES BORROWS,
   CONSTRAINT FK_COPY FOREIGN KEY (COPY)
                    REFERENCES COPIES);

ALTER TABLE BORROWS_LISTS ADD (ARDATE_IN_PAST_IND
AS
   (cast(Day_in_future(To_CHAR(ACTUAL_RETURN_DATE,
   'MM/DD/YYY')) AS VARCHAR2(1))));

ALTER TABLE BORROWS_LISTS ADD CONSTRAINT
           BORROWS_LISTS_CHK3 CHECK
   (ARDATE_IN_PAST_IND = 'Y');

ALTER TABLE BORROWS_LISTS ADD (DRDATE_IN_PAST_IND
AS
   (cast(Day_in_far_future(To_CHAR(DUE_RETURN_DATE,
   'MM/DD/YYY')) AS VARCHAR2(1))));

ALTER TABLE BORROWS_LISTS ADD CONSTRAINT
           BORROWS_LISTS_CHK4 CHECK
   (DRDATE_IN_PAST_IND = 'Y');

CREATE TRIGGER BORROWS_LISTS_AUTON before insert
on BORROWS_LISTS
   for each row
   begin
    if (:new.X is null) then
      select BORROWS_LISTS_SEQ.nextval into :new.X
from dual;
    end if;
   end;

-- DML statements for populating the Library_DB db
```

```
INSERT INTO PERSONS(LName) VALUES ('Homer');
INSERT INTO PERSONS(FName, LName) VALUES (
            'William', 'Shakespeare');
INSERT INTO PERSONS(FName, LName, e_mail) VALUES
   ('Peter', 'Buneman', 'opb@inf.ed.ac.uk');
INSERT INTO PERSONS(FName, LName) VALUES
('Serge', 'Abiteboul');
INSERT INTO PERSONS(FName, LName, e_mail) VALUES
   ('Dan', 'Suciu', 'suciu@washington.edu');

INSERT INTO BOOKS(First_Author, BTitle, BYear)
   VALUES (1, 'Odyssey', -700);
INSERT INTO BOOKS(First_Author, BTitle, BYear)
   VALUES (2, 'As You Like It', 1600);
INSERT INTO BOOKS(First_Author, BTitle, BYear)
   VALUES (4, 'Data on the Web: From Relations to
Semi-structured Data and XML', 1999);

INSERT INTO CO_AUTHORS(Book, Co_Author,
                        Pos_In_List)
   VALUES (3, 3, 2);
INSERT INTO CO_AUTHORS(Book, Co_Author,
                        Pos_In_List)
   VALUES (3, 5, 3);

INSERT INTO PUBLISHERS(Pub_Name) VALUES
                ('Apple Academic Press');
INSERT INTO PUBLISHERS(Pub_Name) VALUES
('Springer Verlag');
INSERT INTO PUBLISHERS(Pub_Name) VALUES (
                    'Morgan Kaufmann');
INSERT INTO PUBLISHERS(Pub_Name) VALUES (
                    'Penguin Books');
INSERT INTO PUBLISHERS(Pub_Name) VALUES (
            'Washington Square Press');
```

```
INSERT INTO EDITIONS(Publisher, First_Book,
                            ETitle, EYear)
  VALUES (4, 1, 'The Odyssey translated by
Robert Fagles', 2012);
INSERT INTO EDITIONS(Publisher, First_Book,
                            ETitle, EYear)
  VALUES (5, 2, 'As You Like It', 2011);
INSERT INTO EDITIONS(Publisher, First_Book,
                            ETitle, EYear)
  VALUES (3, 3, 'Data on the Web: From Relations
to Semi-structured Data and XML', 1999);

INSERT INTO VOLUMES(Edition, VNo, VTitle, ISBN,
Price) VALUES
  (1, 1, NULL, '0-670-82162-4', 12.95);
INSERT INTO VOLUMES(Edition, VNo, VTitle, ISBN,
Price) VALUES
  (2, 1, NULL, '978-1613821114', 9.99);
INSERT INTO VOLUMES(Edition, VNo, VTitle, ISBN,
Price) VALUES
  (3, 1, NULL, '978-1558606227', 74.95);

INSERT INTO VOLUMES_CONTENTS(Volume, Book,
  Book_Pos) VALUES
  (1, 1, 1);
INSERT INTO VOLUMES_CONTENTS(Volume, Book,
  Book_Pos) VALUES
  (2, 2, 1);
INSERT INTO VOLUMES_CONTENTS(Volume, Book,
  Book_Pos) VALUES
  (3, 3, 1);

INSERT INTO COPIES(Inv_No, Volume) VALUES (
                            'H-O-1', 1);
INSERT INTO COPIES(Inv_No, Volume) VALUES (
                            'H-O-2', 1);
```

```
INSERT INTO COPIES(Inv_No, Volume) VALUES (
                            'S-AYL1-1', 2);
INSERT INTO COPIES(Inv_No, Volume) VALUES (
                            'ABS-W-1', 3);
INSERT INTO COPIES(Inv_No, Volume) VALUES (
                            'ABS-W-2', 3);


INSERT INTO BORROWS(Subscriber, Borrow_Date) VALUES
   (5, TO_DATE('10/29/2012', 'MM/DD/YYYY'));
INSERT INTO BORROWS_LISTS(Borrow, Copy,
   Due_Return_Date, Actual_Return_Date)
   VALUES (1, 1, TO_DATE('11/29/2012', 'MM/DD/
YYYY'), TO_DATE('11/23/2012', 'MM/DD/YYYY'));
INSERT INTO BORROWS_LISTS(Borrow, Copy,
   Due_Return_Date) VALUES
   (1, 3, TO_DATE('12/29/2012', 'MM/DD/YYYY'));
```

4.3.3 ORACLE MYSQL 5.7 PUBLIC LIBRARY RDB

> *There is no friend as loyal as a book.*
> —Ernest Hemingway

```
CREATE DATABASE LibraryDB;
-- DDL statements for creating the LibraryDB
tables
CREATE TABLE PERSONS (
   x MEDIUMINT AUTO_INCREMENT PRIMARY KEY,
   FName VARCHAR(128),
   LName VARCHAR(64) NOT NULL,
   'e-mail' VARCHAR(255),
   'Author?' BOOLEAN,
 UNIQUE INDEX PKey (FName, LName, 'e-mail'));

CREATE TABLE BOOKS (
   x MEDIUMINT AUTO_INCREMENT PRIMARY KEY,
```

```
  FirstAuthor MEDIUMINT NOT NULL,
  BTitle VARCHAR(255) NOT NULL,
  BYear INT,
 UNIQUE INDEX BKey (FirstAuthor, BTitle, BYear),
 CONSTRAINT fkFA FOREIGN KEY (FirstAuthor)
      REFERENCES PERSONS (x));

DELIMITER |
CREATE FUNCTION checkBYear(y INTEGER, message
VARCHAR(256))
  RETURNS INTEGER DETERMINISTIC
  BEGIN
    IF NOT y IS NULL AND (y < -5000 OR
                          y > EXTRACT(Year FROM
CurDate())) THEN
      SIGNAL SQLSTATE 'ERR0R'
      SET MESSAGE_TEXT := message;
    END IF;
    RETURN y;
  END;
|
DELIMITER $$
CREATE TRIGGER CHCKI_BYear BEFORE INSERT ON BOOKS
FOR EACH ROW
  BEGIN
    SET @dummy := checkBYear(NEW.Byear,
CONCAT(NEW.Byear, ' is not a year between -5000
and ', EXTRACT(Year FROM
CurDate()),'!'));
  END;
$$

DELIMITER $$
CREATE TRIGGER CHCKU_BYear BEFORE UPDATE ON BOOKS
FOR EACH ROW
  BEGIN
```

```
    SET @dummy := checkBYear(NEW.Byear,
CONCAT(NEW.Byear, ' is not a year between -5000
and ', EXTRACT(Year FROM CurDate()),'!'));
  END;
$$

CREATE TABLE CO_AUTHORS (
  x MEDIUMINT AUTO_INCREMENT PRIMARY KEY,
  Co_Author MEDIUMINT NOT NULL,
  Book MEDIUMINT NOT NULL,
  PosInList TINYINT UNSIGNED NOT NULL,
 UNIQUE INDEX CAAKey (Co_Author, Book),
 UNIQUE INDEX CAPKey (Book, PosInList),
 CONSTRAINT fkP FOREIGN KEY (Co_Author)
      REFERENCES PERSONS (x),
 CONSTRAINT fkB FOREIGN KEY (Book) REFERENCES
BOOKS (x));

DELIMITER |
CREATE FUNCTION checkPosInList(p INTEGER, message
VARCHAR(256))
  RETURNS INTEGER DETERMINISTIC
  BEGIN
    IF p < 2 OR p > 16 THEN
      SIGNAL SQLSTATE 'ERROR';
      SET MESSAGE_TEXT := message;
    END IF;
    RETURN p;
  END;
|

DELIMITER $$
CREATE TRIGGER CHCKI_PosInList BEFORE INSERT ON
CO_AUTHORS
  FOR EACH ROW
  BEGIN
```

```
  SET @dummy := checkPosInList(NEW.PosInList,
     CONCAT(NEW.PosInList, ` is not a position
between 2 and 16!'));
  END;
$$


DELIMITER $$
CREATE TRIGGER CHCKU_PosInList BEFORE UPDATE ON
CO_AUTHORS
  FOR EACH ROW
  BEGIN
   SET @dummy := checkPosInList(NEW.PosInList,
    CONCAT(NEW.PosInList, ` is not a position
  between 2 and 16!'));
  END;
$$


CREATE TABLE PUBLISHERS (
  x SMALLINT AUTO_INCREMENT PRIMARY KEY,
  PubName VARCHAR(128) NOT NULL UNIQUE);

CREATE TABLE EDITIONS (
x INT AUTO_INCREMENT PRIMARY KEY,
Publisher SMALLINT NOT NULL,
FirstBook MEDIUMINT NOT NULL,
ETitle VARCHAR(255) NOT NULL,
EYear INT,
 CONSTRAINT fkPub FOREIGN KEY (Publisher)
         REFERENCES PUBLISHERS(x),
 CONSTRAINT fkBk FOREIGN KEY (FirstBook)
         REFERENCES BOOKS(x),
 UNIQUE INDEX EKey (FirstBook, Publisher,
EYear));
DELIMITER |
```

```
CREATE FUNCTION checkEYear(y INTEGER, message
VARCHAR(256))
  RETURNS INTEGER DETERMINISTIC
  BEGIN
   IF NOT y IS NULL AND (y < -2500 OR
                            y > EXTRACT(Year FROM
CurDate())) THEN
      SIGNAL SQLSTATE 'ERROR';
      SET MESSAGE_TEXT := message;
   END IF;
   RETURN y;
  END;
|
DELIMITER $$
CREATE TRIGGER CHCKI_EYear BEFORE INSERT ON
                      EDITIONS FOR EACH ROW
  BEGIN
   SET @dummy := checkEYear(NEW.Eyear,
CONCAT(NEW.Eyear, ' is not a year between -2500
and ', EXTRACT(Year FROM CurDate()),'!'));
  END;
$$

DELIMITER $$
CREATE TRIGGER CHCKU_EYear BEFORE UPDATE ON
                      EDITIONS FOR EACH ROW
  BEGIN
   SET @dummy:= checkBYear(NEW.Eyear, CONCAT(NEW.
Eyear, ' is not a year between -2500 and ',
EXTRACT(Year FROM CurDate()),'!'));
  END;
$$

CREATE TABLE VOLUMES (
x INT AUTO_INCREMENT PRIMARY KEY,
Edition INT NOT NULL,
VNo TINYINT UNSIGNED NOT NULL,
```

```
VTitle VARCHAR(255),
ISBN VARCHAR(16) UNIQUE,
Price DECIMAL(8,2) NOT NULL,
 CONSTRAINT fkE FOREIGN KEY (Edition)
                REFERENCES EDITIONS(x),
 UNIQUE INDEX (Edition, VNo),
 UNIQUE INDEX (Edition, VTitle));

DELIMITER |
CREATE FUNCTION checkVNo_Price(n INTEGER, p
DECIMAL, messageN VARCHAR(256),
          messageP VARCHAR(256))
  RETURNS INTEGER DETERMINISTIC
  BEGIN
    IF n < 1 THEN
      SET @r := 1;
      SIGNAL SQLSTATE 'ERROR';
      SET MESSAGE_TEXT := messageN;
    ELSEIF p < 0 OR p > 100000 THEN
      SET @r := 2;
      SIGNAL SQLSTATE 'ERROR';
      SET MESSAGE_TEXT := messageP;
    ELSE
      SET @r := 0;
    END IF;
    RETURN @r;
  END;
|

DELIMITER $$
CREATE TRIGGER CHCKI_VNo_Price BEFORE INSERT ON
VOLUMES FOR EACH ROW
  BEGIN
  SET @dummy := checkVNo_Price(NEW.VNo, NEW.
Price, CONCAT(NEW.VNo, ' is not a volume number
```

```
greater than 0!'), CONCAT(NEW.Price, ' is not a
price between 0 and 100,000!'));
   END;
$$


DELIMITER $$
CREATE TRIGGER CHCKU_VNo_Price BEFORE UPDATE ON
VOLUMES FOR EACH ROW
   BEGIN
   SET @dummy := checkVNo_Price(NEW.VNo, NEW.
Price, CONCAT(NEW.VNo, ' is not a volume number
greater than 0!'),
       CONCAT(NEW.Price, ' is not a price between
0 and 100,000!'));
   END;
$$


CREATE TABLE VOLUMES_CONTENTS (
x INT AUTO_INCREMENT PRIMARY KEY,
Volume INT NOT NULL,
Book MEDIUMINT NOT NULL,
BookPos TINYINT UNSIGNED NOT NULL,
CONSTRAINT fkV FOREIGN KEY (Volume)
REFERENCES VOLUMES(x),
 CONSTRAINT fkVB FOREIGN KEY (Book)
REFERENCES BOOKS(x),
 UNIQUE INDEX (Volume, Book),
 UNIQUE INDEX (Volume, BookPos));
DELIMITER |


CREATE FUNCTION checkBookPos(p INTEGER, message
VARCHAR(256))
   RETURNS INTEGER DETERMINISTIC
   BEGIN
    IF p < 1 OR p > 16 THEN
      SIGNAL SQLSTATE 'ERROR';
```

```
      SET MESSAGE_TEXT := message;
    END IF;
    RETURN p;
  END;
|


DELIMITER $$
CREATE TRIGGER CHCKI_BookPos BEFORE INSERT ON
VOLUMES_CONTENTS
  FOR EACH ROW
  BEGIN
    SET @dummy := checkBookPos(NEW.BookPos,
        CONCAT(NEW.BookPos, ' is not a position
between 1 and 16!'));
  END;
$$


DELIMITER $$
CREATE TRIGGER CHCKU_BookPos BEFORE UPDATE ON
VOLUMES_CONTENTS
  FOR EACH ROW
  BEGIN
    SET @dummy: = checkBookPos(NEW.BookPos,
        CONCAT(NEW.BookPos, ' is not a position
between 1 and 16!'));
  END;
$$


CREATE TABLE COPIES (
x INT UNSIGNED AUTO_INCREMENT PRIMARY KEY,
Volume INT NOT NULL,
InvNo VARCHAR(32) NOT NULL UNIQUE,
CONSTRAINT fkS FOREIGN KEY (Volume)
REFERENCES VOLUMES(x));


CREATE TABLE BORROWS (
```

```
x BIGINT UNSIGNED AUTO_INCREMENT PRIMARY KEY,
Subscriber MEDIUMINT NOT NULL,
BorrowDate DATE NOT NULL,
 CONSTRAINT fkSubscriber FOREIGN KEY (Subscriber)
REFERENCES
   PERSONS(x),
 UNIQUE INDEX (BorrowDate, Subscriber));
DELIMITER |

CREATE FUNCTION checkBorrowDate(d DATE, message
VARCHAR(256))
  RETURNS DATE DETERMINISTIC
  BEGIN
   IF d < '2000-01-06' OR d > CurDate() THEN
     SIGNAL SQLSTATE 'ERROR';
     SET MESSAGE_TEXT := message;
   END IF;
   RETURN d;
END;
|
DELIMITER $$

CREATE TRIGGER CHCKI_BorrowDate BEFORE INSERT ON
BORROWS
  FOR EACH ROW
  BEGIN
   SET @dummy := checkBorrowDate(NEW.BorrowDate,
   CONCAT(NEW.BorrowDate, ' is not a date
between 2000-01-06 and ', CurDate(),'!'));
  END;
$$
DELIMITER $$

CREATE TRIGGER CHCKU_BorrowDate BEFORE UPDATE ON
BORROWS
  FOR EACH ROW
```

```
  BEGIN
    SET @dummy := checkBorrowDate(NEW.BorrowDate,
    CONCAT(NEW.BorrowDate, ' is not a date
between 2000-01-06 and ', CurDate(),'!'));
  END;
$$

CREATE TABLE BORROWS_LISTS (
x BIGINT UNSIGNED AUTO_INCREMENT PRIMARY KEY,
Borrow BIGINT UNSIGNED NOT NULL,
Copy INT UNSIGNED NOT NULL,
DueReturnDate DATE NOT NULL,
ActualReturnDate DATE,
 CONSTRAINT fkBw FOREIGN KEY (Borrow)
REFERENCES BORROWS(x),
 CONSTRAINT fkCy FOREIGN KEY (Copy)
REFERENCES COPIES(x),
 UNIQUE INDEX (Copy, Borrow));
DELIMITER |

CREATE FUNCTION checkDue_ActualReturnDates(d
DATE, a DATE, messageD VARCHAR(256), messageA
VARCHAR(256))
  RETURNS INTEGER DETERMINISTIC
  BEGIN
    IF d < '2000-01-06' OR d > DATE_ADD(CurDate(),
INTERVAL 300 DAY)
    THEN
      SET @r := 1;
      SIGNAL SQLSTATE 'ERROR';
      SET MESSAGE_TEXT := messageD;
    ELSEIF a < '2000-01-06' OR a > CurDate() THEN
      SET @r := 2;
      SIGNAL SQLSTATE 'ERROR';
      SET MESSAGE_TEXT := messageA;
    ELSE
```

```
      SET @r := 0;
   END IF;
   RETURN @r;
  END;
|


DELIMITER $$
CREATE TRIGGER CHCKI_Due_ActualReturnDates BEFORE
INSERT ON BORROWS_LISTS FOR EACH ROW
  BEGIN
    SET @dummy := checkDue_ActualReturnDates(NEW.
DueReturnDate,
NEW.ActualReturnDate,
CONCAT(NEW.DueReturnDate,
' is not a date between 2000-01-06 and ',
DATE_ADD(CurDate(),
INTERVAL 300 DAY), '!'),
CONCAT(NEW.ActualReturnDate,
'is not a date between 2000-01-06 and
', CurDate(),'!'));
  END;
$$
DELIMITER $$

CREATE TRIGGER CHCKU_Due_ActualReturnDates BEFORE
UPDATE
  ON BORROWS_LISTS FOR EACH ROW
  BEGIN
    SET @dummy := checkDue_ActualReturnDates(NEW.
DueReturnDate,
NEW.ActualReturnDate,
CONCAT(NEW.DueReturnDate,
' is not a date between 2000-01-06 and ',
DATE_ADD(CurDate(),
INTERVAL 300 DAY), '!'),
CONCAT(NEW.ActualReturnDate,
```

```
` is not a date between 2000-01-06 and `,
CurDate(),'!'));
  END;
$$


-- DML statements for populating the LibraryDB db

INSERT INTO PERSONS (FName, LName, "e-mail",
"Author?")
VALUES (NULL, 'Homer', NULL, true),
  ('William', 'Shakespeare', NULL, true),
  ('Peter', 'Buneman', 'opb@inf.ed.ac.uk',
true),
  ('Serge', 'Abiteboul', NULL, true),
  ('Dan', 'Suciu', 'suciu@washington.edu',
true);

INSERT INTO BOOKS(First_Author, BTitle, BYear)
VALUES (1, 'Odyssey', -700),
  (2, 'As You Like It', 1600),
  (4, 'Data on the Web: From Relations to
  Semi-structured Data and XML', 1999);

INSERT INTO CO_AUTHORS(Book, Co_Author,
                            Pos_In_List)
VALUES (3, 3, 2), (3, 5, 3);

INSERT INTO PUBLISHERS(Pub_Name)
VALUES ('Apple Academic Press'), ('Springer
Verlag'),
  ('Morgan Kaufmann'), ('Penguin Books'),
  ('Washington Square Press');

INSERT INTO EDITIONS(Publisher, First_Book,
                        ETitle, EYear) VALUES
  (4, 1, 'The Odyssey translated by Robert
Fagles', 2012),
```

```
   (5, 2, 'As You Like It', 2011),
   (3, 3, 'Data on the Web: From Relations to
Semi-structured Data and XML', 1999);

INSERT INTO VOLUMES(Edition, VNo, VTitle, ISBN,
Price)
VALUES (1, 1, NULL, '0-670-82162-4', 12.95),
   (2, 1, NULL, '978-1613821114', 9.99),
   (3, 1, NULL, '978-1558606227', 74.95);

INSERT INTO VOLUMES_CONTENTS(Volume, Book,
                                   Book_Pos)
VALUES (1, 1, 1), (2, 2, 1), (3, 3, 1);

INSERT INTO COPIES(Inv_No, Volume) VALUES ('H-O-
1', 1), ('H-O-2', 1), ('S-AYL1-1', 2), ('ABS-W-
1', 3), ('ABS-W-2', 3);

INSERT INTO BORROWS(Subscriber, Borrow_Date)
VALUES (5, '10/29/2012');

INSERT INTO BORROWS_LISTS(Borrow, Copy,
      Due_Return_Date, Actual_Return_Date)
VALUES (1, 1, '11/29/2012', '11/23/2012'),
   (1, 3, '12/29/2012', NULL);
```

4.3.4 MS SQL SERVER 2014 *PUBLIC LIBRARY RDB*

> *Books are a uniquely portable magic.*
> —Stephen King

```
CREATE DATABASE LibraryDB;

-- DDL statements for creating the Library_DB tables

CREATE TABLE PERSONS (
  x INT IDENTITY PRIMARY KEY,
```

```
   FName VARCHAR(128),
   LName VARCHAR(64) NOT NULL,
   [e-mail] VARCHAR(255),
   [Author?] BIT,
   CONSTRAINT PKey UNIQUE ([e-mail], LName,
FName));

CREATE TABLE BOOKS (
x INT IDENTITY PRIMARY KEY,
FirstAuthor INT NOT NULL,
BTitle VARCHAR(255) NOT NULL,
BYear INT CHECK (BYEAR BETWEEN -5000 AND
Year(SysDateTime())),
CONSTRAINT fkFA FOREIGN KEY (FirstAuthor)
     REFERENCES PERSONS,
CONSTRAINT BKey UNIQUE (BTitle, FirstAuthor,
BYear));

CREATE TABLE [CO-AUTHORS] (
x INT IDENTITY PRIMARY KEY,
[Co-Author] INT NOT NULL,
Book INT NOT NULL,
PosInList TINYINT NOT NULL
CHECK (PosInList BETWEEN 2 AND 16),
CONSTRAINT fkCA FOREIGN KEY ([Co-Author])
     REFERENCES PERSONS,
CONSTRAINT fkB FOREIGN KEY (Book)
     REFERENCES BOOKS,
CONSTRAINT CAAKey UNIQUE (Book, [Co-Author]),
CONSTRAINT CAPKey UNIQUE (Book, PosInList));

CREATE TABLE PUBLISHERS (
x INT IDENTITY PRIMARY KEY,
PubName VARCHAR(128) NOT NULL UNIQUE);

CREATE TABLE EDITIONS (
x INT IDENTITY PRIMARY KEY,
```

```
Publisher INT NOT NULL,
FirstBook INT NOT NULL,
ETitle VARCHAR(255) NOT NULL,
EYear INT CHECK (EYEAR BETWEEN -2500 AND
Year(SysDateTime())),
CONSTRAINT fkPub FOREIGN KEY (Publisher)
     REFERENCES PUBLISHERS,
CONSTRAINT fkBk FOREIGN KEY (FirstBook)
     REFERENCES BOOKS,
CONSTRAINT EKey UNIQUE (FirstBook, Publisher,
EYear));

CREATE TABLE VOLUMES (
x INT IDENTITY PRIMARY KEY,
Edition INT NOT NULL,
VNo TINYINT NOT NULL CHECK (VNo BETWEEN 1 AND
255),
VTitle VARCHAR(255),
ISBN VARCHAR(16) UNIQUE,
Price DECIMAL(8,2) NOT NULL CHECK (Price BETWEEN
0 AND 100000),
CONSTRAINT fkE FOREIGN KEY (Edition)
REFERENCES EDITIONS,
  CONSTRAINT VNKey UNIQUE (Edition, VNo),
  CONSTRAINT VTKey UNIQUE (Edition, VTitle));
CREATE TABLE VOLUMES_CONTENTS (
x INT IDENTITY PRIMARY KEY,
Volume INT NOT NULL,
Book INT NOT NULL,
BookPos TINYINT NOT NULL CHECK (BookPos BETWEEN 1
AND 255),
CONSTRAINT fkV FOREIGN KEY (Volume)
REFERENCES VOLUMES,
CONSTRAINT fkVB FOREIGN KEY (Book)
REFERENCES BOOKS,
CONSTRAINT VCBKey UNIQUE (Volume, Book),
```

```
  CONSTRAINT VCPKey UNIQUE (Volume, BookPos));
CREATE TABLE COPIES (
x INT IDENTITY PRIMARY KEY,
Volume INT NOT NULL,
InvNo VARCHAR(32) NOT NULL UNIQUE,
CONSTRAINT fkS FOREIGN KEY (Volume)
REFERENCES VOLUMES);
CREATE TABLE BORROWS (
x BIGINT IDENTITY PRIMARY KEY,
Subscriber INT NOT NULL,
BorrowDate DATETIME NOT NULL CHECK (BorrowDate
BETWEEN
'6/1/2000' AND SysDateTime()),
CONSTRAINT fkSubscriber FOREIGN KEY
(Subscriber) REFERENCES
PERSONS,
CONSTRAINT BRKey UNIQUE (BorrowDate,
Subscriber));
CREATE TABLE BORROWS_LISTS (
x BIGINT IDENTITY PRIMARY KEY,
Borrow BIGINT NOT NULL,
Copy INT NOT NULL,
DueReturnDate DATETIME NOT NULL CHECK
  (DueReturnDate BETWEEN '6/1/2000' AND
  DATEADD (DAY, 300, SysDateTime())),
ActualReturnDate DATETIME CHECK (ActualReturnDate
BETWEEN
'6/1/2000' AND SysDateTime()),
  CONSTRAINT fkBw FOREIGN KEY (Borrow) REFERENCES
BORROWS,
  CONSTRAINT fkCy FOREIGN KEY (Copy) REFERENCES
COPIES,
  CONSTRAINT BLKey UNIQUE (Copy, Borrow));

-- DML statements for populating the LibraryDB db
```

```
INSERT INTO PERSONS(LName) VALUES ('Homer');
INSERT INTO PERSONS(FName, LName) VALUES
('William', 'Shakespeare');
INSERT INTO PERSONS(FName, LName, [e-mail])
VALUES ('Peter', 'Buneman', 'opb@inf.ed.ac.uk');
INSERT INTO PERSONS(FName, LName) VALUES
('Serge', 'Abiteboul');
INSERT INTO PERSONS(FName, LName, [e-mail])
VALUES ('Dan', 'Suciu', 'suciu@washington.edu');
INSERT INTO BOOKS(FirstAuthor, BTitle, BYear)
VALUES (1, 'Odyssey', -700);
INSERT INTO BOOKS(FirstAuthor, BTitle, BYear)
VALUES (2, 'As You Like It', 1600);
INSERT INTO BOOKS(FirstAuthor, BTitle, BYear)
VALUES (4, 'Data on
      the Web: From Relations to Semi-structured
Data and XML', 1999);
INSERT INTO [CO-AUTHORS](Book, [Co-Author],
PosInList)
VALUES (3, 3, 2);
INSERT INTO [CO-AUTHORS](Book, [Co-Author],
PosInList)
VALUES (3, 5, 3);
INSERT INTO PUBLISHERS(PubName) VALUES ('Apple
Academic Press');
INSERT INTO PUBLISHERS(PubName) VALUES ('Springer
Verlag');
INSERT INTO PUBLISHERS(PubName) VALUES ('Morgan
Kaufmann');
INSERT INTO PUBLISHERS(PubName) VALUES ('Penguin
Books');
INSERT INTO PUBLISHERS(PubName) VALUES ('Washing-
ton Square Press');
INSERT INTO EDITIONS(Publisher, FirstBook,
                            ETitle, EYear)
```

```
VALUES (4, 1, 'The Odyssey translated by Robert
Fagles', 2012);
INSERT INTO EDITIONS(Publisher, FirstBook,
                          ETitle, EYear)
VALUES (5, 2, 'As You Like It', 2011);
INSERT INTO EDITIONS(Publisher, FirstBook,
                          ETitle, EYear)
VALUES (3, 3, 'Data on the Web: From Relations to
Semi-structured Data and XML', 1999);
INSERT INTO VOLUMES(Edition, VNo, VTitle, ISBN,
Price)
VALUES (1, 1, NULL, '0-670-82162-4', 12.95);
INSERT INTO VOLUMES(Edition, VNo, VTitle, ISBN,
Price)
VALUES (2, 1, NULL, '978-1613821114', 9.99);
INSERT INTO VOLUMES(Edition, VNo, VTitle, ISBN,
Price)
VALUES (3, 1, NULL, '978-1558606227', 74.95);
INSERT INTO VOLUMES_CONTENTS(Volume, Book,
                                BookPos)
VALUES (1, 1, 1);
INSERT INTO VOLUMES_CONTENTS(Volume, Book,
                                BookPos)
VALUES (2, 2, 1);
INSERT INTO VOLUMES_CONTENTS(Volume, Book,
                                BookPos)
VALUES (3, 3, 1);
INSERT INTO COPIES(InvNo, Volume) VALUES ('H-O-
1', 1);
INSERT INTO COPIES(InvNo, Volume) VALUES ('H-O-
2', 1);
INSERT INTO COPIES(InvNo, Volume) VALUES ('S-
AYL1-1', 2);
INSERT INTO COPIES(InvNo, Volume) VALUES ('ABS-W-
1', 3);
```

```
INSERT INTO COPIES(InvNo, Volume) VALUES ('ABS-W-
2', 3);
INSERT INTO BORROWS(Subscriber, BorrowDate)
VALUES (5, '10/29/2012');
INSERT INTO BORROWS_LISTS(Borrow, Copy,
                            DueReturnDate,
ActualReturnDate) VALUES (1, 1, '11/29/2012',
'11/23/2012');
INSERT INTO BORROWS_LISTS(Borrow, Copy,
                            DueReturnDate)
VALUES (1, 3, '12/29/2012');
```

4.3.5   MS ACCESS 2013 *PUBLIC LIBRARY RDB*

> *If one cannot enjoy reading a book over and over again,*
> *there is no use in reading it at all.*
> —Oscar Wilde

Especially as *Access* is not translating in *SQL* for its users the schemas of the tables it manages, in order to create tables through *SQL* you need to save any `CREATE TABLE` statement in a separate query and to run then each of them.

For example, in order to create table *PERSONS*, you should click on the *Query Design* icon of the *Create* ribbon menu, then click on the *Close* button of the *Show Table* pop-up window (without selecting any table or view), then click on the *SQL View* icon of the *File* ribbon menu, type the corresponding *DDL SQL* statement in the newly open *Query1* tab (see Fig. 4.2), then close it and save it (changing its default *Query1* name in, for example, *createBooks*), and, finally, executing this *DDL* query by double-clicking on its name.

Here is the *DDL SQL* needed for creating all the tables of the *LibraryDB* db:

```
CREATE TABLE PERSONS (
x COUNTER PRIMARY KEY,
FName VARCHAR(128),
```

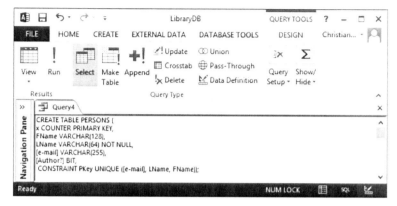

FIGURE 4.2   Creating a *SQL DDL* query in *Access*.

```
LName VARCHAR(64) NOT NULL,
[e-mail] VARCHAR(255),
[Author?] BIT,
  CONSTRAINT PKey UNIQUE ([e-mail], LName, FName));

CREATE TABLE BOOKS (
x COUNTER PRIMARY KEY,
FirstAuthor LONG NOT NULL,
BTitle VARCHAR(255) NOT NULL,
BYear INT,
CONSTRAINT fkFA FOREIGN KEY (FirstAuthor)
                REFERENCES PERSONS,
   CONSTRAINT BKey UNIQUE (BTitle, FirstAuthor,
BYear));

CREATE TABLE [CO-AUTHORS] (
x COUNTER PRIMARY KEY,
[Co-Author] LONG NOT NULL,
Book LONG NOT NULL,
PosInList BYTE NOT NULL,
CONSTRAINT fkP FOREIGN KEY ([Co-Author])
                REFERENCES PERSONS,
  CONSTRAINT fkB FOREIGN KEY (Book) REFERENCES BOOKS,
```

```
     CONSTRAINT CAAKey UNIQUE (Book, [Co-Author]),
     CONSTRAINT CAPKey UNIQUE (Book, PosInList));

CREATE TABLE PUBLISHERS (
x COUNTER PRIMARY KEY,
PubName VARCHAR(128) NOT NULL UNIQUE);

CREATE TABLE EDITIONS (
x COUNTER PRIMARY KEY,
Publisher LONG NOT NULL,
FirstBook LONG NOT NULL,
ETitle VARCHAR(255) NOT NULL,
EYear INT,
CONSTRAINT fkPub FOREIGN KEY (Publisher)
                REFERENCES PUBLISHERS,
CONSTRAINT fkBk FOREIGN KEY (FirstBook)
                REFERENCES BOOKS,
   CONSTRAINT EKey UNIQUE (FirstBook, Publisher,
EYear));

CREATE TABLE VOLUMES (
x COUNTER PRIMARY KEY,
Edition LONG NOT NULL,
VNo BYTE NOT NULL,
VTitle VARCHAR(255),
ISBN VARCHAR(16) UNIQUE,
Price CURRENCY NOT NULL,
CONSTRAINT fkE FOREIGN KEY (Edition)
      REFERENCES EDITIONS,
   CONSTRAINT VNKey UNIQUE (Edition, VNo),
   CONSTRAINT VTKey UNIQUE (Edition, VTitle));
CREATE TABLE VOLUMES_CONTENTS (
x COUNTER PRIMARY KEY,
Volume LONG NOT NULL,
Book LONG NOT NULL,
```

```
BookPos BYTE NOT NULL,
CONSTRAINT fkV FOREIGN KEY (Volume)
REFERENCES VOLUMES,
CONSTRAINT fkVB FOREIGN KEY (Book)
REFERENCES BOOKS,
  CONSTRAINT VCBKey UNIQUE (Volume, Book),
  CONSTRAINT VCPKey UNIQUE (Volume, BookPos));

CREATE TABLE COPIES (
x COUNTER PRIMARY KEY,
Volume LONG NOT NULL,
InvNo VARCHAR(32) NOT NULL UNIQUE,
CONSTRAINT fkS FOREIGN KEY (Volume)
REFERENCES VOLUMES);

CREATE TABLE BORROWS (
x COUNTER PRIMARY KEY,
Subscriber LONG NOT NULL,
BorrowDate DATE NOT NULL,
CONSTRAINT fkS FOREIGN KEY (Subscriber)
              REFERENCES PERSONS,
CONSTRAINT BRKey UNIQUE (BorrowDate,
                        Subscriber));

CREATE TABLE BORROWS_LISTS (
x COUNTER PRIMARY KEY,
Borrow LONG NOT NULL,
Copy LONG NOT NULL,
DueReturnDate DATE NOT NULL,
ActualReturnDate DATE,
CONSTRAINT fkBw FOREIGN KEY (Borrow)
REFERENCES BORROWS,
CONSTRAINT fkCy FOREIGN KEY (Copy)
REFERENCES COPIES,
  CONSTRAINT BLKey UNIQUE (Copy, Borrow));
```

The easiest way to add the check (tuple) and domain constraints is through the *Access* GUI, by using the *Validation Rule* and *Validation Text* properties of the corresponding table and columns, respectively (in the corresponding tables' *Design View*). Figures 4.3–4.11 show how to enforce all the *LibraryDB* domain constraints.

Here is the *DML SQL* needed for populating the tables (that should be stored one by one in distinct queries, just like for the *DDL* ones (see Fig. 4.2), as *Access* is only manipulating single statement *SQL* scripts):

```
INSERT INTO PERSONS(LName) VALUES ("Homer");
INSERT INTO PERSONS(FName, LName) VALUES
("William", "Shakespeare");
INSERT INTO PERSONS(FName, LName, [e-mail])
VALUES ("Peter", "Buneman", "opb@inf.ed.ac.uk");
INSERT INTO PERSONS(FName, LName) VALUES
("Serge", "Abiteboul");
INSERT INTO PERSONS(FName, LName, [e-mail])
VALUES ('Dan', 'Suciu', 'suciu@washington.edu');
```

FIGURE 4.3 Adding *BYear*'s domain constraint through Access GUI.

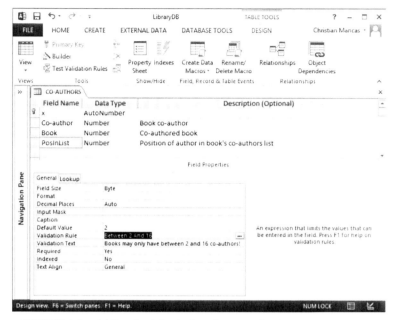

FIGURE 4.4   Adding *PosInList*'s domain constraint through *Access* GUI.

FIGURE 4.5   Adding *EYear*'s domain constraint through *Access* GUI.

FIGURE 4.6   Adding *VNo*'s domain constraint through *Access* GUI.

FIGURE 4.7   Adding *Price*'s domain constraint through *Access* GUI.

FIGURE 4.8 Adding *BookPos*' domain constraint through *Access* GUI.

FIGURE 4.9 Adding *BorrowDate*'s domain constraint through *Access* GUI.

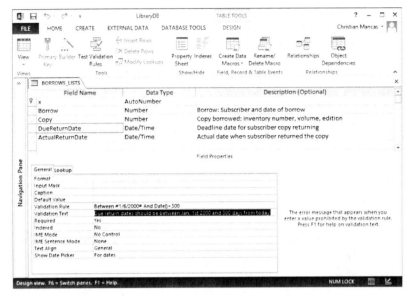

FIGURE 4.10   Adding *DueReturnDate*'s domain constraint through *Access* GUI.

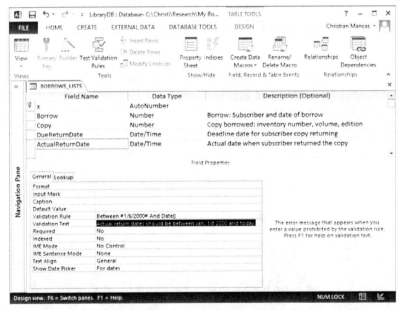

FIGURE 4.11   Adding *ActualReturnDate*'s domain constraint through *Access* GUI.

```
INSERT INTO BOOKS(FirstAuthor, BTitle, BYear)
VALUES (1, "Odyssey", -700);

INSERT INTO BOOKS(FirstAuthor, BTitle, BYear)
VALUES (2, "As You Like It", 1600);
INSERT INTO BOOKS(FirstAuthor, BTitle, BYear)
VALUES (4, "Data on the Web: From Relations to
Semi-structured
    Data and XML", 1999);

INSERT INTO [CO-AUTHORS](Book, [Co-Author],
PosInList)
VALUES (3, 3, 2);
INSERT INTO [CO-AUTHORS](Book, [Co-Author],
PosInList)
VALUES (3, 5, 3);

INSERT INTO PUBLISHERS(PubName) VALUES ("Apple
Academic Press");
INSERT INTO PUBLISHERS(PubName) VALUES ("Springer
Verlag");
INSERT INTO PUBLISHERS(PubName) VALUES ("Morgan
Kaufmann");
INSERT INTO PUBLISHERS(PubName) VALUES ("Penguin
Books");
INSERT INTO PUBLISHERS(PubName) VALUES
("Washington Square Press");

INSERT INTO EDITIONS(Publisher, FirstBook,
                        ETitle, EYear)
VALUES (4, 1, "The Odyssey translated by Robert
Fagles", 2012);
INSERT INTO EDITIONS(Publisher, FirstBook,
                        ETitle, EYear)
VALUES (5, 2, "As You Like It", 2011);
INSERT INTO EDITIONS(Publisher, FirstBook,
                        ETitle, EYear)
```

```
VALUES (3, 3, "Data on the Web: From Relations to
Semi-structured Data and XML", 1999);

INSERT INTO VOLUMES(Edition, VNo, VTitle, ISBN,
Price)
VALUES (1, 1, NULL, "0-670-82162-4", 12.95);
INSERT INTO VOLUMES(Edition, VNo, VTitle, ISBN,
Price)
VALUES (2, 1, NULL, "978-1613821114", 9.99);
INSERT INTO VOLUMES(Edition, VNo, VTitle, ISBN,
Price)
VALUES (3, 1, NULL, "978-1558606227", 74.95);

INSERT INTO VOLUMES_CONTENTS(Volume, Book,
                                  BookPos)
VALUES (1, 1, 1);
INSERT INTO VOLUMES_CONTENTS(Volume, Book,
                                  BookPos)
VALUES (2, 2, 1);
INSERT INTO VOLUMES_CONTENTS(Volume, Book,
                                  BookPos)
VALUES (3, 3, 1);

INSERT INTO COPIES(InvNo, Volume) VALUES ("H-O-
1", 1);
INSERT INTO COPIES(InvNo, Volume) VALUES ("H-O-
2", 1);
INSERT INTO COPIES(InvNo, Volume) VALUES ("S-
AYL1-1", 2);
INSERT INTO COPIES(InvNo, Volume) VALUES ("ABS-W-
1", 3);
INSERT INTO COPIES(InvNo, Volume) VALUES ("ABS-W-
2", 3);

INSERT INTO BORROWS(Subscriber, BorrowDate)
VALUES (5, #10/29/2012#);
```

```
INSERT INTO BORROWS_LISTS(Borrow, Copy,
     DueReturnDate, ActualReturnDate) VALUES
(1, 1, #11/29/2012#, #11/23/2012#);
INSERT INTO BORROWS_LISTS(Borrow, Copy,
     DueReturnDate) VALUES (1, 3, #12/29/2012#);
```

4.4 THE REVERSE ENGINEERING ALGORITHM FOR TRANSLATING *ACCESS 2013* RDB SCHEMAS INTO *SQL* ANSI *DDL* SCRIPTS (*REA2013A0*), A MEMBER OF *REAF0'*

Let's reduce to the minimum our fixations!
—Lucian Blaga

There are very few exceptions among RDBMSs that do not provide facilities for generating *SQL DDL* scripts to create the tables of the dbs they are managing; the only one among the five ones considered in this book is MS *Access*.

This is why we chose *Access 2013* as an example for designing algorithms of the reverse engineering family *REAF0'*. Obviously, as, for example, all other four RDBMS versions considered in this book have *DDL* GUI interfaces too, the algorithm presented in this subsection (let's call it *REA2013A0*, although it also works for many previous *Access* versions) may be an example for designing own similar algorithms for them too, although it is trivially simpler to first use their facilities for generating *DDL* scripts in their own *SQL* idioms and then apply much simpler algorithms to translate these scripts into ANSI standard ones.

Not only because it is a reverse engineering algorithm, but especially in order to first gain experience, we are starting with how to actually do it, before summarizing what has generally to be done in the corresponding algorithm.

In order to simplify things as much as possible, the figures that follow are captured from the *Access 2013 LibraryDB* rdb (see Subsection 4.3.5).

For opening an *Access* db, just double-click on its name; note that if you only have a *Runtime Access* version, you can only manipulate *Access* created db applications and data (if any): you need a full *Access* license in order to also modify db schemas and applications of such dbs.

When double-clicking the *LibraryDB.accdb* file on a computer having a full *Access 2013* license installed, the window shown in Fig. 4.12 is opened. In the db *Navigation* left pane, under the *All Tables* group, you see all the tables belonging to the corresponding db. In order to open the scheme of anyone of them, right-click on its name and then click on the *Design View* option; then, in order to get more screen space, you can minimize the *Navigation* left pane by clicking on its upper rightmost button (the "«" one; to reopen anytime this pane, just click on the same button, which, when this pane is minimized, displays "»").

Figure 4.13 shows the scheme of an *Access* table: its name (in this case *BORROWS_LISTS*) is displayed in the newly opened tab header; its column names are displayed in the *Field Name* column; their corresponding data types are shown in the middle *Data Type* column; the third *Description* (*optional*) column is intended to store the semantics associated to each column.

The columns that are making up the primary key of the table (if any) have to their left a yellow small key icon: in Fig. 4.13, this is displayed only to the left of the *x* column.

You can see the data subtypes and the domain constraints associated to each column in the *General* left subpane under the scheme table; for example, from Fig. 4.13, you can see that the `Autonumber` data type is by default a subtype of the `LONG` one and, if you open the corresponding *Field Size* combo-box, that the only other alternative for it is the `Replication ID` one (see Fig. 4.14).

If you open the *New Values* combo-box, you can see that *Access* generates autonumbers either incrementally (which is the default) or randomly (see Fig. 4.15). From the *Access* documentation, you can see that both the starting and the increment values for its incremental autonumbers are 1.

Finally for single primary keys, just like for any other single key, you can see in Fig. 4.13 that the *Indexed* property is set to *Yes* (*No Duplicates*), which means that a unique index is built on this column and enforces its one-to-oneness.

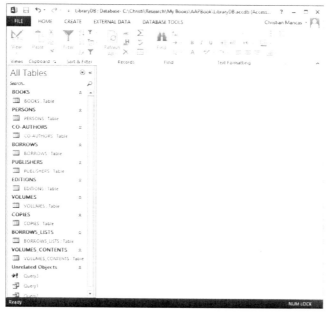

FIGURE 4.12 *Access 2013 Navigation All Tables* pane.

FIGURE 4.13 The scheme (*Design View*) of an *Access 2013* table.

Up to this point, after consulting Tables 4.1 and 4.9 above (where we can see that LONG is NAT(9) in *Access*), we may infer that algorithm *REA2013A0* should output for this table the following *DDL* statement:

```
CREATE TABLE BORROWS_LISTS (
  x INTEGER(9) GENERATED ALWAYS AS IDENTITY(1,1)
  PRIMARY KEY
);
```

If the cursor is then moved on the second (*Borrow*) line of the table from Fig. 4.3, we can see (Fig. 4.16) that this column is a LONG that does not accept null values (because the *Required* property is set to *Yes*). Consequently, the output of *REA2013A0* should now be the following:

```
CREATE TABLE BORROWS_LISTS (
  x INTEGER(9) GENERATED ALWAYS AS IDENTITY(1,1)
  PRIMARY KEY,
  Borrow INTEGER(9) NOT NULL
);
```

When moving the cursor on the third line (*Copy*, see Fig. 4.17), we can see that this column is similar to the previous one, so that the output of *REA2013A0* should now be the following:

```
CREATE TABLE BORROWS_LISTS (
  x INTEGER(9) GENERATED ALWAYS AS IDENTITY(1,1)
  PRIMARY KEY,
  Borrow INTEGER(9) NOT NULL,
  Copy INTEGER(9) NOT NULL
);
```

When moving the cursor on the fourth line (*DueReturnDate*, see Fig. 4.18), we can see that this column is a DATE/TIME one that does not accept nulls either and which has an associated domain constraint, as its *Validation Rule* property is not empty.

FIGURE 4.14 The two types of *Access* autonumbers (LONG and REPLICATION).

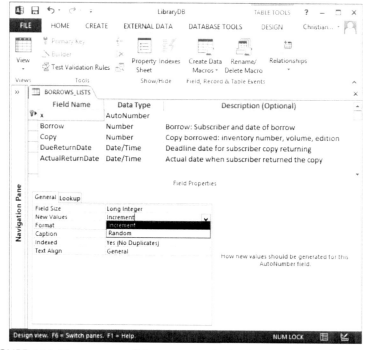

FIGURE 4.15. The two types of *Access* autonumbering (Increment and Random).

FIGURE 4.16   An example of *Access* LONG required column.

FIGURE 4.17   Another example of *Access* LONG required column.

FIGURE 4.18   An example of *Access* domain constraint (`Validation Rule`).

Consequently, with the same *SysDate()* convention used throughout this book[111] and also knowing that in *Access* any integer added to a `DATE` value is expressed in days, the output of *REA2013A*0 should now be the following:

```
CREATE TABLE BORROWS_LISTS (
x  INTEGER(9)  GENERATED  ALWAYS  AS  IDENTITY(1,1)
PRIMARY KEY,
Borrow INTEGER(9) NOT NULL,
Copy INTEGER(9) NOT NULL,
DueReturnDate DATE NOT NULL CHECK (DueReturnDate
BETWEEN
  '01/06/2000' AND SysDate() + 300 DAYS)
);
```

When moving the cursor on the fifth line (*ActualReturnDate*, see Figure 4.19), we can see that this column is similar to the fourth one, except that it accepts nulls (as its *Required* property is set to *No*) and that it has

---

[111] Unfortunately, there is no standard at least for the most frequently needed functions in dbs, although this would be of great help for everybody.

another associated domain constraint; the output of *REA2013A0* should now be the following:

```
CREATE TABLE BORROWS_LISTS (
  x INTEGER(9) GENERATED ALWAYS AS IDENTITY(1,1)
  PRIMARY KEY,
  Borrow INTEGER(9) NOT NULL,
  Copy INTEGER(9) NOT NULL,
  DueReturnDate DATE NOT NULL CHECK (DueReturnDate
  BETWEEN
     '01/06/2000' AND SysDate() + 300 DAYS),
  ActualReturnDate DATE CHECK (ActualReturnDate
  BETWEEN
     '01/06/2000' AND SysDate())
);
```

As inspection of all table's columns is finished, we can proceed by examining the semantic (candidate, unique) keys of the table; in *Access*, they can be browsed (and/or modified) in the popup window that opens

FIGURE 4.19   Another example of an *Access* domain constraint.

when clicking on the *Indexes* icon of the ribbon (see Fig. 4.20). Note that all indexes are defined in this window, but only those that have the property *Unique* set to *Yes* are enforcing keys (the other ones are of no interest in this first volume: they will be discussed in the fourth chapter of the second one).

In this example, there are only two indexes, named *blKey* and *PrimaryKey* (according to the *Index Name* column from Fig. 4.20), both enforcing keys, the first one being made out of the columns *Borrow* and *Copy* (in this order), and the second one, which corresponds to the primary key of the table, only of the column *x* (according to the column *Field Name*).

Consequently, the output of *REA2013A0* should now be the following:

```
CREATE TABLE BORROWS_LISTS (
  x INTEGER(9) GENERATED ALWAYS AS IDENTITY(1,1)
  PRIMARY KEY,
  Borrow INTEGER(9) NOT NULL,
  Copy INTEGER(9) NOT NULL,
  DueReturnDate DATE NOT NULL CHECK (DueReturnDate BETWEEN
    '01/06/2000' AND SysDate() + 300 DAYS),
  ActualReturnDate DATE CHECK (ActualReturnDate BETWEEN
    '01/06/2000' AND SysDate()),
    CONSTRAINT blKey UNIQUE (Borrow, Copy)
);
```

As inspection of all table keys is done, you should close the *Indexes* windows. Next step is to discover the referential integrity constraints associated with this table: the simplest way to start with is to browse the instance of the *Access* metacatalog table *MSysRelationships*, which stores all the referential integrity constraints enforced in the db.

By default, all metacatalog tables are hidden; in order to be able to see in the *Navigation* pane the system tables too, right-click the pane (e.g., to the right of its *All Tables* header) and then click on *Navigation Options*.

Figure 4.21 shows the corresponding popup window that opens and in which you should check the *Show System Objects* check-box (bottom left). When you then click on its *OK* button (bottom right), this window closes

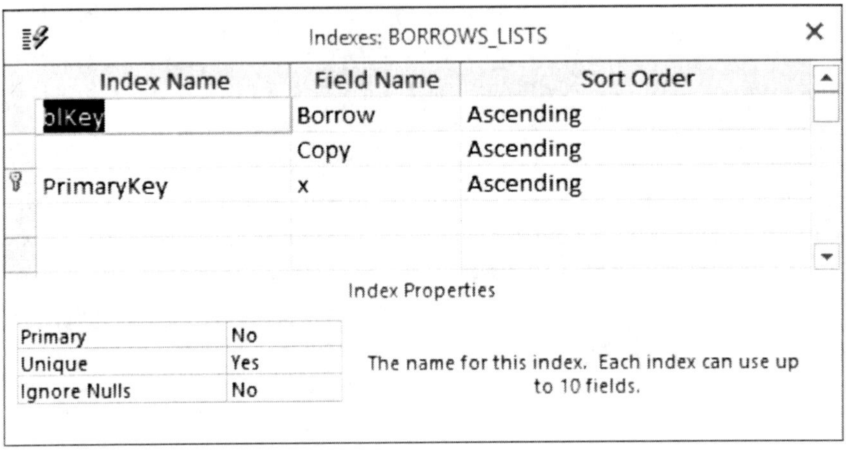

Indexes: BORROWS_LISTS			
Index Name	**Field Name**	**Sort Order**	
olKey	Borrow	Ascending	
	Copy	Ascending	
PrimaryKey	x	Ascending	

Index Properties

Primary	No	
Unique	Yes	The name for this index. Each index can use up
Ignore Nulls	No	to 10 fields.

FIGURE 4.20   An example of Access (unique) keys (and indexes).

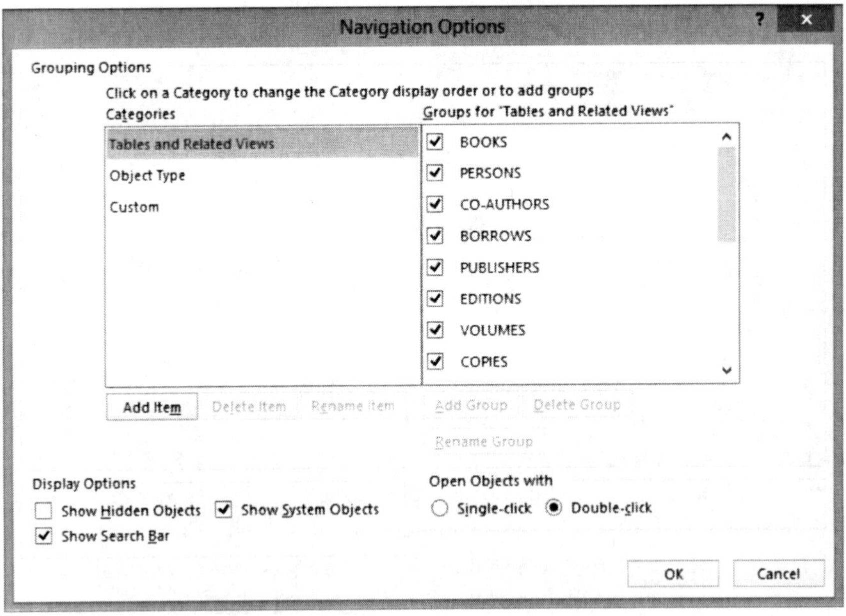

Navigation Options

Grouping Options

Click on a Category to change the Category display order or to add groups

Categories	Groups for 'Tables and Related Views'
Tables and Related Views	☑ BOOKS
Object Type	☑ PERSONS
Custom	☑ CO-AUTHORS
	☑ BORROWS
	☑ PUBLISHERS
	☑ EDITIONS
	☑ VOLUMES
	☑ COPIES

Add Item   Delete Item   Rename Item   Add Group   Delete Group

Rename Group

Display Options

☐ Show Hidden Objects   ☑ Show System Objects

☑ Show Search Bar

Open Objects with

○ Single-click   ● Double-click

OK   Cancel

FIGURE 4.21   *Access Navigation Options* window.

and you can see in the *Navigation* pane the system tables too (with their icons and names, prefixed by *MSys*, slightly dimmed, see Fig. 4.22).

When you double-click on the *MSysRelationships* table icon, its instance opens (see Fig. 4.23); right-click on the name of the *szObject* column and then click on the *Sort A to Z* option; for better viewing the table instance, double-click the dividing lines between columns to size them according to their content.

Column *szRelationship* stores the names of the corresponding referential integrity constraints; column *grbit* stores their subtypes (4352 for ON UPDATE CASCADE and ON DELETE CASCADE; 4096 for ON UPDATE RESTRICT and ON DELETE CASCADE; 256 for ON UPDATE CAS-CADE and ON DELETE RESTRICT; and 0 for the default ON UPDATE RESTRICT and ON DELETE RESTRICT).

Note that whenever the referenced column is an autonumber one[112], cascading updates does not make sense –as values of automatically generated numbers cannot ever be changed–, so that *Access* ignores in such cases the ON UPDATE CASCADE setting); column *szColumn* stores the names of the foreign keys; column *szObject* stores the corresponding table names; column *szReferencedColumn* stores the columns that are referenced by the foreign keys; column *szReferencedObject* stores the corresponding table names.

For example, from the third and the fourth lines of the table instance from Fig. 4.23 we can find out that table *BORROWS_LISTS* has two such referential integrities (both of them of the default subtype, as *grbit* is 0, named *COPIESBORROWS_LISTS* and *BORROWSBORROWS_LISTS*, respectively), the first one associated to its foreign key *Copy* (which references column *x* of table *COPIES*) and the second one to its foreign key *Borrow* (which references column *x* of table *BORROWS*).

Consequently, the output of *REA2013A0* should now be the following:

```
CREATE TABLE BORROWS_LISTS (
   x INTEGER(9) GENERATED ALWAYS AS IDENTITY(1,1)
   PRIMARY KEY,
```

---

[112] As it should always be, except for the cases when the corresponding table represents a subset and its primary key is also a foreign key referencing the primary key of the superset corresponding table (as the primary key of the subset table cannot be an autonumber, but only a LONG number referencing an autonumber); normally, only for such exceptions, you should always enforce the ON UPDATE CASCADE option for all foreign keys referencing the primary keys of all tables that implement subsets.

FIGURE 4.22   *Access Navigation Pane* also showing its system tables.

FIGURE 4.23   *Access MSysRelationships* metacatalog table.

```
Borrow INTEGER(9) NOT NULL,
Copy INTEGER(9) NOT NULL,
DueReturnDate DATE NOT NULL CHECK (DueReturnDate
BETWEEN
```

```
   '01/06/2000' AND SysDate() + 300 DAYS),
ActualReturnDate  DATE  CHECK  (ActualReturnDate
BETWEEN
   '01/06/2000' AND SysDate()),
   CONSTRAINT blKey UNIQUE (Borrow, Copy),
    CONSTRAINT COPIESBORROWS_LISTS FOREIGN KEY
(Copy) REFERENCES
COPIES(x) ON UPDATE RESTRICT ON DELETE RESTRICT,
    CONSTRAINT BORROWSBORROWS_LISTS FOREIGN KEY
(Borrow) REFERENCES
BORROWS(x) ON UPDATE RESTRICT ON DELETE RESTRICT
);
```

As inspection of referential integrity constraints is done for this table, you can close the *MSysRelationships* table (without saving its sorting and column resizing's, as we only have read-only rights on the system tables) and finally check whether the table scheme also includes tuple (check) constraints.

Tuple (check) constraints are browsable/editable in *Access* in the *Validation Rule* property of the *Table Properties Property Sheet* window that you can open in a right subpane by right-clicking the left upper corner of the table header from the *Design View* window of the corresponding db table (immediately at the left of the *Field Name* column header—see Figs. 4.13–4.19) and then click on the *Properties* option. Figure 4.24 shows this *Table Properties Property Sheet* window for the *BORROWS_LISTS* table.

As the *Validation Rule* for the *BORROWS_LISTS* table is empty, the algorithm *REA2013A0* would add no tuple constraint to the above CRE- ATE TABLE statement.

If a tuple constraint *ActualReturnDate ≤ DueReturnDate* were also been part of this table scheme[113], then the corresponding *Validation Rule* property would have look like in Fig. 4.25 and *REA2013A0* would have also added a corresponding tuple constraint of the form:
```
CONSTRAINT BORROWS_LISTS_CHK1 CHECK
      (ActualReturnDate <= DueReturnDate)
```

---

[113] But it is not, as this would prevent storing data that a book copy has actually been returned, when- ever its actual return date were greater than the corresponding due one.

As everything on the *BORROWS_LISTS* table scheme has been found out, you should close its *Design View* window and open the one for a next table. After all table schemas are inspected, you can close the db and/or *Access*.

Figure 4.26 presents the *REA2013A*0 algorithm (designed for all currently used *Access* versions and a member of the *REAF*0′ family of algorithms for translating rdb schemes into corresponding ANSI *SQL DDL* scripts).

As it can be seen, algorithm *REA2013A*0 is assuming, as it should always be the case, that all tables have a primary key and that all primary keys are singletons and that they are either autonumbers (for all tables implementing object sets that are not subsets of other ones) or (for tables implementing subsets) integers referencing autonumbers; Exercise 4.8

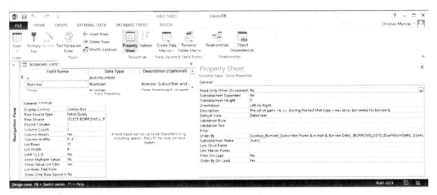

FIGURE 4.24   *Access Table Properties Property Sheet* window (with no tuple constraint).

FIGURE 4.25   *Access Table Properties Property Sheet* window (with a tuple constraint).

Algorithm REA2013A0 (Translation of Access db schemas into SQL ANSI DDL)
Input: an *Access* (all 7 versions between 2007 and 2013) db *S* (.accdb/mdb);
Output: a corresponding *SQL* ANSI *DDL* script *S.sql* for creating *S'* scheme;
Strategy: open db *S*, its *MSysRelationships* table, and script (text file) *S.sql*;
loop for all tables T from S (see *All Tables* subpane of the *Navigation* pane)
  open *T* in *Design View* mode;
  if T's primary key *x* does not appear in *szColumn* of *MSysRelationships* *then*
    add to *S*.sql line "CREATE TABLE " & *T* & "(" & *x* & "NUMERIC(9,
            0) GENERATED ALWAYS AS IDENTITY(1,1) PRIMARY";
  else add to *S*.sql line "CREATE TABLE " & *T* & "(" & *x* & "LONG
    PRIMARY, CONSTRAINT" & *szRelationship(x)* & "FOREIGN KEY ("
    & *x* & ") REFERENCES" & *szReferenceObject(x)* & "(" &
    szReference(x) & ")";
  end if;
  loop for all columns C of T, except for its primary key *x*, having domain *D*
    compute *D'*, the ANSI equivalent of *D*, based on Tables 4.1, 4.11, and 4.12;
    s = ", " & *C* & *D'*;
    if C has *Required* = *Yes* then *s* = *s* & " NOT NULL "; *end if;*
    if C has *Indexed* = *Yes* (*No Duplicates*) then *s* = *s* & " UNIQUE "; *end if;*
    if C has *Validation Rule* Not Null *then s* = *s* & " CHECK (" & *Validation*
                            Rule & ") ";*end if;* add to *S*.sql line *s*;
  end loop for all columns C of T;
  loop for all concatenated unique indexes CK of T (see *Indexes* window)
    s = ", CONSTRAINT " & *IndexName* & " UNIQUE ("; *first* = *true;*
    loop for all CK component attributes A
      if first then *first* = *false; else s* = *s* & ", "; *end if; s* = *s* & *A*;
    end loop;
    add to *S*.sql line *s* & ") ; ";
  end loop (for all concatenated (unique) keys CK of T);
  loop for all MSysRelationships rows y, with *szObject(y)* = *T*
    add to *S*.sql line ", CONSTRAINT " & *szRelationship(y)* & " FOREIGN
         KEY (" & *szColumn(y)* & ") REFERENCES " & *szReferenceObject(y)*
                            & "(" & *szReference(y)* &
")";
  end loop;
  if *T* has *Validation Rule* Not Null *then* add to *S*.sql line ", CONSTRAINT
    CK_" & *T* & " CHECK (" & *Validation Rule* & ") ";
end loop (for all tables T from S);
save and close script *S*.sql; close *Access*;
End REA2013A0 (Translation of Access db schemas into SQL ANSI DDL);

FIGURE 4.26 Algorithm *REA2013A0* (Translation of (all current versions of) *Access*
db schemas into corresponding *SQL ANSI DDL* scripts).

invites readers to consider all possible exceptions to these best practice rules too.

Moreover, *REA2013A*0 is also assuming that there is no foreign keys circle-type cycle made out from at least two tables (i.e., no cycle in the corresponding db E-RD graph has no node in which two of its arrows meet sharp point with sharp point); exercise 4.9 invites readers to consider such cases too.

4.5   THE ALGORITHMS FOR TRANSLATING E-R DATA MODELS INTO RDBS AND ASSOCIATED NONRELATIONAL CONSTRAINT SETS (*AF*1–8)

> *Traditional science is all about finding shortcuts.*
> —Rudy Rucker

As already seen in Section 1.7, the algorithms of the *AF*1–8 family are "shortcut" ones composed out of the following (family of) algorithms (in the order of their application): *A*1–7, *A*7/8–3, *A*8, and *AF*8' (i.e., any of its members g viewed as a mapping can be obtained as $g = f \circ A8 \circ A7/8–3 \circ A1–7$, where f is a member of *AF*8', which means that, for any E-RDM data model m, the corresponding rdb $g(m)$ is computable as $g(m) = f(A8(A7/8–3(A1–7(m))))$).

Of course that the challenge and beauty of any shortcut is to cut as short as advisable (i.e., without jeopardizing the journey), in order to gain time. To begin with in this case, as we have already seen in the previous chapter, it is not at all advisable to skip *A*7/8–3. Fortunately, as *A*7/8–3 can also be applied directly on rdb tables, we can apply it after applying *AF*1–8, so one step is already gained (and, actually, $g = f \circ A8 \circ A1–7$, $g(m) = f(A8(A1–7(m)))$).

What can next be also done to gain more time is to design *AF*1–8 as an extended *A*1–7 (as output) from RDM directly to RDBMS versions, by skipping completely both the RDM and the ANSI standard *SQL DDL* scripts. The price to be paid is only losing the portability of your implementations, but, in general, this is very rarely needed today, so it is generally worthwhile to do it.

As most of the currently available RDBMS versions also have a GUI for *SQL DDL* statements too, two versions of *AF*1–8 algorithms can generally be designed for any RDBMS version: one which generates ANSI standard *SQL DDL* scripts and one which guides you step by step on how to create corresponding rdbs through the corresponding GUI.

In this section, for exemplifying *AF*1–8, we chose one that outputs *SQL DDL* scripts, not only because, generally, all RDBMS vendors are advertising their GUIs in order for their products to be as much attractive as possible, or because the vast majority of the books published in this field and even YouTube is full with similar approaches, but mainly because this process should be completely automatized and thus made transparent to RDBMS users (and the only way to automatize it is to internally generate *SQL DDL* scripts): one day, even if unfortunately not that soon, not only a very few RDBMSes (including the most recent prototype versions of *MatBase*), as it is the case now, will provide to their users only higher level data models and completely hide the RDM one, but all of them will do it.

In particular, we have chosen as target RDBMS version the *MS SQL Server 2014*, so that this section is introducing (as *g*) the *A2014SQLS*1–8 algorithm, a member of the *AF*1–8 family that can be also used for all currently available versions of the *SQL Server*, not only the 2014 one, in order to translate E-RDs and associated restrictions lists directly into *SQL Server DDL* scripts for creating corresponding rdbs and associated non-relational constraint sets. Of course that this algorithm will also help the readers of this book to better understand how was obtained the *DDL* script from Subsection 4.3.4.

Figures 4.27–4.34 show the algorithm *A2014SQLS1–8*. Just like *A*1–7, this algorithm assumes that there are no duplicated object set names or mapping names defined on a same object set; moreover, it assumes that every table has a single numeric surrogate primary key, as it should always be the case, and that static small set elements are string values that do not contain single quotes (see Exercise 4.14). For example, by applying this algorithm to the E-R data model of the public library case study from Subsections 2.8.1 and 2.8.2, the *DDL* script from Subsection 4.3.4 is obtained (see Exercise 4.10.a).

Exercise 4.11 invites readers to design similar algorithms for *DB2 10.5*, *Oracle 12c*, *MySQL 5.7*, *Access 2013*, as well as other RDBMS versions. Exercise 4.12 is similar, but the requested outputs are corresponding

Algorithm A2014SQLS1-8 (Translation of E-R data models into SQL Server rdb schemas and associated non-relational constraint sets)

Input: an E-RD data model *M*

Output: the corresponding *SQL Server* rdb scheme and its associated non-relational constraint set

Strategy:

open scripts (text files) *S.sql* and *non-relS.sql*;

add to *S*.sql line "CREATE DATABASE ["& *M* & "];";

loop for all E-RDs *D* from *M*

  loop for all rectangles *R* in *D* (in bottom-up order, from only referenced, non-referencing object sets to non-referenced, only referencing ones)

    createSQLSTable(R);

  end loop (for all rectangles);

  loop for all diamonds *R* in *D* (in bottom-up order, from only referenced, non-referencing object sets to non-referenced, only referencing ones)

    createSQLSTable(R);

    loop for all R's underlying object set *S* occurrences (having role *r*)

      addSQLSForeignKey(R, r, S, "NN");

    end loop (for all R's underlying object set occurrences);

  end loop (for all diamonds);

  loop for all computed object sets *C* (according to *SQL* expression *E*)

    add to *S*.sql line "CREATE VIEW ["& *C* & "] AS "& *E* & ";";

  end loop (for all computed object sets);

end loop (for all E-RDs);

add all unused (non-relational) restrictions to *non-relS.sql*;

close scripts (text files) *S.sql* and *non-relS.sql*;

End Algorithm A2014SQLS1-8;

FIGURE 4.27   Algorithm A2014SQLS1–8 (Translation of E-R data models into corresponding SQL Server rdb schemas and non-relational constraint sets).

instructions on how to implement the rdbs through the GUIs instead of *SQL DDL* scripts.

4.6   THE REVERSE ENGINEERING FAMILY OF ALGORITHMS FOR TRANSLATING RDB SCHEMAS INTO E-R DATA MODELS (*REAF0–2*)

> *Poet's road is always towards the sources.*
>
> —Lucian Blaga

As already seen in Section 1.7, the algorithms of the *REAF0–2* family are "shortcut" ones composed out of the following (family of) algorithms (in the

method createSQLSTable(R)
if there is no table *R* in rdb *M* and *R* has attributes or is referencing or it is
   not referencing, but it is not static and with small cardinality *then*
    add to *S.sql* line "CREATE TABLE [" & *R* & "] (";
    if R is a subset of *S then*
      AddSQLSForeignKey(R, x, S, "PK");
      loop for all other sets *T* with $R \subseteq T$
        AddSQLSForeignKey(R, xT, T, "NNU");
      end loop;
    else
     N = chooseSQLSnumericDT(maxcard(R));
     add to *S.sql* line "[" & *x* & "]" & *N* & " IDENTITY PRIMARY KEY";
    end if; completeSQLScheme(R); add to *S.sql* line ");";
  end if;
end method createSQLSTable;

FIGURE 4.28   Algorithm A2014SQLS1–8's createSQLSTable method.

method addSQLSForeignKey(R, A, S, o)
if S is a table *then*
  N = chooseSQLSnumericDT(maxcard(S), 0);
  s = s & ", [" & *A* & "] " & *N*;
else
  l = 0;
  loop for all $e \in S$
    if length(e) > l then l = length(e);
  end loop;
  s = s & "[" & *A* & "] VARCHAR(" & *l* & ") CHECK ([" & *A* & "] IN (";
first = true;
  loop for all $e \in S$
    if first then first = false; else s = s & ", "; *end if; s = s* & """ & *e* & """;
  end loop;
end if;
if $o \in$ {"NN", "NNU"} *then s = s* & " NOT NULL "; *if o="NNU"thens=*
                             s & " UNIQUE ";
elseif o = "PK" then s = s & " PRIMARY KEY "; *end if;*
s = s & ", CONSTRAINT [" & *R* & "-" & *A* & "-FK] FOREIGN KEY
([" & *A* & "]) REFERENCES [" & *S* & "]";
add to *S.sql* line *s*;
end method addForeignKey;

FIGURE 4.29   Algorithm A2014SQLS1–8's createSQLSForeignKey method.

method completeSQLScheme(R)
loop for all arrows *A* leaving *R* and heading to object set *S*, except for set
        inclusion-type ones
    if A is compulsory *then addForeignSQLSKey(R, A, S, "NN");*
    else addForeignSQLSKey(R, A, S,); end if;
end loop (for all arrows);
loop for all ellipses *e* connected to *R*, except for the surrogate key *x*
    if e is compulsory *then addSQLSColumn(R, e, "NN"); else*
addSQLSColumn(R, e,); end if;
end loop (for all ellipses);
loop for all uniqueness restrictions *u* associated to *R*
    s = ", CONSTRAINT [" & *u* & "] UNIQUE ("; *first* = *true*;
    loop for all mappings *m* making up *u*
        if first then first = *false; else s* = *s* & ", "; *end if; s* = *s* & " [" & *m* &
        "]";
    end loop;
    add to *S.sql* line *s* & ") ";
end loop (for all uniqueness restrictions);
i = 1;
loop for all tuple-type restrictions *t* associated to *R*
    add to *S.sql* line ", CONSTRAINT CHK_" & *R* & *i* & " CHECK (" &
    t & ") "; *i* = *i* + 1;
end loop (for all other tuple-type restrictions);
end method completeSQLScheme;

FIGURE 4.30   Algorithm A2014SQLS1–8's completeSQLScheme method.

method addSQLSColumn(R, e, o)
s = ", " & *e* & " " & *chooseSQLSDT(R, e);*
if o ∈ {"NN", "NNU"} *then s* = *s* & " NOT NULL ";
if o = "NNU" *then s* = *s* & " UNIQUE ";
if e is computed according to expression *E then s* = *s* & " AS " & *E*;
add to *S.sql* line *s*;
end method addSQLSColumn;

FIGURE 4.31   Algorithm A2014SQLS1–8's addSQLSColumn method.

method chooseSQLSnumericDT(n, m)
if m = 0 then
  select case n
   case < 3: return " TINYINT";
   case ∈ {3, 4}: *return* " SMALLINT";
   case ∈ [5; 9]: *return* " INT";
   case ∈ [10; 18]: *return* " BIGINT";
   case ∈ [19; 38]: *return* " DECIMAL(" & *n* & ", 0) ";
   case ∈ [39; 308]: *return* " FLOAT(53)";
   else return " INTEGER OVERFLOW";
  end select;
else
  select case n
   case < 39: return " DECIMAL(" & *n* & ", " & *m* & ")";
   case ∈ [39; 308]: *return* "FLOAT(53)";
   else return "RATIONAL OVERFLOW";
  end select;
end if;
end method chooseSQLSnumericDT;

FIGURE 4.32 Algorithm A2014SQLS1–8's chooseSQLSnumericDT method.

method chooseSQLStextDT(t, p)
if t = ASCII
  if p ≤ 8000 then return " VARCHAR(" & *p* & ") ";
  elseif p ∈ [8001; $2^{31} - 1$] *then return* " VARCHAR(MAX) ";
  else return "ASCII OVERFLOW"; *end if;*
else
  if p ≤ 4000 then return " NVARCHAR(" & *p* & ") ";
  elseif p ∈ [4001; $2^{30} - 1$] *then return* " VARCHAR(MAX) ";
  else return "UNICODE OVERFLOW"; *end if;*
end if;
end method chooseSQLSDT;

FIGURE 4.33 Algorithm A2014SQLS1–8's chooseSQLStextDT method.

---

method chooseSQLSDT(e)
t = e.Type; p = e.Precision; s = e.Scale; m = e.minVal; M = e.maxVal;
E = e.ComputeExpr;
select case t
  case BOOLEAN: *return* "BIT";
  case ∈ {NAT, INT, RAT}: *return chooseSQLSnumericDT(p, s);*
  case ∈ {NAT, INT, RAT}: *return chooseSQLStextDT(t, p);*
  case CURRENCY : *if p* < 39 *return* " DECIMAL (" & *p* & ", 2) ";
                else return "CURRENCY OVERFLOW"; *end if;*
  case DATE/TIME: *if* 0001-01-01 ≤ *m*, *M* ≤ 9999-12-31 *then*
              if s = 0 *return* " DATE ";
              else return " DATETIME2 "; *end if;*
              else return "DATE/TIME OVERFLOW"; *end if;*
  case Computed: *return* " AS " & *E*;
  case User-defined: *return* " EXEC sp_addtype ";
  case BLOB: *if p* ≤ 8000 *then return* " VARBINARY (" & *p* & ") ";
        elseif p ∈ [8001; 2 GB] *then return* " VARBINARY(MAX) ";
        else return "BLOB OVERFLOW"; *end if;*
  case Transaction UIDs: *return* " ROWVERSION ";
  case Tree nodes: *return* " HIERARCHYID ";
  case XML: *return* " XML ";
  case GPS coordinates: *return* " GEOGRAPHY ";
  case Planar Euclidean coordinates: *return* " GEOMETRY ";
  case Any (Variant): *return* " SQL_VARIANT ";
  else return "UNKNOWN DATA TYPE";
end select;
end method chooseSQLSDT;

FIGURE 4.34   Algorithm A2014SQLS1–8's chooseSQLSDT method.

order of their application): *REAF0'*, *REA0*, *REA1*, and *REA2* (or *REAF0–1* and *REA2*, as *REAF0–1* is the composition of *REAF0'*, *REA0*, and *REA1*).

In other words, any of its members *g* viewed as a mapping can be obtained as *g = f ° REA0 ° REA1 ° REA2*, where *f* is a member of *REAF0'*, which means that, for any rdbs *r*, the corresponding E-R data model *g(r)* is computable as *g(r) = f(REA0(REA1(REA2(r)))))* (or, equivalently, *g = REAF0–1 ° REA2, g(r) = REAF0–1(REA2(r)))*.

You should first of all please note that, as (E)MDM is not tackled in this first volume, neither are *REA1*, *REA2*, nor *REAF0–1*.

Algorithm REA2013A0-2 (Translation of Access db schemas into E-R models)
Input: an *Access* (all 7 versions between 2007 and 2013) db *S* (.accdb/mdb);
Output: a corresponding E-R data model *M* = (*E-RD*, *RS*, *D*);
Strategy: open db *S* and its *MSysRelationships* table;
open empty .txt files *E-RD* (diagram), *RS* (restriction set), and *D* (description);
loop for all tables T from S (see *All Tables* subpane of the *Navigation* pane)
    draw in *E-RD* a rectangle labeled *T*; write *T* in *RS*; *i* = 1;
    open *T* in *Design View* mode;
    loop for all columns C of T having domain *D*;
        if <*T,C*> *not in* {(*szObject, szColumn*)} *then*
            draw in *E-RD* an ellipse labeled *C* and an edge from it to the rectangle *T*;
            if C is (part of) the primary key of *T* then primaryKey(C, 'e', D'); end if*;
            if C has *Indexed* = *Yes* (*No Duplicates*) *then* write in *RS C*'s uniqueness
                restriction, labeled *Ti*; *i* = *i* + 1; draw the one-to-oneness arrow on *C*'s
                edge; *end if*;
        end if;
        compute *D'*, the ANSI equivalent of *D*, based on Tables 4.1, 4.11, and 4.12;
        if C has *Validation Rule* Not Null *then s* = *Validation Rule*;
        elseif <*T,C*> *not in* {(*szObject, szColumn*)} *then s* = *D'*; *else s* = *null*; *end if*;
        if s is Not Null *then* write in *RS C*'s range restriction *s*, labeled *Ti*; *i* = *i* + 1;
        end if; *if C* has *Required* = *Yes then* write in *RS C*'s compulsory restriction,
                labeled *Ti*; *i* = *i* + 1; *end if*;
    end loop (for all columns C of T);
    loop for all concatenated unique indexes CK of T (see *Indexes* window)
        write in *RS CK* uniqueness restriction, labeled *Ti*; *i* = *i* + 1;
        loop for all CK component attributes A
            write *A* in *Ti*, separating it from the previous column (if any) by comma;
        end loop;
    end loop (for all concatenated (unique) keys CK of T);
    if T has *Validation Rule* Not Null *then* write it in *RS*, labeled *Ti*;
end loop (for all tables T from S);
loop for all MSysRelationships rows *y* with *szObject*(*y*) *not* prefixed by "MSys"
    T = *szObject*(*y*); *C* = *szColumn*(*y*); *R* = *szReferenceObject*(*y*);
    draw in *E-RD* an arrow from the rectangle labeled *T* to the one labeled *R* and
    label it *C*; *if* column *C* has *Indexed* = *Yes* (*No Duplicates*) *then*
        make current arrow a double one-to-one one; *end if*;
    if column *C* is (part of) the primary key *then primaryKey(C, 'a',)*;
end loop (for all MSysRelationships rows); *addSemanticsAndClose*;
end REA2013A0-2 (Translation of Access db schemas into E-R data models);

FIGURE 4.35   Algorithm *REA2013A0–2* (Translation of (all current versions of) *Access* db schemas into corresponding E-R data models).

Consequently, we are forced to come up in this volume with algorithms of this family simply by extending, from the output point of view, the composition of *REA0* and *REAF0'* from standard ANSI *SQL DDL* scripts all the way upwards to E-R data models.

Consequently, mainly in order to use the experience and intuition gained in section 4.4 above, we chose to exemplify this family of reverse engineering algorithms also with its member for MS *Access* dbs. Figures 4.35–4.37 present the corresponding algorithm, called *REA2013A0–2*. All other members of this family are left to the readers (see Exercise 4.15).

As it can be seen, just like *REA2013A0*, algorithm *REA2013A0–2* is assuming, as it should always be the case, that all tables have a primary key, that all primary keys are singletons, and that they are either autonumbers (for all tables implementing object sets that are not subsets of other ones) or (for tables implementing subsets) integers referencing autonumbers; Exercise 4.8 invites readers to consider all possible exceptions to these best practice rules too.

Moreover, *REA2013A0–2* is also assuming that there is no foreign keys circle-type cycle made out from at least two tables (i.e., no cycle in the corresponding db E-RD graph has no node in which two of its arrows meet sharp point with sharp point); exercise 4.9 invites readers to consider such cases too.

---

method primaryKey(C, t, D)
underline *C*;
if t = 'e' *then*
    draw the one-to-oneness arrow on the edge from ellipse *C* to its
    rectangle;
    write in *RS* the maximum cardinality restriction for *T*, labeled *Ti*,
    equal to *D*'s cardinal;
    $i = i + 1$;
else
    make the arrow labeled *C* a double one;
end if;
end primaryKey;

FIGURE 4.36   Algorithm *REA2013A0–2*'s method *primaryKey*.

> *method addSemanticsAndClose*
> investigate the resulted *E-RD* and *RS*, and, based on your gained insights, document in *RS* the semantics of all object sets, properties, and restrictions, and write in *D* a first clear, concise, and as much as possible exhaustive description of rdb *S*;
> save and close .txt files *E-RD*, *RS*, and *D*, which make up the E-R data model *M*;
> close *Access*;
> *end addSemanticsAndClose*;

FIGURE 4.37 Algorithm *REA2013A0–2'*s method *addSemanticsAndClose*.

4.7 CASE STUDY: REVERSE ENGINEERING OF AN *ACCESS STOCKS* DB SCHEME INTO BOTH AN ANSI STANDARD *SQL DDL* SCRIPT AND AN E-R DATA MODEL

> *Physics is mathematical not because we know so much about*
> *the physical world, but because we know so little;*
> *it is only its mathematical properties that we can discover.*
> —Bertrand Russell

Figures 4.38–4.74 present a *Stocks Access* db (see section 4.4 above on how such screen captures are generally obtainable and why). Figure 4.75 shows the corresponding ANSI standard *SQL DDL* script for creating this db obtained by applying the algorithm *REA2013A0* to this rdb. Figures 4.76 to 4.88 contain the corresponding E-R data model obtained by applying the algorithm *REA2013A0–2* to it.

4.7.1 THE *STOCKS* DB

> *If stock market experts were so expert, they would be buying stock,*
> *not selling advice.*
> —Norman Ralph Augustine

In Fig. 4.38 we can see that this db is made out of the following 10 tables: *Colors, Customers, Entries, EntryItems, EntryTypes, OrderItems, Orders, Products, Stocks,* and *Warehouses* (as *Switchboard Items* is the standard *Access* table for storing the application menu, which is a table-driven one, and the 6 dimmed tables –whose names start with *MSys*– are making up the standard *Access* metadata catalog).

Figure 4.39 shows the metadata stored by *Access* on the referential integrity constraints of this db in its table *MSysRelationships*. We can see that there are the following 12 foreign keys (as the other two ones belong to and are referencing metacatalog tables: their names are prefixed by "MSys"):

> *fk1068REF548:Entries.EntryTypereferencesEntryTypes.#EntryType*
> *fk1065REF545:Entries.CustomerreferencesCustomers.#Customer*

FIGURE 4.38   The *Stocks* db *Navigation* pane.

FIGURE 4.39 The *Stocks* db *MSysRelationships* table instance.

> ➢ *fk1075REF554*: *EntryItems.Warehouse* references *Warehouses*. *#Warehouse*
> ➢ *fk1073REF551*: *EntryItems.Product* references *Products.#Product*
> ➢ *EntriesEntryItems*: *EntryItems.Entry* references *Entries.#Entry*
> ➢ *OrdersOrderItems*: *OrderItems.Order* references *Orders.#Order*
> ➢ *fk1084REF554*: *OrderItems.Warehouse* references *Warehouses*. *#Warehouse*
> ➢ *fk1082REF551*: *OrderItems.Product* references *Products.#Product*
> ➢ *fk1087REF545*: *Orders.Customer* references *Customers.#Customer*
> ➢ *fk1093REF544*: *Products.Color* references *Colors.#Color*
> ➢ *fk1084REF554*: *Stocks.Warehouse* references *Warehouses.#Warehouse*
> ➢ *fk1101REF551*: *Stocks.Product* references *Products.#Product*

Figure 4.40 presents the scheme of table *Colors*; it can be seen that it has two columns: a surrogate autonumber primary key *#Color* and a string *Color* one, which may have at most 64 characters (*Field Size*), does not accept nulls (*Required = Yes*), and is one-to-one (*Indexed = Yes* (*No Duplicates*)).

Similarly, Fig. 4.41 presents the scheme of table *Customers* that also has two columns: a surrogate autonumber primary key *#Customer* and a string *Customer* one, which may have at most 64 characters, does not accept nulls, and is one-to-one.

Figures 4.42–4.46 present the seven columns of table *Entries*: *#Entry* (surrogate autonumber primary key), *Customer* (long integer, no nulls,

FIGURE 4.40   The *Stocks* db table *Colors* scheme.

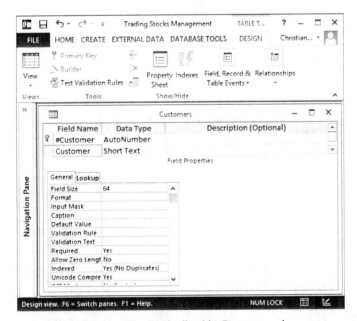

FIGURE 4.41   The *Stocks* db table *Customers* scheme.

FIGURE 4.42   The *Stocks* db table *Entries* column *Customer* definition.

FIGURE 4.43   The *Stocks* db table *Entries* column *EntryNo*.

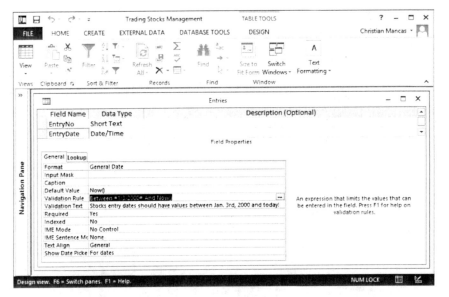

FIGURE 4.44   The *Stocks* db table *Entries* column *EntryDate*.

FIGURE 4.45   The *Stocks* db table *Entries* column *Stocked*.

FIGURE 4.46   The *Stocks* db table *Entries* column *EntryType*.

FIGURE 4.47   The *Stocks* db table *Entries* indexes.

indexed), *EntryNo* (string of at most 64 characters, no nulls, one-to-one), *EntryDate* (date between Jan. 3rd, 2000 and the current system date and time, which is also the default one, no nulls), *Stocked* (Boolean), and *EntryType* (long integer, default 1, no nulls). As seen in Fig. 4.47, this table does not have other keys (as its third index, *CustomerIdx*, is not a unique one).

Figures 4.48–4.51 present the five columns of table *EntryItems*: *#Entry-Item* (surrogate autonumber primary key), *Entry* (long integer, no nulls, indexed), *Product* (long integer, no nulls, indexed), *Qty* (double rational strictly positive and at most 1000, scale 2, not null), and *Warehouse* (long integer, no nulls). As seen in Fig. 4.52, this table has the key *EntryItem-Key*, made up of *Entry* and *Product* (as its other two indexes, *EIProductIdx* and *EIEntryIdx*, are not unique).

Figure 4.53 presents the two columns of table *EntryTypes*: *#EntryType* (surrogate autonumber primary key) and *EntryType* (string of at most 64 characters, no nulls, one-to-one).

Figures 4.54–4.57 present the five columns of table *OrderItems*: *#OrderItem* (surrogate autonumber primary key), *Order* (long integer, no nulls, indexed), *Product* (long integer, no nulls, indexed), *Qty* (double rational strictly positive and at most 1000, scale 2, not null), and *Warehouse* (long integer, no nulls). As seen in Fig. 4.58, this table has the key *OrderItemKey*, made up of *Order* and *Product* (as its other two indexes, *OIProductIdx* and *OIOrderIdx*, are not unique).

FIGURE 4.48   The *Stocks* db table *EntryItems* column *Entry*.

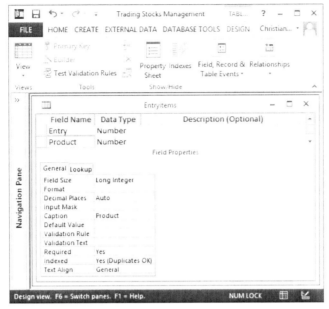

FIGURE 4.49   The *Stocks* db table *EntryItems* column *Product*.

FIGURE 4.50   The *Stocks* db table *EntryItems* column *Qty*.

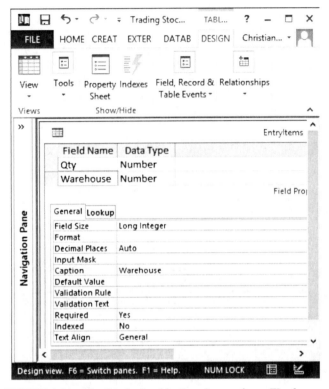

FIGURE 4.51 The *Stocks* db table *EntryItems* column *Warehouse*.

Index Name	Field Name	Sort Order
#Entry	Entry	Ascending
ElProductIdx	Product	Ascending
EntryItemKey	Entry	Ascending
	Product	Ascending

Index Properties

Primary	No
Unique	Yes
Ignore Nulls	No

Indexes: EntryItems

FIGURE 4.52 The *Stocks* db table *EntryItems* keys.

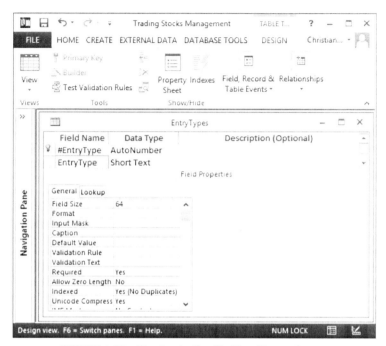

FIGURE 4.53   The *Stocks* db table *EntryTypes* scheme.

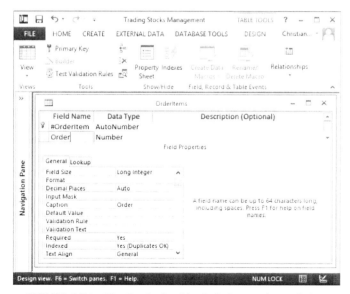

FIGURE 4.54   The *Stocks* db table *OrderItems* column *Order*.

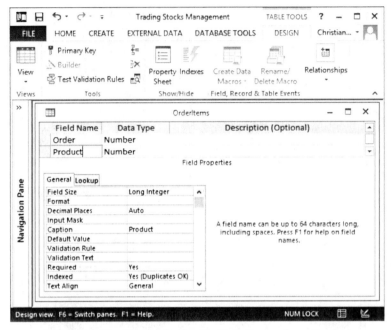

FIGURE 4.55   The *Stocks* db table *OrderItems* column *Product*.

FIGURE 4.56   The *Stocks* db table *OrderItems* column *Qty*.

FIGURE 4.57   The *Stocks* db table *OrderItems* column *Warehouse*.

	Index Name	Field Name	Sort Order	
	OrderItemKey	Order	Ascending	
		Product	Ascending	
		Warehouse	Ascending	
🔑	PrimaryKey	#OrderItem	Ascending	

Indexes: OrderItems

Index Properties

Primary	No	
Unique	Yes	The name for this index. Each index can use up
Ignore Nulls	No	to 10 fields.

FIGURE 4.58   The *Stocks* db table *OrderItems* keys.

Figures 4.59–4.62 present the five columns of table *Orders*: *#Order* (surrogate autonumber primary key), *Customer* (long integer, no nulls, indexed), *OrderNo* (string of at most 64 characters, no nulls, one-to-one), *OrderDate* (date between Jan. 3rd, 2000 and the current system date and time, which is also the default one, no nulls), and *Delivery* (Boolean). As seen in Fig. 4.63, this table does not have other keys (as its third index, *CustomerIdx*, is not a unique one).

Figures 4.64–4.66 present the four columns of table *Products*: *#Product* (surrogate autonumber primary key), *ProductCode* (string of at most 64 characters, no nulls), *Color* (long integer, no nulls), and *Product* (string of at most 64 characters, no nulls. indexed). As seen in Fig. 4.67, this table also has the key *ProductKey* made of *ProductCode* and *Color* (as its third index, *ProductsProduct*, is not unique).

Figures 4.68–4.72 present the six columns of table *Stocks*: *#Stock* (surrogate autonumber primary key), *Warehouse* (long integer, no nulls, indexed), *Product* (long integer, no nulls, indexed), *Stock* (positive double, scale 2, no nulls), *InitialStock* (positive double, scale 2, default 0, no

FIGURE 4.59   The *Stocks* db table *Orders* column *Customer*.

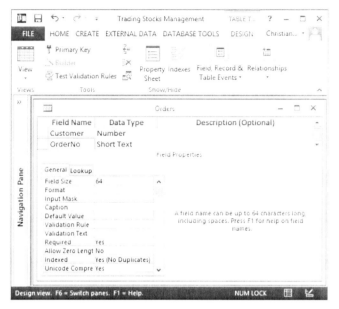

FIGURE 4.60   The *Stocks* db table *Orders* column *OrderNo*.

FIGURE 4.61   The *Stocks* db table *Orders* column *OrderDate*.

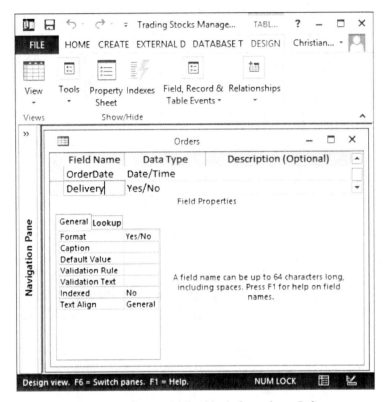

FIGURE 4.62   The *Stocks* db table *Orders* column *Delivery*.

Index Name	Field Name	Sort Order
Customer	Customer	Ascending
OrderNo	OrderNo	Ascending
PrimaryKey	#Order	Ascending

Index Properties

Primary	No
Unique	No
Ignore Nulls	No

If Yes, every value in this index must be unique.

FIGURE 4.63   The *Stocks* db table *Orders* keys.

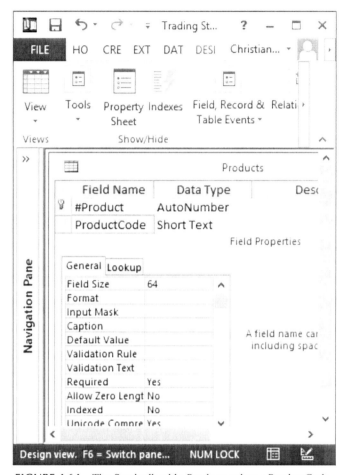

FIGURE 4.64   The *Stocks* db table *Products* column *ProductCode*.

nulls), and *InitialStockDate* (date between Jan. 3rd, 2000 and the current system date and time, no nulls). As seen in Fig. 4.73, this table also has the key *StocksKey* made of *Warehouse* and *Product* (as its other two indexes, *#Product* and *#Warehouse*, are not unique).

Figure 4.74 presents the two columns of table *Warehouses*: *#Warehouse* (surrogate autonumber primary key) and *Warehouse* (string of at most 64 characters, no nulls, one-to-one).

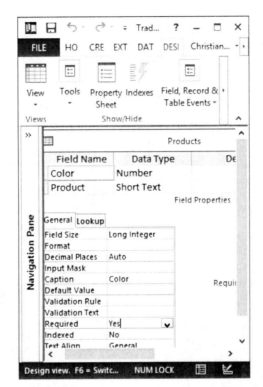

FIGURE 4.65   The *Stocks* db table *Products* column *Color*.

FIGURE 4.66   The *Stocks* db table *Products* column *Product*.

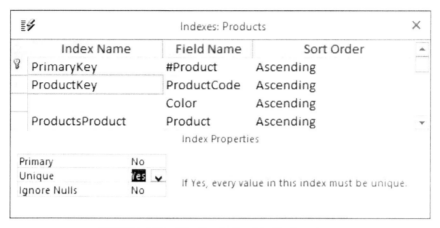

FIGURE 4.67   The *Stocks* db table *Products* keys.

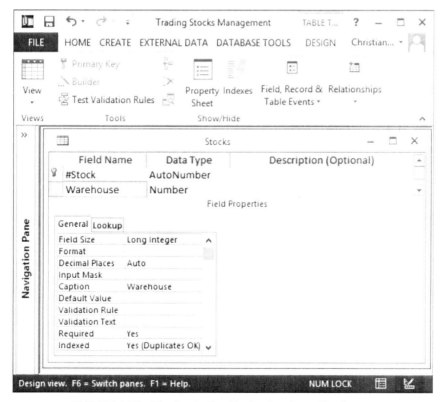

FIGURE 4.68   The *Stocks* db table *Stocks* column *Warehouse*.

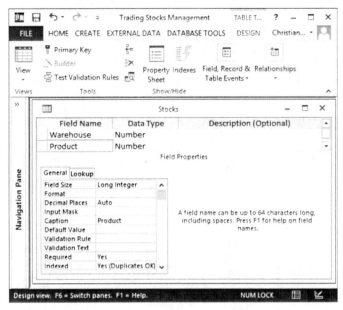

FIGURE 4.69   The *Stocks* db table *Stocks* column *Product*.

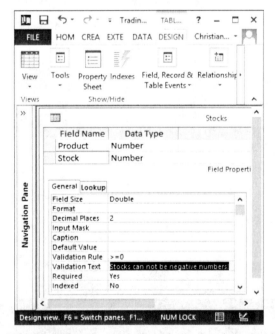

FIGURE 4.70   The *Stocks* db table *Stocks* column *Stock*.

FIGURE 4.71   The *Stocks* db table *Stocks* column *InitialStock*.

FIGURE 4.72   The *Stocks* db table *Stocks* column *InitialStockDate*.

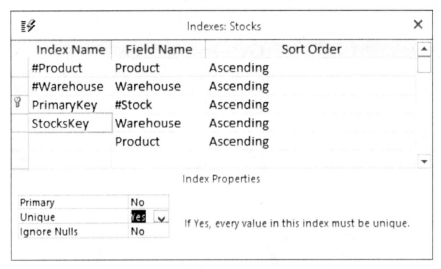

FIGURE 4.73   The *Stocks* db table *Stocks* keys.

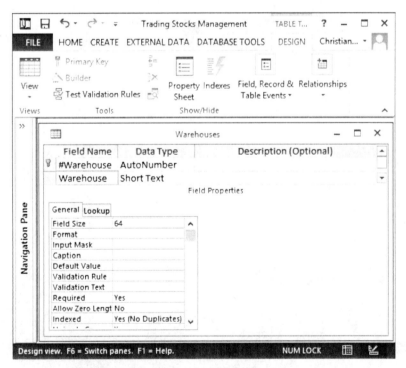

FIGURE 4.74   The *Stocks* db table *Warehouses* scheme.

4.7.2 THE CORRESPONDING ANSI STANDARD SQL DDL SCRIPT

> *The illiterate of the twenty-first century will not be those who cannot read and write, but those who cannot learn, unlearn, and relearn.*
> —Alvin Toffler

Figure 4.75 shows the result of applying the reverse engineering algorithm *REA2013A*0 to the above rdb. Note that for all foreign keys the columns they reference are omitted, as all such columns are primary keys. Also note that, due to their corresponding domain (check) constraints, the precision of the *Qty* columns has been downsized to only 6.

4.7.3 THE CORRESPONDING E-R DATA MODEL

> *Simplicity is prerequisite for reliability.*
> —Edsger Dijkstra

Figures 4.76–4.88 show the result of applying the reverse engineering algorithm *REA2013A*0–2 to the *Stocks* rdb, with a slight variation (see Exercise 4.17): a separate structural E-RD was drawn.

The first ten of them (4.76 to 4.85) present detailed E-RDs per table (object set). Figure 4.86 shows the structural E-RD of this subuniverse. Figure 4.87 is dedicated to the associated set of restrictions.

Finally, Fig. 4.88 contains an informal description of this subuniverse as reversed engineered from the E-RDs and associated restrictions.

Generally, you can get more insight into the semantics of tables, columns, and constraints by also inspecting the db instance, which allows refining the description component of the E-RD model: see, for example, Exercise 4.18.a.

Moreover, after reverse engineering dbs, you can not only discover flaws in their design and/or implementation that need corrections (see, for example, Exercise 4.18.c.), but also get ideas on what extensions to them would be of real help to both their users and applications programmers: see, for example, Exercise 4.19.

Note that the range for *Stock* and *InitialStock* was computed as follows: according to Table 4.11, the DOUBLE *Access* data type may store decimal numbers of up to 308 digits; as these two columns need 2 of them for the scale, the maximum possible mantissa is 306; moreover, given that they both accept only positive values, RAT^+ is their base set (instead of RAT, the set of all rationals).

```
CREATE DATABASE STOCKS;
CREATE TABLE Colors (
  #Color INTEGER(9) GENERATED ALWAYS AS
IDENTITY(1,1) PRIMARY KEY,
  Color VARCHAR(64) NOT NULL UNIQUE
);
CREATE TABLE Customers (
 #Customer INTEGER(9) GENERATED ALWAYS AS
IDENTITY(1,1) PRIMARY KEY,
  Customer VARCHAR(64) NOT NULL UNIQUE
);
CREATE TABLE EntryTypes (
#EntryType INTEGER(9) GENERATED ALWAYS AS
IDENTITY(1,1) PRIMARY KEY,
   EntryType VARCHAR(64) NOT NULL UNIQUE
);
CREATE TABLE Warehouses (
#Warehouse INTEGER(9) GENERATED ALWAYS AS
IDENTITY(1,1) PRIMARY KEY,
   Warehouse VARCHAR(64) NOT NULL UNIQUE
);
CREATE TABLE Products (
  #Product INTEGER(9) GENERATED ALWAYS AS
IDENTITY(1,1) PRIMARY KEY,
   ProductCode VARCHAR(64) NOT NULL,
   Color INTEGER(9) NOT NULL,
   Product VARCHAR(64) NOT NULL,
```

FIGURE 4.75 The *Stocks* db ANSI standard *SQL DDL* creation script.

```
 CONSTRAINT ProductKey UNIQUE (ProductCode,
Color),
 CONSTRAINT fk1093REF544 FOREIGN KEY (Color)
REFERENCES Colors
);
CREATE TABLE Stocks (
   #Stock INTEGER(9) GENERATED ALWAYS AS
IDENTITY(1,1) PRIMARY KEY,
   Warehouse INTEGER(9) NOT NULL,
   Product INTEGER(9) NOT NULL,
   Stock DECIMAL(40,2) NOT NULL CHECK (Stock >=
0),
   InitialStock DECIMAL(40,2) NOT NULL CHECK
(InitialStock >= 0),
   InitialStockDate DATE NOT NULL CHECK
(InitialStockDate BETWEEN 2000-01-03 AND
SysDate()),
 CONSTRAINT StocksKey UNIQUE (Warehouse,
Product),
 CONSTRAINT fk1101REF551 FOREIGN KEY (Product)
REFERENCES Products,
   CONSTRAINT fk1103REF554 FOREIGN KEY
(Warehouse) REFERENCES Warehouses
   );
   CREATE TABLE Entries (
      #Entry INTEGER(9) GENERATED ALWAYS AS
IDENTITY(1,1) PRIMARY KEY,
      Customer INTEGER(9) NOT NULL,
      EntryNo VARCHAR(64) NOT NULL UNIQUE,
      EntryDate DATE NOT NULL CHECK (EntryDate
BETWEEN 2000-01-03 AND SysDate()),
      Stocked BOOLEAN,
      EntryType INTEGER(9) NOT NULL,
    CONSTRAINT fk1065REF545 FOREIGN KEY
(Customer) REFERENCES Customers,
```

FIGURE 4.75   (Continued).

```
    CONSTRAINT fk1068REF548 FOREIGN KEY
(EntryType) REFERENCES EntryTypes
  );
  CREATE TABLE EntryItems (
  #EntryItem INTEGER(9) GENERATED ALWAYS AS
IDENTITY(1,1) PRIMARY KEY,
     Entry INTEGER(9) NOT NULL,
     Product INTEGER(9) NOT NULL,
     Qty DECIMAL(6,2) NOT NULL CHECK (Qty > 0
AND Qty <= 1000),
     Warehouse INTEGER(9) NOT NULL,
   CONSTRAINT EntryItemKey UNIQUE (Entry,
Product, Warehouse),
   CONSTRAINT EntriesEntryItems FOREIGN KEY
(Entry) REFERENCES Entries,
    CONSTRAINT fk1073REF551 FOREIGN KEY
(Product) REFERENCES Products,
    CONSTRAINT fk1075REF554 FOREIGN KEY
(Warehouse) REFERENCES Warehouses
  );
  CREATE TABLE Orders (
    #Order INTEGER(9) GENERATED ALWAYS AS
IDENTITY(1,1) PRIMARY KEY,
     Customer INTEGER(9) NOT NULL,
     OrderNo VARCHAR(64) NOT NULL UNIQUE,
     OrderDate DATE NOT NULL CHECK (OrderDate
BETWEEN 2000-01-03 AND SysDate()),
     Delivery BOOLEAN,
   CONSTRAINT fk1087REF545 FOREIGN KEY
(Customer) REFERENCES Customers
  );

  CREATE TABLE OrderItems (
  #OrderItem INTEGER(9) GENERATED ALWAYS AS
IDENTITY(1,1) PRIMARY KEY,
```

FIGURE 4.75 (Continued).

```
      Order INTEGER(9) NOT NULL,
      Product INTEGER(9) NOT NULL,
      Qty DECIMAL(6,2) NOT NULL CHECK (Qty > 0
AND Qty <= 1000),
      Warehouse INTEGER(9) NOT NULL,
    CONSTRAINT OrderItemKey UNIQUE (Order,
Product, Warehouse),
    CONSTRAINT OrdersOrderItems FOREIGN KEY
(Order) REFERENCES Orders,
    CONSTRAINT fk1082REF551 FOREIGN KEY
(Product) REFERENCES Products,
    CONSTRAINT fk1084REF554 FOREIGN KEY
(Warehouse) REFERENCES Warehouses
    );
```

FIGURE 4.75   (Continued).

FIGURE 4.76   *Colors* E-RD.

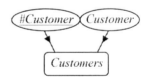

FIGURE 4.77   *Customers* E-RD.

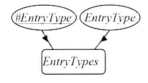

FIGURE 4.78   *EntryTypes* E-RD.

FIGURE 4.79   *Warehouses* E-RD.

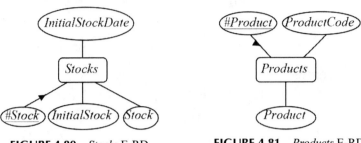

FIGURE 4.80   *Stocks* E-RD.   **FIGURE 4.81**   *Products* E-RD.

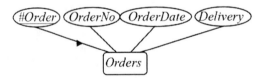

FIGURE 4.82   *Entries* E-RD.   **FIGURE 4.83**   *EntryItems* E-RD.

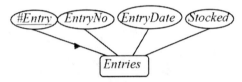

FIGURE 4.84   *Orders* E-RD.   **FIGURE 4.85**   *OrderItems* E-RD.

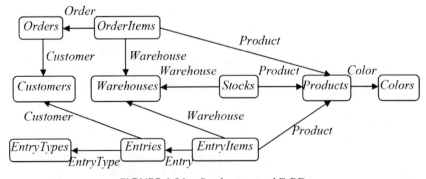

FIGURE 4.86   *Stocks* structural E-RD.

Customers (The set of customers and suppliers of interest.)	
a. Cardinality: max(card(CUSTOMERS)) = 1,000,000,000	(*RC0*)
b. Data ranges:	
#Customer: INT(9)	(*RC1*)
Customer: UNICODE(64)	(*RC2*)
c. Compulsory data: Customer	(*RC3*)
d. Uniqueness: Customer (no two customers may have	
same names)	(*RC4*)
Colors (The set of product colors.)	
a. Cardinality: max(card(Colors)) = 1,000,000,000	(*RK0*)
b. Data ranges:	
#Color: INT(9)	(*RK1*)
Color: UNICODE(64)	(*RK2*)
c. Compulsory data: Color	(*RK3*)
d. Uniqueness: Color (there may not be two colors with	
same name)	(*RK4*)
Products (The set of products of interest.)	
a. Cardinality: max(card(Products)) = 1,000,000,000	(*RP0*)
b. Data ranges:	
#Product: INT(9)	(*RP1*)
Product: UNICODE(64)	(*RP2*)
ProductCode: UNICODE(64)	(*RP3*)
c. Compulsory data: Product, ProductCode, Color	(*RP4*)
d. Uniqueness: Product • Color (there may not be two	
products having same codes and colors)	(*RP5*)
Warehouses (The set of warehouses of interest.)	
a. Cardinality: max(card(Warehouses)) = 1,000,000,000	(*RW0*)
b. Data ranges:	
#Warehouse: INT(9)	(*RW1*)
Warehouse: UNICODE(64)	(*RW2*)
c. Compulsory data: Warehouse	(*RW3*)
d. Uniqueness: Warehouse (no 2 warehouses may have	
same names)	(*RW4*)

FIGURE 4.87   *Stocks* associated restriction set.

Stocks (The set of pairs <*w, p*> meaning that warehouse
w stores product *p*.)
a. *Cardinality: max(card(Stocks))* = 1,000,000,000   (*RS0*)
b. *Data ranges:*
    #Stock: INT(9)   (*RS1*)
    Stock, InitialStock: RAT⁺(306, 2)   (*RS2*)
    InitialStockDate: [2000–01–03, *SysDate*()]   (*RS3*)
c. *Compulsory data: Product, Warehouse, Stock,*
    InitialStock, InitialStockDate   (*RS4*)
d. *Uniqueness: Product • Warehouse* (any valid *Stocks*
    instance should be a set: any pair <*w, p*> should be
    stored only once)   (*RS5*)

EntryTypes (The set of product colors.)
a. *Cardinality: max(card(EntryTypes))* = 1,000,000,000   (*RT0*)
b. *Data ranges:*
    #EntryType: INT(9)   (*RT1*)
    EntryType: UNICODE(64)   (*RT2*)
c. *Compulsory data: EntryType*   (*RT3*)
d. *Uniqueness: EntryType* (there may not be two entry
    types having same name)   (*RT4*)
Entries (The set of product entries into warehouses.)
a. *Cardinality: max(card(Entries))* = 1,000,000,000   (*RE0*)
b. *Data ranges:*
    #Entry: INT(9)   (*RE1*)
    EntryNo: UNICODE(64)   (*RE2*)
    EntryDate: [2000–01–03, *SysDate*()]   (*RE3*)
    Stocked: BOOLE (was entry *p* operated in *Stocks*?)   (*RE4*)
c. *Compulsory data: EntryType, EntryNo, EntryDate,*
    Customer   (*RE5*)
d. *Uniqueness: EntryNo* (there may not be two entries
    having same number)   (*RE6*)
EntryItems (The set of triples <*e, w, p*> meaning that, from
entry *e*, in warehouse *w*, products *p* were stored too.)

FIGURE 4.75   (Continued).

a. Cardinality: max(card(EntryItems)) = 1,000,000,000	*(RI0)*
b. Data ranges:	
#EntryItems: INT(9)	*(RI1)*
Qty: [0, 1000] ⊂ RAT(4, 2)	*(RI2)*
c. Compulsory data: Entry, Product, Warehouse, Qty	*(RI3)*
d. Uniqueness: Entry • Product • Warehouse (any valid	
EntryItems instance should be a set: any triple *<e, w, p>*	
should be stored only once)	*(RI4)*
Orders (The set of customer orders.)	
a. Cardinality: max(card(Orders)) = 1,000,000,000	*(RO0)*
b. Data ranges:	
#Order: INT(9)	*(RO1)*
OrderNo: UNICODE(64)	*(RO2)*
OrderDate: [2000–01–03, *SysDate*()]	*(RO3)*
Delivered: BOOLE (was order *o* delivered?)	*(RO4)*
c. Compulsory data: OrderNo, OrderDate, Customer	*(RO5)*
d. Uniqueness: OrderNo (no two orders may have same	
numbers)	*(RO6)*
OrderItems (The set of triples *<o, w, p>* meaning that, for	
order *o*, from warehouse *w*, products *p* were delivered too.)	
a. Cardinality: max(card(OrderItems)) = 1,000,000,000	*(ROI0)*
b. Data ranges:	
#OrderItems: INT(9)	*(ROI1)*
Qty: [0, 1000] ⊂ RAT(4, 2)	*(ROI2)*
c. Compulsory data: Order, Product, Warehouse, Qty	*(ROI3)*
d. Uniqueness: Order • Product • Warehouse (any valid	
OrderItems instance should be a set: any triple *<o, w, p>*	
should be stored only once)	*(ROI4)*

FIGURE 4.75 (Continued).

A company trades products characterized by compulsory codes, names, and colors (all three of them having at most 64 UNICODE characters). There may not be two distinct products having same code and color. There may not be two colors having same name. Products are bought from and sold to suppliers/customers whose compulsory names have at most 64 UNICODE characters. There may not be two such partners having same names. Products are stored into warehouses whose compulsory names have at most 64 UNICODE characters. There may not be two warehouses having same names.

In any warehouse, stocks for any product are compulsory and positive, having two scale digits; moreover, there should have been initial stocks (possibly 0, also compulsory and positive, two scale digits) in any warehouse, for any product stocked in that warehouse, at a compulsory date between January 3rd, 2000 and the current system date (probably prior to the first corresponding stock entry that was recorded in the db).

The company buys products from (compulsory) suppliers/customers, according to entry documents having compulsory types (at most 64 UNICODE characters) and unique numbers, at compulsory dates between January 3rd, 2000 and the current system date, and stores them in (compulsory) warehouses in (compulsory) strictly positive, two scale digits, and maximum 1000 unit quantities per entry, warehouse, and product. There may not be two entry types having same name.

The company sells products to (compulsory) suppliers/customers, according to order documents having compulsory unique numbers, at compulsory dates between January 3rd, 2000 and the current system date, and delivers them from (compulsory) warehouses in (compulsory) strictly positive, two scale digits, and maximum 1000 unit quantities per order, warehouse, and product.

There may be at most 10^9 elements in each of the following sets: colors, entry types, warehouses, customers, products, stocks, stock entries, stock entry items, orders, and order items.

FIGURE 4.88   *Stocks* db description.

4.8   BEST PRACTICE RULES

> *The laws of nature are but the mathematical thoughts of God.*
> —Euclid

4.8.1   DATABASES

> *We are what we repeatedly do.*
> *Excellence, then, is not an act, but a habit.*
> —Aristotle

R-I-D-0.(Do not add any objects to system tablespaces)

Never place your objects in system tablespaces, but only in user ones: otherwise, you increase data contention and slow both the server and your db applications.

R-I-D-1.(Safely keep up to date DDL scripts)

Always save *SQL DDL* scripts in a secure location; every time you make changes to db objects, be sure to script and check them into some version-control software repository. This is not a vital part of your db documentation, but also makes it simpler to replicate your db (e.g., for new customers) and compare it to other similar ones (including previous versions of your own).

R-I-D-2.(No UNICODE identifiers)

Do not use UNICODE identifiers, even when they are allowed: it complicates your applications whenever conversions are needed.

4.8.2   TABLES

> *When you realize you've made a mistake,*
> *take immediate steps to correct it.*
> —Dalai Lama

R-I-T-1.(Concatenated keys column ordering)

For concatenated keys (generally, for all such indexes, be them unique or not), the order of their columns should be given by the cardinality of their corresponding duplicate values: any column should have more duplicates than its predecessor and fewer than its successor.

For example, as in a *CITIES* table (see, for example, Fig. 3.5) there are much more distinct *ZipCode* values (rare duplicates being possible only between countries) than **Country* name ones (as, generally, there are very many cities in a country), the corresponding key should be defined in the order *<ZipCode, *Country>* and not vice versa.

R-I-T-2.(Don't use triggers to enforce relational constraints)
Triggers should mainly be used for enforcing (event-driven) non-relational constraints and/or to automate db maintenance and auditing. As they are slow, never use them to enforce relational constraints.

For example, you might use triggers to supplement declarative referential integrity too, but never to replace it.

R-I-T-3.(Always use same data types in check constraints)
Always make sure that check constraint expressions compare same data types: otherwise, in the best case, implicit conversions are needed, which is adding to the time spent for constraint enforcing; moreover, in such cases you also risk unpredictable results sometimes.

R-I-T-4.(Place BLOB data from frequently queried tables in subtables)
BLOB data columns should not be defined in frequently queried tables, because of performance issues: it is better to vertically split such tables, keeping all non-BLOB columns in the frequently queried one and the BLOB ones in a separate (sub)table, implemented as if it were corresponding to a subset of it.

For example, split table *PEOPLE* having columns *x*, *FirstName*, *LastName*, *Father*, *Mother*, *Sex*, *BirthPlace*, *BirthDate*, *Notes*, where *Notes* is its only BLOB column, into tables *PEOPLE* having same columns as above except for *Notes* and table *PEOPLENotes* having columns *x* and *Notes*, with *x* referencing the primary key *x* of *PEOPLE*.

4.8.3   COLUMNS

> *Anyone who stops learning is old, whether at 20 or 80.*
> *Anyone who keeps learning stays young.*
> —Henry Ford

R-I-C-0.(Always use correct data types)

Using incorrect data types might decrease the efficiency of the RDBMS optimizers, hurt performance, and cause applications to perform unnecessary data conversions, which are very costly whenever repeated for thousands of times.

For example, don't ever use strings to store dates, times, Booleans, or numbers.

R-I-C-1.(Always use the smallest data types needed)

Always use the smallest data type necessary to implement the required functionality: you are not only saving disk space, but you speed up processing too.

For example, do not use `NUMERIC` or `FLOAT` if you only need integers, do not use `BIGINT` if you only need `SMALLINT`, etc.

R-I-C-2.(Always use (VAR)BINARY instead of (VAR)CHAR BIT)

Although in *DB2* you can also specify the subtype `BIT` for `CHAR` and `VARCHAR` columns that contain `BIT` data, for better performance always use the `BINARY` or `VARBINARY` data types instead.

R-I-C-3.(Always use VARCHAR(n) instead of CHAR(n) for n > 17)

For better processing speed, use `VARCHAR` (variable length strings) instead of `CHAR` (fixed length strings) for all strings having at least 18 characters, unless you know that the entire column is to always be filled (equivalently: always avoid storing many blanks).

For example, in *DB2*, if you store a timestamp in character form (not as the *DB2* `TIMESTAMP` data type), you need a column with some number of characters: in Version 9, that number would be 26 characters; in ASCII, EBCDIC, or UTF-8, that column is 26 bytes; in UTF-16, that column is 52 bytes. Because the timestamp is always the same size, using a varying-length column does not save storage. However, suppose that you have a name field that is in ASCII or EBCDIC and allows for names of 26 characters (just as in ASCII SBCS or EBCDIC SBCS, but in UTF-8 you need 78 bytes, and in UTF-16 52 bytes). In this case, you want to use a varying-length column, because the name field is likely to have many blanks and you should not want to store them.

R-I-C-4.(AS/400 DB2 DECIMAL versus NUMERIC)

In *DB2* on AS/400, if you are going to perform arithmetic operations with a number, it is more efficient to define it as DECIMAL; if a number is going to be used mostly for display purposes, it is more efficient to define it as NUMERIC.

R-I-C-5.(DB2, COBOL, and PL/I best data types matching)

If your application is written in *COBOL* or *PL/I*, store your data in UTF-16 and use the GRAPHIC and VARGRAPHIC data types, in order for the UNICODE format in your application to match the one in your *DB2* db: this setup avoids conversion costs.

4.8.4   CONSTRAINTS

> *The tools we use have a profound (and devious) influence on our*
> *thinking habits, and, therefore, on our thinking abilities.*
> —Edsger W. Dijkstra

R-I-P-0.(Always use all constraint types provided by the target DBMS)

In order to guarantee as much as possible db instances' plausibility, always use all constraint types provided by the host DBMS. All following five best practice rules from this category are particular cases of this one (for the five main relational constraint types).

For example, use MS *Access' Allow Zero Length = No* setting for rejecting dirty nulls for any text type column.

Moreover, always enforce through the host DBMS in your dbs all constraints from your E-R data models that may be enforced by that DBMS: enforcing them only in applications built on top a db leaves that db defenseless against direct data manipulations, which circumvents the applications.

R-I-P-1.(Always enforce all (co-)domain constraints)

You should always enforce for any column its associated (co-)domain constraint.

For example, there are very few types of columns that do not need such constraints (e.g. Boolean, Hyperlink); for text ones, you need to enforce at least its maximum possible length and sometimes also its plausible subset (e.g., for *PEOPLE* or *CREATURES Sex* you only need one character and only a few letters out of the ASCII alphabet); for numbers and dates, only a subset of a corresponding data type (e.g. [0; 1,000.00] for *Stock* or *Price*, [1/1/2010, SysDate()] for *PaymentDate* and *OrderDate*, etc.).

R-I-P-2.(Always enforce all not null constraints, at least one per table additional to the corresponding surrogate primary key one)
You should always enforce for any column that should always be filled with data its associated not null constraint: any table should have at least one such semantic column (i.e. not only the syntactic surrogate key one).

For example, there may be no countries, cities, companies, books, songs, movies, etc. without names (titles), no order, invoice, bill, delivery lines without product, price, and quantity, etc.

R-I-P-3.(Always enforce for all foreign keys their corresponding referential integrity constraints)
Unfortunately, some DBMS GUIs (e.g. MS *Access*) let you define foreign keys without or with only partial referential integrity enforcement: do not ever forget to fully enforce them in order to ban dangling pointers from your db instances.

For example, in MS *Access* you might not leave or turn to *No* the combo-boxes *Limit To List* property (thus allowing users to store anything else than the values provided by the combo-box) or do not check or uncheck the *Enforce Referential Integrity* check-box (thus allowing users to delete rows even if they are referenced by any number of rows).

R-I-P-4.(Always enforce all minimal uniqueness constraints and never superkeys)
Unfortunately, most DBMSes let you both not defining any (unique) key or enforcing superkeys for any db table; you should enforce for any fundamental table all of its (unique) keys and no superkey.

For example, for table *COUNTRIES* from Fig. 3.4, besides its primary key x, you should also enforce its other 3 semantic (candidate) keys (*Country*, *CountryCode*, and *Capital*) and no superkey (e.g. neither *Country* • *CountryCode*, or *Country* • *Capital*, or *CountryCode* • *Capital*, or *Country* • *CountryCode* • *Capital*, or one of their permutations).

R-I-P-5.(Always enforce all tuple/check constraints)

You should use DBMSes for also enforcing any tuple/check constraints existing in your corresponding restriction set.

For example, table *PEOPLE* should have enforced the constraint *BirthDate ≤ PassedAwayDate*, table *SCHEDULES* should have enforced the constraint *StartHour ≤ EndHour*, etc.

4.9   THE MATH BEHIND THE ALGORITHMS PRESENTED IN THIS CHAPTER

> *In most sciences one generation tears down what another has built and what one has established another undoes.*
> *In mathematics alone each generation adds a new story to the old structure.*
> —Hermann Hankel

There is no math behind this chapter, except for the compulsory propositions characterizing the algorithms that it introduces. As they are very simple and similar, for example, to Propositions 3.1 and 4.1 (see Exercise 4.1), they are all left to the reader.

Thus, Exercise 4.1 characterizes *A*8, while Exercise 4.20 invites readers to characterize all of the other algorithms presented in this chapter, as well as all of the extensions to them and all of the similar ones designed by the readers as proposed by the other exercises from subsection 4.11.

4.10   CONCLUSION

> *Machines should work; people should think.*
> —IBM Pollyanna Principle

New dbs or extensions to existing ones can be implemented either from relational schemes or directly from E-R data models. Whenever they are directly implemented by skipping the RDM level, algorithm *A*7/8–3 should immediately be applied on all newly created or modified tables. Skipping the RDM level only affects portability, which is lost in such cases, but may not be needed sometimes.

Database implementation can be done algorithmically, either through running corresponding *SQL DDL* scripts or by using the GUIs of the RDBMSs (which are generally helping by being able to generate at any time corresponding *SQL DDL* scripts).

This chapter provides such algorithms for both implementation paths that are linear in the size of the input RDM schemes and E-R data models, respectively.

Moreover, it also provides dual reverse engineering algorithms of same complexity, for understanding semantics of legacy, not enough and/or up to date documented rdbs.

SQL statements' syntactical differences between RDBMSs are minor; corresponding differences between additionally provided functions (other than the aggregate ones) and, especially, those between available data types, extended *SQL*s, and implementation limitations are much more important and need very carefully design, implementation, and optimization techniques.

For example, even very commonly used built-in data types like integer, date/time, ASCII, and UNICODE strings or functions for getting current system date/time and their components have distinct syntax and semantics; moreover, the limitations of the type no autonumbering, no check constraints, no nondeterministic functions in check constraints, no null values in indexed columns, etc. are imposing workarounds using extended *SQL* stored triggers, functions, and procedures.

Dually, for example, *XML*, enumerated small static sets, file handlers, attachments, hyperlinks, geospatial data, (even recursive) user defined type hierarchies, and all other provided additional data types are easing the tasks of architecting, designing, and implementing rdbs on the corresponding RDBMSs.

RDBMS manufacturers are also providing free editions of their products, but care must be exercised when choosing the best fitted one for your needs, as various severe limitations apply to them.

4.11   EXERCISES

It is exercise alone that supports the spirits, and keeps the mind in vigor.
—Cicero

4.11.1   SOLVED EXERCISES

> *The test and the use of man's education is that he finds*
> *pleasure in the exercise of his mind.*
> —Jacques Barzun

4.0. Apply algorithm *A2014SQLS*1–8 to the E-R data model obtained in the Exercise 2.0 and add to the resulting script the *DML* statements needed in order to populate the db with the instance from Figure 3.26.

Solution:

a.   The corresponding *SQL DDL* script is the following:

```
CREATE DATABASE PAYMENTS;
CREATE TABLE COUNTRIES(
  x smallint IDENTITY PRIMARY KEY,
  Country varchar(128) NOT NULL UNIQUE
);
CREATE TABLE CITIES(
  x int IDENTITY PRIMARY KEY,
  Country smallint NOT NULL,
  City varchar(64) NOT NULL,
  ZipCode varchar(16) NOT NULL,
 CONSTRAINT key_CITIES UNIQUE (ZipCode, Country),
 CONSTRAINT FK_CITIES_Country FOREIGN
KEY(Country) REFERENCES
     COUNTRIES
);
CREATE TABLE CARD_TYPES(
   x smallint IDENTITY PRIMARY KEY,
   CardType varchar(16) NOT NULL UNIQUE
);
CREATE TABLE PARTNERS(
   x int IDENTITY PRIMARY KEY,
   Name varchar(128) NOT NULL,
   [Bank?] bit,
   Location int NOT NULL,
```

```
   Address varchar(255) NOT NULL,
   SubsidiaryOf int,
 CONSTRAINT key_PARTNERS UNIQUE (Name, Address,
Location),
 CONSTRAINT key_PARTNERS_Subsid UNIQUE
(SubsidiaryOf, Address, Location, [Bank?]),
 CONSTRAINT FK_PARTNERS_Location FOREIGN KEY
(Location) REFERENCES
CITIES,
 CONSTRAINT FK_PARTNERS_SubsidiaryOf FOREIGN KEY
(SubsidiaryOf)
REFERENCES PARTNERS
);
CREATE TABLE ACCOUNTS(
   x int IDENTITY PRIMARY KEY,
   Holder int NOT NULL,
   BankSubsidiary int NOT NULL,
   [No.] varchar(32) NOT NULL UNIQUE,
   Amount decimal(32, 2) NOT NULL CHECK (Amount
>= 0),
 CONSTRAINT KEY_ACCOUNTS UNIQUE ([No.],
                            BankSubsidiary),
 CONSTRAINT FK_ACCOUNTS_BankSubsidiary FOREIGN
KEY(BankSubsidiary)
     REFERENCES PARTNERS,
 CONSTRAINT FK_ACCOUNTS_Holder FOREIGN
KEY(Holder) REFERENCES
PARTNERS
);
CREATE TABLE CARDS(
   x int IDENTITY PRIMARY KEY,
   Type smallint NOT NULL,
   [No.] char(16) NOT NULL CHECK ([No.] BETWEEN
'1000000000000000'
AND '9999999999999999'),
```

```
SecurityCode char(3) NOT NULL CHECK (
  SecurityCode BETWEEN '000' and '999'),
   Account int NOT NULL,
   [Month] smallint NOT NULL CHECK ([Month]
BETWEEN 1 AND 12),
   [Year] smallint NOT NULL CHECK ([Year] BETWEEN
Year(GetDate()) -
   2 AND Year(GetDate()) + 2),
   [D/C] char(1) NOT NULL CHECK ([D/C] IN ('D',
'C')),
 CONSTRAINT key_CARDS UNIQUE ([No.], Type, [Year]),
 CONSTRAINT key_CARDS_Account UNIQUE (Account,
[Year]),
 CONSTRAINT FK_CARDS_Account FOREIGN KEY(Account)
REFERENCES
      ACCOUNTS,
CONSTRAINT FK_CARDS_Type FOREIGN KEY(Type)
      REFERENCES CARD_TYPES
);
CREATE TABLE INVOICES(
   x int IDENTITY PRIMARY KEY,
   Supplier int NOT NULL,
   [No.] varchar(16) NOT NULL,
   [Date] datetime NOT NULL CHECK ([Date] BETWEEN
'01/01/2000' AND getdate())
CONSTRAINT def_inv_date DEFAULT getdate(),
   Customer int NOT NULL,
   [*TotAmount] decimal(18,2),
 CONSTRAINT CHK_INVOICES_AntiCommut CHECK (Supplier
<> Customer),
 CONSTRAINT key_INVOICES UNIQUE (Supplier, [No.],
[Date]),
 CONSTRAINT FK_INVOICES_Supplier FOREIGN
KEY(Supplier) REFERENCES
      PARTNERS,
```

```
 CONSTRAINT FK_INVOICES_Customer FOREIGN
KEY(Customer) REFERENCES
PARTNERS
);
CREATE TABLE PAYM_DOCS(
   x int IDENTITY PRIMARY KEY,
   Debtor int NOT NULL,
   [From] int,
   [To] int,
   Card int,
   [No.] varchar(16) NOT NULL,
   [Date] datetime NOT NULL CHECK ([Date] BETWEEN
'01/01/2000' AND getdate())
CONSTRAINT def_paym_docs_date DEFAULT getdate(),
   Type char(1) NOT NULL CHECK (Type IN ('c',
'C', 'w')),
   [*TotAmount] decimal(18,2),
 CONSTRAINT CHK_PAYM_DOCS_AntiCommut CHECK
([From] <> [To]),
 CONSTRAINT key_PAYM_DOCS UNIQUE (Debtor, Type,
[No.], [Date]),
 CONSTRAINT key_PAYM_DOCS_From UNIQUE ([No.],
[Date], [From]),
 CONSTRAINT key_PAYM_DOCS_To UNIQUE ([No.],
[Date], [To]),
 CONSTRAINT key_PAYM_DOCS_Card UNIQUE ([No.],
[Date], Card),
 CONSTRAINT FK_PAYM_DOCS_Debtor FOREIGN
KEY(Debtor) REFERENCES
     PARTNERS,
CONSTRAINT FK_PAYM_DOCS_From FOREIGN KEY([From])
REFERENCES
     PARTNERS,
 CONSTRAINT FK_PAYM_DOCS_To FOREIGN KEY([To])
REFERENCES PARTNERS,
```

```
CONSTRAINT FK_PAYM_DOCS_Card FOREIGN KEY(Card)
REFERENCES CARDS
);
CREATE TABLE PAYMENTS (
    x int IDENTITY PRIMARY KEY,
    PaymDoc int NOT NULL,
    Invoice int NOT NULL,
    Amount decimal(8,2) NOT NULL CHECK (Amount > 0),
  CONSTRAINT key_PAYMENTS UNIQUE (Invoice,
                                        PaymDoc),
  CONSTRAINT FK_PAYMENTS_PaymDoc FOREIGN
KEY(PaymDoc) REFERENCES
    PAYM_DOCS,
  CONSTRAINT FK_PAYMENTS_Invoice FOREIGN
KEY(Invoice) REFERENCES
INVOICES
);
```

b. The corresponding non-relational constraints set is the following (for their enforcement see exercise 3.0 of the second volume of this book):

 ➢ *SubsidiaryOf* should be acyclic. (*P6*)
 ➢ All and only the subsidiaries of a bank should be of bank type. (*P7*)
 ➢ No payment for an invoice can be done before the invoice is established or after more than 100 years after corresponding establishing date. (*PD6*)
 ➢ Whenever *Type* is 'c' (cash), *To*, *From*, and *Card* should be null. (*PD7*)
 ➢ Whenever *Type* is 'C' (card), *From* should be null, whereas *To* and *Card* should not be null. (*PD8*)
 ➢ Whenever *Type* is 'w' (bank wire), *Card* should be null and *To* and *From* should not be null. (*PD9*)
 ➢ Sum of all payments for a same invoice should not be greater than the corresponding invoice total amount. (*PM4*)
 ➢ Bank subsidiaries should always be bank type partners. (*A5*)

> ➤ No card payment may be done with an expired card. (*C*8)

c. The needed *DML* statements are the following (where function *CAST*(*v*, *t*) converts value *v* to data type *t*):

```
INSERT COUNTRIES (Country) VALUES ('France');
INSERT COUNTRIES (Country) VALUES ('Romania');
INSERT COUNTRIES (Country) VALUES ('Holland');
INSERT CITIES (Country, City, ZipCode) VALUES (1,
  'Chouilly', '51530');
INSERT CITIES (Country, City, ZipCode) VALUES (3,
  'Amsterdam', '10');
INSERT CITIES (Country, City, ZipCode) VALUES (2,
  'Bucharest', '0');
INSERT PARTNERS(Name, [Bank?], Location, Address,
                                  SubsidiaryOf)
VALUES ('Champagne Nicolas Feuillate', NULL, 1,
'CD 40 A «Plumecoq», 51530', NULL);
INSERT PARTNERS(Name, [Bank?], Location, Address,
  SubsidiaryOf)
VALUES ('ING Commercial Banking', 1, 3,
'Amsterdamse Poort Building Bijlmerplein 8881102 MG',
  NULL);
INSERT PARTNERS(Name, [Bank?], Location, Address,
  SubsidiaryOf)
VALUES ('ING Bank N.V. Amsterdam Romania', 1, 2,
'Bd. Iancu de Hunedoara nr. 48, Sector 1, 011745',
  2);
INSERT INVOICES (Supplier, [No.], [Date],
                    Customer, [*TotAmount])
VALUES (1, '12', '11/21/2012', 2, CAST(1000.00 AS
  Decimal(18, 2)));
INSERT INVOICES (Supplier, [No.], [Date],
                    Customer, [*TotAmount])
VALUES (1, '13', '11/21/2012', 3, CAST(200.00 AS
                    Decimal(18, 2)));
```

```
INSERT INVOICES (Supplier, [No.], [Date],
                 Customer, [*TotAmount])
VALUES (1, '123', '12/21/2012', 3, CAST(300.00 AS
  Decimal(18, 2)));
INSERT ACCOUNTS (Holder, BankSubsidiary, [No.],
  Amount)
  VALUES (1, 2, 'HO01INGB00009999900111111',
     CAST(10000000.00 AS Decimal(32, 2)));
INSERT ACCOUNTS (Holder, BankSubsidiary, [No.],
  Amount)
  VALUES (2, 2, 'HO01INGB0000999900000000',
     CAST(99999999999.00 AS Decimal(32, 2)));
INSERT ACCOUNTS (Holder, BankSubsidiary, [No.],
  Amount)
  VALUES (3, 3, 'RO81INGB0000999900123456',
     CAST(100000.00 AS Decimal(32, 2)));
INSERT CARD_TYPES (CardType) VALUES ('American
  Express');
INSERT CARD_TYPES (CardType) VALUES ('Visa');
INSERT CARD_TYPES (CardType) VALUES
                 ('Visa Electron');
INSERT CARD_TYPES (CardType) VALUES ('Maestro');
INSERT CARDS (Type, [No.], SecurityCode, Account,
  [Month], [Year], [D/C])
  VALUES (2, '1234567898765432', '123', 1, 1, 2015,
  'C');
INSERT PAYM_DOCS (Debtor, [From], [To], [No.],
  [Date], Type, [*TotAmount])
VALUES (2, 2, 1, '210', '11/30/2012', 'w',
CAST(500.00 AS Decimal(32, 2)));
INSERT PAYM_DOCS (Debtor, [From], [To], [No.],
  [Date], Type, [*TotAmount])
VALUES (2, 2, 1, '320', '12/30/2012', 'w',
CAST(500.00 AS Decimal(32, 2)));
INSERT PAYM_DOCS (Debtor, [From], [To], [No.],
  [Date], Type, [*TotAmount])
```

```
VALUES (3, 3, 1, '101', '12/30/2012', 'w',
CAST(500.00 AS Decimal(32, 2)));
INSERT PAYMENTS (PaymDoc, Invoice, Amount)
VALUES (1, 1, CAST(500.00 AS Decimal(32, 2)));
INSERT PAYMENTS (PaymDoc, Invoice, Amount)
VALUES (2, 1, CAST(500.00 AS Decimal(32, 2)));
INSERT PAYMENTS (PaymDoc, Invoice, Amount)
VALUES (3, 2, CAST(200.00 AS Decimal(32, 2)));
INSERT PAYMENTS (PaymDoc, Invoice, Amount)
VALUES (3, 3, CAST(300.00 AS Decimal(32, 2)));
```

4.1. Prove the following *Proposition* 4.1 (*Algorithm A8 characteriza-tion*): *A8* has the following properties:

(*i*) its complexity is $O(n)$, where $n = t + c + r + c'$, where t is the number of input tables, c is the total one of their columns, r is the total one of their relational constraints, and c' is the total number of column occurrences that make up their concatenated keys;

(*ii*) it is complete (i.e., it is generating *DDL* creation statements for all tables, columns, and associated relational constraints);

(*iii*) it is sound (i.e., it is generating and outputting only necessary *DDL* statements for implementing the input rdb scheme).

Proof:

(*i*) (*Complexity*) Obviously, as it has six embedded loops, all of them finite (hence, it never loops infinitely) and the first one takes t steps, the second one c steps plus part of the r ones –those for the single keys and the not nulls–, the third one takes another part –those for the concatenated keys, which embeds the fourth one that takes c' steps–, and the fifth and sixth ones with the rest of r, for the refer-ential integrities and tuple/check constraints, respectively. Conse-quently, the total number of steps is $n = t + c + r + c'$.

(ii) (*Completeness*) Obviously, all tables, columns, and relational con-straints are considered.

(iii) (*Soundness*) Trivially, no other *SQL* statements than those needed are generated.   *Q.E.D.*

4.2.

a. Apply algorithm *REA2013A0–2* to the MS *Access Northwind* demo db.

b. Apply algorithm *A7/8–3* to the table *Purchase Order Details* of the MS *Access Northwind* demo db and add all discovered semantic keys to its scheme.

Solution:

a. If you did not already downloaded the MS *Access Northwind* db, first do the following:

 1. Open MS *Access 2013*.

 2. In its top screen *Search for Online Templates* text-box type *Northwind* and then press *Enter*: the window shown in Fig. 4.89 will open.

 3. Click on its *Northwind* icon: the window shown in Fig. 4.90 will open.

4. Choose the desired name for the *Northwind* db and the folder of your PC where you want it downloaded, and then click on the *Create* tile: the window shown in Fig. 4.91 will open.

5. Click on the *Enable Content* button from its *SECURITY WARNING* row: the window shown in Fig. 4.92 will open.

6. Either leave the default demo employee "Andrew Cencini" or select any other one and then click on the *Login* button: the window shown in Fig. 4.93 will open.

7. Close the *Home* tab, open the *Navigation Pane*, select its *All Access Objects* container, and open its *Tables* subset one: the window shown in Fig. 4.94 will open.

 As it can be seen from Fig. 4.94, the *Northwind* db has the following 20 tables: *Customers, Employee Privileges, Employees, Inventory Transaction Types, Inventory Transactions, Invoices, Order Details, Order Details Status, Orders, Orders Status, Orders Tax Status, Privileges, Products, Purchase Order Details, Purchase Order Status, Purchase Orders, Sales Reports, Shippers, Strings*, and *Suppliers*.

8. Enable showing of the *Access* system metacatalog tables too (see subsection 4.4, Figs. 4.21 and 4.22): the window shown in Fig. 4.94 will then look like in Fig. 4.95.

9. Double-click on the *MSysRelationships* table, close the *Navigation Pane*, resize *MSysRelationships* columns so that all data is fully

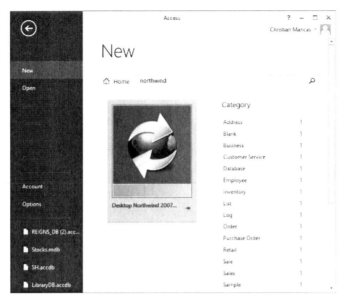

FIGURE 4.89 *Access 2013 Northwind* demo db download selection window.

FIGURE 4.90 *Access 2013 Northwind* demo db download window.

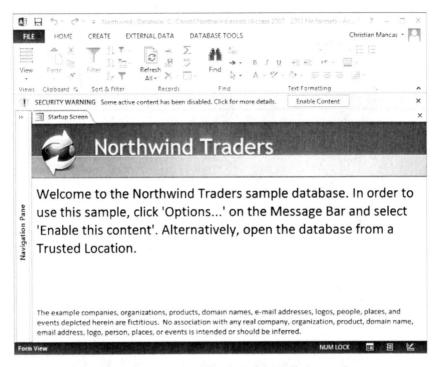

FIGURE 4.91 *Access 2013 Northwind* demo db *Startup Screen*.

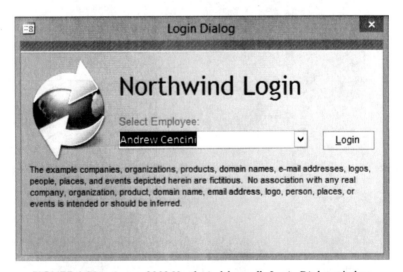

FIGURE 4.92 *Access 2013 Northwind* demo db *Login Dialog* window.

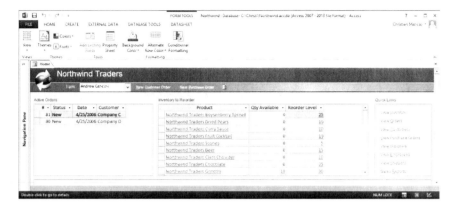

FIGURE 4.93 *Access 2013 Northwind* demo db application *Home* tab.

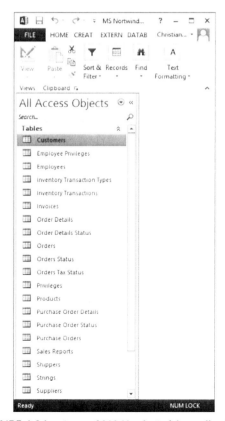

FIGURE 4.94 *Access 2013 Northwind* demo db tables.

FIGURE 4.95   *Access 2013 Northwind* demo db tables, including the system ones.

FIGURE 4.96   *Access 2013 Northwind* demo db *MSysRelationships* table instance.

displayed, and then sort in ascending order this table instance on the data of its column *szObject* (right-click on its name and then click on the *Sort A to Z* option): the window shown in Fig. 4.96 will open.

As it can be seen from Fig. 4.96, the *Northwind* db has the following 21 referential integrity constraints (as the ones of the *MSysNavPaneGroups* and *MSysNavPaneGroupToObjects* tables are system ones):

➤ *New_EmployeePriviligesforEmployees*: column *Employee ID* of table *Employee Privileges* references column *ID* of table *Employees*;

➤ *New_EmployeePriviligesLookup*: column *Privilege ID* of table *Employee Privileges* references column *Privilege ID* of table *Privileges*;

➤ *New_OrdersOnInventoryTransactions*: column *Customer Order ID* of table *Inventory Transactions* references column *Order ID* of table *Orders*;

➤ *New_PuchaseOrdersonInventoryTransactions*: column *Purchase Order ID* of table *Inventory Transactions* references column *Purchase Order ID* of table *Purchase Orders*;

➤ *New_ProductOnInventoryTransaction*: column *Product ID* of table *Inventory Transactions* references column *ID* of table *Products*;

➤ *New_TransactionTypesOnInventoryTransactiosn*: column *Transaction Type* of table *Inventory Transactions* references column *ID* of table *Inventory Transaction Types*;

➤ *New_OrderInvoice*: column *Order ID* of table *Invoices* references column *Order ID* of table *Orders*;

➤ *New_OrderDetails*: column *Order ID* of table *Order Details* references column *Order ID* of table *Orders*;

➤ *New_OrderStatusLookup*: column *Status ID* of table *Order Details* references column *Status ID* of table *Order Details Status*;

➤ *New_ProductsOnOrders*: column *Product ID* of table *Order Details* references column *ID* of table *Products*;

➤ *New_EmployeesOnOrders*: column *Employee ID* of table *Orders* references column *ID* of table *Employees*;

➤ *New_ShipperOnOrder*: column *Shipper ID* of table *Orders* references column *ID* of table *Shippers*;

➤ *New_CustomerOnOrders*: column *Customer ID* of table *Orders* references column *ID* of table *Customers*;

> *New_TaxStatusOnOrders*: column *Tax Status* of table *Orders* refer-
ences column *ID* of table *Orders Tax Status*;
> *New_OrderStatus*: column *Status ID* of table *Orders* references
column *Status ID* of table *Orders Status*;
> *New_ProductOnPurchaseOrderDetails*: column *Product ID* of table
Purchase Order Details references column *ID* of table *Products*;
> *New_PurchaseOrderDeatilsOnPurchaseOrder*: column *Purchase
Order ID* of table *Purchase Order Details* references column *Pur-
chase Order ID* of table *Purchase Orders*;
> *New_InventoryTransactionsOnPurchaseOrders*: column *Inventory
ID* of table *Purchase Order Details* references column *Transaction
ID* of table *Inventory Transactions*;
> *New_PurchaseOrderStatusLookup*: column *Status ID* of table *Pur-
chase Orders* references column *Status ID* of table *Purchase Order
Status*;
> *New_EmployeesOnPurchaseOrder*: column *Created By* of table
Purchase Orders references column *ID* of table *Employees*;
> *New_SuppliersOnPurchaseOrder*: column *Supplier ID* of table
Purchase Orders references column *ID* of table *Suppliers*.

Next steps require to open in *Design View*, one by one, all of the 20
tables, browse through all of their 177 columns (in total), and inspect all of
their 67 non-primary indexes (see, as similar examples, Sections 4.4 and
4.6). In order to save corresponding editorial space, we apologize to the
readers for not having included these additional 244 print screens (which
would need another 122 pages!) and invite them to do it as homework.

Figures 4.97–4.117 present the corresponding E-R data model obtained
after applying *REA20130–2* to this *Northwind* db, followed by the associated restrictions set and subuniverse description.

Note that table *Products*, because of its multivalued column *Supplier
IDs* (represented in Fig. 4.116 as a double-edged "multivalued" arrow), is
not in 1NF.

Please note that, very probably, in fact, there are other 4 referential
integrity constraints missing from this db, as, in *Order Details*, *Purchase
Order ID* and *Inventory ID* are, in fact, foreign keys referencing *Purchase
Orders* and *Inventory Transactions*, respectively, just as in *Purchase*

Orders both *Approved By* and *Submitted By* are, in fact, foreign keys referencing *Employees*.

This is why all these 4 columns are represented in the corresponding E-RDs above both as ellipses (as they are) and as arrows (as they should actually be).

The corresponding associated restrictions set is the following:

1. **Customers** (The set of customers of interest)
   a. *Cardinality:max*(*card*(*Customers*)) = 10^9 (*ID*)   (*C*0)
   b. *Data ranges*:
      Company, Last Name, First Name, E-mail Address,
      Job Title, City, State/Province,
      Country/Region: UNICODE(50)   (*C*1)
      Business Phone, Home Phone, Mobile Phone,
      Fax Number: UNICODE(25)   (*C*2)
      Address, Notes: UNICODE(4096)   (*C*3)
      ZIP/Postal Code: UNICODE(15)   (*C*4)
      Web Page: HYPERLINK   (*C*5)
      Attachments: ATTACHMENT   (*C*6)
   c. *Compulsory data: ID*   (*C*7)
   d. *Uniqueness: ID* (trivial: there may not be two
      customers having same autogenerated id numbers)   (*C*8)

2. **Employees** (The set of employees of interest)
   a. *Cardinality:max*(*card*(*Employees*)) = 10^9 (*ID*)   (*E*0)
   b. *Data ranges*:
      Company, Last Name, First Name, E-mail Address,
      Job Title, City, State/Province,
      Country/Region: UNICODE(50)   (*E*1)
      Business Phone, Home Phone, Mobile Phone,
      Fax Number: UNICODE(25)   (*E*2)
      Address, Notes: UNICODE(4096)   (*E*3)
      ZIP/Postal Code: UNICODE(15)   (*E*4)
      Web Page: HYPERLINK   (*E*5)
      Attachments: ATTACHMENT   (*E*6)
   c. *Compulsory data: ID*   (*E*7)
   d. *Uniqueness: ID* (trivial: there may not be two
      employees having same autogenerated id numbers)   (*E*8)

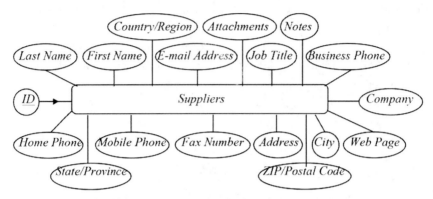

FIGURE 4.97   *Access 2013 Northwind Customers E-RD.*

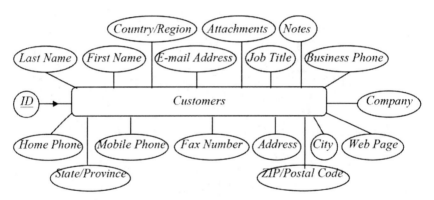

FIGURE 4.98   *Access 2013 Northwind Employees E-RD.*

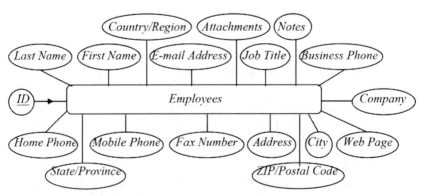

FIGURE 4.99   *Access 2013 Northwind Shippers E-RD.*

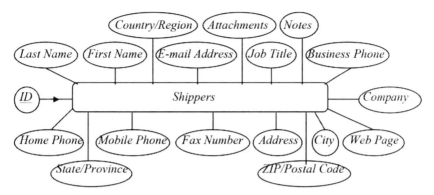

FIGURE 4.100   *Access 2013 Northwind Suppliers* E-RD.

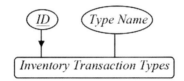

FIGURE 4.101   *Access 2013 Northwind Inventory Transaction Types* E-RD.

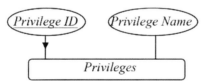

FIGURE 4.102   *Access 2013 Northwind Privileges* E-RD.

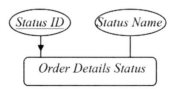

FIGURE 4.103   *Access 2013 Northwind Order Details Status* E-RD.

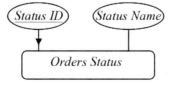

FIGURE 4.104   *Access 2013 Northwind Orders Status* E-RD.

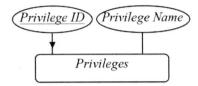

FIGURE 4.105   *Access 2013 Northwind Privileges* E-RD.

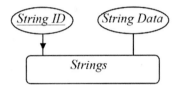

FIGURE 4.106   *Access 2013 Northwind Strings* E-RD.

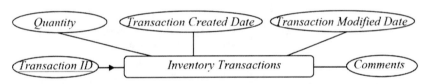

FIGURE 4.107   *Access 2013 Northwind Inventory Transactions* E-RD.

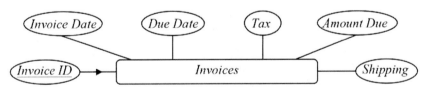

FIGURE 4.108   *Access 2013 Northwind Invoices* E-RD.

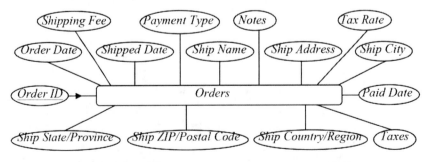

FIGURE 4.109   *Access 2013 Northwind Orders* E-RD.

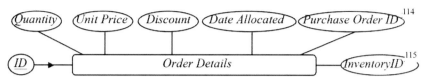

FIGURE 4.110   *Access 2013 Northwind Order Details E-RD.*

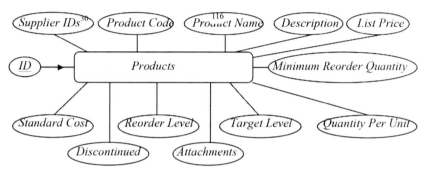

FIGURE 4.111   *Access 2013 Northwind Products E-RD.*

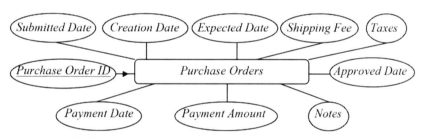

FIGURE 4.112   *Access 2013 Northwind Purchase Orders E-RD.*

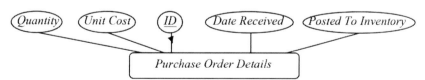

FIGURE 4.113   *Access 2013 Northwind Purchase Order Details E-RD.*

---

[114] In fact, very probably, *Purchase Order ID* should be a foreign key referencing *Purchase Orders*, that is an arrow in Fig. 4.116, not an ellipse.

[115] In fact, very probably, *Inventory ID* should be a foreign key referencing *Inventory Transactions*, that is an arrow in Fig. 4.116, not an ellipse.

[116] Multivalued attribute, taking values in the powerset of *Suppliers*.

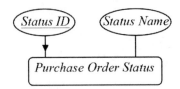

FIGURE 4.114   *Access 2013 Northwind Purchase Order Status E-RD.*

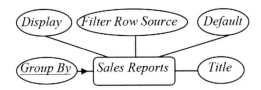

FIGURE 4.115   *Access 2013 Northwind Sales Reports E-RD.*

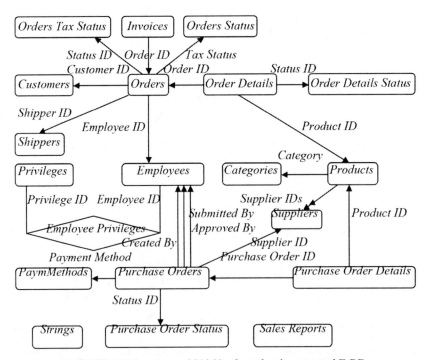

FIGURE 4.116   *Access 2013 Northwind main structural E-RD.*

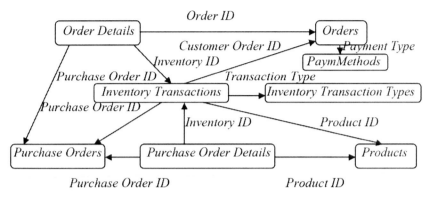

FIGURE 4.117 *Access 2013 Northwind* additional structural E-RD.

3. ***Suppliers*** (The set of suppliers of interest)
    a. *Cardinality:max(card(Suppliers))* = 10^9 *(ID)*   *(S0)*
    b. *Data ranges*:
    Company, Last Name, First Name, E-mail Address,
    Job Title, City, State/Province,
    Country/Region: UNICODE(50)   *(S1)*
    Business Phone, Home Phone, Mobile Phone,
    Fax Number: UNICODE(25)   *(S2)*
    Address, Notes: UNICODE(4096)   *(S3)*
    ZIP/Postal Code: UNICODE(15)   *(S4)*
    Web Page: HYPERLINK   *(S5)*
    Attachments: ATTACHMENT   *(S6)*
    c. *Compulsory data: ID*   *(S7)*
    d. *Uniqueness: ID* (trivial: there may not be two suppliers
        having same autogenerated id numbers)   *(S8)*
4. ***Shippers*** (The set of shippers of interest)
    a. *Cardinality:max(card(Shippers))* = 10^9 *(ID)*   *(H0)*
    b. *Data ranges*:
    Company, Last Name, First Name, E-mail Address,
    Job Title, City, State/Province,
    Country/Region: UNICODE(50)   *(H1)*

Business Phone, Home Phone, Mobile Phone,

Fax Number: UNICODE(25)   (*H2*)

Address, Notes: UNICODE(4096)   (*H3*)

ZIP/Postal Code: UNICODE(15)   (*H4*)

Web Page: HYPERLINK   (*H5*)

Attachments: ATTACHMENT   (*H6*)

   c. Compulsory data: ID   (*H7*)

   d. Uniqueness: ID (trivial: there may not be two

     shippers having same autogenerated id numbers)   (*H8*)

5. **Inventory Transaction Types** (The set of inventory

  transaction types of interest)

   a. Cardinality:max(card(Inventory Transaction

     Types)) = 10^9 (*ID*)   (*T0*)

   b. Data ranges:

   Type Name: UNICODE(50)   (*T1*)

   c. Compulsory data: ID, Type Name   (*T2*)

   d. Uniqueness: ID (trivial: there may not be two inventory

     transaction types having same autogenerated id numbers) (*T3*)

6. **Order Details Status** (The set of order details status of interest)

   a. Cardinality:max(card(Order Details Status)) = 10^9

     (*Status ID*)   (*ODS0*)

   b. Data ranges:

   Status Name: UNICODE(50)   (*ODS1*)

   c. Compulsory data: Status ID, Status Name   (*ODS2*)

   d. Uniqueness: Status ID (trivial: there may not be two

     order details status having same autogenerated id

     numbers)   (*ODS3*)

7. **Orders Status** (The set of orders status of interest)

   a. Cardinality:max(card(Orders Status)) = 10^9

     (*Status ID*)   (*OS0*)

   b. Data ranges:

   Status Name: UNICODE(50)   (*OS1*)

   c. Compulsory data: Status ID, Status Name   (*OS2*)

   d. Uniqueness: Status ID (trivial: there may not be two

     orders status having same autogenerated id numbers) (*OS3*)

8. ***Orders Tax Status*** (The set of orders tax status of interest)
   a. *Cardinality:max(card(Orders Tax Status))* = 10^9 (*ID*)   (*OTS0*)
   b. *Data ranges*:
   Tax Status Name: UNICODE(50)   (*OTS1*)
   c. *Compulsory data: Status ID, Tax Status Name*   (*OTS2*)
   d. *Uniqueness: Status ID* (trivial: there may not be two
      orders tax status having same autogenerated id numbers) (*OTS3*)
9. ***Purchase Order Status*** (The set of purchase order status
   of interest)
   a. *Cardinality:max(card(Purchase Orders Status))* = 10^9
      (*Status ID*)   (*POS0*)
   b. *Data ranges*:
   Status: UNICODE(50)   (*POS1*)
   c. *Compulsory data: Status ID*   (*POS2*)
   d. *Uniqueness: Status ID* (trivial: there may not be two
      purchase order status having same autogenerated id
      numbers)   (*POS3*)
10. ***Privileges*** (The set of types of privileges of interest that
    employees may have on the MS *Access 2013 Northwind*
    db application objects)
    a. *Cardinality:max(card(Privileges))* = 10^9 (*Privilege ID*) (*R0*)
    b. *Data ranges*:
    Privilege Name: UNICODE(50)   (*R1*)
    c. *Compulsory data: Privilege ID*   (*R2*)
    d. *Uniqueness: Privilege ID* (trivial: there may not be two
       types of privileges having same autogenerated id numbers) (*R3*)
11. ***Categories*** (The set of product categories of interest)
    a. *Cardinality:max(card(Categories))* = 128   (*G0*)
    b. *Data ranges*:
    Category: {"Baked Goods & Mixes", "Beverages",
    "Candy", "Canned Fruit & Vegetables",
    "Canned Meat", "Cereal", "Chips", "Snacks", "Condiments",
    "Dairy Products",
    "Dried Fruit & Nuts", "Grains", "Jams", "Preserves",
    "Oil", "Pasta", "Sauces", "Soups"}   (*G1*)
    c. *Compulsory data: Category*   (*G2*)

    d. Uniqueness: Category (there may not be two product
       categories having same names)   (*G*3)

12. **PaymMethods** (The set of purchase orders payment
    methods of interest)
    a. Cardinality:max(*card*(*PaymMethods*)) = 8   (*M*0)
    b. Data ranges:
    Payment Method: {"Credit card", "Check", "Cash"}   (*M*1)
    c. Compulsory data: Payment Method   (*M*2)
    d. Uniqueness: Payment Method (there may not be two
       payment methods having same names)   (*M*3)

13. **Employee Privileges** (The set of pairs <*e*, *p*> storing the
    fact that employee *e* has privilege *p* on the objects of the
    MS Access 2013 Northwind db application objects.)
    a. Cardinality:max(*card*(*Employee Privileges*)) = 130,000   (*EP*0)
    c. Compulsory data: Employee ID, Privilege ID   (*EP*1)
    d. Uniqueness: Employee ID • Privilege ID (there is no
       use in storing more than once any pair <*e*, *p*>)   (*EP*2)

14. **Invoices** (The set of all issued invoices by the *Nortwhind*
    Traders to their customers since the beginning of the
    twentieth century.)
    a. Cardinality:max(*card*(*Invoices*)) = 10^9 (*Invoice ID*)   (*I*0)
    b. Data ranges:
    Invoice Date, Due Date: [1/1/1900, 12/31/9999]   (*I*1)
    Tax, Shipping, Amount Due: CURRENCY(14)   (*I*2)
    c. Compulsory data: Invoice ID   (*I*3)
    d. Uniqueness: Invoice ID (trivial: there may not be two
       invoices having same autogenerated id numbers)   (*I*4)

15. **Orders** (The set of all orders placed to the *Nortwhind*
    Traders by their customers since the beginning of the
    twentieth century.)
    a. Cardinality:max(*card*(*Orders*)) = 10^9 (*Order ID*)   (*O*0)
    b. Data ranges:
    Order Date, Shipped Date, Paid Date:
    [1/1/1900, 12/31/9999]   (*O*1)
    Shipping Fee, Taxes: CURRENCY(14)   (*O*2)

Ship Name, Ship City, Ship State/Province,
Ship ZIP/Postal Code, Ship Country/Region:
     UNICODE(50)   *(O3)*
Ship Address, Notes: UNICODE(4096)   *(O4)*
c. *Compulsory data: Order ID*   *(O5)*
d. *Uniqueness: Order ID* (trivial: there may not be two
    orders having same autogenerated id numbers)   *(O6)*

16. **Purchase Orders** (The set of all purchase orders placed
    by the *Nortwhind Traders* to their suppliers since the
    beginning of the twentieth century.)
    a. *Cardinality:max(card(Purchase Orders))* = 10^9
      (Purchase Order ID)   *(PO0)*
    b. *Data ranges*:
    Submitted Date, Creation Date, Expected Date,
    Payment Date, Approved Date: [1/1/1900, 12/31/9999] *(PO1)*
    Shipping Fee, Taxes, Payment Amount:
    CURRENCY(14)   *(PO2)*
    Notes: UNICODE(4096)   *(PO3)*
    c. *Compulsory data: Purchase Order ID*   *(PO4)*
    d. *Uniqueness: Purchase Order ID* (trivial: there may
      not be two purchase orders having same
      autogenerated id numbers)   *(PO5)*

17. **Products** (The set of products that the *Nortwhind*
    Traders ever bought from their suppliers, stored in their
    stocks, and/or sold to their customers.)
    a. *Cardinality:max(card(Products))* = 10^9 *(ID)*   *(P0)*
    b. *Data ranges*:
    Product Code: UNICODE(25)   *(P1)*
    Product Name, Quantity Per Unit: UNICODE(50)   *(P2)*
    Standard Cost, List Price: CURRENCY(14)   *(P3)*
    Minimum Reorder Quantity, Reorder Level (Inventory
    quantity that triggers reordering),
    Target Level (Desired Inventory level after a purchase
    reorder): INT(9)   *(P4)*
    Description: UNICODE(4096)   *(P5)*

Discontinued: BOOLEAN   (*P6*)

Attachments: ATTACHMENT   (*P7*)

c. *Compulsory data: ID, List Price*   (*P8*)

d. *Uniqueness: ID* (trivial: there may not be two

   products having same autogenerated id numbers)   (*P9*)

18. **Inventory Transactions** (The set of all *Nortwhind Traders* inventory transactions since the beginning of the twentieth century.)

   a. *Cardinality:* $max(card(Inventory\ Transactions)) = 10^9$

      (*Transaction ID*)   (*IT0*)

   b. *Data ranges*:

   Transaction Created Date, Transaction Modified Date:

   [1/1/1900, 12/31/9999]   (*IT1*)

   Quantity: INT(9)   (*IT2*)

   Comments: UNICODE(255)   (*IT3*)

   c. *Compulsory data: Transaction ID, Transaction Type, Product ID, Quantity*   (*IT4*)

   d. *Uniqueness: Transaction ID* (trivial: there may not be two inventory transactions having same autogenerated id numbers)   (*IT5*)

19. **Order Details** (The set of all details of the orders placed to the *Nortwhind Traders* by their customers since the beginning of the twentieth century.)

   a. *Cardinality:* $max(card(Order\ Details)) = 10^9$ (*ID*)   (*OD0*)

   b. *Data ranges*:

   Date Allocated: [1/1/1900, 12/31/9999]   (*OD1*)

   Unit Price: CURRENCY(14)   (*OD2*)

   Quantity: RAT(18, 4)   (*OD3*)

   Discount: $[0, 1] \subset RAT(0, 308)$ (*OD4*)

   c. *Compulsory data: ID, Order ID, Quantity, Discount*   (*OD5*)

   d. *Uniqueness: ID* (trivial: there may not be two order details having same autogenerated id numbers)   (*OD6*)

20. **Purchase Order Details** (The set of all details of the purchase orders placed by the *Nortwhind Traders* to their suppliers since the beginning of the twentieth century.)

a. *Cardinality:max*(*card*(*Purchase Order Details*)) =
   10^9 (*ID*) (*POD*0)
b. *Data ranges*:
Date Received: [1/1/1900, 12/31/9999] (*POD*1)
Unit Cost: CURRENCY(14) (*POD*2)
Quantity: RAT(18, 4) (*POD*3)
Posted To Inventory: BOOLEAN (*POD*4)
c. *Compulsory data: ID, Purchase Order ID, Quantity,*
   Unit Cost (*POD*5)
d. *Uniqueness: ID* (trivial: there may not be two purchase
   order details having same autogenerated id numbers) (*POD*6)
21. **Sales Reports** (The set of all sales reports provided by
   the *Nortwhind Traders* db application.)
   a. *Cardinality:max*(*card*(*Sales Reports*)) = 10^{15}
      (*Group By*) (*SR*0)
   b. *Data ranges*:
   Group By, Display, Title: UNICODE(50) (*SR*1)
   Filter Row Source: UNICODE(4096) (*SR*2)
   Default: BOOLEAN (*SR*3)
   c. *Compulsory data: Group By* (*SR*4)
   d. *Uniqueness: Group By* (there may not be two sales
      reports having same group by clauses) (*SR*5)
22. **Strings** (The set of all strings used by the *Nortwhind Traders*
   demo db application for its titles, messages, etc.)
   a. *Cardinality:max*(*card*(*Strings*)) = 10^9 (*String ID*) (*S*0)
   b. *Data ranges*:
   String Data: UNICODE(255) (*S*1)
   c. *Compulsory data: String ID* (*S*2)
   d. *Uniqueness: String ID* (trivial: there may not be two
      strings having same autogenerated id numbers) (*S*3)
Figure 4.118 contains the corresponding description of the MS *Access
Nortwhind Traders* demo db.

b. Figure 4.119 shows the scheme of the MS *Access Nortwhind Traders* demo db *Purchase Order Details* table and Fig. 4.120 its indexes.

Besides its surrogate primary key *ID*, the table has the following 7
columns: *Purchase Order ID, Product ID, Quantity, Unit Cost, Date*

Received, Posted To Inventory, and *Inventory ID.* Besides the *PrimaryKey* index, the table has no other unique index, so no other key.

$i = 1$:

✓ *Purchase Order ID* is prime, but it is not a key, as any purchase order may have several detail lines (products);

✓ *Product ID* is prime, but it is not a key, as any product may be ordered in several purchase orders and even partially received for a same such order;

✓ *Inventory ID* is a key, as no inventory transaction may correspond to several purchase order details (products): $K = \{$ *ID, Inventory ID* $\}$;

✓ *Quantity* is not a key, as there may be, even in a same purchase order, several products ordered in a same quantity and it is not prime either: from the uniqueness point of view, it does not matter in what quantity a product is ordered and/or received, so *Quantity* cannot participate in any uniqueness of purchase order details;

✓ *Unit Cost* is not a key, as there may be, even in a same purchase order, several products having a same unit cost, and it is not prime either: from the uniqueness point of view, it does not matter which is the unit cost of a product, so *Unit Cost* cannot participate in any uniqueness of purchase order details;

✓ *Date Received* is prime, but not a key, as there may be, even in a same purchase order, several products having a same received date;

✓ *Posted To Inventory* is not a key, as there may be, even in a same purchase order, several products having been posted to inventory and several other ones having not yet been posted to inventory, and it is not prime either: from the uniqueness point of view, it does not matter whether an ordered product received quantity was posted to inventory or not (the ones not yet received cannot of course be posted), so *Posted To Inventory* cannot participate in any uniqueness of purchase order details.

$i = 2$:

In this step we are left with only three prime attributes: *Purchase Order ID, Product ID,* and *Date Received.*

✓ *Purchase Order ID* • *Product ID* is not a key, as even in a same purchase order, some amounts of an ordered product may have been received (at different dates) and the rest of the ordered amount not yet.

The MS *Access Northwind Traders* sample db consists of 2 unrelated and 20 related object sets. The unrelated sets are:

➤ *Sales Reports*, which stores all sales reports provided by the *Northwind Traders* db application (grouped by column names –at most 50 characters, compulsory and unique–, title, and displayed name –both at most 50 characters–, corresponding *SQL* statements for computing the row source –at most 4096 characters–, and whether it is the default sales report or not);

➤ *Strings*, which stores the set of all strings used by the *Northwind Traders* sample db application for its titles, messages, etc. (an autonumber and the corresponding string data of at most 255 characters).

The related sets store data since 1/1/1900 on the following items (all currency-type ones may have at most 14 mantissa digits; all sets may have at most 10^9 elements and have an autonumber id, which is their only uniqueness):

➤ *Customers*, *Suppliers*, *Shippers*, and *Employees* (company, last and first name, e-mail address, job title, city, state/province, and country/region –at most 50 characters each of them–, business, home, mobile phone, and fax number –at most 25 characters each of them–, address and notes –at most 4096 characters each of them–, zip/postal code –at most 15 characters–, a web page hyperlink, and file attachments);

➤ *Products* traded (code –at most 25 characters–, suppliers, category –one of 18 predefined ones–, name and quantity per unit –at most 50 characters each of them–, standard cost and compulsory list price –currency values of at most 14 digits–, minimum reorder quantity, inventory quantity that triggers reordering, and desired inventory level after a purchase reorder –integers of at most 9 digits–, description –at most 4096 characters–, whether discontinued or not, and file attachments);

➤ *Invoices* (customer order, issue and due dates, tax, shipping, and amount due currency values);

➤ *Inventory transactions* (type, product, purchase/customer order, create and modified dates, quantity –an integer of at most 9 digits–,

FIGURE 4.118 *Access 2013 Northwind db description.*

and comments –at most 255 characters–; compulsory are only the type, product, and quantity);

➢ *Purchase orders* (supplier, status, employees that submitted, created, and approved it, payment type –one of 3 predefined ones–, submitted, creation, expected, payment, approved, and (per product) received dates, shipping fee, taxes, payment amount, and (per product) unit cost currencies, products and corresponding inventory transactions, per product quantities – rationals of at most 18 mantissa and 4 scale digits–, per product notes –at most 4096 characters–, and whether received products were posted to inventory or not; compulsory are only the id, purchase order id, quantity, and unit cost);

➢ *Customer orders* (customer, status, tax status, and per product detail status, employee in charge, shipper, payment type –one of 3 predefined ones–, order, shipped, paid, and (per product) allocated dates, shipping fee, taxes, and (per product) unit cost currencies, products and corresponding inventory transactions and purchase orders, per product quantities –rationals of at most 18 mantissa and 4 scale digits–, ship name, city, state/province, zip/postal code, and country/region –at most 50 characters–, ship address and notes –at most 4096 characters–, and per product discount percentage; compulsory are only the order id, quantity, and discount);

➢ *Purchase* and *Customer Order* and *Details Status*, including tax ones, and inventory transaction types (compulsory name, at most 50 characters);

➢ *Employee Privilege Types* (name, at most 50 characters);

➢ *Employee Privileges* (unique compulsory pairs of employees and privilege types).

FIGURE 4.118   (Continued)

✓ *Purchase Order ID • Date Received* is not a key, as even in a same purchase order, some amounts of different ordered products may have been received at a same date.

✓ *Product ID • Date Received* is not a key, as there may happen that same products were received at same dates, but for different purchase orders.

$i = 3$:

✓ *Purchase Order ID • Product ID • Date Received* is a key, as no product should appear more than once in any purchase order as being received at a same date.

Consequently, the algorithm stops (as no other combinations of prime attributes are possible) and the *Purchase Order Details* table should have

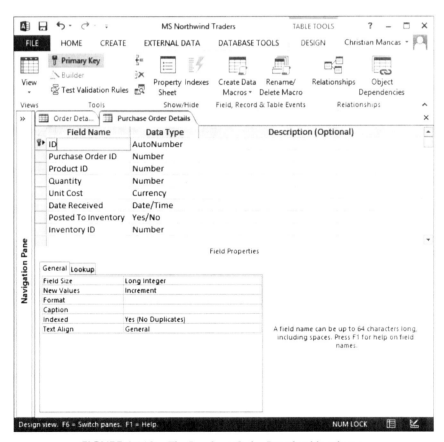

FIGURE 4.119 *The Purchase Order Details* table scheme.

Index Name	Field Name	Sort Order
Indexes: Purchase Order Details		
ID	ID	Ascending
Inventory ID	Inventory ID	Ascending
InventoryTransactionsOnPurchaseOrders	Inventory ID	Ascending
OrderID	Purchase Order ID	Ascending
PrimaryKey	ID	Ascending
ProductID	Product ID	Ascending
ProductOnPurchaseOrderDetails	Product ID	Ascending
PurchaseOrderDeatilsOnPurchaseOrder	Purchase Order ID	Ascending

Index Properties

Primary	No	
Unique	No	The name for this index. Each index can use up to 10 fields.
Ignore Nulls	No	

FIGURE 4.120   *The Purchase Order Details* table initial indexes.

the following three keys: $K = \{$ *ID, Inventory ID, Purchase Order ID • Product ID • Date Received* $\}$.

The corresponding *SQL DDL* statements for enforcing these keys are the following:

```
ALTER TABLE [Purchase Order Details]
ADD CONSTRAINT keyPODInventoryID UNIQUE ([Inventory
ID]);

ALTER TABLE [Purchase Order Details]
ADD CONSTRAINT keyPODProductID UNIQUE ([Purchase
Order ID],
  [Date Received], [Product ID]);
```

Figure 4.121 shows the *Index* window of this table after declaring the *keyPODInventoryID* above key, while Fig. 4.122 shows it after declaring the *keyPODProductID* too.

Index Name	Field Name	Sort Order
ID	ID	Ascending
Inventory ID	Inventory ID	Ascending
InventoryTransactionsOnPurchaseOrders	Inventory ID	Ascending
keyPODInventoryID	Inventory ID	Ascending
OrderID	Purchase Orde	Ascending
🔑 PrimaryKey	ID	Ascending
ProductID	Product ID	Ascending
ProductOnPurchaseOrderDetails	Product ID	Ascending
PurchaseOrderDeatilsOnPurchaseOrder	Purchase Orde	Ascending

Index Properties

Primary	No	
Unique	Yes	
Ignore Nulls	No	

If Yes, every value in this index must be unique.

FIGURE 4.121 *The Purchase Order Details keyPODInventoryID* key.

Index Name	Field Name	Sort Order
ID	ID	Ascending
Inventory ID	Inventory ID	Ascending
InventoryTransactionsOnPurchaseOrders	Inventory ID	Ascending
keyPODInventoryID	Inventory ID	Ascending
keyPODProductID	Purchase Order ID	Ascending
	Date Received	Ascending
	Product ID	Ascending
OrderID	Purchase Order ID	Ascending
🔑 PrimaryKey	ID	Ascending
ProductID	Product ID	Ascending
ProductOnPurchaseOrderDetails	Product ID	Ascending
PurchaseOrderDeatilsOnPurchaseOrder	Purchase Order ID	Ascending

Index Properties

Primary	No	
Unique	Yes	
Ignore Nulls	No	

If Yes, every value in this index must be unique.

FIGURE 4.122 *The Purchase Order Details keyPODProductID* key.

4.11.2 PROPOSED EXERCISES

Many difficulties which nature throws in our way may be smoothed
away by the exercise of intelligence.
—Titus Livius

4.3. Based on Section 4.2, for automatizing translations of ANSI *SQL DDL* scripts into corresponding scripts of various RDBMS versions:
   a. design the algorithms of the family *AF'8'* for *DB2 10.5, Oracle 12c, MySQL 5.7, SQL Server 2014*, and *Access 2013*;
   b. design such algorithms for at least one current version of another RDBMS vendor or open source consortium (e.g., *PostgreSQL, MongoDB*, etc.);
   c. based on Tables 4.3–4.12, establish a consolidated table for pinpointing both equivalences and differences between the data types provided by the five RDBMS versions considered in this book.

*4.4. Extend algorithm *A*8 with the following features:
   a. structures and abstract data types;
   b. *XML*;
   c. geospatial data types.

4.5. For automatizing translations of various RDBMS versions' *SQL DDL* scripts into corresponding ANSI standard ones:
   a. design the algorithms of the family *REAF0'* for *DB2 10.5, Oracle 12c, MySQL 5.7*, and *SQL Server 2014*;
   b. design such algorithms for at least one current version of another RDBMS vendor or open source consortium (e.g., *PostgreSQL, MongoDB*, etc.).

4.6. Design the reverse engineering algorithm *REA*0, for translating *SQL* ANSI *DDL* scripts into relational db schemas.

4.7. Modify algorithm *A*8 in order for it to also treat the totally not recommended concatenated foreign keys (*hint*: take as an example its loop for generating concatenated (unique) keys).

4.8. Modify algorithm *REA2013A*0 such that it also copes with tables having no primary keys, and/or having not autonumber or integer primary keys referencing other autonumber ones, and/or having primary keys made up of several columns, and/or having primary keys with other data types than the integer ones.

4.9. Consider the case of rdbs in which there is at least one cycle in their corresponding E-RD graph in which no two arrows meet sharp point to sharp point:

a. should such cycles be allowed by RDBMSs? Why?

b. if yes, design needed (if any) modifications to algorithm *REA2013A0* such that it also copes with such cases.

4.10. Apply algorithm *A2014SQLS*1–8 to:

a. the E-R data model of the public library case study from Subsections 2.8.1 and 2.8.2; compare its result with the *DDL* script from Subsection 4.3.4

b. the E-R data metamodel of the E-RDM presented in Section 2.9, in order to obtain a MS *SQL Server DDL* script for generating new metacatalog tables for it (that might be used when designing and developing an E-RDM GUI)

c. the RDM metamodel of itself presented in subsection 3.7.2

*d. compare the results of c) above with the *SQL Server 2014*'s corresponding metacatalog tables.

4.11.

a. Design algorithms similar to *A2014SQLS*1–8 for *DB2 10.5, Oracle 12c, MySQL 5.7, Access 2013*.

*b. Same thing as above for the latest versions of *PostgreSQL* and *MongoDB*.

4.12.

a. Design algorithms similar to *A2014SQLS*1–8 for *SQL Server 2014, DB2 10.5, Oracle 12c, MySQL 5.7, Access 2013*, but for corresponding instructions on how to implement rdbs through their GUIs (not through *SQL DDL* scripts).

*b. Same thing as above for the latest versions of *PostgreSQL* and *MongoDB*.

4.13

a. Design extensions to both *REA2013A0* and *REA2013A0*–2 so that they also generate *DML* scripts to populate dbs with the source rdb instances.

b. Apply these extensions to the public library case study db instance from Figure 3.13 and compare their output with the *DML* script from Subsection 4.3.5.

4.14 Design extensions to *A2014SQLS*1–8 so that it also accepts:

a. the totally not recommended tables without primary keys, and/or having not autonumber or integer primary keys referencing other autonumber ones, and/or having primary keys made up of several columns, and/or having primary keys with other data types than the integer ones;

b. small static lookup sets whose elements might contain single quotes.

4.15. For automatizing translations of various RDBMS managed rdbs into corresponding E-R data models:

a. design the algorithms of the family *REAF*0–2 for *DB2 10.5, Oracle 12c, MySQL 5.7*, and *SQL Server 2014*;

b. design such algorithms for at least one current version of another RDBMS vendor or open source consortium (e.g., *PostgreSQL, MongoDB*, etc.).

4.16. Based on the intuition gained from Section 4.7:

a. extend the algorithm *REA2013A*0 so that it also includes in the resulting ANSI standard *SQL DDL* scripts the default column values and the table indexes (*hint*: the corresponding syntaxes are `colName data-type` … `DEFAULT expression` … and `INDEX idxName`
`(colName1, …, colNamen))`;

b. what would be the corresponding *SQL DDL* script from Section 4.7 if it were obtained with this extended version of the algorithm?

4.17. Slightly modify algorithm *REA2013A*0–2 such that for more than 8 tables it draws separate structural E-RDs, each of which having at most 16 object sets (just like in Fig. 4.86).

4.18. Fig. 4.123 shows the instance of the *EntryTypes* table from the db *Stocks* (see Section 4.7):

a. based on it, improve the description of this db;

b. generalize the above and correspondingly improve the algorithms of the *REAF*0–2 family, by also taking into account rdb instances;

c. given the instance from Fig. 4.123, would you implement the *EntryTypes* object set with a table? If not, explain why and correspondingly correct both the E-R data model and the *DDL* script from Subsection 4.7.2.

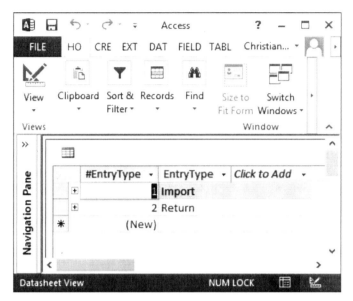

FIGURE 4.123. Instance of the *EntryTypes* table from the db *Stocks*.

4.19.

a. Would you add to the *Customers* table of the *Stocks* db (see Section 4.7) two Boolean columns *Customer?* and *Supplier?* Why?

b. Make all necessary changes, both in the ANSI standard *SQL DDL* script from Subsection 4.7.2 and in the E-R data model from Subsection 4.7.3 in order to add these two columns to the *Stocks* db.

4.20. Based on the examples provided by Proposition 3.1 and Exercise 4.1, characterize and prove the corresponding properties of the algorithms *REA2013A0*, *REA2013A0–2*, and *A2014SQLS1–8*, as well as of all of the other algorithms that you designed from the scratch or extended as asked by the exercises 4.3.a and b, 4.4 to 4.8, 4.9.b, 4.11, 4.12, 4.13a, 4.14, 4.15, 4.16a, 4.17, and 4.18b.

4.21. Apply the algorithm *A8* to the relational schemas from Figs. 3.12 and 3.20, Tables 3.7–3.13, as well as to all those obtained by solving Exercises 3.9, 3.28.a, and 3.76–3.81.

4.22 Apply the algorithm *A2014SQLS1–8* to the E-R data models from subsections 2.9.1 and 2.9.2, 3.7.3.1 and 3.7.3.2, Figs. 3.28–3.30, as well as to all those obtained by solving the exercises from Chapter 2.

4.23. Apply the algorithm *REA2013A0* to the MS *Access Northwind* db (see Exercise 4.2).

4.24. Consider the E-R data model obtained by the Exercise 4.23:

a.   is it a good one? Why? (*hints*: for example, only referring to its *Customers* object set: what's the difference between *Customers* and *Employees* and *Shippers* and *Suppliers*? Are all of these four distinct object sets needed, or could we consolidate them in only two: *PEOPLE* and *COMPANIES*? Are *Job Title*, *City*, *State/Province*, and *Country/Region* only strings? Is it ok that all of its properties, except for the primary key, are not compulsory? Is it enough to uniquely identify customers only through the surrogate key ID? Is this table at least in 3NF? Why? After inspecting their instances too, do you think that the sets *Inventory Transaction Types, Order Detail Status, Orders Status, Orders Tax Status*, and *Purchase Order Status* should be thought of as object sets? Why?);

b.   correct this E-R data model;

c.   apply to this corrected E-R data model the algorithms *A2014SQLS1–8* and *A2013A1–8* (obtained by solving Exercise 4.11.a) and run the resulted *SQL DDL* scripts in MS *SQL Server* and *Access* in order to obtain the corresponding correctly designed *Northwind* rdbs.

4.25. Consider the MS *SQL Server Northwind* demo db (that can be downloaded   from   http://www.microsoft.com/en-us/download/details.aspx?id=23654):

a.   obtain its corresponding E-R data model (through reverse engineering);

b.   analyze it: is it a good one? Why?

c.   correct this E-R data model;

d.   apply to this corrected E-R data model the algorithms *A2014SQLS1–8* and *A7/8–3*, and then run the resulted *SQL DDL* script in MS *SQL Server* in order to obtain the corresponding correctly designed *Northwind* rdb.

4.26. Consider the solved problem 4.0 above:

a.   add tables *PRODUCTS* and *INVOICE_DETAILS* (see Exercise 2.8);

    b. add triggers in order to automatically compute total amounts for both invoice detail lines and invoices;

*c. enforce the nonrelational constraints through *SQL Server 2014* triggers;

**d. generalize the experience you got when solving *a.* above and extend the algorithm *A2014SQLS*1–8 such that it also enforces at least the types of nonrelational constraints that you encountered in this exercise.

    4.27. Extend the algorithm *REA2013A0*–2 such that it also takes into consideration *Value List* type combo-boxes (*hint*: for example, in the *MS Access 2013 Northwind* demo db, column *Category* of table *Products* is referencing such a value list; consequently, this value list was considered in Fig. 4.116 as a (value object) set *Categories* (static and small) and *Category* is considered as an arrow in Fig. 4.116, instead of an ellipse from Fig. 4.111; same thing goes for the *Payment Method* of *Purchase Orders*).

    4.28. Extend the algorithm *REA2013A*0–2 such that for tables having only one double key (primary or not) whose both columns are foreign keys to generate corresponding diamonds instead of rectangles (*hint*: for example, in the MS *Access 2013 Northwind* demo db, table *Employee Privileges* only has such a double primary key—see Fig. 4.117).

    4.29. Extend the algorithm *REA2013A*0–2 such that for tables having no primary key the maximum cardinality of the corresponding object set is 2,147,483,648 (i.e., 2 GB, the maximum size of an *Access* db) divided by the average row length in bytes and for tables having a compound primary key made out of two foreign keys (see Exercise 4.28) it is this size minus the sum of half of the average maximum sizes of the two referenced table rows, everything divided by the average size of the referencing table rows (see, for example, the maximum cardinal of the *Employee Privileges* object set in the solution of problem 4.2 above).

    4.30. Extend the algorithm *REA2013A*0–2 such that for tables having non-numeric primary keys the maximum cardinality of the corresponding object set is computed from the corresponding maximum cardinal of the primary key domain, limited to 2 GB (the maximum size of an *Access* db) divided by the average row length (*hint*: for example, table *Sales Reports* from Exercise 4.2 above has a UNICODE(50) primary key; $card(\text{UNICODE}) = 65536$; $max(card(\text{UNICODE}(50))) = 65536^{50}$, which is immensely larger than 2 GB; $max(length(Sales\ Reports\ \text{rows})) = 4096 + 154 = 4250$ bytes; $avg(length(Sales\ Reports\ \text{rows})) = 2125$ bytes $= 0.000001.979060471057891845703125$ GB; consequently, $max(card(Sales\ Reports)) = 2$ GB$/0.00000197906047105789184570 31\ 25 = 1085102592571150.0950588235294118$, which is approximately $1,085,102,592,571,150$—that is a little bit more than 10^{15}).

4.31. Apply Algorithm $A7/8–3$ to all of the tables:

a.   of the dbs *Stocks* (see subsection 4.7.1) and MS *Access Northwind* (see Exercise 4.0);

b.   obtained when solving Exercises 4.22 and 4.25.d.

4.32. Apply algorithms $A7/8–3$ and $RESQLS2014A0–2$ (designed in Exercise 4.15) to the *MS SQL Server Pubs* sample db.

4.33. Apply algorithms $A7/8–3$ and $REO12cA0–2$ (designed in Exercise 4.15) to the *Oracle Express HR* sample db.

4.34. Apply algorithms $A7/8–3$ and $REDB2105A0–2$ (designed in Exercise 4.15) to the *DB2 Express-C Sample* db.

4.35. Apply algorithms $A7/8–3$ and $REMySQL57A0–2$ (designed in Exercise 4.15) to the *MySQL Sakila* and *World* sample dbs.

4.36. Apply at least one algorithm of the $AF1–8$ family to each E-R data model obtained by solving Exercise 2.10.

*4.37. Extend the RDM metamodel of itself from Subsection 3.7.2 with all tables and columns that it does not have, but you can find in the visible parts of the metacatalogs of:

a.   MS *Access 2013* (*hint*: its system "MSys" prefixed tables);

b.   MS *SQL Server 2014 Express* (*hint*: its *master* db tables);

c.   *Oracle MySQL 5.7* (*hint*: its *information_schema* tables).

4.38. Apply algorithm $REA1–2$ to all extensions added in Exercise 4.37 so that to obtain corresponding extended E-R data models of the RDM (see subsection 3.7.3).

*4.39. Apply algorithms of the family $REAF0–2$ to the views provided by the metacatalogs of:

a.   MS *SQL Server 2014 Express* (*hint*: its *master* db *system views*);

b.   *Oracle 12c Express* (*hint*: only the most frequently ones indicated by http://www.techonthenet.com/oracle/sys_tables/ would be enough for this exercise);

    c.   IBM *DB2 10.5 Express-C* (*hint*: only the most frequently *SYS-CAT* ones indicated by the following web page and its next (3) one http://www.devx.com/dbzone/Article/29585/0/page/2 would be enough for this exercise).

**4.40.

    a.   Reapply algorithm *A*0 to the descriptions from the E-R data models obtained in Exercise 4.39, in order to correct these E-R data models (i.e., conceptually reverse engineer the corresponding views, in order to discover the metacatalog tables from which they were computed).

    b.   Apply corresponding algorithms of the family *AF*1–8 to the corrected E-R data models obtained at *a*. above, in order to obtain *SQL DDL* scripts for creating the corresponding metacatalog tables, run them in the corresponding RDBMS versions, and then design for them the views similar to those in Exercise 4.39 (your starting point to this journey; *hint*: trivially, you have to do it in user dbs of your own, not in the corresponding system ones!).

4.12   PAST AND PRESENT

> *The past is the shadow of the future.*
> —Lucian Blaga
> *Look closely at the present you are constructing. It should look like the future you are dreaming.*
> —Alice Walker

All of the best practice rules and algorithms presented in this section, as well as their characterization propositions, application examples and exercises come from (Mancas, 2001). All these algorithms are also included in *MatBase* (see, e.g., Mancas and Mancas (2005)).

Algorithms similar to *A*8 are included in all RDBMSes that provide translation of the rdbs they manage into *SQL DDL* scripts.

Algorithms similar to *REA*0 are included in all RDBMSes for creating or modifying rdbs when executing *SQL DDL* scripts.

Different variants of algorithms from the *AF*1–8 family are embedded in the data modeling tools provided by *DB2*, *Oracle*, *SQL Server*, as well as by third party solutions (see, for example, the universal *Toad* one, now owned by the Dell Corp.: http://www.quest.com/toad-data-modeler/).

There exist other third party tools as well for generating corresponding *SQL DDL* scripts from *Access* dbs: for example, see the *DBWScript* from DBWeigher.com (http://dbweigher.com/dbwscript.php).

For obtaining E-RD variants from rdbs there are also tools both RDBMS embedded (like, for example, the MS *Access Relationships* and *SQL Server Database Diagrams*, the *Oracle SQL Developer Data Modeler*, and the IBM *DB2 Data Studio* or *InfoSphere Data Architect*) and third party (see, for example, the CA *ERwin Data Modeler*, http://erwin.com/products/data-modeler, the *Visual Paradigm*, http://www.visual-paradigm.com/features/database-design/ or the *SmartDraw*, http://www.smartdraw.com/resources/tutorials/entity-relationship-diagrams/). Unfortunately, since its 2013 version, the MS *Visio* does not have this facility anymore.

Excellent *DB2 10.5*, *Oracle 12.c*, *MySQL 5.7*, *SQL Server 2014*, and *Access 2013* documentation is provided by IBM, Oracle, and Microsoft, respectively (see references in the next subsection).

Even if only in the fourth chapter of the next volume will we tackle rdb optimizations, some initial aspects of it were also introduced in this chapter, both from own experience, RDBMS documentations, and from Bradford (2011), Feuerstein (2008), MacDonald (2013), and Vaughn and Blackburn (2006).

All URLs mentioned in this section were last accessed on September 30th, 2014.

KEYWORDS

- (z)Linux
- @@DBTS
- @@IDENTITY
- A1-7
- A2014SQLS1-8
- A7/8-3
- A8

- Access
- addSemanticsAndClose
- addSQLSColumn
- ADO
- ADOX
- AF1-8
- AF8'
- AFTER UPDATE
- AIX
- ALL_OBJECTS
- Allow Zero Length
- ALTER SESSION
- ALTER TABLE
- ANCHOR
- ANSI
- ANYDATA
- ANYDATASET
- ANYTYPE
- Array
- ASCII
- AS ROW BEGIN
- AS ROW CHANGE
- AS ROW END
- associative array
- ATTACHMENT
- AUTO_INCREMENT
- Azure
- BasicFile LOB
- BEFORE INSERT
- BEFORE UPDATE
- BFILE
- BIGINT
- BINARY

- **BINARY_DOUBLE**
- **BINARY_FLOAT**
- **BIT**
- **BITEMPORAL**
- **BLOB**
- **BLU**
- **BOOL**
- **BOOLEAN**
- **built-in data type**
- **BUSINESS_TIME**
- **BYTE**
- **cardinality**
- **CASCADE DELETE**
- **CASCADE UPDATE**
- **CAST**
- **CCSID**
- **CHAR**
- **check constraint**
- **chooseSQLSDT**
- **chooseSQLSnumericDT**
- **chooseSQLStextDT**
- **CLOB**
- **CODEUNITS16**
- **CODEUNITS32**
- **COMMENT**
- **completeSQLScheme**
- **compulsory data**
- **computed column**
- **CONCAT**
- **COUNT**
- **COUNTER**
- **CREATE DATABASE**
- **CREATE FUNCTION**

- **CREATE SEQUENCE**
- **createSQLSForeignKey**
- **createSQLSTable**
- **CREATE TABLE**
- **CREATE TRIGGER**
- **CREATE TYPE**
- **CREATE USER**
- **CurDate**
- **CURRENCY**
- **CURRENT_DATE**
- **CURRENT_TIME**
- **CURRENT_TIMESTAMP**
- **CURRENT_TIMEZONE**
- **CURRENT DATE**
- **CURRENT LOCALE LC_TIME**
- **CURRENT TIME**
- **CURRENT TIMESTAMP**
- **CURRENT TIMEZONE**
- **CurTime**
- **DANGLING**
- **DAO**
- **Data Architect**
- **database**
- **Database Partitioning Feature (DPF)**
- **data event**
- **DATALINK**
- **Data Links Manager**
- **data range**
- **Data Studio**
- **data type**
- **DATE**
- **DATE_ADD**
- **DATE_SUB**

- **DATETIME2**
- **DATETIMEOFFSET**
- **DB2**
- **DB2SECURITYLABEL**
- **DB2SQLSTATE**
- **DBA_OBJECTS**
- **DBCLOB**
- **DBCS**
- **DBMS_XMLSCHEMA**
- **DBTimeZone**
- **DBURIType**
- **DBWScript**
- **DDL**
- **DECFLOAT**
- **DECIMAL**
- **DELETE RESTRICT**
- **delimited identifier**
- **DEREF**
- **DESCRIBE**
- **Design View**
- **DETERMINED BY**
- **DETERMINISTIC**
- **Developer Data Modeler**
- **directly usage**
- **direct recursive type**
- **DISTINCT**
- **DML**
- **domain constraint**
- **DO statement**
- **DOUBLE**
- **DOUBLEPRECISION**
- **DROP**
- **dummy date**

- **E-R data model**
- **EBCDIC**
- **Enforce Referential Integrity**
- **ENUM**
- **ERWin Data Modeler**
- **Excel**
- **EXPLAIN**
- **Express-C**
- **extended data type**
- **Extract**
- **FIPS Flagger**
- **FLOAT**
- **FLUSH**
- **FOR BIT DATA**
- **FOR EACH ROW**
- **foreign key**
- **functional dependency**
- **GENERATE_UNIQUE**
- **GEOGRAPHY**
- **GEOMETRY**
- **geospatial data type**
- **GetDate**
- **GetUTCDate**
- **global temporary table**
- **global transaction identifier**
- **GRAPHIC**
- **grbit**
- **GROUP BY**
- **GTID**
- **GUID**
- **Hadoop**
- **HEX**
- **HIERARCHYID**

- **HP-UX**
- **HTTPURIType**
- **IBM**
- **identifier**
- **IDENTITY**
- **indexed**
- **index file**
- **indirectly usage**
- **indirect recursive type**
- **infinity**
- **information_schema**
- **InfoSphere Data Replication**
- **InfoSphere Warehouse**
- **INSERT INTO**
- **INT**
- **INTEGER**
- **INTERVAL DAY**
- **INTERVAL YEAR**
- **iSeries**
- **key**
- **LAST_INSERT_ID**
- **LIMIT**
- **Limit to List**
- **Linux**
- **LOAD DATA INFILE**
- **LOB technology**
- **Local_Timestamp**
- **local temporary table**
- **LOGICAL**
- **LONG**
- **LONGBLOB**
- **LONG RAW**
- **LONGTEXT**

- **Lookup**
- **Lookup Wizard**
- **Management Studio**
- **MatBase**
- **materialized view**
- **MAX_STRING_SIZE**
- **MAXDB**
- **MBCS**
- **MEDIUMBLOB**
- **MEDIUMINT**
- **MEDIUMTEXT**
- **Microsoft**
- **minimal uniqueness**
- **Mixed Based Replication (MBR)**
- **MIXED DATA**
- **MIXED DECP**
- **MONEY**
- **MongoDB**
- **MSysRelationships**
- **MULTILINESTRING**
- **MULTIPLE BYTE**
- **MULTIPOINT**
- **MULTIPOLYGON**
- **MyISAM**
- **MySQL**
- **NaN**
- **NCHAR**
- **NCLOB**
- **NDBCLUSTER**
- **NESTED TABLE**
- **NEWID**
- **Northwind**
- **NOT NULL**

- not unique index
- Now
- number
- NUMBER
- number sequence
- NVARCHAR
- NVARCHAR2
- OBJECT
- ODBC
- Office
- OID
- OLEDB
- ON COMMIT PRESERVE ROWS
- ON DELETE
- ON UPDATE
- OpenSolaris
- OpenVMS
- Oracle
- Oracle Express
- Oracle Golden Gate
- ORDER BY
- ordinary array
- ordinary identifier
- OS X
- PARAMETERS
- PIVOT
- PL/SQL
- POINT
- POLYGON
- PostgreSQL
- primaryKey
- primary key
- pureScale

- **QBE**
- **QMF**
- **Query Design**
- **Query Management Facility**
- **quiet NaN**
- **QUOTED_IDENTIFIER**
- **R-I-C-0**
- **R-I-C-1**
- **R-I-C-2**
- **R-I-C-3**
- **R-I-C-4**
- **R-I-C-5**
- **R-I-D-0**
- **R-I-D-1**
- **R-I-D-2**
- **R-I-P-0**
- **R-I-P-1**
- **R-I-P-2**
- **R-I-P-3**
- **R-I-P-4**
- **R-I-P-5**
- **R-I-T-1**
- **R-I-T-2**
- **R-I-T-3**
- **R-I-T-4**
- **RAW**
- **REA0**
- **REA1**
- **REA2**
- **REA2013A0**
- **REA2013A0-2**
- **REAF0-1**
- **REAF0-2**

- **REAF0'**
- **REAL**
- **recursive type**
- **REF**
- **REFERENCE**
- **REFERENCES**
- **referential integrity**
- **REGEXP**
- **relational scheme**
- **REPLACE**
- **Required**
- **RESET**
- **reverse engineering**
- **root table**
- **root type**
- **Row Based Replication (RBR)**
- **ROW BEGIN**
- **ROW CHANGE TIMESTAMP**
- **ROW END**
- **ROWGUIDCOL**
- **ROWID**
- **ROWVERSION**
- **Runtime**
- **Sakila**
- **SBCS**
- **SDO_GEOMETRY**
- **SDO_GEORASTER**
- **SDO_TOPO_GEOMETRY**
- **SecureFile LOB**
- **SELECT**
- **Sequence**
- **SERIAL**
- **Server Pubs**

- **SessionTimeZone**
- **SET**
- **SharePoint**
- **SHORT**
- **SHORT TEXT**
- **SHOW**
- **Show System Objects**
- **SI_AverageColor**
- **SI_Color**
- **SI_ColorHistogram**
- **SI_FeatureList**
- **SI_PositionalColor**
- **SI_StillImage**
- **SI_Texture**
- **signaling NaN**
- **SINGLE**
- **SMALLDATETIME**
- **SMALLINT**
- **SMALLMONEY**
- **Smart Draw**
- **Solaris**
- **Spatial and Graph**
- **Spatial Extender**
- **SQL**
- **SQL/MM StillImage**
- **SQL_VARIANT**
- **SQL Server**
- **Statement Based Replication (SBR)**
- **STDDT**
- **STRAIGHT_JOIN**
- **strongly typed data type**
- **STRUCTURED**
- **subnormal number**

- subtable
- SUM
- Supertable
- supertype
- SYSCAT
- SysDate
- SysDateTime
- SysDateTimeOffset
- SYSDUMMY
- SYSFUN
- SYSIBM
- SYSPROC
- SYSTEM_TIME
- system views
- SysTimestamp
- SysUTCDateTime
- szColumn
- szObject
- szReferencedColumn
- szReferencedObject
- szRelationship
- table
- tablespace
- target type
- temporary table
- TEXT
- Time
- TIMESTAMP
- TIMESTAMP WITH LOCAL TIME ZONE
- TIMESTAMP WITH TIME ZONE
- TINYBLOB
- TINYINT
- TINYTEXT

- **TOP**
- **TRANSACTION START ID**
- **TRANSFORM**
- **TRANSLATE**
- **Trigger**
- **TRIM**
- **tuple constraint**
- **typed table**
- **typed view**
- **type hierarchy**
- **UDT**
- **UNICODE**
- **UNION ALL**
- **UNIQUEIDENTIFIER**
- **unique index**
- **uniqueness**
- **Universal Naming Convention (UNC)**
- **Unix**
- **UNPIVOT**
- **UNSIGNED**
- **UPDATE OF**
- **UPDATE RESTRICT**
- **URIFactory**
- **URIType**
- **UROWID**
- **USER_OBJECTS**
- **user defined data type**
- **UTF-16**
- **UTF-8**
- **VALIDATED**
- **VALIDATING**
- **Validation Rule**
- **VARBINARY**

- **VARCHAR**
- **VARCHAR2**
- **VARGRAPHIC**
- **VARRAYS**
- **VBA**
- **virtual column**
- **Visio**
- **Visual Paradigm**
- **VM/VSE**
- **weakly typed data type**
- **Windows**
- **WTDDT**
- **XML**
- **XMLEXISTS**
- **XMLQUERY**
- **XMLType**
- **XPath**
- **XQuery**
- **Year**
- **Yes (No Duplicates)**
- **YESNO**
- **zOS**

REFERENCES

So all of the music has reference, or is inspired by something of the dharma[117]
that I've come in contact with.
—Joseph Jarman

Bradford, R. (2011). *Effective MySQL: Optimizing SQL Statements*. McGraw-Hill Osborne Media: New York, NY.
Feuerstein, S. (2008). *Oracle PL/SQL Best Practices. 2nd edition.* O'Reilly Media: Sebastopol, CA.

[117] *Dharma*, in Sanskrit, means the "Law that upholds, supports, or maintains the regulatory order of the universe".

IBM Corp. (2014). *DB2 database product documentation.* (http://www01.ibm.com/support/docview.wss?uid=swg2700.9474).

IBM Corp. (2014). *DB2 for Linux, UNIX, and Windows Best Practices.* (https://www.ibm.com/developerworks/community/wikis/home?lang=en#!/wiki/Wc9a068d-7f6a6_4434_aece_0d297ea80ab1/index?sort=updated&tag=best_practices).

MacDonald, M. (2013). *Access 2013: The Missing Manual.* O'Reilly Media: Sebastopol, CA.

Mancas, C. (2001). *Databases for Undergraduates Lecture Notes*: Computer Science Dept., *Ovidius* State University, Constanta, Romania.

Mancas, C., Mancas, S. (2005). *MatBase* E-R Diagrams Subsystem Metacatalog Conceptual Design. In Proc. IASTED DBA 2005. Conf. on DB and App., 83–89, Acta Press, Innsbruck, Austria.

Microsoft Corp. (2013). *Access 2013. (http://msdn.*microsoft.com/en-us/library/office/fp179695(v=office.15). aspx).

Microsoft Corp. (2014). *SQL Server Database Engine* (http://msdn.microsoft.com/en-us/library/ms1878.75.aspx).

Oracle Corp. (2014). *MySQL 5.7 Reference Manual.* (http://dev.mysql.com/doc/refman/5.7/en/).

Oracle Corp. (2014). *Oracle Database Documentation Library 12c Release 1.* (http://docs.oracle.com/database/121/index.htm).

Vaughn, W. R., Blackburn, P. (2006). *Hitchhiker's Guide to Visual Studio and SQL Server: Best Practice Architectures and Examples, 7th edition.* Addison-Wesley: Boston, MA.

CHAPTER 5

CONCLUSION

The possession of knowledge does not kill the sense of wonder and mystery.
There is always more mystery.
—Anaïs Nin

CONTENTS

5.1   Database Axioms 595

5.2   Why Do We Need Another Conceptual Level for Expert DB
      Design?.. 604

5.3   What Are the Most Important Things That We Should Be
      Aware of in DBs? 609

Keywords 611

5.1   DATABASE AXIOMS

God does not care about our mathematical difficulties;
He integrates empirically.
—Albert Einstein

Generally, in order to be able to compare several solutions to a same problem, but also to judge a solution for correctness, elegance, optimality, etc., the existence of a widely accepted set of guiding axioms is a must.

Unfortunately, for database design, implementation, manipulation and optimization there is no such established set of axioms. This section presents my corresponding set proposal.

5.1.1 DESIGN AXIOMS

We cannot solve our problems with the same thinking we used when we created them.
—Albert Einstein

The *relational db scheme design problem* is, essentially, two folded: what table schemes should be included in a db scheme? what columns and constraints should contain every of its table schemes?

The following db design axioms provide criteria for comparing and hierarchizing alternative design solutions:

A-DA0. Non-relationally rdb design axiom: rdb schemes should not be designed relationally; instead, they should first be designed by using E-RDs and associated restriction sets; then, conceptual design should be refined by using a more powerful data model, able to heavily assist us in discovering and designing all existing constraints, as well as in correcting and refining all errors in E-RDs; finally, resulted conceptual schemes should be (preferably, automatically) translated into corresponding relational ones, plus associated non-relational constraint sets. As an emergency temporary shortcut, you could skip the intermediate data model level, but never the E-RDM one.

A-DA01. Data plausibility axiom: any db instance should always store only plausible data; implausible ("garbage") data might be stored only temporarily, during updating transactions (i.e., any time before start and after end of such a transaction all data should be plausible).

A-DA02. No semantic overloading axiom: no fundamental object set or property should be semantically overloaded (i.e., any such object set or property should have only one simple and clear associated semantics).

A-DA03. Non-redundancy axiom: in any db, any fundamental data should be stored only once, such that inserting new data, as well as updating or deleting existing one should be done in/from only one table row.

A-DA04. Unique objects axiom: just like, generally, for set elements, object sets do not allow for duplicates (i.e., each object for which data is stored in a db should always be uniquely identifiable through its corresponding data).[118]

---

[118] Consequently, no db fundamental table should ever contain duplicate rows (the so-called *table syntactic horizontal minimality*). Moreover, for such tables, except for some very few exceptions (namely, tables corresponding to subsets or to object sets of the type poultry/rabbit/etc. cages), their rows should

A-DA05. Best possible performance axiom: db design, implementation and optimization should guarantee obtaining the maximum possible performance (i.e., the overall best possible execution speed for critical queries and updates, as well as the best possible average execution speed for the noncritical ones).

A-DA06. Controlled-redundancy axiom: only for satisfying the best possible performance axiom above, dbs might store information too (i.e., computed, thus redundant data), provided that redundancy is strictly controlled: users should at most read information (i.e., never write or delete it) and information should be automatically recomputed immediately after corresponding fundamental data has changed.

A-DA07. No trivial constraints axiom: no trivial constraint (e.g., any set includes both the empty set and itself, any set is equal to itself, etc.) should ever be added to a db scheme.

A-DA08. No constraints, but keys on redundant data axiom: no constraint should be enforced on redundant data, except for keys: as redundant data is automatically computed, it should always be valid, so adding constraints on it would only slow down updates speed. The exception for keys is justified as, sometimes, redundant data is needed in order to correctly model all existing keys.[119]

A-DA09. Constraints discovery axiom: for any nontrivial and noncontradictory and not implied (i.e., computable) restriction existing in the modeled subuniverse, any db should enforce a corresponding constraint.

A-DA10. No implausible constraints axiom: no db should enforce a constraint that does not correspond to a restriction existing in the modeled subuniverse.

A-DA11. Constraint sets optimality axiom: for any db scheme, no implied constraint should be enforced and the constraint set should be enforceable in the minimum possible time (i.e., there should not exist another equivalent constraint set whose enforcement takes less time in the given context).

A-DA12. Constraint sets coherence axiom: for any db scheme, the set of its constraints should be coherent (i.e., it should not be contradictory,

---

not differ between them only in their surrogate key column, but in other columns as well (the so-called *table semantic horizontal minimality*).

[119] see, for example, keys *Country • Zip* from Fig. 3.5 (Subsection 3.2.3) and *ConstrName • *DB* from Table 3.11 (Subsection 3.7.2)

thus only allowing for some object sets or properties the empty set as their instances).

A-DA13. Minimum inclusions axiom: in any db model, for any inclusion between object sets $T \subseteq S$, S should be the smallest set including T.

A-DA14. Inclusions acyclicity axiom: no fundamental object set inclusions chain should be a cycle (dually, for any db scheme, its fundamental inclusions graph should be acyclic).

A-DA15. Sets equality axiom: no equality of two fundamental object sets should ever be added to a conceptual db scheme (i.e., such equalities, if any, are always to be interpreted simply as declaring object set aliases).[120]

A-DA16. Mappings equality axiom: no equality of two fundamental mappings should ever be added to a db scheme (i.e., such equalities, if any, are always to be interpreted simply as declaring mappings aliases).[121]

A-DA17. ER-Ds axiom: except for an overview one, no ER-D should have more than one letter/A4 page and its nodes and edges should not intersect between them, except for tangent points of edges to their connected nodes.

A-DA18. ER-D cycles axiom: all ER-D cycles should be thoroughly analyzed in order to discover whether or not they should be broken or they need declaring additional constraints in the corresponding db scheme.

A-DA19. Naming axiom: all dbs and their components (object sets, mappings, constraints, tables, columns, queries, stored procedures, indexes, etc.) should be consistently named (i.e., strictly adhering to some naming conventions), always embedding as much unambiguous semantics as possible in each name. Names given during data analysis should be kept identical on all design levels and down to corresponding db implementations. Names for elements of any such sets should be unique within their corresponding sets.

A-DA20. Documentation axiom: Document all of your software (consequently, data analysis, db design, implementation, manipulation and

---

[120] Dually, rdb implementations may also contain, for optimality reasons, several tables corresponding to a same object set obtained by vertical and/or horizontal splitting, provided that all of them are correctly linked and processed: instances of those obtained horizontally should always remain pairwise disjoint, whereas any of the $n-1$ schemas ($n > 1$, natural) obtained vertically should be implemented as corresponding to subsets of the remaining (master) one.

[121] That is no table should ever contain two columns whose values should always be equal (the so-called *table vertical minimality*).

optimization too) as if tomorrow you will be struck by total amnesia, but do not overdocument.

Obviously, axioms *A-DA*02, *A-DA*05, *A-DA*09, *A-DA*10, *A-DA*17, and *A-DA*19 cannot be guaranteed by any data model, but only by us, as their users.

5.1.2 IMPLEMENTATION AXIOMS

Beware of the problem of testing too many hypotheses; the more you torture the data, the more likely they are to confess, but confessions obtained under duress may not be admissible in the court of scientific opinion.
—Stephen M. Stigler

A-DI0. Instances consistence axiom: any possible rdb instance should be consistent (not only with respect to relational-type constraints, but to the non-relational ones too).[122]

A-DI01. Tables' semantics axiom: in any rdb, each fundamental table should correspond to an object set (be it actual or virtual) in the modeled subuniverse of discourse.

A-DI02. Non-static fundamental objects representation axiom: for each nonstatic fundamental object set of the modeled subuniverse there should correspond a fundamental table in the modeling rdb (dually, only static fundamental object sets, like, for example, rainbow colors, sex types, etc., could be represented not by tables, but, for example, by enumerated sets).

A-DI03. Tables indivisibility axiom: any fundamental object set should always be modeled with only one table; these tables may be horizontally and/or vertically splitted in actual implementations only when needed, but under strict db applications control (see footnote of axiom *A-DA*15 above).

A-DI04. Columns' semantics axiom: in any table, each fundamental column should correspond to a property of the objects in the object set corresponding to that table.

A-DI05. Empty instances axiom: any fundamental table should accept both empty and nonempty instances.

---

[122] That is all existing constraints (in the corresponding subuniverse of discourse), be them relational or not, should be enforced in the db.

A-DI06. Constraints enforcement axiom: for each fundamental table, any of its constraints whose type is provided by the target RDBMS has to be enforced through that RDBMS; all other db constraints should be enforced by all software applications built on top that rdb.[123]

A-DI07. Domain constraints axiom: any fundamental column should have an associated domain constraint restricting its range to a plausible finite subset of the corresponding data type.

A-DI08. Rows uniqueness axiom: any fundamental table row should be always uniquely identifiable according to all existing uniqueness constraints in the corresponding subuniverse and it should correspond to a unique object from the corresponding objet set.

A-DI09. No void rows axiom: for any fundamental table, at least one semantic fundamental column should not accept null values.

A-DI10. Referential integrity axiom: any foreign key value should always reference an existing value (dually, no rdb should ever contain dangling pointers).

A-DI11. Primary keys axiom: any fundamental table should have a surrogate integer primary key whose range cardinality should be equal to the maximum possible cardinality of the corresponding modeled object set; except for tables corresponding to subsets, primary keys should store auto generated values.

A-DI12. Foreign keys axiom: any foreign key should be simple (i.e., not concatenated), reference a surrogate integer primary key and have as range exactly the range of the referenced primary key.

A-DI13. Key Propagation Principle axiom: any mapping (i.e., many-to-one or one-to-many relationship) between two fundamental object sets should be implemented according to the *Key Propagation Principle*: a foreign key referencing its codomain (i.e., the "one" side) is added to the table corresponding to its domain (i.e., the "many" side).

A-DI14. Set inclusion relational implementation axiom: any inclusion between two object sets $T \subseteq S$ should be represented by a dedicated foreign key of T referencing the primary key of S; one of them may be the primary key of T.

---

[123] That is never leave relational constraints enforcement to db applications: otherwise, first of all, developers might forget to enforce them in some particular cases (if not in all of them…) and, secondly, db instances might violate even those enforced by the applications, whenever they are updated directly, not through application control.

A-DI15. No superkeys axiom: no superkey should ever be enforced; generally, no other constraints than those declared in the corresponding conceptual db scheme should ever be enforced.[124]

A-DI16. No modification anomalies axiom: tables of any rdb should be completely anomaly free.[125]

A-DI17. Minimal data storing axiom: no data should be stored more than once in any db.

Obviously, axiom *A-DI06* cannot be guaranteed by any data model or DBMS, but only by us, as their users.

5.1.3 INSTANCES MANIPULATION AXIOMS

> *The competent programmer is fully aware of the limited size of his own skull.*
> *He therefore approaches his task with full humility,*
> *and avoids clever tricks like the plague.*
> —Edsger W. Dijkstra

A-DM0. Positive data axiom: all data stored in conventional dbs is positive.

A-DM01. Closed world axiom: facts that are not known to be true (i.e., for which there is no stored data) are implicitly considered as being false.

A-DM02. Closed domains axiom: there are no other objects than those for which there is stored data.

A-DM03. Null values set axiom: there is a distinguished countable set of *null values*.

A-DM04. Null values axiom: *Null values* may model either (temporary) unknown or inapplicable or no information data.

A-DM05. Available columns axiom: only the columns of the tables, views and subqueries of the FROM clause are available for all other clauses of the corresponding *SQL* statements.

A-DM06. GROUP BY (golden rule) axiom: in the presence of a GROUP BY clause, corresponding SELECT clause can freely contain only column expressions also present in the GROUP BY one; all other available columns may be used in SELECT only as arguments of aggregation functions.

---

[124] Obviously, this is a particularization of axiom *A-DA*11 above.

[125] That is not only anomaly-free with respect to the domain and key constraints, but with respect to all other constraints, regardless of their type.

A-DM07. Relevant data and processing axiom: any query should consider and minimally process, in each of its steps, only relevant data; for example,

- ✓ the `SELECT` and `FROM` clauses should contain only needed columns and tables, respectively;
- ✓ use the `GROUP BY` clause only when needed (e.g., computing groups always containing only one row is senseless);
- ✓ in the presence of a `GROUP BY` clause, all logical conditions that may be evaluated before grouping (i.e., not containing aggregate functions) should be placed in the corresponding `WHERE` clause and only those that contain aggregated columns should be placed in the corresponding `HAVING` one;
- ✓ ordering should be done only on needed columns (expressions);
- ✓ joins, groupings, and as much as possible of the needed logical conditions should be performed on numeric columns (surrogate keys and foreign keys referencing them): only in the last step of a query should surrogate keys and pointers (i.e., foreign keys) be replaced by corresponding desired semantic columns.

A-DM08. ORDER BY axiom: except for well-founded exceptions, as part of a minimal courtesy towards our customers, query results should be presented ordered by the most interesting possible order (even if not explicitly asked for); dually, ordering should be done only once per query, as its last step.

A-DM09. Data importance axiom: as data is one of the most important assets of our customers, preserving its desired values, as well as consistence of the corresponding db instances is a must (e.g., do not ever insert, update or delete data that should not be inserted/updated/deleted).

A-DM10. Query semantics correctness axiom: any syntactically correct query computes a table, but neither its instance, nor even its header is guaranteed to be the result expected by our customers (e.g., some of our queries are, in fact, equivalent possible definitions of the empty set).[126]

A-DM11. Fastest possible manipulation axiom: at least in production environments, data manipulations should be done with best possible

---

[126] In order to guarantee their semantic correctness (the only one relevant to our customers), if you cannot formally prove it, at least informally read each query meaning immediately after you consider it finalized and compare reading with corresponding initial request: if they do not match exactly, then reconsider query design and/or development.

algorithms and technologies, so that processing speed be the fastest possible.

A-DM12. Regularly backup fundamental data intelligently axiom: fundamental data has to be regularly backed-up, by intelligently using all corresponding RDBMS's facilities (incremental, hot, etc.).

A-DM13. Always process sets (not elements) axiom: in data manipulation, do not think like in imperative (procedural, algebraic) programming, in terms of elements of sets, but in terms of sets of elements.[127]

Obviously, axioms *A-DM07*, *A-DM08*, *A-DM09*, *A-DM10*, and *A-DM13*, cannot be guaranteed by any data model or DBMS, but only by us, as their users.

5.1.4 OPTIMIZATION AXIOMS

> *We must develop knowledge optimization initiatives*
> *to leverage our key learnings.*
> —Scott Adams

Even if no intuitive support may be given for the time being for optimization, which heavily depends on current corresponding rdb instances, workload, and data access paths (see second volume of this book), for having all axioms grouped together, optimization ones are already presented in this section too.[128]

Optimization also heavily depends on the used DBMS versions; as such, first are presented here only the very general ones.

Finally, specific optimization axioms are presented in the second volume for the five RDBMS versions considered as examples in this book.

A-DO0. Reduce data contention axiom: intelligently use hard disks, big files, multiple tablespaces, partitions, and segments with adequate block sizes, separate user and system data, avoid constant updates of a same row, etc. in order to always reduce data contention to the minimum possible.

---

[127] Data is not like trees (that you have to cut one by one), but rather like grains (that you may crop one by one, but then all of us would eventually starve).

[128] Consequently, beginner readers might skip it at a first sequential read of this volume and return to it after reading and understanding the second volume.

A-DO1. Minimize db size axiom: regularly shrink tablespaces, tables, and indexes for maintaining high processing speeds and minimal db and backup sizes.

A-DO2. Maximize use of RDBMS statistics axiom: regularly gather and intelligently use statistics provided by RDBMSs for db fine-tuning.

A-DO3. Follow RDBMS advisors recommendations axiom: regularly monitor and apply all recommendations of RDBMSs' advisors.

A-DO4. Commit as infrequently as is practical axiom: for performance reasons, RDBMSs are generally writing asynchronously to the db data files, but committing requires synchronous writes, because they must be guaranteed at their occurring moments of time.[129]

Obviously, axioms *A-DO*0, *A-DO*3 and *A-DO*4 cannot be guaranteed by any data model or DBMS, but only by us, as their users.

5.2 WHY DO WE NEED ANOTHER CONCEPTUAL LEVEL FOR EXPERT DB DESIGN?

Whenever you set out to do something, something else must be done first.
—(Corollary 6 to the first Murphy's Law:
If anything can go wrong, it will.)

Essentially, this first volume advocates using at least three levels from Fig. 1.1 (namely the E-RDM, RDM, and RDBMS ones) and five corresponding FE algorithms (*A*0, *A*1–7, *A*7/8–3, A8, and one of the *AF*8′), whenever having to create a new rdb or to extend an existing one. Dually, when having to understand a not enough documented rdb, it advocates using same three levels and three corresponding RE algorithms (namely, one of the *REAF*0', *REA*0, and *REA*1–2).

Obviously, there is an even shorter possible path presented in this book too, namely using one of the composed algorithms from the *AF*1–8 family, instead of the composition of one of the *AF*8 and *A*1–7. This should be considered only by exception, when under time pressure: as we have already seen, skipping the RDM level means losing portability (sometimes even between different versions of a same RDBMS).

---

[129] The need for synchronous write is one of the reasons why the online redo log can become a bottleneck for applications.

Anyhow, even in such cases, do not forget to also apply $A7/8–3$ in the end, for all new and/or modified fundamental tables. Please note that, although it is presented in the RDM framework, you can straightforwardly apply it to any RDBMS version instead.

Dually, it is generally preferable not only under time pressure to directly use the needed member of the $REAF0–2$ shortcut family, instead of composing $REA1–2$ and the needed one of the $REAF0$ family: when working with a legacy not documented rdb, it is very rarely that you would also come across another equivalent rdb implemented in another RDBMS version.

5.2.1 IS E-RDM NEEDED? IS IT ENOUGH?

> *Science does not know its debt to imagination.*
> —Ralph Waldo Emerson

Of course that, at "co-volume", one might reduce this minimum advisable forward path to only two algorithms: the needed one from a family, let's say, $AF0–8$, obtained by composing the corresponding one from $AF1–8$ and $A0$, and $A7/8–3$—thus eliminating the E-RDM. Personally, I wouldn't ever recommend such an extreme shortcut, at least from the following three fundamental reasons:

- customers are very rarely experts in mastering the considered RDBMS version and even too rarely in correctly and fully bridging between their subuniverse of interest and that RDBMS version (think that, generally, among your decision maker customers are also CFOs, CEOs, stakeholders, etc., not only CTOs): skipping the E-RDM level results, from this point of view, in replacing its business-oriented jargon with, let's say, an *Oracle 12c* one, whenever you communicate with your customers, which is obviously not an advantage, but rather a major disadvantage;
- very few RDBMS versions have satisfactory graphical capabilities for showing (foreign key) links between the tables of a rdb (e.g., MS *SQL Server* is showing only one line between any two tables, regardless of how many links there are between them, and even when there is only one such link, you cannot determine which are

the corresponding connected columns; not even MS *Access*, which is far more capable from this point of view, showing any link apart, exactly between the corresponding columns, is not satisfactory: for one-to-one relationships you can only guess which is the foreign key and which the corresponding referenced column, for linked tables belonging to other dbs than the current one the type of the relationship is not showed, and for each link between two tables or same table instances they use a different table instance for the referenced table, which is generally confusing very many beginners, at least): skipping the E-RDM level results, from this point of view, in losing most of the very powerful graphical advantages of the E-RDs;

• even when very experienced, only using a technological level inevitably distorts your thinking habits, which tends to favor technology over conceptual data analysis and db architecturing; sooner or later, you will start neglecting, for example, also considering the business rules that are formalizable only as non-relational constraints, as well as the informal descriptions of the corresponding subuniverses of discourse: skipping the E-RDM level results, from this point of view, in losing both the accuracy of your data models and their informal descriptions.

Of course that, regarding this latter above reason, one might argue that informal descriptions may be also obtained directly from the corresponding rdbs and that, anyhow, at this level, we do not care at all about non-relational constraint enforcement (so what's the use of discovering and including such constraints in our models?) and not even about making sure that there are no other ones (apart those already discovered) that govern the corresponding subuniverses as well.

To the first objection, regarding the informal descriptions, I agree, provided it is actually done every time and also including all of the non-relational constraints that govern the corresponding subuniverse.

To the second one, regarding the usefulness of the non-relational constraints, I would stress again that any business rule governing a subuniverse of interest to our data models and corresponding dbs should be discovered, included in the model, and enforced: otherwise, data plausibility is compromised. Do, for example, civil engineering architects or car designers ignore any constraint in their fields, hoping that the workers who implement their models will discover and enforce them during manufacturing or, even worse, that not the workers, but the users of those products will do it?

Moreover, generally, not only me would always stress that, for its sake, as well as for our's (workers in this field or/and users of its artifacts), the IT industry should rapidly mature at least at the level of, let's say, the car manufacturing's one (as building has already more than 5,000 years and is very, very hard to narrow such gap rapidly...): would you buy or even only drive or use a car (driven by somebody else) that was artisanally manufactured, without proper design, without taking into consideration all physics and safety rules that govern this subuniverse?

To conclude the above considerations, I am only one of the very many who are convinced that we should not skip the E-RDM, but, on the contrary, we should always make full and intelligent use of it in the first step of the conceptual data analysis and modeling process.

Dually, unfortunately, even extended with restriction sets, E-RDM is not enough for data modeling, at least from the following four points of views:

- E-RDM is totally ignoring the following db design axioms (see Subsection 5.1.1): *A-DA*03, *A-DA*05, *A-DA*07, *A-DA*08, *A-DA*11 to *A-DA*16, *A-DA*18, and *A-DA*20.
- E-RDM is only partially taking into account the following db design axioms (see Subsection 5.1.1): *A-DA*01 (as discovery of all existing business rules in any subuniverse of discourse is not even an aim of restriction sets), *A-DA*04 (as for restriction sets declaring only one uniqueness per object set is enough), and *A-DA*20 (as only informal descriptions of the subuniverses of discourse to be modeled are compulsory).
- E-RDM has no means to check at least the syntactical correctness of E-R data models (e.g., are all nonfunctional relationship-type object sets and all functional relationships and all attributes correctly defined mathematically?).
- Moreover, there are very few commercial (and even prototype) DBMSs that provide E-R user interfaces which make completely transparent to users the RDM (or any other) level of db scheme

implementation, and even much more fewer the ones that also provide E-R query languages at least as powerful as *SQL*.

All these four considerations imply that E-RDM alone is not enough for accurate, expert conceptual data analysis, db design, implementation, and usage.

To conclude this subsection, the answers to the questions in its title are: "definitely *yes*" and "definitely *no*", respectively.

5.2.2   IS RDM NEEDED? IS IT ENOUGH?

> *What is now proved was only once imagined.*
> —William Blake

Due to the fact that RDBMSs are still the norm in this field and that they are so close to RDM (and, hence, to one another), there is a great temptation to skip RDM completely, in the processes of both db design and usage.

However, as already discussed in details in Section 3.13, the answers to the questions in the title of this subsection are: "*yes*" (at least for portability and RLs reasons) and "definitely *no*", respectively.

I would only add here to the pros that it is always much better not to try to solve any problem from the beginning in technological terms, but in higher, more abstract ones. In particular, in the RDM case, despite the great number of fundamental compatibilities between nearly all RDBMS versions, there are, however, enough technological differences too between them, sometimes even only between different versions of a same RDBMS.

Dually, for the cons, I would also add here the following two other reasons:

- RDM is totally ignoring the following db design axioms (see Section 5.1): *A-DA*07, *A-DA*08, *A-DA*11 to *A-DA*16, *A-DA*18, *A-DI*01 to *A-DI*05, *A-DI*08, *A-DI*09, and *A-DI*12 to *A-DI*17.
- RDM is only partially taking into account the following db design axioms (see Section 5.1): *A-DA*01, *A-DI*0 (as it does not take into account non-relational constraints), and *A-DA*04 (as users are not coerced to declare at least one key for any fundamental table).

5.2.3 WHAT SHOULD AN INTERMEDIATE MODEL BETWEEN E-RDM AND RDM BRING?

> *The important thing in science is not so much to obtain new facts as to discover new ways of thinking about them.*
> —Sir William Bragg

Not only from my point of view, first of all, such a data model should fully take into account all db axioms from Section 5.1: even for those that it cannot automatically guarantee, it should at least provide a framework and tools for its users to help them comply with.

Next, it should also incorporate:

- an algorithm of type $A3$, for the (at least syntactical) validation of the initial E-R data models;
- one of type $A4$, for assisting E-RD cycles analysis (because, as shown in the second chapter of the second volume of this book, they are very frequently embedding lot of non-relational constraints);
- one of type $A5$, for assisting the not always obvious task of guaranteeing the coherence of the constraint sets;
- one of type $A6$, for assisting the much more difficult task of guaranteeing the minimality of the constraint sets.

Finally, above all, in order to guarantee data plausibility, such a model should also incorporate all needed constraint types, not only the relational ones: ideally, it should provide at least the subclass of closed Horn clauses[130].

There are hundreds of higher (than RDM and E-RDM) level conceptual data models, but, to my knowledge, none of them incorporates all the above as well, except the (E)MDM that is presented and used in the second volume of this book.

5.3 WHAT ARE THE MOST IMPORTANT THINGS THAT WE SHOULD BE AWARE OF IN DBS?

> *The most important of my discoveries have been suggested to me by my failures.*
> —Humphrey Davy

---

[130] Recall that the Horn clauses are the biggest class of first order predicate formulas for which the implication problem is decidable.

As we have already seen, no data models and/or DBMS might take into account all above db axioms and best practice rules. Consequently, I would like to take this final (in the realm of this volume) opportunity to summarize once more the main ideas behind the crucial ones in the following dodecalog:

1. *Data is* increasingly becoming *the most important asset* of customers: we always must commit to keep it plausible, safe, and available at the maximum possible speed.

2. *Only humans can understand semantics* and break it down into atomic ones: algorithms and/or machinery can only assist us with it.

3. It is not only in our customers', but also in our own interest to *never semantically overload* either object sets or mappings.

4. *Discovering all existing business rules* that govern any subuniverse of interest, adding them to our data models, and enforcing them in any corresponding db *is crucial in guaranteeing data plausibility*: *missing anyone of them allows for storing implausible data.*

5. Adding *any implausible constraint* to a data model *should be always banned*, as it would not allow for storing plausible data in the corresponding dbs.

6. For any db, its associated *constraint set should be coherent and minimal.*

7. *Without proper data analysis and conceptual db design*, resulting dbs have *very few chances to satisfy customers.*

8. Just like, for example, in the case of aircrafts, *following prescriptions of algorithms and best practice rules is the safest thing to do* in the db field too.

9. *Redundant data should always be strictly controlled*: corresponding values should always be read-only for users and should be immediately calculated automatically when needed.

10. *Object elements represented in dbs only by null values or/and which are not uniquely identifiable semantically* (i.e., not only by surrogate keys) within their object sets are useless, confusing, taking up storage space and slowing down query execution for nothing: consequently, they *should be banned from any db instance.*

11. When executed, a query always returns the correct answer to the corresponding customer question if and only if it was also semantically

(not only syntactically) correctly designed and coded (dually: only *syntactically correct queries are always returning correct answers to a question, but not necessarily to the desired one*).

12. When manipulating data, *thinking in terms of sets of elements is far better than thinking in terms of elements of sets*, as it results in much greater query execution speeds (just like in the case of parallel versus sequential programming) and much higher *SQL* (be it pure, extended, or embedded) developing productivity.

KEYWORDS

- **A-DA0**
- **A-DA1**
- **A-DA2**
- **A-DA3**
- **A-DA4**
- **A-DA5**
- **A-DA6**
- **A-DA7**
- **A-DA8**
- **A-DA9**
- **A-DA10**
- **A-DA11**
- **A-DA12**
- **A-DA13**
- **A-DA14**
- **A-DA15**
- **A-DA16**
- **A-DA17**
- **A-DA18**
- **A-DA19**
- **A-DA20**
- **A-DI0**
- **A-DI1**
- **A-DI2**
- **A-DI3**
- **A-DI4**
- **A-DI5**
- **A-DI6**

- **A-DI7**
- **A-DI8**
- **A-DI9**
- **A-DI10**
- **A-DI11**
- **A-DI12**
- **A-DI13**
- **A-DI14**
- **A-DI15**
- **A-DI16**
- **A-DI17**
- **A-DM0**
- **A-DM1**
- **A-DM2**
- **A-DM3**
- **A-DM4**
- **A-DM5**
- **A-DM6**
- **A-DM7**
- **A-DM8**
- **A-DM9**
- **A-DM10**
- **A-DM11**
- **A-DM12**
- **A-DM13**
- **closed world**
- **constraint enforcement**
- **constraint set coherence**
- **constraint set minimality**
- **controlled redundancy**
- **data analysis**
- **data contention**
- **data plausibility**
- **database design**
- **database design axiom**
- **database implementation axiom**
- **database instance manipulation axiom**
- **database optimization axiom**
- **(E)MDM**
- **E-RDM**

- **GROUP BY golden rule**
- **HAVING clause**
- **key propagation principle**
- **non-relational constraints**
- **null value**
- **ORDER BY clause**
- **positive data**
- **RDM**
- **referential integrity**
- **relational constraints**
- **SQL**
- **WHERE clause**

APPENDIX

MATHEMATIC PREREQUISITES FOR THE MATH BEHIND

Do not worry about your difficulties in Mathematics.
I can assure you mine are still greater.
—Albert Einstein

A *set* is a collection/ensemble/etc. of unique elements (i.e., sets do not allow for duplicates) whose order is irrelevant. Examples: the set of web sites, the set of your friends, the set of e-messages you received/sent this year, etc.

The number of elements of a set A is called its *cardinal* and is denoted by $card(A)$. For example, the cardinal of the set of rainbow colors ({red, orange, yellow, green, blue, indigo}) is 6, the one of the world's countries affiliated to the United Nations is 193, etc.

The *empty set* (denoted $\emptyset$) is the set with no elements; trivially, $card(\emptyset) = 0$. For example, the set of (natural) satellites of the Mercury planet is empty (since it has no satellites).

The *union* ($\cup$) of two sets is the set made out of all elements from all corresponding sets (and as sets do not allow for duplicates, the school union definition adds "taken only once"): $A \cup B = \{x \mid x \in A \vee x \in B\}$.

For example, if $A = \{a, b, c\}$ and $B = \{b, c, d\}$, then $A \cup B = \{a, b, c, d\}$; *CHILDREN $\cup$ PARENTS = PEOPLE*.

The *intersection* ($\cap$) of two sets is the set containing all and only the elements belonging to both of them: $A \cap B = \{x \mid x \in A \wedge x \in B\}$.

For example, if $A = \{a, b, c\}$ and $B = \{b, c, d\}$, then $A \cap B = \{b, c\}$; *SINGLES $\cap$ MARRIED_AT_LEAST_ONCE = WIDOWED $\cup$ DIVORCED*.

Two sets are *disjoint* if they don't share any element (i.e., their intersection is the empty set; for example, A and B are disjoint if $A \cap B = \emptyset$).

For example, if $A = \{a, b, c\}$ and $B = \{1,2,3\}$, then $A \cap B = \emptyset$; *NEVER_MARRIED $\cap$ MARRIED_AT_LEAST_ONCE = $\emptyset$*.

A set is *distinguished* if it is disjoint of any other set, i.e. none of its elements belongs to any other set.

A set A is *included* in a set B (denoted $A \subseteq B$ or $B \supseteq A$) if all elements of A (called the *subset*) also belong to B (called the *superset*). Trivially, any set includes itself and the empty set (both of them called *improper subsets*), and equal sets are each included in the other (i.e., $A \subseteq B \wedge B \subseteq A \Leftrightarrow A = B$). For subsets that may not be equal to their supersets, symbols $\subset$ and $\supset$ are used instead. $A \not\subseteq B$ denotes noninclusion. For example,

- EMPLOYEES $\subset$ PEOPLE,
- CHILDREN $\subseteq$ PEOPLE,
- TAKEOFF_AIRPORTS $\subseteq$ LANDING_AIRPORTS,
- LANDING_AIRPORTS $\subseteq$ TAKEOFF_AIRPORTS
- (so, TAKEOFF_AIRPORTS = LANDING_AIRPORTS = AIRPORTS),
- OCEANS $\not\subseteq$ SEAS, etc.

The *complement* (^{C}) of a set is the set of all elements not belonging to that set: $A^C = \{x \mid x \notin A\}$. For example, if $A = \{1\}$, A^C contains the whole Universe, except for the digit 1.

The *relative complement* of a set A with respect to a set B (or the *difference* between B and A) is the set of all elements of B that do not also belong to A: $B \setminus A = \{x \mid x \in B \wedge x \notin A\}$.

If $A \subseteq B$, then $B \setminus A$ is called the *absolute complement* of A with respect to B.

For example, if $A = \{1,2,3\}$, $B = \{1\}$, then $B \setminus A = \emptyset$ and $A \setminus B = \{2,3\}$; *PEOPLE \ CHILDREN = ADULTS* (so *ADULTS* is the absolute complement of *CHILDREN* with respect to *PEOPLE*); etc.

An *axiom* (*postulate*, *assumption*) is any premise mathematical statement, so evident as to be accepted as true without controversy, which serves as a starting point from which other statements are logically derived. Within the system they define, axioms (unless redundant) cannot be derived by principles of deduction, nor are they demonstrable by mathematical proofs, simply because they are starting points; there is nothing else from which they logically follow: otherwise, they would be classified as *theorems*. For example, in Euclidian geometry, an axiom states that all right angles are equal to one another, while Thales' theorem states that, in any circle, any triangle formed between three distinct points of the circle is a right one if and only if two of the points are on a circle's diameter.

Especially in metalogic and computability (recursion) theory (but not only), methods (procedures, algorithms[131]) taking some class of problems and reducing the solution to a finite set of steps that, for all class problems instances, always computes the right answers are called *effective*.

Functions (see their definition below) with an effective method are called *effectively calculable*. Several independent efforts to provide a formal characterization of effective calculability led to a variety of proposed definitions[132] (that were ultimately shown to be equivalent): the notion captured by these definitions is called (*recursive*) *computability*. The *Church–Turing thesis*[133] states that these two notions coincide: any number-theoretic function that is effectively calculable is recursively computable and vice-versa.

Generally, any function defined on and taking values from a set of formulas is called an *inference (derivation, deductive, transformation) rule*. In classical logic (as well as the semantics of many other nonclassical logics) inference rules are also truth preserving[134]: if the premises are true (under an interpretation) then so is the conclusion. Usually, only rules that are recursive (i.e., for which there is an effective procedure for determining whether, according to the rule, any given formula is the conclusion of a given set of formulae) are important.

For example, "if x is an ancestor of y and z is x's father, then z is an ancestor of y too" is a (transitive recursive) inference rule. *Modus ponens* $((p \wedge (p \Rightarrow q)) \Rightarrow q)$, *modus tollens* $((\neg q \wedge (p \Rightarrow q)) \Rightarrow \neg p)$, *resolution* $((p \vee q) \wedge (\neg p \vee r)) \Rightarrow (q \vee r))$, *hypothetical* $((p \Rightarrow q) \wedge (q \Rightarrow r)) \Rightarrow (p \Rightarrow r))$ and *disjunctive* $((p \vee q) \wedge \neg p) \Rightarrow q)$ *syllogisms* are well-known examples of propositional logic (see below) inference rules.

In linguistics, computer science, and mathematics, a *formal language* is a set of strings of symbols that may be constrained by rules that are

---

[131] Some authors consider only effective methods for calculating the values of a function as being *algorithms*.

[132] Namely, general recursion, Turing machines, and λ-calculus.

[133] Note that this is not a mathematical statement, so it cannot be proven mathematically.

[134] In many-valued logics, they preserve a general designation.

specific for it. The *alphabet* of a formal language is the set of symbols, letters, or tokens from which the strings of the language may be formed; frequently it is required to be finite. The strings formed from this alphabet are called *words*, and the words that belong to a particular formal language are sometimes called *well-formed words/formulas*. For example, all computer programming languages (and not only) are studied using the formal languages theory.

A formal language is often defined by means of a *formal grammar*[135]: a set of formation rules for rewriting strings (*rewriting rules*), along with a *start symbol* from which rewriting must start. The rules describe how to form strings from the language's alphabet that are valid according to the language's syntax. A grammar is usually thought of as a language generator. However, it can also sometimes be used as the basis for a *"recognizer"* — a tool that determines whether a given string belongs to the language or is grammatically incorrect.

A *formal system* is a mathematical well-defined system of abstract thought consisting of finite sets of symbols (its *alphabet*), axioms and inference rules (the latter two constituting its *deductive system*), plus a grammar for generating and/or accepting the well-formed formulas forming its formal language. For example, Euclid's *Elements* (or even Spinoza's *Ethics*, its apparently nonmathematical imitation), the predicate, propositional, lambda, and domain relational calculi are formal systems.

Deductive systems should be *sound* (i.e., all provable statements are true) and *complete* (i.e., all true statements are provable). Similarly, algorithms are said to be sound and complete (if they analyze/compute only and all possible valid cases/values of the corresponding subuniverse).

Propositional (or *sentential*) *logic* (or *calculus*) is a formal system in which formulas of a formal language may be interpreted as representing propositions. A system of inference rules and axioms allows certain formulas (called *theorems*) to be derived, which may be interpreted as *true propositions*. The series of formulas which is constructed within such a system is called a *derivation* and the last formula of the series is a theorem, whose derivation may be interpreted as a proof of the truth of the proposition represented by the theorem.

---

[135] For example, regular, context-free, deterministic context-free, etc.

Truth-functional propositional logic is a propositional logic whose interpretation limits the *truth* values of its propositions to two, usually denoted as (*Boolean*) *true* and *false*. Its *logical expressions* (yielding either *true* or *false*) operands are logical ones: the constants *true* and *false*, (in) equalities ($=, \neq, <, \leq, >, \geq$—also called the *standard math operators/relations*), logical functions, subexpressions enclosed in parenthesis, etc.; their fundamental operators are *not* ($\neg$), *and* ($\wedge$), and *or* ($\vee$); most important derived operators are *implication* ($\Rightarrow, \Leftarrow$) and *equivalence* ($\Leftrightarrow$). Truth-functional propositional logic is considered to be *zeroth-order logic*.

Zeroth-order logic *truth tables* are the following:
$\neg\,true = false$; $\neg\,false = true$.
$false \wedge true = false$; $false \wedge false = false$; $true \wedge true = true$;
$false \vee false = false$; $false \vee true = true$; $true \vee true = true$;
$a \Rightarrow b = \neg\,a \vee b = b \Leftarrow a$; $a \Leftrightarrow b = a \Rightarrow b \wedge b \Rightarrow a$.

First-order logic (*first-order predicate calculus, lower predicate calculus, quantification theory, predicate logic*) (*1OL*) is a formal system used in mathematics, philosophy, linguistics, computer science, etc., which extends propositional logic with the *existential* ("there is/are", denoted '∃') and *universal* ("for any", denoted '∀') *quantifiers* that are applied to predicates' variables.

A theory about some topic is usually first-order logic together with a specified domain of discourse (over which the quantified variables range), finitely many functions (which map from that domain into it), finitely many predicates (defined on that domain), and a recursive set of axioms (which are believed to hold for those things). Note that, sometimes, "theory" is understood in a more formal sense, which is just a set of sentences in first-order logic, and that, in first-order theories, predicates are often associated with atomic sets (i.e., not sets of sets).

In 1OL there are two classes of formulas: *open* ones, containing at least one variable occurrence not bound to any quantifier, and *closed* ones (*propositions*), in which all variable occurrences are bound to a quantifier. In any fixed interpretation, propositions are either true or false, whereas open formulas truth values depend on their actual variables values.[136]

---

[136] Note that db *constraints* are closed while *queries* are open formulas.

An operator θ over some set A is said to be:

✓ *associative*, if $a\ \theta\ (b\theta\ c) = (a\ \theta\ b)\ \theta\ c$, $\forall a, b, c \in A$ (case in which parenthesis may be dropped, and simply write $a\ \theta\ b\ \theta\ c$ instead and, generally, $a_1\theta...\theta\ a_n$);

✓ *commutative*, if $a\theta\ b = b\theta\ a$, $\forall a, b \in A$;

✓ *anticommutative*, if $a\theta\ b \neq b\theta\ a$, $\forall a, b \in A$;

✓ *distributive over another operator* τ, if $a\ \theta\ (b\ \tau\ c) = a\ \theta\ b\ \tau\ a\ \theta\ b$, $\forall a, b, c \in A$.

For example, for all number sets, addition is associative, commutative, and distributive over multiplication, while matrix multiplication is anticommutative.

Given any two sets A and B, their *Cartesian product*, denoted $A \times B$, is the set of all possible pairs $\langle x, y \rangle$, with x from A and y from B: $A \times B = \{(x, y) \mid x \in A \wedge y \in B\}$. Obviously, $card\ (A \times B) = card\ (A) * card(B)$. As $\times$ is associative, it may also be considered to be a *n-ary* operator, $n > 1$ natural. For example, if $A = \{1,2\}$, $B = \{a, b\}$, and $C = \{true, false\}$, then $A \times B \times C = \{\langle 1, a, true \rangle, \langle 1, a, false \rangle, \langle 1, b, true \rangle, \langle 1, b, false \rangle, \langle 2, a, true \rangle, \langle 2, a, false \rangle, \langle 2, b, true \rangle, \langle 2, b, false \rangle\}$.

$\times$ is anticommutative: for any sets A and B, $A \times B \neq B \times A$; the only commutative Cartesian products are the homogenous ones that is those defined on a same set: $A \times A = A \times A$, for any set A. For example, if $A = \{a, b\}$ and $B = \{1, 2, 3\}$, there are two (distinct!) Cartesian product sets on them: $A \times B = \{(a,1), (a,2), (a,3), (b,1), (b,2), (b,3)\}$ and $B \times A = \{(1, a), (2, a), (3, a), (1, b), (2, b), (3, b)\}$.

Given any two sets A and B, a *binary* or *dyadic* (*mathematical*) *relation* R on them ("from A to B") is any nonvoid subset of their Cartesian product. For example, let $A = \{a, b\}$, $B = \{1, 2, 3\}$, and $R = \{(a,1), (a,3), (b,2)\} \subset A \times B$. A scheme and instance type notation is also used: $R = (A, B; G_R)$, where A and B are called R's *underlying sets* (forming its *scheme*) and $G_R \subseteq A \times B$ is called R's *graph* (its *instance*). If $A = B$, R is said to be *homogenous*. Relations may have any natural number of underlying sets (i.e., they may be ternary, quaternary, etc., generally *n-ary*).

A *mapping* (*function*) is a binary relation satisfying the following two properties (constraints): total definition and functionality. A mapping $f \subseteq A \times B$ is denoted $f : A \rightarrow B$, with A being called its *domain* and B its *codomain*.

A binary relation $f \subseteq A \times B$ is *totally defined* if its graph contains at least one pair for each element of its domain (A); it is *functional* if its graph contains at most one pair for each element of its domain (A). Obviously, these two properties may be consolidated into only one: f is a *mapping* (*function*) if its graph contains exactly one pair (denoted $(x, f(x))$) for each element of its domain (A). For example:

- R above is totally defined, but not functional (as there are two pairs for $a \in A$)
- $f = \{(a,1), (b,2)\} \subset A \times B$ is a mapping $f : A \rightarrow B$, with $f(a) = 1$ and $f(b) = 2$, where $A = \{a, b\}$, $B = \{1,2\}$.
- *BirthPlace* : *PEOPLE* $\rightarrow$ *CITIES* is a mapping, as anybody is/was born in one and only one city.

Two mappings are *equal* if they have same domain, codomain, and graph. For example, $sin : (-\infty, \infty) \rightarrow (-\infty, \infty)$ is not equal to $sin : (-\infty, \infty) \rightarrow [-1, 1]$, as they have different codomains.

For any Cartesian product $A \times B$ there are two unique distinguished mappings $pr_A : A \times B \rightarrow A$ and $pr_B : A \times B \rightarrow B$, called its *canonical* (*Cartesian*) *projections*, which are mapping each pair to its corresponding pair element: $pr_A(x, y) = x$, $pr_B(x, y) = y$, $\forall (x, y) \in A \times B$. For example, if $A = \{a, b\}$ and $B = \{1, 2, 3\}$, $pr_A(a,1) = pr_A(a,2) = pr_A(a,3) = a$, $pr_A(b,1) = pr_A(b,2) = pr_A(b,3) = b$, $pr_B(a,1) = pr_B(b,1) = 1$, $pr_B(a,2) = pr_B(b,2) = 2$, $pr_B(a,3) = pr_B(b,3) = 3$.

For any mapping $f : A \rightarrow B$, its *image*, denoted $Im(f) = \{y \in B \mid \exists x \in A, y = f(x)\} \subseteq B$, is the set of all of its values. For example, if $f : \{1,2,3\} \rightarrow \{a, b, c\}$, with $f(1) = b$, $f(2) = b$, $f(3) = c$, then $Im(f) = \{b, c\}$.

The *mapping composition* $h = g \circ f$ of mappings $f : A \rightarrow B$ and $g : B \rightarrow C$ is the unique mapping $h : A \rightarrow C$ computed as $h(x) = g(f(x))$, $\forall x \in A$. For example, if $f : \{1,2,3\} \rightarrow \{a, b, c\}$, with $f(1) = b$, $f(2) = b$, $f(3) = c$, $g : \{a, b, c\} \rightarrow \{true, false\}$, with $g(a) = true$, $g(b) = false$, $g(c) = true$, then $h : \{1,2,3\} \rightarrow \{true, false\}$, with $h(1) = g(f(1)) = g(b) = false$, $h(2) = g(f(2)) = g(b) = false$, $h(3) = g(f(3)) = g(c) = true$.

A mapping $f : A \rightarrow A$ is called an *automapping* (*autofunction*). For example, *ReportsTo* : *EMPLOYEE* $\rightarrow$ *EMPLOYEE* is an automapping.

An automapping is said to be *idempotent* if $f \circ f = f^2 = f$. For example, $f : \{1,2,3\} \rightarrow \{1,2,3\}$, with $f(1) = 3$, $f(2) = 2$, $f(3) = 3$, is idempotent, as $f^2(1) = f(f(1)) = f(3) = 3 = f(1)$, $f^2(2) = f(f(2)) = f(2) = 2 = f(2)$, $f^2(3) = f(f(3)) = f(3) = 3 = f(3)$.

A mapping $f : A \rightarrow B$ is *one-to-one* (*injective*) if its graph contains at most one pair for each element of its codomain (B). For example, $g : A \rightarrow B$, with $g(a) = 2$ and $g(b) = 1$ is one-to-one, whereas $g : A \rightarrow B$, with $g(a) = 1$ and $g(b) = 1$ is not; $MAC : NETWORK\_ADAPTERS \rightarrow GUIDS$ is one-to-one (as any network adapter has a unique GUID), whereas $BirthPlace : PEOPLE \rightarrow CITIES$ is not (as several people are/were born in any city). Obviously, one-to-oneness is a dual of functionality.

A mapping $f : A \rightarrow B$ is *onto* (*surjective*) if $Im(f) = B$.

For example, $f : \{1,2,3\} \rightarrow \{a, b, c\}$, with $f(1) = b, f(2) = b, f(3) = c$ is not onto, whereas $f : \{1,2,3\} \rightarrow \{a, b, c\}$, with $f(1) = b, f(2) = a, f(3) = c$ it is; $sin : (-\infty, \infty) \rightarrow (-\infty, \infty)$ is not onto, but $sin : (-\infty, \infty) \rightarrow [-1, 1]$ it is; $BirthPlace : PEOPLE \rightarrow CITIES$ is onto (as, in any city, at least one person is/were born), whereas $Capital : COUNTRIES \rightarrow CITIES$ it is not (as not all cities are capitals).

A mapping $f : A \rightarrow B$ is *bijective* if it is both one-to-one and onto; sometimes, bijective mappings are denoted $f : A \leftrightarrow B$; such A and B are called *equipotent* (as $card(A) = card(B)$) or *isomorphic*.

For example, $f : \{1,2,3\} \rightarrow \{a, b, c\}$, with $f(1) = b, f(2) = a, f(3) = c$ is bijective, whereas $f : \{1,2,3\} \rightarrow \{a, b, c\}$, with $f(1) = b, f(2) = b, f(3) = c$ it is not; $sin : (-\infty, \infty) \rightarrow [-1, 1]$ is not bijective, as it is not one-to-one (e.g., $sin(0) = sin(\pi) = sin(2\pi) = \ldots = sin(k\pi) = 0$, k integer), whereas $sin : [-\pi/2, \pi/2] \rightarrow [-1, 1]$ it is; the real linear function ($ax + b$) is bijective, whereas the squaring one ($ax^2 + bx + c$, $a \neq 0$) it is not.

For any set A there is a unique distinguished bijective auto-mapping $1_A : A \rightarrow A$ called A's *unity function*, which is mapping each of its elements to itself: $1_A(x) = x$, $\forall x \in A$.

A *mapping (Cartesian) product* $h = f \bullet g$ of mappings $f : A \rightarrow B$ and $g : A \rightarrow C$ is the unique mapping $h : A \rightarrow B \times C$ computed as $h(x) = (f(x), g(x))$, $\forall x \in A$.

For example, if $f = FirstName : PEOPLE \rightarrow$ ASCII(255) and $g = LastName : PEOPLE \rightarrow$ ASCII(128), $h = FirstName \bullet LastName : PEOPLE \rightarrow$ ASCII(255) × ASCII(128); if $f(x) =$ 'Dinu' and $g(x) =$ 'Lipatti', then $h(x) =$ ('Dinu', 'Lipatti').

If $f : A \rightarrow B$ is one-to-one then, $\forall g : A \rightarrow C, f \bullet g$ is one-to-one too.

For example, as $SSN : PEOPLE \rightarrow$ NAT(9) is one-to-one (as there may not be two persons having same SSN), $SSN \bullet BirthDate : PEOPLE \rightarrow$ NAT(9) $\times$ [1/1/1900, $SysDate()$] is one-to-one too.

A mapping (Cartesian) product $p = f_1 \bullet \ldots \bullet f_n$, $n > 1$ natural, is *minimally one-to-one* if it is one-to-one but none of its proper subproducts is one-to-one; in particular (for $n = 2$), $f \bullet g$ is *minimally one-to-one* if it is one-to-one and neither f, nor g are.

For example, $SSN \bullet BirthDate : PEOPLE \rightarrow$ NAT(9) $\times$ [1/1/1900, $SysDate()$] is not minimally one-to-one (as SSN is one-to-one), whereas *Country* $\bullet State : STATES \rightarrow COUNTRIES \times$ ASCII(64) it is: not only it is one-to-one (as there may not be two states with a same name in a same country), but none of its component is (*Country* as there are several states in a country, e.g. 50 + a federal district in the U.S., 41 in Romania, etc., and *State* as there may be two states of different countries having same name, e.g. Cordoba, in Spain and Argentina; La Rioja, in Spain, Peru, and Argentina, etc.).

$[x]$, x real, denotes the *integer part* of x. For example, $[\pi] = 3$, $[5.9] = 5$.

$n! = 1 * 2 * \ldots * (n—1) * n$ is the *factorial* of n, n natural.

For example, $4! = 1 * 2 * 3 * 4 = 24$.

x^y denotes the y-th *power* of x, x, y being complex (real, integer, natural, etc.) numbers. For example, $2^4 = 2 * 2 * 2 * 2 = 16$.

Applying Newton's binomial theorem for $x = y = 1$, $2^n = $ C(n, 0) $+ \ldots +$ C(n, n), n natural, where C(n, j) denotes the (natural) number of "*n choose j*", j natural, $n \geq j$ (i.e., *combinations of j elements taken from n ones*: C(n, j) = $n!/(j! * (n—j)!)$).

For example, $\forall n$ natural, C(n, 0) = C(n, n) = 1; C(4,3) = $4!/(3! * 1!) = 24/6 = 4$; etc.

The number of operands of an operator or of arguments of a function is called its *arity*. For example, constants have arity 0 (are *nullar*), the logical *not* has arity 1 (is *unary*), just like the functions *card* and *Im*, while *and* and *or* have arity 2 (are *binary*), just like the function (Cartesian) product. As this product is associative, $f_1 \bullet \ldots \bullet f_n$ has arity n (or is *n-ary*), n natural. Notationally, arity is denoted just like the absolute value operator: for example, $|f_1 \bullet \ldots \bullet f_n| = n$.

A unary operator f is *monotonic* (*monotone*) if it preserves ordering that is if $\forall x, y$, $(x \leq y \Rightarrow (f(x) \leq f(y)))$, where $\leq$ is an order (e.g., $\leq$ on numbers, but also $\subseteq$ on sets, etc.).

For example, the linear numeric function $f(x) = ax + b$, $a \neq 0$ is monotonic (and, in particular, increasing, when $a > 0$, and decreasing, when $a < 0$), whereas the squaring one ($ax^2 + bx + c$, $a \neq 0$) it is not,

A binary homogenous relation R over a set A is:

✓ *total* (or *complete*), if $aRb \vee bRa$, $\forall a, b \in A$;

✓ *connected*, if $aRb \vee bRa$, $\forall a, b \in A$, $a \neq b$;

✓ *partial* (*not total*), if $\exists a, b \in A$ such that $a \neg Rb \wedge b \neg Ra$;

✓ *reflexive*, if aRa, $\forall a \in A$;

✓ *irreflexive*, if $a \neg Ra$, $\forall a \in A$; (sometimes called *antireflexive*)

✓ *symmetric*, if $aRb \Rightarrow bRa$, $\forall a, b \in A$;

✓ *antisymmetric*, if $aRb \wedge bRa \Rightarrow a=b$, $\forall a, b \in A$;

✓ *asymmetric*, if $aRb \Rightarrow \neg(bRa)$, $\forall a, b \in A$;

✓ *transitive*, if $aRb \wedge bRc \Rightarrow aRc$, $\forall a, b, c \in A$;

✓ *intransitive*, if $aRb \wedge bRc \Rightarrow \neg(aRc)$, $\forall a, b, c \in A$; (sometimes called *antitransitive*)

✓ *Euclidean*, if $aRb \wedge aRc \Rightarrow bRc$, $\forall a, b, c \in A$;

✓ *cyclic*, if $\exists \{a_1, \ldots, a_n\} \subseteq A$, $n > 0$, natural, such that $a_1 Ra_2 \wedge \ldots \wedge a_{n-1} Ra_n \wedge a_n Ra_1$; such a $\{a_1, \ldots, a_n\}$ subset of A is called a (*simple*) *cycle* (*circuit, circle, polygon*) (in R's graph);

✓ *acyclic*, if it has no cycles.

For example, if $A = \{a, b, c\}$, then:

✓ $R = \{(a, b), (b, c), (c, a)\}$ is total, irreflexive, asymmetric, intransitive, and cyclic;

✓ $R = \{(a, a)\}$ is partial, reflexive, antisymmetric, and cyclic;

✓ $R = \{(a, b), (b, a)\}$ is partial, irreflexive, symmetric, and cyclic;

✓ $R = \{(a, b), (b, c), (a, c)\}$ is total, irreflexive, asymmetric, transitive, Euclidean, and acyclic;

✓ $R = \{(a, b), (a, c)\}$ is total, irreflexive, asymmetric, intransitive, and acyclic.

Trivially, if $A = \mathbf{R}$ (the set of real numbers), then, for example:

✓ $\geq$ is total, reflexive, antisymmetric, transitive, and not asymmetric;

✓ $>$ is partial, irreflexive, asymmetric, antisymmetric, transitive, and acyclic;

✓ $\neq$ is irreflexive and symmetric.

Obviously, auto-mappings are particular types of homogenous binary (dyadic) relations, so they might also be reflexive, irreflexive, etc.

A binary homogenous relation R is a *partial order* if it is reflexive, antisymmetric and transitive.

A *partially ordered* (by a partial order *po*) *set S* is abbreviated *poset* and denoted (S, po). Inclusions are the most well-known examples of partial orders.

For each inclusion $A \subseteq B$ there is a unique one-to-one distinguished mapping $i_A: A \rightarrow B$ called the *canonical inclusion* function, which maps each subset element to the corresponding superset one: $i_A(x) = x$, $\forall x \in A$. Trivially, when $B = A$, $i_A = 1_A$.

An order is called *total* if it is also total. $<, \leq, >, \geq$ over numeric sets are the most well-known examples of total orders.

A binary homogenous relation R is a *preorder* (*quasiorder*) if it is reflexive and transitive.

Functional database dependency is the best rdb example of a preorder.

A binary homogenous relation is an *equivalence* one if it is reflexive, symmetric and transitive.

Two sets are *equal* (=) if they have exactly same elements; trivially, $A \subseteq B \wedge B \subseteq A \Leftrightarrow A = B$, for any sets A and B. Equality (not only of sets) is the most well-known example of an equivalent relation.

Other well-known examples of equivalence relations are geometric equivalences (e.g., triangle, rectangle, etc.); in rdb, all keys of any relation are equivalent (see Exercise 3.42(iv)), all attributes that mutually depend on each other (see Exercise 3.42(i)) and all constraint sets having same closure are equivalent too, etc.

Two or more subsets of a set are called its *partitions* if they are pair-wise disjoint and their union is equal to the set. Equivalently, a set is partitioned into several partitions if each of its elements belongs to one and only one partition.

Most often, partitions of a set are indicated with the derived *direct sum* ($\oplus$) operator: $A = P_1 \oplus \ldots \oplus P_n$, $n > 1$, natural, where P_i are A's partitions, is equivalent with $A = P_1 \cup \ldots \cup P_n \wedge P_i \cap P_j = \emptyset$, $\forall i \neq j$, $1 \leq i, j \leq n$.

For example, *PEOPLE = ADULTS $\oplus$ CHILDREN, PEOPLE = ELECTED $\oplus$ NOT_ELECTED, ELECTED = PRESIDENT $\oplus$ SENATORS $\oplus$ CONGRESSMEN $\oplus$ MAYORS $\oplus$ COUNCILLORS*, etc.

When a set is partitioned by an equivalence relation, then its partitions are members of the corresponding *quotient set*. For any quotient set $A/\sim$ (generated by equivalence relation $\sim$ over set A) there is a unique onto mapping $\rho_{\sim} : A \rightarrow A/\sim$ called its *canonical surjection*, which maps any element of A to the partition to which it belongs. Sometimes, partitions are identified by a (unique) representative element from A, in which case $A/\sim \subseteq A$ is called a *representative system* (*system of representatives*).

For example, let A = *CITIZENS* and $\sim \subseteq$ *CITIZENS* $\times$ *CITIZENS*, defined by "$\forall x, y \in$ *CITIZENS*, $x\sim y$ iff x and y have the same supreme head of state"; then *CITIZENS*$/\sim$ includes, for example, the classes of all subjects of HM Queen Elisabeth II of the U.K. (also including Canadians, Australians, New-Zealanders, etc.), of all citizens of the U.S. and their territories, of all German citizens, etc. The corresponding representative system would now include the Japanese Emperor Akihito, Queens Elisabeth II of the U.K., Margrethe II of Denmark, presidents Barack Obama, Xi Jinping, Vladimir Putin, François Hollande (of the U.S., China, Russia, and France, respectively), etc. The canonical surjection *SupremeStateHead* : *CITIZENS* $\rightarrow$ *CITIZENS*$/\sim$ has, for example, the following values: *SupremeStateHead*(Seiji Ozawa) = Emperor Akihito, *SupremeStateHead*(Sir Elton Hercules John) = Queen Elisabeth II of the U.K., *SupremeStateHead*(Céline Marie Claudette Dion) = Queen Elisabeth II of the U.K., *SupremeStateHead*(Eldrick Tont "Tiger" Woods) = President Barack Obama, *SupremeStateHead*(Dalai Lama) = President Xi Jinping, etc.

The *kernel* (*nucleus*) of a mapping $f : A \rightarrow B$, denoted $ker(f)$, is an equivalence relation over its domain, partitioning it into classes grouping together all of its elements for which f takes same values: $ker(f) = \{ (x, x') \in A^2 \mid f(x) = f(x') \}$. Sometimes, instead of the functional notation $ker(f)$, an infix one is used: $x =_f x' \Leftrightarrow f(x) = f(x')$. The corresponding induced partition (and quotient set) is $\{\{w \in A \mid f(x) = f(w)\} \mid x \in A \} = A/_{ker(f)}$, which is also called the function *coimage* (denoted $coim(f)$). Note that coimages are naturally isomorphic to images, as there is a bijection from $coim(f)$ to $Im(f)$. It is very easy to show that $ker(f \bullet g) = ker(f) \cap ker(g)$, for any function product $f \bullet g$.

For example, with *Country* : *STATES* $\rightarrow$ *COUNTRIES* and *Region* : *STATES* $\rightarrow$ *REGIONS*, $ker(Country)$ = {(Alabama, Alabama), (Alabama, Alaska), ..., (Wisconsin, Wyoming), (Wyoming, Wyoming), (Alba, Alba), (Alba, Arad), ..., (Vâlcea, Vrancea), (Vrancea, Vrancea), ...}, $ker(Region)$ = {(Connecticut, Connecticut), (Connecticut, Maine),

..., (Oregon, Washington), (Washington, Washington), (Argeş, Argeş), (Argeş, Brăila), ..., (Covasna, Harghita), (Harghita, Harghita), ...}, and *ker(Country•Region)* = *ker(Country)* ∩ *ker(Region)* = *ker(Region)* = {(Connecticut, Connecticut), (Connecticut, Maine), ..., (Oregon, Washington), (Washington, Washington), (Argeş, Argeş), (Argeş, Brăila), ..., (Covasna, Harghita), (Harghita, Harghita), ...}.

A set together with an associative binary operation on it forms a *semigroup*. Semigroups are one of the simplest *algebraic structures*, that is an arbitrary set (called *underlying* (*carrier*) *set*) with one or more *finitary operations* (i.e., operations having a finite arity) defined on it. For example, the set of integers forms semigroups both with addition and multiplication.

Given a set and a binary operation on it, the element of the set that leaves all set elements (including itself) unchanged when combined with them through the binary operation, if any, is called the *identity* (or *neutral*) *element* (of that set, with respect to that binary operation) or, simply, the *identity* (e.g., for any number sets, 0 for sum and subtraction, 1 for multiplication and division).

A semigroup with identity is called *monoid*. For example, obviously, any number set (i.e., complex, real, rational, integers, naturals) forms monoids both with addition and multiplication.

A set *S* of symbols (called the *alphabet*) equipped with the associative binary operation called *string concatenation* (the operation of joining two character strings end-to-end) and augmented with a unique distinguished identity element (called the *empty string*, generally denoted by ε or λ, and representing, by convention, the unique sequence of zero elements from *S*) is called the *freely*[137] (*generated*) *monoid* over *S*, denoted *S**. Obviously, *S** elements are character strings of various lengths made up with the symbols of *S*.

Examples: ASCII*, UNICODE*, etc. ASCII(*n*) ⊂ ASCII*, where *n* is a natural, is the subset made out only of strings having maximum length *n*.

The *closure* of a homogeneous binary relation *R* is the relation $R^+ = \{(x, z) \mid (x, y) \in R \wedge (y, z) \in R\}$.

R' is the *transitive closure* of a relation *R* if and only if:

---

[137] Generally, an abstract monoid (or semigroup) *S* is described as *free* if it is isomorphic to the free monoid (or semigroup) on some set, where, informally, a free algebraic object over a set *S* can be thought as being a "generic" algebraic structure over *S*: the only equations that hold between elements of a free object are those that follow from the defining axioms of the corresponding algebraic structure.

1. R' is transitive,
2. $R \subseteq R'$, and
3.   for any relation R'', if $R \subseteq R''$ and R'' is transitive, then $R' \subseteq R''$ (i.e., R' is the smallest relation that satisfies (1) and (2)).

Similarly, any other such closure (e.g., reflexive, symmetric, etc.) can be defined.

Obviously, the transitive closure of a relation can be obtained by closing it, closing the result, and continuing to close the results of the previous closures until no further elements are added. Note that the *digraph of the transitive closure* of a relation is obtained from the digraph of the relation by adding for each directed path the arc that shunts the path if one is not already there.

Let R be a relation from X to Y and S one from Y to Z;

The *composition* (or *join or concatenation*) of R and S, written $R.S$, is the relation $R.S = \{(x, z) \in X \times Z \mid xRy \text{ and } ySz, y \in Y\}$.

A function-style notation $S°R$ is also sometimes used, although it is quite inconvenient for relations. The notation $R.S$ is easier to deal with, as the relations are named in the order that leaves them adjacent to the elements that they apply to (thus, $x(R.S)z$, because xRy and ySz for some y).

Transitivity and composition may seem similar, but they are not: both are defined using x, y, and z, for one thing, but transitivity is a property of a single relation, while composition is an operator on two relations that produces a third relation (which may or may not be transitive).

The *product* of two relations R and S is the relation $R \times S = \{(w, x, y, z) \mid wRx \wedge ySz\}$.

The *converse* (or *transpose*) of R, written R^{-1}, is the relation $R^{-1} = \{(y, x) \mid xRy\}$.

Symmetric and converse may also seem similar, but they are actually unrelated: both are described by swapping the order of pairs, but symmetry is a property of a single relation, while converse is an operator that takes a relation and produces another relation (which may or may not be symmetric). It is trivially true, however, that the union of a relation with its converse is a symmetric relation.

Examples:

Let $X = \{$Airplane, Pool, Restaurant$\}$, $Y = \{$Goggles, Heels, Seatbelt, Tuxedo$\}$, $Z = \{$Buckle, Pocket, Strap$\}$, and $R = \{$(Airplane, Seatbelt), (Airplane, Tuxedo), (Pool, Goggles), (Restaurant, Heels), (Restaurant, Tuxedo)} be the relation "*is where one wears*", and S = {(Goggles, Strap), (Heels, Buckle), (Heels, Strap), (Seatbelt, Buckle), (Seatbelt, Strap), (Tuxedo, Pocket)} the relation "*can have a*"; then:

- $R.S$ = {(Airplane, Buckle), (Airplane, Pocket), (Airplane, Strap), (Pool, Buckle), (Pool, Strap), (Restaurant, Buckle), (Restaurant, Pocket), (Restaurant, Strap)}.
- R^{-1} = {(Goggles, Pool), (Heels, Restaurant), (Seatbelt, Airplane), (Tuxedo, Airplane), (Tuxedo, Restaurant)}.

As relations are sets (even if of pairs), all the operations and relations on sets also apply to relations as well; for example:

The *intersection* of any R, S, is the relation $R \cap S = \{x(R \cap S)y \mid xRy \wedge xSy\}$.

The *union* of R and S, is the relation $R \cup S = \{x(R \cup S)y \mid xRy \vee xSy\}$.

The *difference* of R and S, is the relation $R - S$ (or $R \backslash S$) = $\{x(R-S)y \mid xRy \wedge \neg xSy\}$.

Let U be a set and R and S be relations on U (their common *underlying* set);

$R = S$ iff $(\forall x, y \in U)(xRy \Leftrightarrow xSy)$.

$R \subseteq S$ iff $(\forall x, y \in U)(xRy \Rightarrow xSy)$.

Originally[138], a sentence is said to be *undecidable* when it can neither be proved, nor refuted. For example, the so-called *axiom of choice* (equivalent to the statement "*the Cartesian product of a collection of nonempty sets is nonempty*") is undecidable in the standard axiomatization set theory except for it.

A *Turing machine* (*TM*) is a theoretical device that manipulates symbols on an infinite strip of tape according to a set (table) of rules, by reading and/or writing one symbol at a time. A TM whose set of rules prescribes at most one action to be performed for any given situation[139] is called *deterministic* (*DTM*) or *ordinary*, whereas when there are several actions for at least one situation it is called *nondeterministic* (*NTM*).

A *complexity class* is a set of problems of related resource-based complexity. A typical complexity class definition has the form "the set of problems that can be solved by an abstract machine M using $O(f(n))$ of resource

---

[138] This first sense arouse from the Gödel's theorems; in the computability theory, there is a second sense for undecidability, which applies to *decision problems* (countably infinite sets of questions each requiring a *yes* or *no* answer) rather than to sentences: there is no algorithm (computable *or effectively calculable* function) that correctly answers every question in the problem set. In this sense, for example, the problem of whether a Turing machine halts or not on any given program is undecidable.

[139] That is a pair of current <state, symbol>.

R", where n, natural, is the size of the input, $f(n)$ is a function of n (e.g., n, n^2, 2^n, etc.), $O(x)$ is the *complexity operator* (called *big O*), and, for example, resources are the number of steps (e.g., needed to be performed by an algorithm in the corresponding worst case), execution time, memory space, etc.

For example, the *NC class* is a set of decision problems decidable in polylogarithmic time on a parallel computer with a polynomial number of processors—a class of problems having highly efficient parallel algorithms.

L (also known as LOGSPACE, LSPACE or DLOGSPACE) is the complexity class containing decision problems that can be solved by a DTM using a logarithmic amount of memory space. L contains precisely those languages expressible in first-order logic with an added commutative transitive closure operator. This result has applications to database query languages: the *data complexity of a query* is defined as the complexity of answering a fixed query, considering the data size as the variable input. For this measure, queries against relational databases with complete information (i.e., having no notion of nulls) as expressed, for instance, in relational algebra are in L.

The main idea of logspace is that you can store a polynomial-magnitude number in logspace and use it to remember pointers to a position of the input. The logspace class is therefore useful to model computation where the input is too big to fit in the internal memory (RAM) of a computer. Long DNA sequences and databases are good examples of problems where only a constant part of the input will be in RAM at a given time and where we have pointers to compute the next part of the input to inspect, thus using only logarithmic memory.

L is a subclass of NL (NLOGSPACE), which is the class of languages decidable in logarithmic space on a nondeterministic Turing machine.

It has been proved that $NC \subseteq L \subseteq NL \subseteq NC^2$, that is: given a parallel computer C with a polynomial number $O(n^k)$ of processors for some constant k, any problem that can be solved on C in $O(\log n)$ time is in L, and any problem in L can be solved in $O(\log^2 n)$ time on C. Still open problems include whether $L = NL$.

The *polynomial complexity class* (P, or *PTIME*, or $DTIME(n^{O(1)})$) contains all decision problems that can be solved by a deterministic Turing machine using a polynomial amount of time. Cobham's thesis holds that P is the class of computational problems that are "efficiently solvable" or "tractable"; however, in practice some problems not known to be in P have practical solutions, and some that are in P do not. A decision problem is *P-complete* if it is in P.

The most famous example of a P-complete problem is the *Turing machine halting* one: given a Turing machine, an input for it, and a number T written in unary, does that machine halt on that input within the first T steps? Other examples include: the *Lexicographically First Depth First Search Ordering* (given a graph with fixed ordered adjacency lists and nodes u and v of it, is vertex u visited before vertex v in a depth-first search induced by the order of the adjacency lists?), the *Context Free Grammar Membership* (given a context-free grammar and a string, can that string be generated by that grammar?), and the *Horn-satisfiability* (given a set of Horn clauses, is there a variable assignment which satisfies them?). Dually, there are problems in P that are not known to be either P-complete or NC, but are thought to be difficult to parallelize. Examples include the decision problem of *finding the greatest common divisor of two numbers* and the one for *determining what answer the extended Euclidean algorithm would return when given two numbers*.

It has been shown that $NL \subseteq P$. An open problem is whether $L = P$?

The *nondeterministic polynomial* complexity class (*NP*) is the set of decision problems whose solutions can be determined by a NTM in polynomial time (i.e., in $O(n^k)$, where k is a natural constant). For example, the *traveling salesman problem* (given an input matrix of distances between n cities, determine if there is a route visiting all cities with total distance less than some k) is in *NP*.

Informally, the *nondeterministic polynomial-time hard* complexity class (*NP-hard*) is the set of decision problems that are at least as hard as the hardest problems in NP. Formally, a problem P is NP-hard when every problem R in NP can be reduced in polynomial time to P. As a consequence, finding a polynomial algorithm to solve any NP-hard problem would give polynomial algorithms for all the problems in NP[140], which is unlikely, as many of them are considered as hard. For example, the

---

[140] A common mistake is to think that the *NP* in *NP-hard* stands for *non-polynomial*: although it is widely suspected that there are no polynomial-time algorithms for NP-hard problems, this has not been proven yet. Moreover, NP also contains all problems that can be solved in polynomial time.

decision *subset sum problem* (given a set of integers, does any nonempty subset of them add up to zero?) is NP-hard.

A decision problem is *NP-complete* when it is both in NP and NP-hard. The set of NP-complete problems is denoted by *NP-C* (or *NPC*). Although any given solution to an NP-complete problem can be verified in polynomial time, there is no known efficient way to find such a solution in the first place (the most notable characteristic of NP-complete problems is that no fast solution to them is known): the time required to solve such problems using any currently known algorithm increases very quickly as the size of the problem grows. As a consequence, determining whether or not it is possible to solve these problems quickly, called the *P versus NP problem*, is one of the main still unsolved problems in computer science. NP-complete problems are generally addressed by using heuristic methods and approximation algorithms.

For example, the above NP-hard subset sum problem is also NP-complete. Another famous NP-hard example is the *Boolean Satisfiability Problem* (also called the *Propositional Satisfiability Problem* and abbreviated as *SATISFIABILITY* or *SAT*): there exists an interpretation that satisfies a given Boolean formula? In other words, can the variables of a given Boolean formula be consistently replaced by *true* or *false* in such a way that the formula evaluates to *true*?[141] Despite the fact that no algorithms are known to solve SAT efficiently, correctly, and for all possible input instances, many instances of SAT-equivalent problems that occur in practice (e.g., in AI, circuit design, automatic theorem proving, etc.) can actually be solved rather efficiently using heuristical SAT-solvers. Although such algorithms are not believed to be efficient on all SAT instances, they tend to work well for many practical applications.

There are decision problems that are NP-hard but not NP-complete, such as the *programs halting problem*: given a program and its input, will it run forever?[142] For example, the Boolean satisfiability problem can be reduced to the programs halting problem by transforming it to the description of a Turing machine that tries all truth value assignments and, when

---

[141] If this is the case, the formula is called *satisfiable*. On the other hand, if no such assignment exists, the function expressed by the formula is identically *false* for all possible variable assignments and the formula is *unsatisfiable*. For example, the formula $a \wedge \neg b$ is satisfiable, because $a = true$ and $b = false$ make it *true*; dually, $a \wedge \neg a$ is unsatisfiable.

[142] That's a *yes/no* question, so this is a decision problem.

it finds one that satisfies the formula, it halts; otherwise, it goes into an infinite loop. The halting problem is not in NP since all problems in NP are decidable in a finite number of operations, while the halting problem is generally undecidable.

There are also NP-hard problems that are neither NP-complete nor undecidable. For instance, the language of *True quantified Boolean formulas (TQBF)*[143] is decidable in polynomial space, but not nondeterministic polynomial time (unless NP = PSPACE).

The *polynomial space* complexity class (*PSPACE*) is the set of decision problems that can be solved by a DTM only using a polynomial amount of (memory) space.

A decision problem is *PSPACE-complete* if it can be solved using an amount of memory that is polynomial in the input length and if every other problem that can be solved in polynomial space can be transformed to it in polynomial time. The problems that are PSPACE-complete can be thought of as the hardest problems in *PSPACE*, because a solution to any such problem could easily be used to solve any other problem in *PSPACE*. The PSPACE-complete problems are thought to be outside of the *P* and *NP* complexity classes, but this has not been proved.[144]

For example, the canonical PSPACE-complete problem is the *quantified Boolean formula problem (QBF)* one, a generalization of the Boolean satisfiability problem, in which both existential and universal quantifiers can be applied to each variable: it asks whether a quantified sentential form over a set of Boolean variables is *true* or *false*; for example, the following is an instance of QBF: $\forall x\, \exists y\, \exists z\, ((x \lor z) \land y)$.

The complexity class *EXPSPACE* is the set of all decision problems solvable by a DTM in $O(2^{p(n)})$ space, where $p(n)$ is a polynomial function of n.

NEXPTIME (or *NEXP*) is the class of problems solvable in exponential time by a nondeterministic Turing machine.

---

[143] A (fully) quantified Boolean formula is a formula in quantified propositional logic where every variable is quantified (or *bound*), using either existential or universal quantifiers at the beginning of the sentence. Such a formula is equivalent to either *true* or *false* (since there are no free variables). If such a formula evaluates to *true*, then it belongs to *TQBF* (also known as *QSAT* or *Quantified SAT*).

[144] It is, however, known that they lie outside of the *NC* class, because problems in *NC* can be solved in an amount of space polynomial in the logarithm of the input size, and the class of problems solvable in such a small amount of space is strictly contained in *PSPACE* (which is proved by a *space hierarchy theorem*).

The complexity class *EXPTIME* (also called *EXP* or *DEXPTIME*) is the set of all decision problems solvable by a DTM in $O(2^{p(n)})$ time, where $p(n)$ is a polynomial function of n.

A decision problem is *EXPTIME-complete* if it is in EXPTIME and every problem in EXPTIME has a polynomial-time many-one reduction to it (i.e., there is a polynomial-time algorithm that transforms instances of one to instances of the other with the same answer). Problems that are EXPTIME-complete might be thought of as the hardest problems in EXPTIME. Notice that although we don't know if NP is equal to P or not, we do know that EXPTIME-complete problems are not in P; it has been proven that these problems cannot be solved in polynomial time (by the time hierarchy theorem).

One of the basic undecidable problems is that of deciding whether a DTM halts. One of the most fundamental EXPTIME-complete problems is a simpler version of this, which asks if a DTM halts in at most k steps: it is in EXPTIME because a trivial simulation requires $O(k)$ time, and the input k is encoded using $O(\log k)$ bits; it is EXPTIME-complete because, roughly speaking, we can use it to determine if a machine solving an EXPTIME problem accepts in an exponential number of steps; it will not use more.[145]

We know that $P \subseteq NP \subseteq PSPACE \subseteq EXPTIME \subseteq NEXPTIME \subseteq EXPSPACE$, with[146] $P \subset EXPTIME \wedge P \neq EXPTIME$, $NP \subset NEXPTIME \wedge NP \neq NEXPTIME$, and $PSPACE \subset EXPSPACE \wedge PSPACE \neq EXPSPACE$, which means that at least one of the first three inclusions and at least one of the last three inclusions must be proper, but it is not known which ones are. Most experts believe all the inclusions are proper.

It is also known that if $P = NP$, then EXPTIME = NEXPTIME.

---

[145] Recall that the same problem with the number of steps written in unary is P-complete.

[146] By the *time hierarchy theorem* and the *space hierarchy theorem*.

INDEX

Π. See function product operator symbol
∃. See there exists a unique
&. See string concatenation operator
(3,3)-NF, 306, 314, 316
(3,3)-NF ⇒ BCNF, 306
(E)MDM, xx, xxi, 8, 23, 24, 97, 239, 324,
    See (elementary) mathematical data
    model constraint, xx
(elementary) mathematical data model, 8
(S, po)
    S's poset. See set S partially ordered
    by po
,
    SQL Cartesian product operator symbol
        Example, 202
.NET, 322
    CLI, 351
.QL, 28, 32
[x]. See integer part (of x)
    example, 166, 173
\. See difference set operator
|>. See left anti-semijoin (RA) operator
|><. See left semi-join (RA) operator
|><|. See join (RA) operator family
|><|R. See right (outer) join (RA) operator
|P|. See arity (of P)
<|. See right anti-semijoin (RA) operator
><|. See right semi-join (RA) operator
÷. See division (RA) operator
1NF, 316, See First Normal Form, See
    First Normal Form
1OL. See first-order logic
2nd NF, 229
2NF, 229, See 2nd NF
3.5NF. See BCNF
3NF, 229, See 3rd NF
3rd NF, 229
4NF, 231, 314, See 4th NF
4NF ⇒ BCNF, 306

4th NF, 231
5NF, 314, 316, See 5th NF
5NF ⇒ 4NF, 307
5th NF, 233
A.I.Cuza University, 32
A0, 23, 21, 23, 111
A1, 21, 22, 23
A1-7, 22, 23, 115, 324
    complexity, 255
A2, 21, 23, 24
A2014SQLS1-8, 489, 536
A3, 21, 23, 24, 609
A3', 21, 23, 24
A4, 21, 23, 24, 609
A5, 21, 23, 24, 609
A6, 21, 23, 24, 609
A7, 21, 22, 23, 24
A8, 21, 22, 23, 115, 428, 570
    characterization proposition, 543
A9', 23
aberrant constraint. See implausible
    constraint
Abiteboul, S., 31, 322, 323, 324, 332, 333
absolute complement, 616
abstract
    data
        type
            Oracle, 388
Access, 22
Access 2013, 428, 578
Achilles, A.-C., 29
ACID, 222
ACM, 31, 32, 33, See Association for
    Computing Machinery
    SIGMOD, 32, 33
    TODS, 31, 33, 112
Acta Press, 33, 113, 338, 593
Actian Corporation, 25
ActiveX Data Objects, 223
acyclic

binary homogenous relation, 624
inclusion, 145
Adams, S., 603
Addison-Wesley, 31, 32, 593
ADO, 223, 351, 352, 389, 402, 418, See
ActiveX Data Objects
Adobe, 26
ADOX, 420
Advantage Database Server, xviii, 16, 26
AF10, 21, 23, 24
AF11, 21, 23
AF1-8, 21, 23, 24
AF8', 21, 23, 24
AF9, 21, 23, 24
AF9', 21, 23, 24
AFIPS, 33, See American Federation of
Information Processing Societies
aggregation, 7, 25
function, 195, 205, 321
Aho, A.V., 314, 322, 333
AI. See Artificial Intelligence
AIX, 366, 374
algebraic structure, 627
algorithm, 617
A0, 24, 604, 605
A1, xx, 20, 24, 604, 605
A1-7, 20, 24, 604, 605
A2, 20, 24
A3, xx, 20, 24, 324
A3', 20, 24, 324
A4, 20, 24
A5, 20, 24
A6, 20, 24
A7, xx, 20, 24,
A7/8-3, xx, 20, 21, 23, 24, 311, 312,
324, 604
characterization proposition, 251
A7/8-3 (table keys discovery assis-
tance), 166
A8, 20, 24, 604
AF0-8, 605
AF10, xxi, 22, 23
AF11, xxi, 22, 23
AF1-8, 20, 22, 24
AF8, 604
AF8', 604
AF8', 20, 24

AF8", 22
AF9, xxi, 22
AF9', xxi, 22
REA0, 604
REA1, xx
REA1-2, 24, 604
REA2, xx
REAF0', 604
REAF0', 22, 24
REAF0-1, 22
REAF0-2, 22, 24
Algorithm A0 (Data analysis and concep-
tual modeling process assistance), 69
Algorithm A1-7 (Translation of E-R data
models into relational schemes and non-
relational constraint sets), 147
Algorithm A2014SQLS1-8 (Translation
of E-R data models into SQL Server
schemas and non-relational constraint
sets), 490
Algorithm A8 (Translation of RDM sche-
mas into SQL ANSI-92 scripts), 349
Algorithm REA1-2 (Reverse engineering
of RDM schemes into E-R data models),
159
Algorithm REA2013A0 (Translation of
Access db schemas into SQL ANSI
DDL), 487
Algorithm REA2013A0-2 (Translation of
Access schemas into E-R data models),
495
alias. See renaming (RA) operator
ALL SQL predicate, 210
Allison, C., 333
allowable values. See attribute range
ALPHA, 321
Alphabet, 627
formal language, 618
formal system, 618
ALPHANUMERIC
MS Access. See MS Access TEXT
ALTER TABLE SQL statement, 196
American National Standards Institute, 16
ANCHOR:
DB2, 361
and
zeroth-order logic, 619

Android, 26
anomaly
    deletion, 234
    free, 235
    insertion, 324
ANSI. See American National Standards
    Institute
ANSI/X3/SPARC, 333
anti-commutative
    Cartesian product, 620
    operator, 620
antisymmetric
    binary homogenous relation, 624
Antonellis, V., de, 324, 333
ANY SQL predicate, 210
ANYDATA
    Oracle, 384
ANYDATASET
    Oracle, 384
ANYTYPE
    Oracle, 384
Apache, 16, 26, 28
Apple, 390
Apple Academic Press, xxiv
Apress Media, LLC, 32
Aristotle, 529
arity
    function product, 120
    function product, operator, 623
    relation, 120
    table, 119
Armstrong, W. W., 318, 333
array
    associative, 366
    cardinality, 366
        maximum, 366
    ordinary, 366
ARRAY
    DB2, 366
AS ROW BEGIN, 368
AS ROW END, 368
AS SQL operator. See renaming (RA)
    operator
    example, 201, 202
AS TRANSACTION START ID, 368
AS/400, 355
ASC(ENDING) SQL predicate, 209

ASCII, 74
    Access, 424
    DB2, 371
    MySQL, 400
    Oracle, 387
    SQL Server, 415
ASCII(n), 52, 627
Asentinel Intl.,iii
Ashdown, L., 31
ASSM. See Oracle Automatic Segment
    Space Management
associative
    array, 366
    Cartesian product, 620
    operator, 620
assumption. See axiom
    closed world, 151
Astrahan, M. M., 31, 323, 333, 339
asymmetric
    binary homogenous relation, 624
    example, 285
Athanasiu, I., xxiii
atomic
    function product, 120
ATTACHMENT
    MS Access, 422, 425
Attribute, 38, 95
    compulsory
        restriction, 50, 96
            examples, 55, 56
    E-RDM, 75
    list. See canonical (Cartesian) projec-
    tion: product
    metacatalog, 186
    range
        restriction, 50,
            examples, 51, 53
    RDM, 119, See:table column
Atzeni, P. , 315, 318, 325, 333
augmentation (RA) operator, 322
Augustine, N. R., 498
AUTO_INCREMENT, 393
autofunction, 44 , 621
AUTOINCREMENT
    MS Access. See MS Access COUN-
    TER
automapping. See auto-function

automorphism, 303
autonumber, 195
AUTONUMBER
    Access,424
    DB2, 429
    MySQL, 399
    Oracle, 387
    SQL Server, 415
auto-numbers
    example, 121
AVG SQL aggregation function, 195
axiom,616
    A-DA0. Non-relationally rdb design,
    596
    A-DA01, 607, 608
    A-DA01. Data plausibility, 596
    A-DA02, 599
    A-DA02. No semantic overloading,
    596
    A-DA03, 607
    A-DA03. Non-redundancy,596
    A-DA04, 607, 608
    A-DA04. Unique objects,596
    A-DA05, 599, 607
    A-DA05. Best possible performance,
    597
    A-DA06. Controlled-redundancy,597
    A-DA07, 607, 608
    A-DA07. No trivial constraints, 597
    A-DA08, 607, 608
    A-DA08. No constraints, but keys on
    redundant data, 597
    A-DA09, 599
    A-DA09. Constraints discovery, 597
    A-DA10, 599
    A-DA10. No implausible constraints,
    597
    A-DA11, 607, 608
    A-DA11. Constraint sets optimality,597
    A-DA12, 607, 608
    A-DA12. Constraint sets coherence,
    597
    A-DA13, 607, 608
    A-DA13. Minimum inclusions, 598
    A-DA14, 607, 608
    A-DA14. Inclusions acyclicity, 598
    A-DA15, 599, 607, 608

A-DA15. Sets equality, 598
A-DA16, 607, 608
A-DA16. Mappings equality, 598
A-DA17, 599
A-DA17. ER-Ds, 598
A-DA18, 607, 608
A-DA18. ER-D cycles, 598
A-DA19, 599
A-DA19. Naming axiom, 598
A-DA20, 607
A-DA20. Documentation, 598
A-DI0, 608
A-DI0. Instances consistence, 599
A-DI01, 608
A-DI01. Tables' semantics, 488
A-DI02. Non-static fundamental
objects representation, 599
A-DI03. Tables indivisibility, 599
A-DI04. Columns' semantics, 599
A-DI05, 608
A-DI05. Empty instances, 599
A-DI06, 607
A-DI06. Constraints enforcement, 600
A-DI07. Domain constraints, 600
A-DI08, 608
A-DI08. Rows uniqueness, 600
A-DI09, 608
A-DI09. No void rows, 600
A-DI10, 123
A-DI10. Referential integrity, 600
A-DI11. Primary keys, 600
A-DI12, 608
A-DI12. Foreign keys, 600
A-DI13, 607
A-DI13. Key Propagation Principle,
600
A-DI14, 607
A-DI14. Set inclusion relational imple-
mentation, 600
A-DI15, 607
A-DI15. No superkeys, 607
A-DI16, 608
A-DI16. No modification anomalies,
601
A-DI17, 120, 601
A-DI17. Minimal data storing, 601
A-DM0. Positive data, 601

A-DM01. Closed world, 601
A-DM02. Closed domains, 601
A-DM03. Null values set, 601
A-DM04. Null values, 601
A-DM05. Available columns, 601
A-DM06. Group by golden rule, 601
A-DM07, 601
A-DM07. Relevant data and process-
ing, 602
A-DM08, 603
A-DM08. Order by, 602
A-DM09, 603
A-DM09. Data importance, 602
A-DM10, 603
A-DM10. Query semantics correctness,
602
A-DM11. Fastest possible manipula-
tion, 602
A-DM12. Regularly backup fundamen-
tal data intelligently, 603
A-DM13, 603
A-DM13. Always process sets (not
elements), 603
A-DO0, 604
A-DO0. Reduce data contention, 603
A-DO1. Minimize db size, 604
A-DO2. Maximize use of RDBMS
statistics, 604
A-DO3, 604
A-DO3. RDBMS advisors recommen-
dations, 604
A-DO4, 604
A-DO4. Commit as infrequently as is
practical axiom, 604
closed domains, 4, See A-DM02
closed world, 4, See A-DM01
of choice, 629
unique objects, 4, See A-DI09
axiomatization
complete, 318
FD ∪ IND
implication
finite and unrestricted, 319
Bacon, R., 18
Bagui, S., 112
Bancilhon, F., 321, 333
Barker, R., 41

E-RD notation, 40, 41 49
Barksdale, J., 129
Barron, T. M., 334
Barzun, J., 536
Băutu, E., 23
BCNF, 230, 314, See Boyce-Codd NF
Beeri, C., 314, 316, 317, 319, 320, 333
Before Insert
trigger, 376
BEGIN TRAN. See START TRANSAC-
TION SQL statement
BEGIN TRANSACTION. See START
TRANSACTION SQL statement
Ben-Gan, I., 333
Bentley, 26
Berkeley University, 25
Berkowitz, N., 333
Bernstein, P. A., 314, 317, 319, 333, 334
best practice rule, xxi, 17
BFILE
Oracle, 381
BI. See Business Intelligence
big data, 26
big O. See complexity operator
BIGINT
DB2, 355
MySQL, 26
SQL Server, 22
BigTable, 26
bijective function, 622
binary
function product, operator
arity, 623
BINARY
MS Access, 421,424
MySQL, 396
SQL Server, 409
binary (mathematical) relation, 620
composition, 628
join. See binary (mathematical)
relation:composition
transpose. See binary (mathematical)
relation:converse
Binary Large Oject, 421
BINARY_DOUBLE
Oracle, 377
BINARY_FLOAT

Oracle, 377
Biskup, J., 317, 334
BIT, 359
  DB2, 375
  MS Access, 419, 424
  MS SQL Server, 415
  SQL Server, 405
BL. See business logic
Blackburn, P., 593
Blaga, L., v, xvii, 71, 97, 146, 188, 341,
  348, 425, 442, 433, 443, 466, 471, 472,
  473, 490, 577
Blaha, M., 32
Blake, W., 608
BLOB. See Binary Large Object
  Access, 425
  DB2, 358, 372
  MySQL, 396, 408
  Oracle, 379, 387
  SQL Server, 415
Bloomberg, 26
BLU, 353, See big data, lightning fast,
  ultra-easy
BMP. See Basic Multilingual Plane
Bohr, N., 240
BOOLEAN
  Access, 415
  DB2, 371
  MySQL, 391, 399
  Oracle, 387
  SQL Server, 415
Boolean Satisfiability Problem, 632
Borges, J.L., 63
Borland, 28
Boyce, R. F., 32, 228, 230, 322, 334, 337,
  339
Boyce-Codd NF, 228, 230, 334, 337, 339
BPR. See best practice rule
  R-DA-0.(No semantic overloading), 76
  R-DA-1.(Naming conventions), 76
  R-DA-10.(No implausible restrictions),
  86
  R-DA-11.(No restrictions on computed
  objects, except for keys, 87
  R-DA-12, 87, 91, 239
  R-DA-12.(Maximum cardinalities), 87
  R-DA-13, 91

R-DA-13.(Range restrictions), 88
R-DA-14, 91
R-DA-14.(Mandatory restrictions), 88
R-DA-15, 91, 239
R-DA-15.(Uniqueness restrictions), 89
R-DA-16, 91
R-DA-16.(Other restriction types), 91
R-DA-17.(Object sets and their restric-
tions ordering and naming), 91
R-DA-18.(Object sets and their proper-
ties and restrictions documentation), 91
R-DA-19.(E-RDs architecture for com-
plex data models), 92
R-DA-2.(Object set types), 77
R-DA-3.(The Key Propagation Prin-
ciple), 78
R-DA-4.(Object sets abstraction), 78
R-DA-5.(Object set basic properties),
80
R-DA-6.(Object set property types), 81
R-DA-7.(Data controlled redundancy),
82
R-DA-8.(Data computability), 85
R-DA-9.(Restrictions necessity), 85
R-E-1.(Use named constants), 246
R-FK-1.(Foreign keys), 240
R-G-0.(Think in sets of elements, not
elements of sets), 241
R-G-1.(Use parallelism), 241
R-G-2.(Minimize I/Os), 241
R-G-3.(Reuse db connections), 242
R-G-4.(Avoid dynamic SQL), 242
R-G-5.(Minimize undo segments que-
rying), 243
R-G-6.(Unit testing), 243
R-I-C-0.(Always use correct data
types), 531
R-I-C-1.(Always use the smallest data
types needed), 531
R-I-C-2.(Always use (VAR)BINARY
instead of (VAR)CHAR BIT), 531
R-I-C-3.(Always use VAR(n) instead of
CHAR(n) for n > 17), 531
R-I-C-4.(AS/400 DB2 DECIMAL
versus NUMERIC), 531
R-I-C-5.(DB2, COBOL, and PL/I best
data types matching), 531

R-I-D-0.(Do not add any objects to system tablespaces), 529
R-I-D-1.(Safely keep up to date DDL scripts), 529
R-I-D-2.(No UNICODE identifiers), 529
R-I-T-1.(Concatenated keys column ordering), 529
R-I-T-2.(Don't use triggers to enforce relational constraints), 530
R-I-T-3.(Always use same data types in check constraints), 530
R-I-T-4.(Place BLOB data from frequently queried tables in subtables), 530
R-K-1, 240
R-K-1.(Primary keys), 239
R-K-2.(Semantic keys), 239
R-M-0.(Backup), 243
R-M-1.(Test SELECTs first), 243
R-M-2.(Use rollback transactions), 244
R-M-3.(Caution with the ON DELETE CASCADE), 244
R-M-4.(Rule out dirty nulls), 244
R-M-5.(Immediately purge unneeded old data), 244
R-Q-0.(Data and processing minimality), 244
R-Q-1.(WHERE-HAVING rule), 245
R-Q-2.(WHERE rules), 245
R-Q-3.(Cautiously use static subqueries), 245
R-Q-4.(Avoid dynamic subqueries), 245
R-Q-5.(Always sort intelligently), 246
R-T-1.(Table creation, update, and drop), 237
R-T-2, 237, 238
R-T-2.(Relational set inclusion implementation), 238
R-T-3.(DKRINF – Domain-Key Referential Integrity NF), 239
Bradford, R., 592
Bradshaw, J., 227
Bragg, W., 313
Brandon, D., 91
Breazu, V.. See Tannen, V.

Bronowski, J., 200
Buddha, 118
Buneman, S., 417
Burghelea, A., xxiii
business rule, xviii, 11, 50, See constraint
    discovery, 610
    enforcement, 610
    modeling, 610
Butte, A., 17
BYTe
    MS Access, 419, 422
C, 28, 351, 616, See set complement
C#, 19, 28
C(n, j). See n choose j
C+. See closure of constraint set C
C++, 322, 351
CA, 111, See Computer Associates
    Erwin, 111
    ERwin
        Data Modeler, 578
CACM, 32, See Communications of the ACM
Caesar, J., 227
calculus
    object, 316
Calude, C., 325, 334
Campbell, J., vii, 32, 85
Camus, A., 25
candidate key. See semantic key
canonical
    Cartesian projection
        example, 95
    injection
        example, 95
canonical (Cartesian) projection, 621
    co-domain, 124
    product, 124
canonical inclusion function, 625
canonical surjection, 626
card. See cardinal
cardinal, 641
    function product image, 120
    relation instance, 120
    table, 119
cardinality, 37
    array, 366
        maximum, 366

carrier set. See underlying set
Cartesian product, 620
   properties, 301
Casals, P., 127
Casanova, M. A., 315, 319, 334
Cascalog, 28
CASE, xvii, See Computer Aided Software
   Engineering
case study, xxi
Cassandra, 16, 26
CAST, 367, 541
categorization, 7
Cattell, R. G. G., 323, 334
CC. See concurrency control
CCSID, 360
CCSID UNICODE, 358
CE. See constraint enforcement
CERN, 27, See European Council for
   Nuclear Research
chain. See function (Cartesian) product
   chain
Chamberlin, D. D., 32, 323, 334, 339
Chamfort, N. de, 126
Chan, E. P. F., 314, 334
Chandra, A. K., 319, 321, 322, 334
Chang, C. L., 320, 334
Chang, C.C., 32
CHAR
   DB2, 357
   MS Access. See MS Access TEXT
   MySQL, 396
   Oracle,379
   SQL Server, 346
chase. See chasing tableaux algorithm
   IND, 320
   JD, 320
   JD ∪ FD, 320
chasing, 319
chasing tableaux algorithm, 320
CH-complete RL, 322
CH-completeness, 322
check
   constraint, 315
check (relational) constraint
   example, 121
check constraint, 145
check RDM constraint, 120

Chen, P. P., 32, 112
Chen, P.P., 41
   E-RD notation, 41
Chiang, R. H. L., 334
Chimenti, D., 32
Church, A., 617
Churcher, C., 32, 335
Churchill, W., 136, 230, 244, 247
Church–Turing thesis, 617
Cicero, 60, 535
Ciobanu, G., xxiii
classification, 7
Clausewitz, C. von, 112
CLOB
   DB2, 357
   Oracle, 379
Clojure, 28
closed
   Horn clause, 609
closed domains axiom, 4, See A-DM02
closed formula, 120, 619
closed world assumption, 151
closed world axiom, 4, See A-DM01
closure
   binary (mathematical) relation, 620
closure of constraint set, 13
closure operator
   properties, 308
Cloud SQL, 26
Cobham's thesis, 631
Cobol, 351
COBOL, 351
Codd, E. F., 32, 228, 230, 314, 321, 334,
   335, 337, 339
CODEUNITS32, 357
co-domain, 621
Cognitect, 28
coherence, 610
coherent constraint set, 15
coim. See coimage
coimage, 626
column, 120
   range. See domain (relational) constraint
columns concatenation. See canonical
   (Cartesian) projection: product

combinations of j elements taken from n
    ones. See n choose j
COMMIT
    SQL statement, 222
commutative
    operator, 620
compatible tuple, 234
complete
    algorithm, 618
        example, 176
    axiomatization, 318
        JD, 320
        limited, 319
    binary homogenous relation. See
        total:binary homogenous relation
    deductive system, 618
    inner join, 218
    join
        properties, 304
    relational language, 221, 303
completeness
    of RL, 321
complexity
    class, 629
    operator, 630
composed key. See minimally one-to-one
    function (Cartesian) product
composition
    binary (mathematical) relation, 628
compulsory
    restriction, 75
compulsory attribute restriction, 50
compulsory constraint. See not-null con-
    straint
computability. See recursive computability
computable queries. See CH-completeness
computed
    column
        Access, 400
        DB2, 371
        MS Access, 401, 423
        MS SQL Server, 413
        Oracle, 387
        SQL Server, 415
    table
        MS Access, 419

Computer Aided Software Engineering,
    xvii
Computer Science Press, 32, 33, 34, 339
concatenated
    foreign key, 347
concatenated key, 128
conceptual db design, 610
concurrency control, 16
conjunctive
    query, 319
connected
    binary homogenous relation, 624
Connelly, T. M. Jr., 100
consistent
    instance
        database, table, 12
constraint, 12, 120, 248, See closed
    formula
    (E)MDM, xxi
    check, 145, 315
    disjunctive existence, 315
    domain, 125, 248
    enforcement, 16
    existence, 249, 315
    fundamental, 13
    holds. See satisfied constraint
    homogeneous binary relation, xxi
    implausible, 15
    implication, 12
    implied, 12
    key, 128, 249
    mapping, xxi
    metacatalog, 186
    non-relational
        enforcement, xxi
    not null, 127
    NOT NULL, 127
    object, xxi, 315
    plausible, 15
    preservation, 317
    primitive, 314
    primitive, 314
    RDM
        check, 120
        domain, 122
        NOT NULL, 123
        referential integrity, 122

uniqueness, 119
redundancy, 14
referential integrity, 138, 248
relational, 119
representation. See constraint preservation
satisfaction, 12
set, xxi
    closure, 13
    coherent, 15
    incoherent, 15
    minimality, 14
trivial, 14
tuple, 249, 315
    MS
        Access, 425
violation, 12
CONSTRAINT SQL clause, 196
    example, 194
content
    table. See instance: table
controlled redundancy, 610
conventional database, 10
converse
    binary (mathematical) relation, 520
CORAL, 27, 33
Coral++, 16
CORAL++, 27, 33
Corbusier, Le, 91
Cosmadakis, S., 318, 320, 335, 337
COUNT SQL aggregation function, 195, 201
    example, 201
COUNTER
    MS Access, 420, 424
Coupland, D., 89
cover
    equivalent constraint sets, 13
CP, 317, See constraint preservation
CR, 321
Crasovschi, L., xxiii, 33, 324, 325
CRC Press, 112
CREATE TABLE SQL statement, 195
    example, 194
cross tabulation, 118
crow's foot notation. See Barker E-RD notation

CurDate(), 395
CURRENCY
    Access, 424
    DB2, 371
    MS Access, 421, 424
    MySQL, 399
    Oracle, 387
    SQL Server, 415
CURRENCY(n), 52
CURRENT DATE, 357
CURRENT LOCALE LC_TIME, 357
CURRENT TIME, 357
CURRENT TIMESTAMP, 357
CURRENT TIMEZONE, 357
Current_Date, 378
CURRENT_DATE, 357
CURRENT_TIME, 367
Current_Timestamp, 378
Current_Timestamp(), 408
CURRENT_TIMEZONE, 357
cursor, 120
CurTime(), 395
cyclic
    binary homogenous relation, 624
DAC. See data access control
Dalai Lama, 529
dangling
    Oracle REF, 381
dangling pointer, 138
DAO, 418, 420
data, 2, 610
    access control, 15
    analysis, 8, 610
    anomaly, 314
    big, 26
    definition language, 15
    dependencies, 320
    event, 346
    heterogeneous, 24
    homogeneous, 5
    implausible, 610
    incomplete, 5
    manipulation
        anomalies, 227
    manipulation language, 15
    model, 6
        (elementary) mathematical, 8

entity-relationship, 8
E-R, 20, 36, 98
hierarchical, 7
network, 7
relational, 7
semantic, 25
modeling, xviii
E-R, xxi
functional, xxi
negative, xxi
plausibility, xxi, 12, 610
redundancy
controlled, 82
redundant, 610
semi-structured, 29
type, 74
data and programs backup and restore, 16
data complexity of a query, 630
data model
logical, 9
database
instance
invalid. See
database:instance:inconsistent
valid. See
database:instance:consistent
database, 9
conventional, 10
deductive, 25
derived, 10
fundamental, 10
instance, 10
consistent, 12
inconsistent, 12
management system, 16
object-oriented, 9
relational, 10
management system, 9
scheme, 10
decomposition, 317
synthesis, 317
software application, 23
Database Partitioning Feature, 351
DATALINK, 366
Datalog, xx, 16, 27, 28
Datalog¬, xx, 28

RA and SQL equivalences, 221
DATASIS ProSoft, iii
DATE
DB2, 356
MS Access. See MS Access DATE-
TIME
MySQL, 393
Oracle, 377
SQL Server, 407
Date(), 356
Date, C. J., 32, 313, 335,
DATE/TIME
Access, 424
DB2, 371
interval
Oracle, 388
MySQL, 399
Oracle, 387
SQL Server, 415
DATETIME
MS Access, 421, 424
MS SQL Server, 415
MySQL, 393
SQL Server, 407
DATETIME2
SQL Server, 407
DATETIMEOFFSET
SQL Server, 407
Datomic, 28
Davila, N. G., 44
Davy, H., 610
Dayal, U., 334
db. See database
deductive, 25
design
conceptual, 610
instance
manipulation
optimization, xxi
math
instance, xx
scheme, xx
object-relational, 27
scheme
optimization, xxi
db standard operators, 138

DB2, xvii, 16, 17, 22, 23, 25, 26, 32, 127, 186, 203, 323, 340, 353, 375, 426, 428, 593, 489
  10.5, 352
  Data Links Manager, 366
  data type
    DISTINCT
      strongly typed, 362
      weakly typed, 363
    delimited identifier, 353
    ordinary identifier, 353
    Query Management Facility, 323, 354
    Spatial Extender, 366
    SQL, 224
DB2 10.5, 351, 578
DBA, 353, 375
DBCLOB. See double-byte character large object
  DB2, 357
DBCS. See Double-Byte Character Set
DBLP, 29
DBMS, xxi, 16, See database management system
  heterogeneous, 24
  NoSQL, 24
DBMS_ROWID, 380
DBTimeZone, 378
DBURIType, 382
DBWeigher.com, 578
DBWScript, 578
DDL, 15, 193, See data definition language
de Antonellis, V., 318
De Moor, O., 33
DECFLOAT
  DB2, 355
decidability, 318
DECIMAL
  DB2, 356
  MySQL, 392
  SQL Server, 406
decision problem, 629
  example, 319
decomposition
  db scheme, 317
deductive
  database, 25

db, xx
deductive system, 618
delete, 323
delete anomaly, 227
DELETE SQL statement, 199
deletion
  anomaly, 235
Dell Corp., 578
Delobel, C., 314, 335, 336
Delphi, 351
Deming, W. E., 55, 142
Deng, Y., 339
dependencies
  data, 320
dependency. See functional dependency, See constraint
  encapsulated implicational, 319
  equalities generating, 317
  functional, 254
  inclusion, 315
  join, 233
  multivalued, 231
  preservation, 259
  theory, 316
  tuples generating, 317
  typed template, 319
derivation, 618
  rule. See inference rule
derivation rule. See inference rule
derived db, 10
DES, 28
DESC(ENDING) SQL predicate, 209
Design View, 323
  Access, 419, 474
deterministic Turing machine, 629
DEXPTIME. See EXPTIME
Di Paola, R., 321, 321
dictatorial constraint. See implausible constraint
Dicu, A., xxiii
difference (RA) operator
  properties, 304
difference set operator, 213, See relative complement
  IsNull relational equivalent, 301
Dijkstra, E., 519
Dijkstra, E. W., 98, 212, 223, 224, 601

Dimo, P., xxiii
Dincă, A., xxiii
direct sum set operator, 625
directly
    recursive
        usage, 365
dirty null, 127
disjoint sets, 641
disjunctive
    existence
        constraint, 315
disjunctive syllogism, 617
Disraeli, B., 125
DISTINCT
    data type
        strongly typed, 362
        weakly typed, 363
    DB2, 362
    MS Access, 425
    SQL predicate, 327
DISTINCT SQL predicate, 202, 214
    example, 208
DISTINCTROW
    SQL predicate, 301
DISTINCTROW SQL predicate, 202, 214
distinguished set, 616
distributive operator (over another operator), 620
division (RA) operator, 221
    properties, 301
DKNF, 234, 260, 314, See Domain-Key NF
DKNF
    the highest RDM normal form theorem, 307
DKNF ⇒ (3,3)-NF, 307
DKNF ⇒ PJNF, 307
DKRINF, xxi, 236, 258, 261, 316, 325, See Domain-Key-Referential-Integrity Normal Form
DLOGSPACE. See L
DML, 15, See data manipulation language
domain
    (relational) constraint, 124
        example, 121
    constraint, 248
        enforcement, 346

function, 620
independence, 316, 321
independent
    expression, 321
    language, 321
    metacatalog, 180
    RDM, 247
    RDM constraint, 122
    relational calculus, 212
domain constraint, 125
Domain-Key NF, 234
Domain-Key-Referential-Integrity Normal Form, 258
DOUBLE
    DB2, 355
    MS Access, 420, 424
    MySQL, 392
double arrow sign
    in the (E)MDM. See one-to-one function
    in the RDM. See functional equivalence
Doyle, A.C., 9
DPBR. See data and programs backup and restore
DPF. See Database Partitioning Feature
Dragomir, S., 33, 338
DRC, 212, See domain relational calculus
DROP SQL statement, 196
DTIME(nO(1)). See polynomial complexity class
DTM. See deterministic Turing machine
dyadic (mathematical) relation. See binary (mathematical) relation
Dylan, B., 77
dynamic SQL, 224
    example, 226
e.g., See for example
Earp, R., 112
eBay, 27
Eclipse, 351
effective method. See effective procedure
effective procedure, 617
effectively calculable function, 617
Eifrem, E., 33
Einstein, A., 15, 35, 80, 92, 143, 196, 221, 241, 256, 596, 615

EKNF, 306, 314, See elementary keys NF
elementary key, 306
elementary keys NF, 306
Ellison, E., v
Ellison, L., 351, 374, 375, 376, 386
Elsevier B.V., 334
Emax, 355
embedded SQL, 223
EMDM, 25, 28
Emerson, R. W., 605
Emin, 355
Emmick, D. J., v
empty set, 615
empty string, 627
encapsulated implicational dependency,
    319
entity, 37, 74, 94
    examples, 37
entity-relationship
    data model, 8
    diagram, 8
ENUM
    MySQL, 396
equal
    functions, 621
    sets, 625
equality
    generating dependencies, 317
equijoin (RA) operator, 218
equipotent sets, 622
equivalence
    binary homogenous relation, 625
    relation
        kernel, 215
        zeroth-order logic, 619
equivalent
    constraint sets, 30
    relation
        example, 625
E-R
    data
        metamodel, 147, 571
        model
            example, 188
        modeling, xxi
    data model, 19, 36, 98
    diagram, xviii

E-R data metamodel of the E-RDM, 111
E-R diagram, 37, 40
    example, 40, 42, 43, 45, 46, 48, 60, 62,
        63, 64, 78, 80, 81, 83, 84, 86
ERD. See E-RD
E-RD, 8, See E-R diagram
E-RD
    example, 190
E-RD, 606
E-RDM
    E-R data metamodel, 111
E-RDM, xx, xxi, 8, 22–25, 28, 148, 571,
    604–606, 609, See entity-relationship
    data model
    attribute, 75
Erhard, W., 131
Etiny, 356
ETL. See Extract, Transform, and Load,
    See Extract, Transform, and Load
Euclid, 529, 618
Euclidean
    binary homogenous relation, 624
event. See data event
    example
        AfterUpdate, 346
        BeforeUpdate, 346
        Validated, 346
        Validating, 346
existence
    constraint, 249, 315
existential quantifier, 619
EXISTS SQL predicate, 210
EXP. See EXPTIME
explicit transaction, 222
expressive power
    query language, 321
EXPTIME, 633
EXPTIME-complete, 634
extended
    SQL, 346
extended SQL, 223
external (natural) join. See full outer join
    (RA) operator
EXTRACT, 378, 395
Facebook, 16, 26
factorial, 623
factorization, 7, 48

Fagin, R., 314, 315, 316, 317, 321, 333,
    334, 335
false
    Boolean constant, 619
FD, 228, 259, 314, 315, 316, 317, 318,
    319, See functional dependency
    implication problem, 319
    preservation, 317
FD ∪ IND
    implication
        problem
            finite
                undecidability, 319
        undecidability
            finite implication, 319
FD ∪ IND, 319
FD ∪ IND
    inite and unrestricted implication axi-
    omatization, 319
FD ∪ IND, 319
FD ∪ MVD, 319
FE. See forward engineering
FED. See functional equivalence
Felea, V., xxiii
Feuerstein, S., 592
field. See column
Figure 1.0, 22
Figure 1.1, 23
Figure 2.1, 40
Figure 2.10, 45
Figure 2.11, 46
Figure 2.12, 48
Figure 2.13, 60
Figure 2.14, 62
Figure 2.15, 64
Figure 2.16, 70
Figure 2.17, 71
Figure 2.18, 78
Figure 2.19, 80
Figure 2.2, 40
Figure 2.20, 81
Figure 2.21, 81
Figure 2.22, 83
Figure 2.23, 84
Figure 2.24, 84
Figure 2.25, 85
Figure 2.26, 86

Figure 2.27, 93
Figure 2.28, 93
Figure 2.29, 93
Figure 2.3, 41
Figure 2.30, 93
Figure 2.31, 93
Figure 2.32, 93
Figure 2.33, 94
Figure 2.34, 94
Figure 2.35, 94
Figure 2.36, 94
Figure 2.37, 95
Figure 2.38, 97
Figure 2.39, 102
Figure 2.4, 41
Figure 2.5, 42
Figure 2.6, 42
Figure 2.7, 43
Figure 2.8, 43
Figure 2.9, 44
Figure 3.1, 120
Figure 3.10, 149
Figure 3.11, 149
Figure 3.12, 150
Figure 3.13, 156
Figure 3.14, 159
Figure 3.15, 161
Figure 3.16, 162
Figure 3.17, 163
Figure 3.18, 164
Figure 3.19, 165
Figure 3.2, 122
Figure 3.20, 178
Figure 3.21, 189
Figure 3.22, 190
Figure 3.23, 194
Figure 3.24, 198
Figure 3.25, 264
Figure 3.26, 265
Figure 3.27, 284
Figure 3.28, 294
Figure 3.29, 294
Figure 3.3, 124
Figure 3.30, 295
Figure 3.31, 312
Figure 3.32, 316
Figure 3.4, 133

Figure 3.5, 135
Figure 3.6, 144
Figure 3.7, 147
Figure 3.8, 148
Figure 3.9, 148
Figure 4.1, 349
Figure 4.10, 470
Figure 4.100, 553
Figure 4.101, 553
Figure 4.102, 553
Figure 4.103, 553
Figure 4.104, 553
Figure 4.105, 554
Figure 4.106, 554
Figure 4.107, 554
Figure 4.108, 554
Figure 4.109, 554
Figure 4.11, 470
Figure 4.110, 555
Figure 4.111, 555
Figure 4.112, 555
Figure 4.113, 555
Figure 4.114, 556
Figure 4.115, 556
Figure 4.116, 556
Figure 4.117, 557
Figure 4.118, 565
Figure 4.119, 567
Figure 4.12, 475
Figure 4.120, 568
Figure 4.121, 569
Figure 4.122, 569
Figure 4.123, 573
Figure 4.13, 475
Figure 4.14, 477
Figure 4.15, 477
Figure 4.16, 478
Figure 4.17, 478
Figure 4.18, 479
Figure 4.19, 480
Figure 4.2, 463
Figure 4.20, 482
Figure 4.21, 482
Figure 4.22, 484
Figure 4.23, 484
Figure 4.24, 486
Figure 4.25, 486

Figure 4.26, 487
Figure 4.27, 490
Figure 4.28, 491
Figure 4.29, 491
Figure 4.3, 466
Figure 4.30, 492
Figure 4.31, 492
Figure 4.32, 493
Figure 4.33, 493
Figure 4.34, 494
Figure 4.35, 495
Figure 4.36, 496
Figure 4.37, 497
Figure 4.38, 498
Figure 4.39, 499
Figure 4.4, 467
Figure 4.40, 500
Figure 4.41, 500
Figure 4.42, 501
Figure 4.43, 501
Figure 4.44, 502
Figure 4.45, 502
Figure 4.46, 503
Figure 4.47, 503
Figure 4.48, 504
Figure 4.49, 505
Figure 4.5, 467
Figure 4.50, 505
Figure 4.51, 506
Figure 4.52, 506
Figure 4.53, 507
Figure 4.54, 507
Figure 4.55, 508
Figure 4.56, 508
Figure 4.57, 509
Figure 4.58, 509
Figure 4.59, 510
Figure 4.6, 468
Figure 4.60, 511
Figure 4.61, 511
Figure 4.62, 512
Figure 4.63, 512
Figure 4.64, 513
Figure 4.65, 514
Figure 4.66, 514
Figure 4.67, 515
Figure 4.68, 515

Figure 4.69, 516
Figure 4.7, 468
Figure 4.70, 516
Figure 4.71, 517
Figure 4.72, 517
Figure 4.73, 518
Figure 4.74, 518
Figure 4.75, 520
Figure 4.76, 523
Figure 4.77, 523
Figure 4.78, 523
Figure 4.79, 523
Figure 4.8, 469
Figure 4.80, 524
Figure 4.81, 524
Figure 4.82, 524
Figure 4.83, 524
Figure 4.84, 524
Figure 4.85, 524
Figure 4.86, 524
Figure 4.87, 525
Figure 4.88, 528
Figure 4.89, 545
Figure 4.9, 469
Figure 4.90, 545
Figure 4.91, 546
Figure 4.92, 546
Figure 4.93, 547
Figure 4.94, 547
Figure 4.95, 548
Figure 4.96, 548
Figure 4.97, 552
Figure 4.98, 552
Figure 4.99, 552
file
    handler
        Access, 425
        DB2, 372
        Oracle, 388
    index
        non-unique, 345
        unique, 345
finitary operation, 627
Finkelstein, L., 50
Fiorina, C., 2
First Normal Form, 116, 118, 120, 250
first-order logic, 619

first-order predicate calculus. See first-
    order logic
Fischer, P. C., 317, 339
FLOAT
    DB2, 355
    MS Access. See MS Access DOUBLE
    MySQL, 392
    Oracle, 377
    SQL Server, 406
FLOAT4
    MS Access. See MS Access SINGLE
FLOAT8
    MS Access. See MS Access DOUBLE
FOR EACH ROW ON UPDATE AS ROW
    CHANGE TIMESTAMP, 368
Ford, H., 390, 391, 398, 400, 531
foreign key, 124, 137, 141, 248, See func-
    tion images inclusion proposition
    column, 122
        example, 121
    concatenated, 347
        RDM constraint. See referential integ-
        rity constraint
formal grammar, 618
formal language, 617
formal system, 618
FORTRAN, 351
forward engineering, xix, 70
FoundationDB, 28
Free Software Foundation, 28
FreeBSD, 390
freely (generated) monoid, 627
FROM SQL clause
    DELETE SQL statement, 199
    SELECT SQL statement, 202
        example, 202
FTP. See File Transfer Protocol
full outer join (RA) operator, 205
    properties, 301
function, 620
    (Cartesian) product, 622
        chain, 250
    aggregation, 321
    co-domain, 620
    coimage, 626
    composition, 621
    domain, 620

effectively calculable, 617
graph, 621
image, 138, 621
kernel, 626
maximal, 305
minimal, 305
structural, 95
tuple, 321
functional
    binary (mathematical) relation, 621
    data
        modeling, xxi
    dependency, 228
        characterization proposition, 254
        definition, 254
        partial, 229
        transitive, 229
    equivalence, 305
    relationship, 75
    type, 43
functional relationship, 95
fundamental constraint, 13
fundamental db, 10
Gallaire, H., 32, 316, 322, 336
Gamboa, R., 32
Garcia-Molina, H., 32, 336
Garey, M. R., 325, 336
Gartner, 352
Gates, B., v, 402, 403, 404, 405, 408, 409,
    413, 414, 417, 418, 419, 421, 423
Gemstone, 27
GENERAL
    MS Access. See MS Access LONGBI-
    NARY
generalization, 7, 25
GENERATE_UNIQUE(), 370
GEOGRAPHY
    SQL Server, 412
Geometry
    MySQL, 398
GEOMETRY
    MySQL, 401
    SQL Server, 412
GEOMETRYCOLLECTION
    MySQL, 398
GetDate(), 408
GetDescendant(), 410

GIS. See geographic information system
Giumale, C., xxiii
global transaction identifier, 395
Goldstein, B.S., 315, 336
Goodman, N., 314, 334
Google, 26
    App Engine, 26
Gould, S.J., 115
GPS coordinates
    DB2, 372
    Oracle, 388
    SQL Server, 417
Graefe, G., 323, 336
Graham, M. H., 337
Grahne, G., 322, 332, 336
GRANT SQL statement, 196
Grant, J., 315, 336
graph
    binary (mathematical) relation, 620
GRAPHIC
    DB2, 357
Groff, J. R., 336
group
    SQL SELECT statement GROUP BY
    clause. See partition
GROUP BY SQL clause, 206
    example, 206, 209
    golden rule, 207
GTID. See global transaction identifier
GUI. See graphic user interface
GUID, 405, See Global Unique Identifier
    MS Access, 420
Gurevich, Y., 319, 336
Haderle, D., 32
Hadoop, 16, 27
Halpin, T., 336
Hammer, M., 32
Hankel, H., 534
Hannan, D., 237
Harel, D., 321, 334
HAVING SQL clause, 207
    example, 208, 209
HBase, 16, 27
HDBMS. See Heterogeneous DBMS
header
    table, 119
Heath, I., 314, 336

Heine, H., 229
Held, G., 33, 339
Hemingway, E., 444
Hernandez, M.J., 32, 336
heterogeneous
    data, 24
    DBMS, 24
Hewlett-Packard, 26, 28
hierarchical data model, 7
HIERARCHYID
    SQL Server, 410
Hive, 16, 27
Hoberman, S., 32
Hodgman, J., 332
homogeneous binary relation
    constraint, xxi
homogeneous data, 5
homogenous binary (mathematical) relation, 620
Honeyman, P., 315, 317, 333, 336
Horn clause, 96
    closed, 609
Horn, A., 336
Horton, D., 92
House, S., 48
Howard, J. H., 333
HP. See Hewlett-Packard
    UX, 374
HR. See Human Resources
HTTPURIType, 383
Hugo, V., 78
Huie, J. L., 258, 321
Hull, R. B., 31, 322, 332, 336
Huxley, A., 133, 234
Huxley, T.H., 11
HYPERLINK
    MS Access, 422, 425
hypothetical syllogism, 617
i.e.. See that is
i5/OS, 340
iA. See canonical inclusion function (of set A)
IASTED, 33, 113, 338, 593, See International Association of Science and Technology for Development

IBM, xvii, xviii, 16, 17, 25, 26, 27, 32, 322, 323, 340, 366, 372, 428, 534, 578, See International Business Machines
    Data Studio, 110, 353
    DB2
        Data Studio, 578
        Express-C, 352
        object
            name, 352
    Infosphere
        Data Architect, 578
        InfoSphere Data Replication, 370
IBM Corp., 593
idempotent, 219, 621
    auto-function, 621
    operator, 304
identifier
    delimited
        DB2, 353
    ordinary
        DB2, 353
identity. See identity element
IDENTITY, 356, 405
identity element, 627
IDENTITY SQL data type. See auto-number
IEEE, 32, See Institute of Electrical and Electronics Engineers
    TKDE. See Transactions on Knowledge and Data Engineering
IEEEDOUBLE
    MS Access. See MS Access DOUBLE
IEEESINGLE
    MS Access. See MS Access SINGLE
iff. See if and only if, See if and only if
Illustra, 27
Im. See image: function, See image: function
image
    function, 621
    function (Cartesian) product, 120
Imielinski, T., 315, 316, 322, 336
implausible
    constraint, 15, 610
    data, 610
implication
    problem, 14, 320

FD, 319
FDs, 229, 236
FDs and MVDs, 232
JD |— MVD, 320
JDs, 233
MVD, 319
zeroth-order logic, 619
implication problem, 318
implicit transaction, 222
implied constraint, 12
example, 294, 308
improper subsets, 616
IN SQL predicate, 210, 301
inclusion
dependency, 315
unary, 319
inclusion (typed) dependency. See function
images inclusion proposition
inclusion dependency. See referential
integrity constraint
incoherent constraint set, 15
incomplete data, 5
inconsistent
instance
database, table, 12
IND, 261, 315, 319, See inclusion dependency
chase, 320
equational theories, 320
implication problem
complexity, 319
unary, 319
index, 354
index file
non-unique, 345
unique, 345
indirectly
recursive
usage, 365
inference
rule, 319
inference rule, 5, 10
disjunctive syllogism, 617
hypothetical syllogism, 617
modus ponens, 617
modus tollens, 617
recursive, 617

resolution, 617
Infinity, 355
information, 2
Informix, 27
Database, 26
InfoSphere Warehouse, 352
Ingres, 25, 33
INGRES, 25, 323
injective function. See one-to-one: func-
tion
in-memory table, 353
Inmon, B., 32
inner join (RA) operator, 202, 216
insert, 323
insert anomaly, 227
INSERT INTO SQL statement, 197
example, 198
insertion
anomaly, 234
instance
binary (mathematical) relation. See
graph:binary (mathematical) relation
database, 9
relation, 120
table, 119, 120
INT
Access, 424
DB2, 371
MS Access. See MS Access LONG
MySQL, 392, 399
Oracle, 387
SQL Server, 392, 415
INT(n), 52
INT+(n), 52
INTEGER
DB2, 355
MS Access. See MS Access LONG
integer part, 623
example, 165
INTEGER1
MS Access. See MS Access BYTE
INTEGER4
MS Access. See MS Access LONG
intersection set operator, 615
relational equivalent, 301
interval
Oracle data types, 377

INTERVAL DAY TO SECOND
  Oracle, 377
INTERVAL YEAR TO MONTH
  Oracle, 377
intransitive
  binary homogenous relation, 624
IR. See inference rule
irreflexive
  binary homogenous relation, 624
  example, 285
IS [NOT] DANGLING, 381
IS-A, 7
iSeries, 352
IsNull
  SQL predicate, 301
IsNULL, 117, 322
ISO 3166, 325
ISO/IEC, 336
isomporhic sets. See equipotent sets
Istrail, S., xxiii
IT. See Information Technology
Ivanov, V., xxiii
Jagadish, H. V., 32, 337
Java, 19, 28, 28, 322, 351
  Hibernate, xvii
JD, 232, 259, 314, 317, 317, See join
  dependency
  chase, 330
  complete
    axiomatization, 320
JD ∪ FD, 320
  chase, 320
JD |— MVD
  implication problem, 320
Jefferson, T., 25
Jena, 28
Jepsen, C. R., 76
Johnson, D. S., 319, 325, 336,336
Johnson, S.,137
join (RA) operator, 202
  properties, 301, 304
join (RA) operator family, 216
join dependency, 232
joinable row, 218
Jung, C., 231
Kaiser, H. J., 263
Kalantri, A., 82

Kandzia, P., 322, 337
Kanellakis, P. C., 319, 319, 320, 332, 335
Karlin, S., 177
kb. See knowledge base
KBMS. See knowledge:base:management
  system
KDBMS. See knowledge and DBMS, See
  knowledge and DBMS
  example, 27
Kent, W., 325, 337
ker. See kernel (equivalence relation)
  operator
ker(f•g), 305, 626
  example, 292
kernel
  equivalence relation, 215, 262
  properties, 305
Kerouac, J., 240
key. See one-to-one function or minimally
  one-to-one function (Cartesian) product
  concatenated, 373
  constraint, 128, 249
  foreign, 137
    metacatalog, 186
  primary, 119, 129
  propagation principle, 140
  Propagation Principle, 325
    axiom, 600
    example, 141
    formalization, 252
  properties, 304, 305
  semantic, 131
  simple, 128
  surrogate, 38, 130
Key Propagation Principle, 43
King, S., 456
Klein, H., 337
Klug, A., 319, 321, 336
Knowledge, 20, 5
  Base, 10, 16, 25
knowledge and DBMS, 16
knowledge base management system, 16
Knuth, D., 146, 152, 156, 160, 177
Koestler, A., 313
Korth, H. F., 315, 337
KPP. See Key Propagation Principle
Kramer, H., 87

Kreps, P., 33, 339
Krishnamurti, R., 32
Kumar, A., xxiv
Kyte, T., 25, 31
L, 630
L|><|. See left (outer) join (RA) operator
L|><|R. See full outer join (RA) operator
Lady Starlight. See Martin, C.
Lakatos, I., 237, 324
Lammel, R., 32
Lao Tzu, xxiii, 110, 156, 162, 232
Large Hadron Collider, 27
Lasseter, J, 37
Lautréamont, C. de, 213
Laver, K., 337
LD, 219, 317, See lossless(-join) decom-
    position
LDL, 16, 27, 32
LeDoux, C. H., 314, 337
Lee, R. M., 320, 334
left (outer) join (RA) operator, 203
    example, 203
left anti-join. See left anti-semijoin (RA)
    operator
left anti-semijoin (RA) operator, 221
left semi-join (RA) operator, 221
Levene, M., 324, 337
Lewis, H. R., 334
Ley, M., 29
Lien, Y. E., 315, 337
Lightstone, S. S., 32, 337
LIMIT SQL predicate, 202
LIMIT TO n ROWS
    MS Access, 425
limited complete axiomatization, 319
LINESTRING
    MySQL, 398
Ling, Y., 339
Linux, 340, 352, 374, 390
Lipski, W., 315, 316, 322, 336
LNCS, 32, See Lecture Notes in Computer
    Science
Local_Timestamp, 378
Lochovsky, F., 33
logic
    implication, 12
logical

data model, 9
expression, 619
LOGICAL
    MS Access. See MS Access BIT
LOGICAL1
    MS Access. See MS Access BIT
LOGSPACE. See L
Loizou, G., 324, 337
LONG, 420
    MS Access, 420, 424
    MS SQL Server, 415
LONG RAW
    Oracle, 379
LONGBINARY
    MS Access, 421, 425
LONGBLOB
    MySQL, 396
LONGCHAR
    MS Access. See MS Access LONG-
    TEXT
LONGTEXT
    MS Access, 421, 424
    MySQL, 396
Lookup, 418
Lookup Wizard, 420
Lorentz, D., 337
Lorie, R. A., 339
lossless
    decomposition, 306
lossless(-join) decomposition, 219
lower predicate calculus. See first-order
    logic
LSPACE. See L
Lungu, A., 23
M.I.T.. See Massachussets Institute of
    Technology
    Press, 33
Mac. See Apple:Macintosh
MacDonald, M., 593
Mach, E., 237
Maier, D., 32, 315, 319
Makowsky, J. A., 334
Mancas, 111
Mancas, C., 32, 33, 41, 111, 113, 315, 324,
    325, 334, 337, 338
    E-RD notation, 41
Mancas, E., xxiii

Mancas, I., xxiii
Mancas, S., 33, 113, 593
mandatory (required) column. See totally
    defined function
mandatory constraint. See not-null con-
    straint
mandatory values restriction. See compul-
    sory attribute restriction
many-to-many relationship, 37, 96, See
    non-functional relationship
many-to-one relationship, 37
mapping. See function
    constraint, xxi
    non-prime, 249
    prime, 249
    semantic overloading, 610
mapping (Cartesian) product. See function
    (Cartesian) product
mapping composition. See function com-
    position
Marcus Aurelius, 318
Marden, O.S., 49
marked null, 315
Markowitz, V., xxiii
Martin, C., 87
Marx, G., 433
Masalagiu, C., xxiii
MatBase, xxi, 28, 33, 325, 338, 489, 577
    Datalog¬ Subsystem, 33
    E-RD Subsystem, 33, 593
    Relational Import Subsystem, 33
math
    db
        instance, xx
        scheme, xx
MAX SQL aggregation function, 195
maximal
    column, 305
MBCS. See Multi-Byte Character Set
MBR. See Mixed Based Replication
MC Press Online, LLC, 32
MCC, 27, See Microelectronics and Com-
    puter Technology Corporation
McGraw-Hill Osborne Media, 592
McLeod, D., 32
Mead, M., 56
Mealy, G.H., 33

MEDIUMBLOB
    MySQL, 396
MEDIUMINT
    MySQL, 392
MEDIUMTEXT
    MySQL, 396
Melkanoff, M. A., 317, 340
MEMO
    MS Access. See MS Access LONG-
    TEXT
Mendelzon, A. O., 337
metacatalog. See metadata catalog
    attribute, 186
    constraint, 186
    domain, 535
    key
        foreign, 186
    rdb, 179
    RDBMS, 324
metadata, 3
    catalog, 149, 177
    example, 10
Michael I of Romania, v
Microsoft, 16, 26, 323, 578
    Research, 28
Microsoft Corp., 593
Miller, A., xxiii
MIN SQL aggregation function, 195
minimal
    column, 305
    constraint set, 14
minimality, 610
minimally one-to-one function product,
    623
    example, 121
Minker, J., 32, 316, 322, 336
Minskz, M., 33
Mitchell, J. C., 319, 338
MIXED, 359
Mixed Based Replication, 395
model, 6
modeling
    data, xviii
modus ponens, 617
modus tollens, 617
MONEY

MS Access. See MS Access CUR-
RENCY
   SQL Server, 406
monoid, 627
monotone. See monotonic
monotonic operator, 623
monotonicity, 304
Moore, S., 304
Morfuni, N. M., 315
Morgan Kaufman, 31
Morgan Kaufmann, 32
Morgan, T., 336
Morgan-Kaufmann Publishers, 33
Mozilla, 26
MS, xvii, xviii, 428, See Microsoft
   .NET, 402
   Access, xviii, 16, 17, 26, 28, 54, 118,
   127, 129, 136, 138, 140, 143, 144, 186,
   187, 201, 203, 212, 214, 249, 290, 323,
   346, 402, 605
      constraint
         tuple, 425
      db, 419
      Northwind db, 544
      object
         name, 418
      query. See view (computed table)
      Relationships, 110, 578
      reserved characters, 418
      SQL
         comments, 425
      table
         temporary, 419
      view, 419
   Access 2013, 593
   Excel, 323
   Query, 323
   Query Designer, 212
   SQL Server, xvii, xviii, 16, 17, 26, 28,
   54, 127, 129, 136, 187, 203, 212, 323,
   593, 605
      2014, 373
      Database Diagrams, 111, 578
      Express, 403
      object
         name, 403
   Visio, 578

   Visual Studio .NET, 402
   Windows, 352, 366, 374, 403
      NT Services, 402
MS Access
   ATTACHMENT, 422, 425
   BINARY, 421, 424
   BIT, 419, 424
   BYTE, 420, 424
   column
      computed, 401, 423
   COUNTER, 420, 424
   CURRENCY, 420, 424
   DATETIME, 421, 424
   DISTINCT, 425
   DOUBLE, 420, 427
   HYPERLINK, 422, 425
   LIMIT TO LIST n ROWS, 425
   LONG, 420, 424
   LONGBINARY, 421, 425
   LONGTEXT, 421, 424
   MULTIPLE
      BYTE, 422, 425
      DECIMAL, 422, 425
      DOUBLE, 422, 425
      GUID, 422, 425
      INT16, 422, 425
      INT32, 422, 425
      SINGLE, 422, 425
      TEXT, 422, 425
   query. See view
   SHORT, 420, 424
   SINGLE, 420, 424
   table
      computed, 419
      row
         Total, 423
   TEXT, 421, 424
   TOP n, 425
   TRANSFORM, 425
   wildcard, 423
MS Access 2013
   Runtime, 418
MS Excel
   pivoting, 425
MS Office 365
   website, 418
MS SharePoint 2013

server, 418
MS SQl Server
  column
    computed, 413
MS SQL Server, 27
  BIT, 415
  DATETIME, 415
  INT, 405
  LONG, 415
MS Windows, 390, 418
MULTILINESTRING
  MySQL, 398
MULTIPLE
  BYTE
    MS Access, 422, 425
  DECIMAL
    MS Access, 422, 425
  DOUBLE
    MS Access, 422, 425
  GUID
    MS Access, 422, 425
  INT16
    MS Access, 422, 425
  INT32
    MS Access, 422, 425
  SINGLE
    MS Access, 422, 425
  TEXT
    MS Access, 422, 425
MULTIPOINT
  MySQL, 398
MULTIPOLYGON
  MySQL, 398
multivalued dependency, 231
  complementary property, 232, 306
Murach, J., 338
Murakami, H., 428
Murdock, M., 40
MVD, 231, 314, 316, 317, 319, See multi-
  valued dependency
  implication problem, 319
MVDs, 259
MySQL, xvii, xviii, 15, 24, 26, 54, 338,
  428
  Enterprise Edition, 389
  NDBCLUSTER, 389
  object

name, 390
Replication
  Mixed Based, 395
  Row Based, 395
  Statement Based, 395
  Workbench, 391
MySQL 5.6, 593
MySQL 5.7, 578
n choose j, 165, 623
n!. See factorial (of n)
Nadeau, T., 32, 337
NaN, 356, See Not a Number
  quiet, 343, 355
  signaling, 355
Naqvi, S., 32, 33
n-ary
  function product, operator
    arity, 623
NAT
  Access, 424
  DB2, 371
  MySQL, 399
  Oracle, 328
  SQL Server, 487
NAT(n), 52
natural join (RA) operator, 217, 301
Nazarian, V., 14
NC class, 633
NCHAR
  Oracle, 379
  SQL Server, 409
NCLOB
  Oracle, 379
negative
  data, 5
Negri, M., 323, 338
Nehru, J., 42
nested
  table
    DB2, 372
    Oracle, 388
nested table, 118
NESTED TABLE
  Oracle, 382
network data model, 7
Neumann, J. L. von, 92
neutral element. See identity element

NEWID(), 405
    example, 405
NEXP. See NEXPTIME
NEXPTIME, 633
NF, 314, 317
Nicolas, J., 32, 316
Nicolas, J. M., 314, 338
Nin, A., 595
NL, 630
NLOGSPACE. See NL
no-information
    null, 315
no-information null value, 126
    characterization, 304
Nokia, 26
non-applicable null value. See non-existent
    null value
non-deterministic polynomial, 631
non-deterministic polynomial-time hard,
    631
non-deterministic Turing machine, 629
non-essential attribute (column, function).
    See non-prime attribute (column, func-
    tion)
non-existent
    null, 315
non-existent null value, 126
non-functional
    relationship
        type, 126
non-nested table. See First Normal Form
non-prime attribute (column, function),
    135
non-prime mapping, 249
non-relational
    constraint
        enforcement, 346
non-relational constraint
    enforcement, xxi
Non-relational constraint
    example, 189
NonStop SQL, 26
NonStop SQL Server, 16
normalization
    algorithm, 317
    problem, 317

NoSQL, 24, 26, See Not Only SQL
    DBMS, See NoSQL DBMS
NoSQL DBMS, 26
not
    zeroth-order logic, 619
NOT EXISTS SQL predicate, 210
NOT IN SQL predicate, 210
NOT NULL
    (relational) constraint, 121, See totally
    defined function
    RDM constraint, 123
    SQL predicate, 195
        example, 194
    values, 123
NOT NULL constraint, 249
not null value restriction. See compulsory
    attribute restriction
not total binary homogenous relation. See
    partial:binary homogenous relation
NOTE
    MS Access. See MS Access LONG-
    TEXT
not-null constraint, 127
Now(), 395, 421
NP. See non-deterministic polynomial
NPC. See NP-complete
NP-C. See NP-complete
NP-complete, 632
    example, 320
NP-completeness, 325
NP-hard. See non-deterministic polyno-
    mial-time hard
    example, 320
NTM. See non-deterministic Turing
    machine
n-tuple. See tuple
nucleus. See kernel equivalence relation,
    See kernel (equivalence relation)
null, 126, 322, See null value
    marked, 315
    no-information, 315
    non-existent, 315
    unknown, 315
    value, 123, 610
null value, 126, See null
    dirty, 127
    no-information, 126

non-applicable. See null value: non-
    existent
non-existent, 126
    unknown, 126
nullar
    function product, operator
        arity, 623
nulls. See null values
NULLS, 95, 123, 247
nulls substitution principle, 314
number
    subnormal, 356
NUMBER
    MS Access. See MS Access DOUBLE
    Oracle, 376
NUMBER SQL data type, 195
NUMERIC
    DB2, 355
    MS Access. See MS Access DOUBLE
NVARCHAR
    SQL Server, 409
NVARCHAR2
    Oracle, 379
O(x). See complexity operator
O'Reilly Media, 592, 593
O'Reilly Media Inc., 32, 33, 333, 335
O2SQL, 323
OBELisQ, 323
object
    calculus, 315
    constraint, xxi, 315
    identifier
        Oracle, 381
    set
        semantic overloading, 610
    uniqueness, 610
OBJECT
    Oracle, 381
object-oriented
    database, 9
object-relational db, 27
Object-Relational Mapping, xvii
OCI. See Oracle Call Interface
ODBC, 351, 352, 289, 402, 418, 419
OID. See Oracle object identifier
OLAP, 352, 353, 374, 402, See Online
    Analytical Processing

OLE, 421
OLEDB, 351, 374, 389, 402, 418, 419
OLEOBJECT
    MS Access. See MS Access LONGBI-
        NARY
Olson, E.C., 193
OLTP, 351, 375
ON DELETE CASCADE SQL predicate,
    140
ON DELETE RESTRICT SQL predicate,
    140
ON UPDATE CASCADE SQL predicate,
    139
ON UPDATE NO ACTION SQL predi-
    cate, 139
ON UPDATE RESTRICT SQL predicate,
    139
ON UPDATE SET NULL SQL predicate,
    139
one-to-many relationship, 37
one-to-one function, 622
one-to-one relationship, 37
onto function, 622
O-O. See object-oriented
O-ODBMS, 24, 27, 29
O-ODM, 24
open formula, 120, 619
OpenSolaris, 390
OpenVMS, 374
optimization
    db
        instance
            manipulation, xxi
            scheme, xxi
or
    zeroth-order logic, 619
O-R. See object-relational
Oracle, xvii, xviii, 16, 17, 22, 23, 25, 26,
    41, 127, 129, 144, 146, 186, 187, 195,
    202, 203, 223, 230, 238, 245, 249, 290,
    322, 337, 375, 426, 428, 578, See Oracle
    Database
    11g, 605
    12c, 373, 374, 593
    autonumber, 376
    Corporation, xviii, 16, 25, 26, 31
    dangling REF, 381

data
    type
        interval, 377
    Enterprise Console, 375
    E-RD notation. See Barker E-RD nota-
        tion
    Express Edition, 403
    FIPS Flagger, 374
    function
        SysDate, 386
        To_Date, 374
    GoldenGate, 376
    object
        identifier, 381
        name, 374
    schema object, 381
    sequence, 376
    Spatial and Graph, 383
    SQL
        comments, 375
    SQL Developer, 375
        Data Modeler, 111, 578
    table
        column
            virtual, 385
    text
        string, 374
    trigger, 386
    user, 375, See Oracle user
    versions, 22
    XE. See Oracle Express Edition
Oracle 12c, 578
Oracle Corp., 593
ORDER BY
    RA operator, 215
        example, 216
    SQL clause, 209
        example, 209
ordinary
    array, 366
ordinary Turing machine. See determinis-
    tic Turing machine
Orford Walpole, H. Earl of, 140
Orlowska, M., 338
ORM. See Object-Relational Mapping
OS. See operating system
OS X, 352, 390, 403, 418

OS/400. See iSeries
outer join (RA) operators, 202, 203
Ovidius State University, iii, xx, xxiii, 113,
    337, 593
Ovidius University Press, 338
Ovidius, P. N., v
OWL. See Web Ontology Language
Ozsoyoglu, G., 338
P. See polynomial complexity class
P versus NP problem, 632
package, 222
Packt Publishing, 33
Page, L., 256
Papadimitriou, C. H., 316, 334, 340
Paradox, 28
paramodulation resolution, 320
Paredaens, J., 321
Parekh, S., 32
Parker, D. S., 314, 337, 339
partial
    binary homogenous relation, 624
    functional dependency, 226
    order, 625
partially ordered set, 625
partially specified values, 315
partition, 205, 625
PB. See petabyte
P-complete, 631
Peale, N. V., 106
Pearson Education Ltd., 32, 336
Pelagatti, S., 338
Penney, J. C., 75
Penzias, A., 85
per se. See in its own
Perl, 351
permitted values. See attribute range
personal numeric code, 51
petabyte, 26
Peter, L. J., 298
Petrov, S. E., 338
Petrov, S. V., 320
PHP, 351
Picasso, P., 78
Pirotte, A., 321, 338
PIVOT
    SELECT clause, 403
PJNF. See projection-join NF

PL. See programming language
PL/I, 351
PL/SQL, 223, 322, 338, 374
plausible
    constraint, 31
    data, 610
Plenum Press, 32
POINT
    MySQL, 398
Politehnica University, iii, xx, xxiii, 33, 113, 337
POLYGON
    MySQL, 398
polynomial complexity class, 630
polynomial space, 633
poset. See partially ordered set
Post, E. L., 319, 338
Postgres, 26, 27
PostgreSQL, 26, 323
postulate. See axiom
pr. See canonical (Cartesian) projection
Pragmatic Bookshelf, 33
predicate
    IsNULL, 322
predicate logic. See first-order logic
Prentice-Hall, 33
preorder, 625
    example, 625
    partial, 291
Price, T. G., 339
primary key, 119, 129
    example, 121
PRIMARY KEY SQL predicate, 195
prime attribute (column, function), 134
prime mapping, 249
primitive
    constraint, 314
primitive constraints, 234
problem
    implication, 318, 320
        FD, 319
    normalization, 317
product
    binary (mathematical) relation, 628
programs halting problem, 632
projection (RA) operator, 213
    properties, 304

projection-join NF, 233
Prolog, 10, 27
proposition. See closed formula
propositional calculus. See propositional logic
propositional logic, 618
Propositional Satisfiability Problem. See Boolean Satisfiability Problem
PSPACE. See polynomial space
PSPACE-complete
    example, 319
PTIME. See polynomial complexity class
    example, 319
Purcell, T., 32
pureScale, 351
pyDatalog, 28
Pythagoras, 376, 378, 381, 384, 386, 391, 396, 398
Python, 28, 351
QBE, 212, 323, 353, 375, 391, See Query-by-Example
QBF. See quantified Boolean formula problem
QL, 322
QMF. See Query Management Facility, See Query Management Facility
QSAT. See True quantified Boolean formulas
quantification theory. See first-order logic
quantified Boolean formula problem, 633
Quantified SAT. See True quantified Boolean formulas
quasiorder. See preorder
Quel, 25, 322, 323
query. See open formula
    conjunctive, 319
    data complexity, 630
    relational, 120
    semantic
        correctness, 611
Query-by-Example, 212
quiet NaN, 355
Quillian, H.R., 33
quotient set, 215, 625
RA, 213, 314, 321, 322, See relational algebra operator

derived
  augmentation, 322
  division, 221
  join
    inner, 216
      complete, 218
      equi, 218
      natural, 217
      θ, 218
    outer, 216
      full, 205
      left, 203
        semi, 221
      right, 204
        semi, 221
  projection
    total, 322
  fundamental
  ORDER BY, 215
  projection, 213
  renaming, 217
  selection, 213
  SQL and Datalog¬ equivalences, 221
RA expression
  Properties, 304
RAC. See Oracle Real Application Clusters
Raicu, A., xxiii
Ramakrishnan, R., 27, 33
range, 75
  restriction, 75
range constraint. See domain constraint
Raspall, F.-V., 318
RAT
  Access, 424
  DB2, 371
  MySQL, 399
  Oracle, 387
  SQL Server, 415
RAT(n.m), 52
RAT+(n.m), 52
RAW
  Oracle, 379
RBR. See Row Based Replication
RC, 314, 321, See relational calculus
rdb, 9, See relational database
  example, 177

instance, 248
metacatalog, 179
scheme, 247, 248
RDBMS, xviii, 9, 23, 24, 604, 608, See
  relational:database management system
  core, 16
  example, 25
  GUI, 426
  metacatalog, 324
  version, 428, 605, 608
RDM, xx, xxi, 7, 23, 25, 28, 188, 257, 314,
  321, 347, 604, 608, 609, See relational
  data model
  attribute, 119, 247
  domain, 247
  portability, 608
  rdb
    instance, 248
    scheme, 248
  relation, 119, 247
    instance, 247
    scheme, 247
  scheme
    design
      decomposition, 259
      synthesis, 259
    theory, 257
    tuple, 119
RE. See reverse engineering
REA0, 22–24, 577
REA1, 22, 23
REA1-2, 22, 24, 605
  Complexity, 256
REA2, 22, 23
REA2013A0, 473, 519
REA2013A0-2, 519, 544
REAF0, 605
REAF0', 22, 24, 344, 494, 604
REAF0-1, 22, 23, 344, 494
REAF0-2, 22, 23, 605
REAL
  DB2, 355
  MS Access. See MS Access SINGLE
  SQL Server, 406
recognizer, 618
recordset, 120, 223
recursive

computability, 617
relationship, 44
    exam, 44
type
    definition, 365
    usage
        direct, 365
        indirect, 365
recursive inference rule, 617
Redmond, E., 27, 33
redundant
    data, 610
redundant constraint, 14
REF
    Oracle, 381
REFERENCE
    DB2, 365
referential integrity, 137, 248, 261, 315,
    See referential integrity constraint
    (relational) constraint. See function
    images inclusion proposition
    RDM constraint, 123
referential integrity constraint, 138
reflexive
    binary homogenous relation, 624
relation
    instance, 120
    mathematical, 620
    RDM, 116, 119, 247
        Instance, 247
        scheme, 247
    relational instance. See table instance
    universal, 317
relational
    algebra, 213
    calculus, 212
    constraint, 119
        check, 120
        domain, 122
        NOT NULL, 123
        referential integrity, 122
        uniqueness, 119
    data model, 7
    database, 10, 116, 118
        instance, 119
        scheme, 119
    language

example, 25
    query, 120
relational database management system, 9
relationship, 37, 95
    example, 37
    functional, 43, 75
    hierarchy
        example, 45
    many-to-many, 37, 96
    many-to-one, 37
    non-functional, 74
    one-to-many, 37
    one-to-one, 37
    recursive, 44
    type
        functional, 43
        non-functional, 43
relative complement, 616
renaming (RA) operator, 214
    properties, 304
representative system, 626
required constraint. See not-null constraint
resolution
    paramodulation, 320
resolution, 617
restriction, xviii, See business rule
    attribute
        compulsory, 50
        range, 50
    compulsory, 75
    other types, 50
        examples, 59
    range, 75
    set, 50, 96
    uniqueness, 50, 75
reverse engineering, xix, 70
REVOKE SQL statement, 196
rewriting rules, 618
REXX, 351
right (outer) join (RA) operator, 203
right anti-join. See right anti-semijoin
    (RA) operator
right anti-semijoin (RA) operator, 221
right semi-join (RA) operator, 221
Rishe, N., 339
Rissanen, J., 314, 338
RL, 608, See relational language

CH-complete, 322
completeness, 321
RL limitations
    transitive closure, 322
Robinson, I., 28, 33
Rockoff, L., 323, 338
Rodin, A., 228
Roeser, M. B., 337
role, 38, 45, 74, 95
    examples, 38
ROLLBACK
    SQL statement, 222
Roosevelt, T., 99
root type, 364
Row Based Replication, 395
ROWID
    Oracle, 379
rowids
    Oracle, 379
ROWVERSION
    SQL Server, 407
RPG, 351
Ruby, 351
Rucker, R., 488
rule
    business, xviii
    derivation. See rule: inference
    inference, 319
Russell, B., 497
S*. See freely (generated) monoid over S
SA. See algebra of sets
Şabac, G., xxiii
Sadri, F., 319, 339
Sagiv, Y., 315, 319, 333, 337, 339
SAP, 26
Saraiva, J., 32
Sarkar, D., 27, 33
Sartre, J.-P., 31
SAT. See Boolean Satisfiability Problem
SATISFIABILITY. See Boolean Satisfi-
    ability Problem
satisfied
    constraint, 12
Sava-Segal, E., xxiii
Sbattella, L., 338
SBCS, 358, See Single-Byte Character Set
SBR. See Statement Based Replication

scheme
    binary (mathematical) relation, 620
    database, 10
    rdb, 247
    relation, 120
    table, 3
Schweitzer, A., 14
Sciore, E., 315, 320, 339
SDO_GEOMETRY
    Oracle, 383
SDO_GEORASTER
    Oracle, 383
SDO_TOPO_GEOMETRY
    Oracle, 383
SELECT
    SQL statement, 420
SELECT @@IDENTITY
    MS Access SQL statement, 201
SELECT LAST_INSERT_ID(), 393
selection (RA) operator, 213
    properties, 304
Selinger, P., 339
semantic
    data
        model, 25
    query correctness, 610
semantic key, 131
semigroup, 627
semi-structured data, 29
Semmle Ltd., 28
sentential calculus. See propositional logic
sentential logic. See propositional logic
SEQUEL, 322
sequence
    Oracle, 376
Şerbănaţi, L.D., xxiii
SessionTimeZone, 378
Set, 615
    closed with respect to implication, 13
    complement, 616
    constraint, xxi
    inclusion, 616
    quotient, 255
    union, 210, 219
SET
    MySQL, 396
Shaw, G. B., 90

Sheshadri, P., 33
SHORT
    MS Access, 420, 388
SI_AverageColor, 383, 384
SI_Color, 383, 388
SI_ColorHistogram, 382, 388
SI_FeatureList, 384, 388
SI_PositionalColor, 384, 388
SI_StillImage, 383, 388
SI_Texture, 384, 388
Sickels, S., xxiv
SIGMOD. See Special Interest Group on
    Management of Data
signaling NaN, 355
simple key, 128
Simsion, G., 32
Simsion, G. C., 33
SINGLE
    MS Access, 420, 421, 424
single key. See one-to-one function
SMALLDATETIME
    SQL Server, 407
SMALLINT
    DB2, 355
    MS Access. See MS Access SHORT
    MySQL, 392
    SQL Server, 405
SMALLMONEY
    SQL Server, 406
SmartDraw, 578
Smith, D. P. C., 33
Smith, J., 239
Smith, J. M., 33, 314, 339
Snicket, L., 63
Social Security Number, 51
Socrates, 110
Solaris, 352, 366, 374, 390
Solis, B., 75
SOME SQL predicate, 210
sound
    algorithm, 557
        example, 179
    deductive system, 557
Sowell, T., 314
sp_addtype, 413, 415
space hierarchy theorem, 633
Spencer, H., 15

Spinoza, B., 618
SPJ (RA) expression, 220
    example, 218
SPJR expression, 311
Springer-Verlag, 33, 113
Spyratos, N., 335
SQL, 16, 25, 27, 257, 322, 338, 346,
    348, 351, 426, See Structured Query
    Language
    engine, 16
    extended, 346
    formal semantics, 323
    NonStop Server, 26
    programming, 323
    RA and Datalog¬ equivalences, 221
    recursive, xx
    statement
        SELECT, 420
SQL PL. See SQL Procedural Language
SQL Server, 22, 426, 428
    index
        null, 414
    Management Studio, 404, 413
    Query Design, 404
SQL Server 2014, 578
SQL stored programs, 10
SQL/MM StillImage, 383
SQL_VARIANT
    SQL Server, 412
SQL-99, 323
SQLite, xvii, 26
Srere, P.A., 6
Srivastava, D., 33
SSN, 7, See Social Security Number
Stănăşilă, O., xxiii
standard math operators, 619
standard math relations. See standard math
    operators
start symbol, 618
START TRANSACTION
    SQL statement, 222
Statement Based Replication, 395
static SQL, 224
    example, 226
STDDT. See Strongly Typed DISTINCT
    Data Type
Stigler, S.M., 599

Stoll, C., 1
Stonebraker, M., 33, 323, 339
stored
    function, 222
    procedure, 222
Storey, V. C., 334
Stravinsky, I., 145
STRING
    MS Access. See MS Access TEXT
string concatenation, 627
strongly typed DISTINCT data type, 362
structural function, 95
STRUCTURED
    DB2, 364
subnormal number, 356
subquery, 209, 210
    dynamic, 210
    static, 210
subset, 616
subtype, 364
Suciu, D., xxiii, 31
Sudarshan, S., 33
Sullivan, L. H., 314
SUM SQL aggregation function, 195
Sun, W., 339
superkey, 124, 128, See one-to-one func-
    tion (Cartesian) product
superset, 616
supertype, 364
surjective function. See onto: function
surrogate key, 38, 96, 130
    example, 38
Sybase, xviii, 16, 26, 136
    SQL
        Server, 26
    SQL Server, xviii, 26
symmetric
    binary homogenous relation, 624
symmetrical difference set operator
    IsNull relational equivalent, 301
syntactic key. See surrogate key
synthesis
    db scheme, 317
SysDate, 378
    Oracle
        Function, 386
SysDate() function, 195

SysDateTime(), 408
SYSFUN, 370
system of representatives. See representa-
    tive system
System R, 25, 31, 323
SystemR, 323
SysTimestamp, 378
Szent–Gyorgyi, A., 96
table
    arity, 119
    cardinal, 119
    column, 119
        name, 122
    content. See table instance
    header, 3, 119, 120
        example, 121
    hierarchies, 354
    instance, 3, 119, 120
        consistent, 12
        example, 121
        inconsistent, 12
        invalid. See
        table:instance:inconsistent
        valid. See table:instance:consistent
    in-memory, 353
    nested, 118
    row
        joinable, 218
    scheme. See table header
    temporary, 353, 375, 391, 404
        local, 354, 375, 404
        global, 354, 375, 404
    typed, 364
Table 1.0, 3
Table 3.10, 181
Table 3.11, 183
Table 3.12, 185
Table 3.13, 187
Table 3.14, 204
Table 3.15, 204
Table 3.16, 206
Table 3.17, 207
Table 3.18, 228
Table 3.19, 230
Table 3.1a, 118
Table 3.1b, 118
Table 3.2, 142

Table 3.20, 230
Table 3.21, 232
Table 3.22, 233
Table 3.23, 236
Table 3.24, 220
Table 3.25, 317
Table 3.3, 142
Table 3.4, 143
Table 3.5, 168
Table 3.6, 172
Table 3.7, 179
Table 3.8, 179
Table 3.9, 180
Table 4.1, 350
Table 4.10, 417
Table 4.11, 424
Table 4.12, 425
Table 4.2, 360
Table 4.3, 371
Table 4.4, 372
Table 4.5, 387
Table 4.6, 388
Table 4.7, 400
Table 4.8, 401
Table 4.9, 415
table scheme. See table header
tableaux, 320
tablespace, 345
Tagore, R., 243
Tandem, 16, 26
Tannen, V., xxiii, 314, 325, 334
target list
    projection (RA) operator, 213
target type, 366
TB. See terrabyte
Technics Publications LLC, 32, 33
temporary
    table, 353, 375, 391, 404
        global, 353, 375
        local, 353, 375
        MS
            Access, 419
Ţene, L., xxiii
Teorey, T. J., 32, 337
terrabyte, 26
TEXT
    MS Access, 421, 424

MySQL, 396
Thalheim, B., 113, 316, 339
The Weather Channel, 27
theorem, 616
    propositional logic, 618
Thue, A., 338
TIME
    DB2, 356
    MS Access. See MS Access DATE-
    TIME
    MySQL, 333, 334
    SQL Server SQL Server, 407
Time(), 421
TIMESTAMP
    DB2, 356
    MS Access. See MS Access DATE-
    TIME
    MySQL, 393
    Oracle, 377
TIMESTAMP WITH LOCAL TIME
ZONE
    Oracle, 377
TIMESTAMP WITH TIME ZONE
    Oracle, 377
TINYBLOB
    MySQL, 396
TINYINT, 399, 415, 493, 494
    MySQL, 392
    SQL Server, 405
TINYTEXT
    MySQL, 396
Titus Livius, 570
TKDE, 32
TM. See Turing machine
To_Date
    Oracle
        Function, 374
Toad, 577
TODS. See Transactions on Database
Systems
Toffler, A., 134, 519
TOP n
    MS Access, 425
TOP SQL predicate, 202
total
    binary homogenous relation, 624
    order, 625

total projection (RA) operator, 322
Total table row
  MS Access, 423
totality constraint. See not-null constraint
totally defined binary (mathematical) relation, 621
totally defined function, 124
TQBF. See True quantified Boolean formulas
transaction, 222
  atomicity, 222
  consistency, 222
  durability, 222
  explicit, 222
  implicit, 222
  isolation, 222
  serialization, 222
TRANSFORM, 418
  MS Access, 425
transitive
  binary homogenous relation, 624
  closure, 223, 303
    example, 302, 303
  functional dependency, 229
transitive closure
  binary (mathematical) relation, 627
  digraph, 628
  RL limitations, 322
TRC, 212, 321, See tuple relational calculus
Trigger, 222, 346, 354
  Before Insert, 376
  Oracle, 386
trivial
  constraint, 14
  FD, 305
  MVD, 306
    characterization theorem, 306
true
  Boolean constant, 518
  Proposition, 618
True quantified Boolean formulas, 633
truth
  preserving, 617
truth tables
  propositional logic, 619
truth-functional propositional logic, 619

Tsichritzis, D. C., 33
Tsou, D. M., 317, 339
T-SQL, 223, 322, 374, 405
Tsur, S., 33, 335
tuple, 120
  constraint, 145, 249, 315, See check constraint
    enforcement, 346
  function, 322
  generating dependencies, 317
  RDM. See:table row
  relational calculus, 212
Turing machine, 617, 629
Turing, A., 617, 623, 629
Twain, M., 69, 197
type
  domain, 321
  hierarchy, 364
  root, 364
  target, 365
typed table, 364
typed template dependency, 319
UDT. See User-defined Data Type
Ullman, J. D., 32, 34, 313–315, 317, 319, 321–323, 333, 336, 337, 339
UML. See Unified Modeling Language
unary
  function product, operator
    arity, 623
  IND, 319
unary inclusion dependency, 319
UNC, 422, See Universal Naming Convention
undecidable, 319
  problems
    finite monoid words, 319
    monoid words, 319
undecidable problem, 14, 629
underlying set, 620, 627
UNICODE, 74, 358
  Access, 400
  DB2, 371
  MySQL, 400
  Oracle, 371
  SQL Server, 416
UNICODE(n), 52

unikey FD schemes characterization theorem, 306
union
  set operator
    example, 13, 234
union (RA) operator
  properties, 304
UNION ALL SQL operator, 211
union set operator, 221, 615
  example, 119, 203–205
UNION SQL operator, 210
  Example, 301
unique columns concatenation (composition). See one-to-one function (Cartesian) product
unique composed key. See minimally one-to-one function (Cartesian) product
unique objects axiom, 4, See A-DI09
UNIQUEIDENTIFIER, 405, 415
  example, 405
uniqueness
  object, 610
  restriction, 50, 75, 96
    examples, 56, 58
uniqueness constraint. See one-to-one function or minimally one-to-one function (Cartesian) product
unity function, 622
universal
  relation, 317
Universal Naming Convention, 422
universal quantifier, 619
University of Wisconsin at Madison, 27
Unix, 340, 352, 390
unknown
  null, 233
unknown null value, 126
UNPIVOT
  SELECT clause, 403
updatable query, 211
update, 323
update anomaly, 227
UPDATE SQL statement, 197, 241
URIType
  Oracle, 382
URL, 422, 578
  Oracle, 388

UROWID, 380
  Oracle, 379
Urowids, 380
UTC. See Coordinated Universal Time
UTF-16, 358, 360
UTF-32, 357
UTF-8, 358, 359, 390
Vadim Leandru, xxiii
value
  domain, 321
VALUES SQL clause, 197
  example, 198
VARBINARY
  MS Access. See MS Access BINARY
  MySQL, 396
  SQL Server, 409
VARCHAR
  DB2, 357
  MS Access. See MS Access TEXT
  MySQL, 396
  SQL Server, 409
VARCHAR2
  Oracle, 378
VARCHAR2 SQL data type, 195
Vardi, M. Y., 316, 317, 319–321, 333–337, 339
VARGRAPHIC
  DB2, 357
VARRAYS
  Oracle, 381
Vassiliou, Y., 315, 339
Vaughn, W. R., 593
VBA, 322, 418, 420
Vianu, V., 31, 323, 332, 333
Vidal, V. M., 315
Vidal, V. M. P., 334
Viesturs, E., 88
view, 221, 353, 354, 375, 391
Vinci, L. da, 38
violated
  constraint, 12
virtual
  column
    Oracle, 385
virtual column. See Oracle computed column
Visser, J., 33

Visual Paradigm, 578
Visual Studio, 351
VLDB, 33, See International Conferences
  on Very Large DataBases
VM/VSE, 352
Walker, A., 577
Wallace, H. A., 86
Wang, H., 338
Watson, T. J., v, 351–354, 357, 361, 370
Watson, T. Jr., 347, 367, 370
weakly typed DISTINCT data type, 363
Webber, J., 33
well-formed formula. See word: formal
  language
well-formed word. See word: formal
  language
WHERE SQL clause
  DELETE SQL statement, 199
  SELECT SQL statement, 202
    example, 208, 209, 216
  UPDATE SQL statement, 197
Whitman, W., 125
Widom, J., 32, 34, 336, 339
Wiener, N., 58
wildcard
  MS Access, 423
Wilde, O., 67, 229, 462
Williams, T. T., 88
Wilson, J. R., 33
Windows, 340
  8, 312
WITH RECURSIVE
  SQL statement, 223
Witt, G. C., 33
Wong, E. A., 33, 315, 339
word
  formal language, 618
WTDDT, 363, See Weakly Typed DIS-
  TINCT Data Type
XDBURIType, 383
XML, 29, 351, See eXtended Markup
  Language
  DB2, 361
  Oracle, 388

SQL Server, 411, 417
XMLType
  Oracle, 382
XPath, 390
XQuery, 351
Yahoo, 27
Yannakakis, M., 316, 340
YEAR
  MySQL, 394
Year(), 408, 421
Year() function, 195
YESNO
  MS Access. See MS Access BIT
z/OS, 340
Zamil, J., 340
Zaniolo, C., 32, 314, 315, 317, 340
zeroth-order logic. See truth-functional
  propositional logic
Zhan, Y., 338
zLinux, 374
Zloof, M., 323, 340
zOS, 352
ε. See empty string
λ. See empty string
∀. See universal quantifier
∃. See existential quantifier
π. See projection (RA) operator
θ-join (relational) operator, 218
ρ. See renaming (RA) operator
σ. See selection (RA) operator
|—. See logic implication
°. See function composition
×. See Cartesian product, See Cartesian
  product, See Cartesian product
•. See function (Cartesian) product, See
  function (Cartesian) product
—. See difference set operator
⊕. See direct sum, See direct sum set
  operator
∅. See empty set
∩. See intersection set operator
∪. See union set operator
⊆. See set inclusion